Rediscover Algebra and Trigonometry

Foundations for High School and Beyond

for Grade 9 Learners on AP, IB & STEM Pathways

Sharpen your skills. Shape the future

Anil Kumar Khandelwal

Copyright Page

Title: *Rediscover Algebra and Trigonometry*
Author: Anil Kumar Khandelwal
ISBN: 978-1-0693862-3-6
First Edition: 2025
Published in Canada
Printed on Demand (Amazon KDP / IngramSpark)
Imprint: Independently Published

Contact:
globalmathinstitute@gmail.com
www.globalmathinstitute.com
www.youtube.com/@MathematicsTutor

Preface

Rediscover Algebra and Trigonometry

A STEM Journey for Every Learner

This book was born from a simple realization:
Many bright and capable students—despite years of learning—struggle to connect the pieces of algebra and trigonometry. They may have learned formulas, but not the framework; they solved problems, but missed the purpose. Others, years later, wish to return and finally see the subject as it was meant to be seen—logical, visual, and profoundly connected to the world around us.

Rediscover Algebra and Trigonometry was written to close that gap. It bridges the space between school-level mathematics and the conceptual depth demanded by **AP Mathematics**, **IB Mathematics**, and **STEM preparatory courses**. It builds readiness for higher learning by revisiting key ideas with clarity, structure, and purpose.

For learners, educators, and parents alike, this book serves two pathways:

- **A Launchpad** — for those preparing to excel in advanced programs, where conceptual insight and problem-solving agility truly matter.
- **A Leadership Path** — for future STEM innovators and thinkers building the mindset, discipline, and creativity to lead in science, technology, and mathematics.

Each chapter blends guided reasoning, visual models, and carefully sequenced examples. The **Challenge Zones** and **STEM Projects** extend learning beyond the textbook, encouraging creativity, exploration, and design thinking. The final **Assessment**, spanning nearly 74 pages, features more than **300 SAT/ACT-style tests** and **Olympiad-level problems**—allowing learners to apply concepts in context and test their readiness for competitive excellence.

Every topic is supported with **QR codes** that connect to my **YouTube channel**, providing over **17,000 free video explanations**—a complete learning ecosystem designed to help students anywhere, anytime.

Let this not be just another math book,
but a *journey to rediscover*:
the **joy of understanding**,
the **satisfaction of solving**, and
the **confidence of knowing** that with curiosity and practice—
you can learn, grow, and thrive,
in school and beyond.

— Anil Kumar Khandelwal
Founder, Global Math Institute Inc.

About the Book & the Author

Rediscover Algebra and Trigonometry – Foundations for High School and Beyond
For Grade 9 Learners on AP, IB & STEM Pathways

Mathematics is not just about solving problems—it is about recognizing patterns, building logic, and connecting ideas with clarity and confidence.

Rediscover Algebra and Trigonometry presents algebra, geometry, and trigonometry as a **connected system of reasoning**, rather than as isolated formulas or procedures. Designed for today's STEM-driven learners, the book strengthens the bridge between school mathematics and the higher-level conceptual thinking required in **AP, IB, and university-prep programs**.

Each chapter builds purposefully, guiding students from foundational skills to advanced applications. Learners develop structural thinking in algebra, explore multiple representations of linear and quadratic relationships, solve systems and inequalities, work with functions and transformations, and apply geometry and trigonometry to real-world contexts. Throughout, emphasis is placed on understanding *why* methods work—not just how to apply them.

The book includes **300+ carefully sequenced questions**, supported by **dedicated video explanations**, allowing learners to revisit concepts, clarify misconceptions, and build confidence independently. The journey concludes with a comprehensive **Assessment and Challenges** chapter featuring test sets, exam-thinking application problems, extended exam-preparation questions, and STEM exploration tasks—preparing students for **AP, IB, SAT, AMC, Olympiad, and university-prep mathematics**. Some advanced problems are intentionally open-ended to encourage deeper reasoning and independent thinking.

Anil Kumar Khandelwal is a mathematics educator, mentor, and author with over **15 years of teaching experience**, supporting students across AP, IB, SAT, Olympiad, and STEM pathways. He is the founder of **Global Math Institute Inc.** and the creator of one of the world's largest independent mathematics learning libraries, with **17,000+ instructional videos** used by learners in over **80 countries**. His work emphasizes clarity, structure, and deep understanding—helping students develop the habits of thinking required for long-term success in mathematics and STEM disciplines.

— Anil Kumar Khandelwal
Founder, Global Math Institute Inc.

Foreword

By Bharat Khandelwal

Opening Connection – Personal Memory

Nearly fifty years ago, when I had just entered high school, I faced my first serious challenge in mathematics: trigonometry.
I vividly recall going to my cousin, Anil Kumar, with a problem I couldn't solve. He helped me work through it, but soon I was back with another. It didn't take him long to realize that my difficulty wasn't only with the *new* material—it was rooted in the fundamental concepts that support trigonometry.

Realization & Lesson Learned – Importance of Foundations

That moment was a turning point. I discovered that trigonometry is not an isolated topic—it depends on a solid command of:

- Algebra and its manipulations
- Ratios and proportions
- The Pythagorean theorem
- Similar triangles and geometry fundamentals

Once these building blocks were strengthened, trigonometry became not just manageable, but fascinating. I began solving more complex problems, including advanced trigonometric identities that I wouldn't have imagined attempting before.

Why This Book Matters – Connection to the Present Work

This book, *Rediscover Algebra and Trigonometry – A STEM Journey for Every Learner*, follows that same philosophy.
It doesn't drop the learner into trigonometry unprepared. Instead, it builds the essential skills first, then progresses to right triangle trigonometry, trigonometric identities, and real-world STEM applications.
The approach mirrors the journey I went through, but in a structured, interactive way—supported by examples, challenges, and even QR code video explanations.

Broader Impact – Lifelong Skills

Strong mathematical foundations are not only for passing exams—they shape problem-solving ability, logical reasoning, and confidence. These are skills that extend far beyond the classroom into professional and everyday life. I've experienced this firsthand in my career and continue to value the clarity and discipline that math teaches.

Closing Endorsement – Personal & Motivational

I am proud to see how Anil has transformed our early learning experiences into a resource that will guide the next generation. Whether you are a student, educator, or lifelong learner, this book will help you rediscover mathematics—not as a set of rules to memorize, but as a tool to explore, connect, and create.

Bharat Khandelwal
Managing Director and Partner, Alvarez and Marsal
Ex-Managing Director and Partner, BCG

Acknowledgments

This book was inspired by two kinds of learners who embody true courage:

- those who return to mathematics to rebuild confidence and understanding, and
- those young, ambitious minds preparing for AP, IB, and advanced STEM programs who seek deeper clarity before leaping ahead.

While teaching thousands of students across the world, I noticed a recurring truth: brilliant learners often struggle not because of a lack of intelligence, but because foundational connections were never fully built. **Rediscover Algebra and Trigonometry** was created to close that gap — to strengthen readiness for higher mathematics and nurture confidence through conceptual clarity.

I owe my deepest gratitude to my father, whose *Saturday math sessions at IIT Delhi* planted the earliest seeds of this journey. His quiet discipline, depth of understanding, and love for structured thinking shaped the teacher I am today.

To my students — from middle schoolers discovering their potential to IB and AP learners aiming high — you are living proof that curiosity is the true starting point of excellence.
Your persistence and enthusiasm make every lesson meaningful.

To parents, educators, and well-wishers across continents who continue to support this mission — thank you for believing in purposeful education.
Learning, after all, is a journey we walk together.

— Anil Kumar Khandelwal
Founder, Global Math Institute Inc.

Contents

Chapter 1: Core Skills for Success

Building the Foundation for Algebra, Trigonometry, and Real-World Problem Solving

Every successful journey into algebra and trigonometry begins with strong foundational skills. In this chapter, you will sharpen your understanding of numbers, operations, patterns, and reasoning — the essential tools used everywhere in STEM, from engineering and finance to physics and computer science.

Big Questions to Explore

As you progress through this chapter, keep these guiding questions in mind:

- What happens if two people simplify the same expression using different rules?
- Can every decimal be expressed as a fraction?
- How do ratios and fractions help us understand motion, speed, and real-life rates?
- If motion can be divided infinitely, can an object ever truly reach its destination? *(Explore this surprising idea through Zeno's Paradox.)*
- Why do governments and businesses track percent change so closely?

Skills You Will Build

By the end of this chapter, you will be able to:

- Apply **Order of Operations** with complete confidence
- Simplify **fractions** and **rational expressions**, including multi-layered forms
- Use **LCM**, **ratio**, and **unitary methods** to solve real-life problems
- Model everyday situations using **speed**, **time**, and **percent** calculations
- Understand **percent change**, **tax**, and **interest** as practical math
- Evaluate your mastery through practice questions and a short test

STEM Insight

Whether you are building bridges, programming robots, planning experiments, or managing budgets, these foundational skills form the everyday language of problem solving. Once mastered, they unlock the door to deeper mathematics — algebra, trigonometry, calculus, and advanced STEM challenges.

Order of Operation

Why We Need Order

A simple expression can lead to many answers:

2 + 3 × 4 − 1

Solved freely, students might get **11, 13, 15, or 19** – but only one is correct.

Mathematics isn't about guessing; it's about clarity and logic.
To ensure everyone simplifies expressions the same way, we follow a universal rule called the **Order of Operations**.

It's not just a shortcut – it's the *grammar of mathematics*, helping us read and simplify expressions consistently.

Use BEDMAS (or PEMDAS) to remember the order:

- **B** – **Brackets** *(Parentheses)* first
- **E** – **Exponents** *(Powers and Roots)*
- ÷ × **D / M** – **Division and Multiplication** *(from left to right)*
- + − **A / S** – **Addition and Subtraction** *(from left to right)*

SCAN ME

> Multiplication and division are done in the order they appear from left to right – same with addition and subtraction.

Let's Revisit Our Example

2 + 3 × 4 − 1

Using BEDMAS:

1. No brackets or exponents
2. Do the multiplication first: $3 \times 4 = 12$
3. Then go left to right: $2 + 12 - 1 = 14 - 1 = 13$

Final Answer: 13

> By following **BEDMAS**, we bring order to operations—and logic to learning.
> It's how mathematics speaks clearly.

Watch Video **Arithmetic Lesson** SCAN ME

Example 1: Evaluate Arithmetic Expressions

Please provide reasons for each step as shown in the first example.

a.	$4 + 3 \times 5$ $= 4 + 15$ Multiplication first $= 19$ Then add	b.	$(4 + 3) \times (-5)$ $= 7 \times (-5)$ $= -35$
c.	$2^2 + (7 + 3) \div (-2) \times (-5)$ $= 4 + (10) \div 2 \times 5$ $= 4 + 5 \times 5$ $= 4 + 25 = 29$	d.	$4 - 4 \times 2 + (-3)^2$ $= 4 - 4 \times 2 + 9$ $= 4 - 8 + 9$ $= -4 + 9 = 5$

Guided Practice 1: Evaluate Arithmetic Expressions

Students are encouraged to provide reasons for each step.

a.	$1 - (-2) + (-3) + 4 - 5 + 6$	b.	$3 \times 4 \div 6 \times 2 \div 4$
c.	$10^2 - 4 \div 2 + 3^2 \times 5$	d.	$1 + 3 \times (4 - 2) \div 6$

Guided Practice 2: Insert brackets to make each statement true.

a.	$8 + 2 \times 3^2 - 2 = 22$	b.	$8 + 2 \times 3^2 - 2 = 42$
c.	$8 - 2 \times 3^2 - 2 = 2$	d.	$8 - 2 \times 3^2 + 2 = 56$

Guided Practice 3: Insert brackets to make the largest number

a.	$4 - 2 \times 3^2 - 5 \div 2$	b.	$2 \times 3 - 8 \div 2 + 4 \times 3^2$

Wonderful Palindromic Numbers with Simple Formula: $(10^n - 1)^2 \div 81$

Use formula $(\mathbf{10^n - 1})^2 \div \mathbf{81}$ to create a series for $\boldsymbol{n} = \mathbf{1, 2, 3, 4, 5, 6}, \ldots$

Record the numbers you get and look for patterns — do they read the same backward and forward?
Note how the number of digits changes and where the middle digits appear.
What do you notice as n increases?

Number Facts — Discovering Patterns in Simplicity

Numbers aren't just for counting — they tell stories, form patterns, and reveal deep connections. Let's rediscover some fascinating truths that make math come alive!

1 The Power of Zero

Zero means *nothing*, yet it changes everything.
Add or subtract it — no change.
Multiply anything by it — all disappears.
Divide by zero? ❌ Undefined — because nothing can be divided into "nothing parts."

2 The Magic of One

One is identity.
Any number multiplied or divided by **1** stays the same.
It's the foundation of the *identity property* — the anchor of all numbers.

3 Opposites Attract

Every positive number has a negative partner:
$+7 \leftrightarrow -7$
Together they balance to **zero**.
That's symmetry — the essence of algebra and geometry.

4 Powers, Indices, or Exponents

Repeated multiplication can be written in a compact form using **exponents**:

$$2^3 = 2 \times 2 \times 2 = 8$$

The small, raised number (called the **index** or **exponent**) shows **how many times** the base is multiplied by itself.
Patterns in powers reveal how numbers grow — doubling, squaring, or soaring to infinity!

5 Patterns Hidden in 9

The number 9 holds remarkable symmetry in our base-ten system: The sum of digits of any multiple of 9 is always 9 or another multiple of 9.
e.g. $18 \rightarrow 1 + 8 = 9$ $63 \rightarrow 6 + 3 = 9$

6 Patterns in Place Value

Each place represents a power of 10: 1, 10, 100, 1000...
A simple shift in place changes everything — 3 becomes 30, 300, 3000.
That's why our number system is called *positional.*

7 Reciprocal Relationships

Every number (except 0) has a **reciprocal** — its *flip partner*.
$\frac{1}{2} \leftrightarrow 2$ $5 \leftrightarrow \frac{1}{5}$
Multiply a number by its reciprocal, and you always get **1.**

8 Power of Ten

Moving the decimal one place to the right multiplies by 10; one place to the left divides by 10.
That simple shift powers our metric system and scientific notation.

9 Prime Foundations

Every whole number is built from *prime factors*.
Like atoms in matter, primes are the **building blocks** of numbers:
e.g. $60 = 2 \times 2 \times 3 \times 5$.

Scientific Notation

Writing real numbers that are too large or too small in decimal form as $a \times 10^n$ where value of a **is at least 1 but less than 10** $(1 \le a < 10)$ and the exponent of 10 is an integer. The coefficient a is called **significand**. It can have negative values. The exponent n is called order of **magnitude**. It helps to compare numbers.

Note: $10^0 = 1;\ 10^1 = 10;\ 10^{-1} = \frac{1}{10} = 0.1$

Example 1: Numbers in Scientific Notation

Numbers	Scientific Notation	Remarks
5	5×10^0 or 5	10^0 is normally omitted
365	3.65×10^2	Days in a year
12742	1.2742×10^4	Earth diameter in km
0.000089	8.9×10^{-5}	Human hair diameter in m
37 trillion	3.7×10^{13}	Cells in human body
1 nanometer	1×10^{-9}	0.000 000 001 m

Significant Digits (or Significant Figures)

Significant digits are the digits in a number that carry meaning about its precision. The digits of a in $a \times 10^n$ in scientific notation **are the significant digits.**

Rules for Significant Digits:

1. All non-zero digits are significant. Example: 345 has 3 significant digits.
2. Zeros between non-zero digits are significant. Example: 1005 has 4 significant digits.
3. Leading zeros (before the first non-zero digit) are not significant. Example: 0.0048 has 2 significant digits.
4. Trailing zeros in a number with a decimal point are significant. Example: 45.00 has 4 significant digits.
5. Trailing zeros in a whole number without a decimal may or may not be significant (depending on context). Example: 1500 could have 2, 3, or 4 significant digits.

The 1500 Dilemma: How Many Significant Digits?

The number **1500** often causes confusion when counting significant digits.

- Written simply as **1500**, it may be taken as **2 significant digits** (the trailing zeros are not counted).
- But in a precise measurement, it could mean **3 or even 4 significant digits**.

The issue: **we cannot know the precision from 1500 alone.**

That's where **scientific notation** removes ambiguity:

- $1.5 \times 10^3 \rightarrow$ **2 significant digits**
- $1.50 \times 10^3 \rightarrow$ **3 significant digits**
- $1.500 \times 10^3 \rightarrow$ **4 significant digits**

Takeaway:

Scientific notation tells us both the size of a number and the precision of its measurement.

Problem: Write the number of significant figures (s.f.) for each number.

a.	10.10 4	b.	0.001 1	c.	4.60×10^3 3	d.	1.100 4

Rounding Numbers

Rounding is the process of reducing the number of digits in a number while keeping its value close to the original. It makes numbers easier to work with in estimation and reporting.

Rounding Rules:

SCAN ME

1. Identify the place value to round off (tens, tenths, hundredths, etc.).
2. Look at the next digit (the one to the right):
   - If it is 5 or more, increase the rounding digit by 1.
   - If it is less than 5, keep the rounding digit the same.
3. Replace all digits to the right with zeros (if rounding whole numbers) or drop them (if decimals).

Examples:

- Round 487 to the nearest tens: 490
- Round 5.846 to the nearest hundredth: 5.85

Rounding to Significant Digits (Significant Figures, s.f.)

When asked to round to a certain number of significant digits:
1. Count digits from the first non-zero digit.
2. Keep only the required number of significant digits.
3. Round off accordingly.

Examples:

$0.003456 \rightarrow 2\ s.f. = 0.0035$

$48{,}726 \rightarrow 3\ s.f. = 48{,}700$

$6.789 \rightarrow 2\ s.f. = 6.8$

$9.10938356 \times 10^{-31} \rightarrow 3\ s.f. = 9.11 \times 10^{-31}$

$1.8987 \times 10^{27} \rightarrow 3\ s.f. = 1.90 \times 10^{27}$

> **Tips:**
>
> Don't confuse decimal places with significant figures, s.f.
>
> ➤ 45.678 rounded to 3 decimal places is 45.678, but to 3 significant figures, it's 45.7.

Upper and Lower Bound

Approximation involves estimating a value that cannot be determined exactly.
The **upper bound** gives the greatest possible value, while the **lower bound** gives the least possible value of a measured or calculated quantity.

Example: Measurement Bound

A metal rod is measured as **25 cm**, correct to the nearest **0.1 cm**.

- **Lower bound:** $25 - 0.05 = 24.95$ cm
- **Upper bound:** $25 + 0.05 = 25.05$ cm
  So, the true length lies between **24.95 cm** and **25.05 cm**.

SCAN ME

> **Application Problem**
>
> Atoms are so small that 6.022×10^{23} gold atoms have a mass of approximately 150 gm. How many atoms of gold are there in 1 kg gold bar? Round your answer to 3 significant figures. 4.01×10^{24}

> **Challenge Problem**
>
> A car travels a distance $55.5\ km$, correct to the nearest $0.05\ km$. It takes $48.7\ minutes$, correct to one decimal place. Find the lower and upper bound for the average speed of the car in km/h, correct to one decimal place. $68.0; 68.7\ km/h$

Fractions and Rational Expressions

Fractions and the Infinite Race

A Paradox to Begin With

The Story of Achilles and the Tortoise

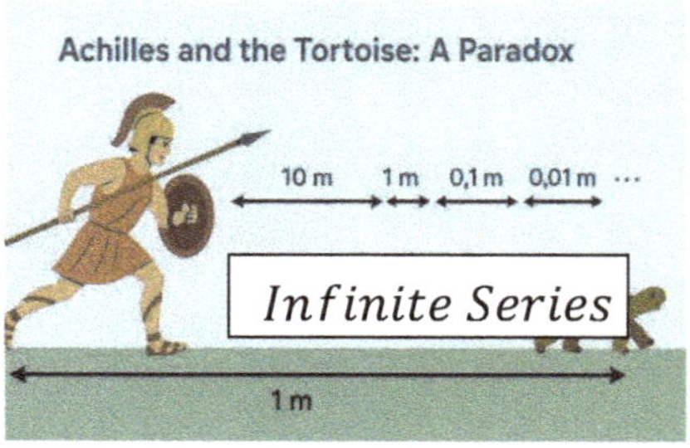

Imagine this:

Achilles — the fastest Greek warrior — challenges a slow-moving tortoise to a race. To be fair, he gives the tortoise a 10-meter head start.

But something strange happens...

- Achilles runs the 10 meters to catch up — but by then, the tortoise has moved 1 more meter.
- Achilles runs that extra meter — and the tortoise is now 0.1 meters ahead.
- He closes that gap — but the tortoise has moved another 0.01 meters.
- And this goes on... and on... and on.

Every time Achilles reaches the tortoise's last spot, the tortoise is slightly ahead.

This ancient puzzle is known as **Achilles' Paradox**, first posed by the Greek philosopher **Zeno**. It challenges our very idea of motion, time — and numbers.

✨ **A Glimpse Ahead...**

As we dive into fractions, you'll learn the secret behind this race — and calculate **exactly** how far the tortoise runs before being overtaken.

Spoiler: It's **not infinite**. It's a **rational number**!

So, let's begin the chapter, not with confusion — but with curiosity.
Because sometimes, chasing a tortoise teaches more than catching it.

Watch Video **Achilles' Paradox** **SCAN ME**

Rational Numbers

Rational numbers can be written in the form $\frac{p}{q}, q \neq 0$

Since **fractions** are exactly this — a numerator divided by a non-zero denominator — **every fraction is a rational number**.umb

SCAN ME

Definition & Key Concepts

- A fraction represents a part of a whole or a division of two quantities.
- It is written as: $\frac{a}{b}$, where a is numerator, b is denominator and $b \neq 0$

Types of Fractions

Type	Example	Description
Proper Fraction	$\frac{3}{4}$	Numerator < Denominator
Improper Fraction	$\frac{7}{4}$	Numerator > Denominator
Mixed Number	$1\frac{1}{2}$	Whole Number + Proper Fraction

Example 1

Write $\frac{7}{4}$ as a mixed number.

Divide:77 $\div 4 = 1$ remainder 3

So, $\frac{7}{4} = 1\frac{3}{4}$

Visual Explanation

A circle divided into 4 equal parts. Shade 7 parts (requires two circles).

$\frac{7}{4}$

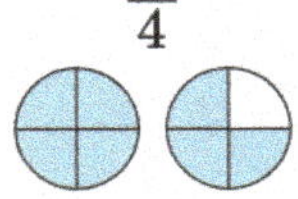

Example 2

Convert $2\frac{2}{3}$ into an improper fraction.

Multiply & add:

$(2 \times 3) + 2 = 8$

So,

$2\frac{2}{3} = \frac{8}{3}$

Practice: Change the form of the following fractions:

a.	$\frac{5}{2}$	b.	$3\frac{2}{5}$	c.	$\frac{6+1}{6}$

Explore Strategy: Find the remainder when $31 \times 43 \times 29 \div 5$ 2

Decimal Equivalents of Fractions

When a fraction is written as a decimal, the result may either terminate (end after a few digits) or repeat endlessly in a pattern.

SCAN ME

Terminating Decimals

A terminating decimal ends after a fixed number of digits.
This happens when the **denominator of the simplified fraction** (in lowest terms) has only 2 and/or 5 as prime factors. In other words, prime factorization of the denominator should be of the form $\mathbf{2^m} \times \mathbf{5^n}$ for terminating decimals.

Example 3: Write Fractions as Terminating Decimals

$\frac{3}{4} = 0.75$	$\frac{17}{25} = 0.68$	$\frac{9}{8} = 1.125$

💡 *These decimals "terminate" because 2 and 5 are the only primes that divide evenly into powers of 10 (like 10, 100, 1000).*

Example 4: Write Terminating Decimals as Rational Numbers (Fractions)

Multiply and divide the number with $\mathbf{10^d}$ where $\boldsymbol{d}$ is number of digits after decimal.

a.	$\mathbf{1.2} = \frac{12}{10} = \frac{6}{5}$	b.	$\mathbf{1.25} = \frac{125}{100} = \frac{5}{4}$	c.	$\mathbf{0.05} = \frac{5}{100} = \frac{1}{20}$

Example 5: Are the following fractions terminating or non-terminating decimals.

a.	$\mathbf{\frac{21}{120}}$	b.	$\mathbf{\frac{26}{128}}$	c.	$\mathbf{\frac{10}{75}}$

Steps: Simplify and then check the prime factors of denominator

#1: Simplify	$\frac{21}{120} = \frac{7}{40}$	$\mathbf{\frac{26}{128}} = \frac{13}{64}$	$\mathbf{\frac{10}{75}} = \frac{2}{15}$
#2: Prime-factor denominator	$\frac{7}{40} = \frac{7}{2 \times 2 \times 2 \times 5}$	$\frac{13}{64} = \frac{13}{2^6}$	$\frac{2}{15} = \frac{2}{3 \times 5}$
#3: Terminating/ Non-terminating	Terminating	Terminating	Non-Terminating since 3 is factor.

Repeating Decimals

A repeating decimal goes on forever, following a repeating pattern of digits. This occurs when the denominator has prime factors other than 2 or 5.

Example 6: Write Fractions as Repeating Decimals

$\frac{2}{3} = 0.6666 \ldots = 0.\overline{6}$	$\frac{11}{6} = 1.8\overline{3}$	$\frac{9}{7} = 1.\overline{285714}$

*The repeating block is called the "**repetend**," and it can be one digit or more. In the fraction $\frac{9}{7} = 1.\overline{285714}$, 285714 is the **repetend**. This repeating block is of 6 digits.*

This shows that **rounding may result in approximation**. For **exact answers**, they should be written in **fractions**.

SCAN ME

Converting Repeating Decimals to Rational Numbers in Simplest Form

Example 7: Write Repeating Decimals as Rational Numbers (Fractions)

Notation $0.\overline{3}$ is short form of writing $0.333 \ldots$

Steps Required:

- Let the decimal number be equal to a variable, x
- Let p be the number of digits in repeating block of x.
- Multiply both sides by 10^p. Now repetend is same for x and $x \times 10^p$
- $x \times 10^p - x$ is a terminating decimal.
- Solve: $x = \frac{x(10^p - 1)}{10^p - 1}$

Illustrated Example:

a.		b.	
	$\mathbf{0.333 \ldots = 0.\overline{3}}$		$\mathbf{0.\overline{47}}$
	$let\ x = 0.333 \ldots$		$let\ x = 0.\overline{47}$
	$10x = 3.333 \ldots$		$100x = 47.\overline{47}$
	$10x - x = 3.\overline{3} - 0.\overline{3}$		$100x - x = 47.\overline{47} - 0.\overline{47}$
	$9x = 3$		$99x = 47$
	$x = \frac{3}{9} = \frac{1}{3}$		$x = \frac{47}{99}$

Achilles' Paradox: Unique Solution

Table below shows the distance travelled by Achilles and Tortoise.

Steps	Achilles Runs	Tortoise Moves
1	10 m	1 m
2	1 m	0.1 m
3	0.1 m	0.01 m
4	0.01 m	0.001 m
...	...	...
Total	11.111... m	1.111... m

This seems like an **infinite process**.
So here's the big question:

Can Achilles ever catch the tortoise?
And how far will he run?

Fractions to the Rescue

Let's look at the **distances**: Achilles runs 11.111...= $11.\overline{1}$

Repeating decimal. Show that $11.\overline{1} = 11\frac{1}{9}$

Exact distance run by

- Achilles is $11\frac{1}{9}$ m
- Tortoise is $1\frac{1}{9}\ m$, (10 m less than Achilles).

That's the power of **fractions** and **mathematical thinking**.

In just a few pages, you have discovered:

- How repeating decimals like 1.111... can be written as **exact fractions**
- How infinite steps can lead to **finite answers**
- And how paradoxes like this are not tricks — they're gateways to deeper understanding

🎯 Conceptual / Deep Thinking: **Infinity meets reason — where endless steps add up to understanding.**

STEM Project: Explore and Learn

$$\frac{1}{7} = 0.\overline{142857}$$

Explore the number **142857**, known as the ***repetend*** *(repeated block)* of $\frac{1}{7}$.

Multiply it by **2, 3, 4, 5,** and **6** to discover why it's called a **cyclic number** — where digits repeat in a rotating pattern.

📘 **Reference:** Explore of cyclic numbers, *Inspired by Math – Volume I: Kaprekar's World – Numbers That Think*, **page 122** by **Anil Kumar Khandelwal**.

Add/Subtract Fractions: Basic Skills Review

Example 8: Add/Subtract Fractions with Same or Common Denominator

a.	$\frac{3}{7}+\frac{2}{7}=\frac{5}{7}$	b.	$\frac{4}{11}-\frac{3}{11}=\frac{1}{11}$

Example 9: Add/Subtract Fractions with Different Denominator

- Find LCM (Lowest Common Multiple) as the Common Denominator.
- Write equivalent fractions.
- Perform operation
- Simplify result, if required

Explore 4 Methods to find LCM next

c.	$\frac{1}{2}+\frac{1}{3}$ $=\frac{3}{6}+\frac{2}{6}$ $=\frac{5}{6}$ Write equivalent fractions with LCM of 6 as denominator Add numerators to get the sum of fractions	d.	$1+\frac{2}{5}-\frac{3}{10}$ $LCM=10$ $=\frac{10}{10}+\frac{4}{10}-\frac{3}{10}$ $=\frac{11}{10}=1\frac{1}{10}$

Guided Practice: Add/Subtract Fractions with Different Denominator

e.	$\frac{1}{4}+\frac{5}{6}$	f.	$1+\frac{4}{5}-\frac{3}{7}$
g.	$\frac{5}{12}+\frac{5}{18}$	h.	$\frac{3}{8}+\frac{1}{6}+\frac{3}{10}$
i.	$\frac{1}{12}-\frac{5}{8}+1$	j.	$2-\frac{2}{3}-\frac{3}{4}$
k.	A tank is **1/5 full**. How much water should be added so that only **1/10 of the tank remains empty**?		

Challenge Question

$\frac{1}{8}$ of a container is filled. How much water should be added so that only $\frac{1}{12}$ of this container remains empty.

Watch Video **Find LCM** **SCAN ME**

Four Methods to LCM and GCF

There are **four methods** to find the LCM **(Lowest Common Multiple)** and GCF (Greatest Common Factor) of two or more numbers. Each method has its own approach and advantages. Let's begin with the first method:

Method 1 for LCM: Listing Multiples for LCM

This method is good for small numbers.

Example 10: Write Multiple to Find LCM

Find LCM of 4 and 6. In short it is also written as **LCM(4, 6)**

Step 1: Write the first few multiples of each number.

- Multiples of 4: 4, 8, **12**, 16, 20, 24, 28, 32, 36
- Multiples of 6: 6, **12**, 18, 24, 30, 36, 42, 48

Step 2: Find the **lowest common multiple** in both lists.
LCM(4, 6) = 12

Note: Other common multiples like 24, 36, etc. are all **multiples of the LCM**.

Method 1 for GCF: Listing Factors for GCF

Example 11: Write Factors to Find GCF

Find GCF of 24 and 16

Step 1: Write the factors for both the numbers.

- Factors of 24: 1, 2, 3, 4, 6, **8**, 12, 24
- Factors of 16: 1, 2, 4, **8**, 16

Step 2: Find the **greatest common factor** in both lists.
GCF(24, 16) = 8

✍ Practice Question

Find LCM and GCF between the following numbers:

a.	12, 18	b.	16, 24	c.	8, 9

Method 2: Prime Factorization

Example 12: Use Prime Factorization to Find LCM and GCF of 12 and 18

Step 1: Write the prime factorization of each number

$12 = 2 \times 2 \times 3 = 2^2 \times 3$

$18 = 2 \times 3 \times 3 = 2 \times 3^2$

Step 2 For LCM: Take **each prime factor** that appears and use the **highest power**

LCM = $2^2 \times 3^2 = 36$

Step 2 For GCF: Take **each prime factor** that appears and use the **lowest power**

GCF = $2 \times 3 = 6$

NOTE: LCM $\times$ GCF = 12 $\times$ 18

Method 3: Venn Diagram with Prime Factorization

Break numbers into their prime factors. Multiply all prime factors with highest powers to get the LCM. Use a Venn diagram to clearly see what's shared and what's unique — helping us calculate both GCF and LCM with ease.

Example 13: Find and Visualize LCM: Prime Factorization and Venn Diagram

Find LCM and GCF of 18 and 24

Step 1: Prime Factorization:

$18 = 2 \times \mathbf{3^2}$

$24 = \mathbf{2^3} \times 3$

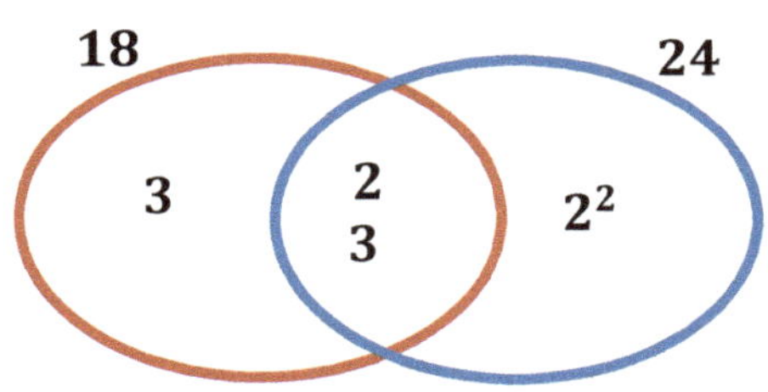

Step 2: Complete Venn Diagram

Step 3: Result of GCF and LCM

GCF is the intersection of prime factors	LCM is the union of prime factors
$\boldsymbol{GCF = 3 \times 2 = 6}$	$\boldsymbol{LCM = 3 \times 2 \times 3 \times 2^2 = 72}$

✍ Practice Question

Find LCM and GCF between the following numbers using Venn Diagram

a.	25, 30	b.	96, 120	c.	1001, 91

Method 4: Find the GCF of 84 and 126 using repeated division.

Step 1 – Divide the larger number by the smaller number

$126 \div 84 = 1\ remainder\ 42$

Step 2 – Replace the larger number with the smaller number, and the smaller number with the remainder
Now work with **84** and **42**:

$84 \div 42 = 2\ remainder\ 0$

Step 3 – Stop when the remainder is 0
When the remainder becomes **0**, the **divisor** at that step is the GCF.

So: GCF=42

Final Answer:
The **GCF of 84 and 126 is 42**.

Method 4: LCM by Ladders Division

Example 14: Add More than One Fraction $\frac{5}{12} + \frac{9}{16} + \frac{7}{18}$

To add the fractions, we need to find the LCM of 12, 16, and 18.

Here is the process of finding LCM

With description on the right side.

2	12	16	18
2	6	8	9
3	3	4	9
	1	4	3

LCM = 2 x 2 x 3 x 4 x 3 = 144

Step-by-Step Process:

1. **Write all numbers side by side.**
2. Start dividing by the **smallest prime number** that divides at least two of them.
3. **Write the quotient below** each number (if divisible), or repeat the number (if not divisible).
4. Repeat the process with the next line until **no two numbers have any common prime factor**.
5. Multiply all the prime numbers to calculate the **LCM**.

$$\frac{5}{12} + \frac{9}{16} + \frac{7}{18} = \frac{60}{144} + \frac{81}{144} + \frac{56}{144} = \frac{197}{144}$$

Add equivalent fractions with denominator of 144

💡 What is the LCM and GCF between two consecutive numbers?

Guided Practice: Calculate each of the following.

a.	$\frac{9}{14}+\frac{3}{20}+\frac{5}{24}$	b.	$\frac{7}{12}+\frac{9}{10}+\frac{8}{15}$	c.	$\frac{5}{8}+\frac{8}{15}+\frac{11}{40}$

Watch Video **HCF 3 Methods** **SCAN ME**

Example 15: Multiply and Simplify Fractions

Simplify means to divide the numerator and denominator with common factors. Multiply numerators together and the denominator together for fractions.

First Multiply and then Simplify		First Simplify and then Multiply.	
$\frac{3}{4}\times\frac{2}{9}$ $=\frac{3\times 2}{4\times 9}$ $=\frac{6}{36}$ $=\frac{1}{6}$	#1 Multiply numerators and denominators #2 The common factor is 6. Divide both numerator and denominator by 6 to write simplified answer.	$\frac{16}{25}\times\frac{15}{24}$ $=\frac{16\times 15}{25\times 24}$ $=\frac{2\times 3}{5\times 3}$ $=\frac{2}{5}$	#1 Do not Multiply but simplify first to avoid working with big numbers. #2 This is a better method as can be seen from the example.

Communication: Explain why division before multiplication is better.

Example 16: Order of Operations with Fractions

Calculate $\left(\frac{2}{3}\right)^2-\frac{4}{9}\times\frac{3}{8}+\frac{1}{2}$ Bracket first	$=\frac{4}{9}-\frac{4}{9}\times\frac{3}{8}+\frac{1}{2}$ $=\frac{4}{9}-\frac{1}{6}+\frac{1}{2}$ Multiply then Add	$=\frac{8}{18}-\frac{3}{18}+\frac{9}{18}$ $=\frac{14}{18}=\frac{7}{9}$ Simplify

Guided Practice

$\frac{1}{2}+6\left(\frac{1}{3}\right)^2-\frac{2}{3}$	$\frac{3}{4}-\left(3-\frac{5}{6}\right)\times\frac{13}{2}$	$\frac{1}{2}+6\div\left(\frac{1}{4}-\frac{1}{3}\right)^2$

Example 17: Evaluate Expression

$$\frac{\frac{1}{2}+\frac{1}{3}}{\frac{1}{2}-\frac{1}{3}}$$

Strategy:

Multiply each term by LCM of 6

$$\frac{\frac{1}{2}+\frac{1}{3}}{\frac{1}{2}-\frac{1}{3}}=\frac{3+2}{3-2}=\frac{5}{1}=5$$

Example 18: Evaluate Expression

$$\frac{\frac{1}{2}+\frac{2}{3}}{\frac{3}{4}-\frac{1}{6}}$$

Strategy:

Multiply each term by LCM of 12

$$\frac{\frac{1}{2}+\frac{2}{3}}{\frac{3}{4}-\frac{1}{6}}=\frac{6+8}{9-2}=\frac{14}{7}=\mathbf{2}$$

Example 19: Evaluate Expression

$$\frac{1}{3-\frac{1}{3-\frac{1}{3}}}$$

Strategy: Start from bottom and move up

$$\frac{1}{3-\frac{1}{3-\frac{1}{3}}}=\frac{1}{3-\frac{1}{\frac{9-1}{3}}}=\frac{1}{3-\frac{1}{\frac{8}{3}}}=\frac{1}{3-\frac{3}{8}}=\frac{1}{\frac{24-3}{8}}=\frac{1}{\frac{21}{8}}=\mathbf{\frac{8}{21}}$$

Guided Practice Evaluate each of the following

$\frac{\frac{1}{2}+\frac{2}{3}}{\frac{3}{4}-\frac{1}{6}}$	$\frac{\frac{2}{3}+\frac{3}{4}}{1-\frac{5}{6}}$	$\frac{1}{2-\frac{3}{4-\frac{1}{5}}}$

Watch Video **Multilayer Fraction** SCAN ME

Example 20: Application of LCM

There are 3 lamps:

- 🔴 Lamp A flashes every 20 seconds
- 🟠 Lamp B flashes every 45 seconds
- 🟢 Lamp C flashes every 120 seconds

They all flash together at 12:00 noon.
Question: How many times will all three lamps flash together in the next hour?

Solution 20:

Step-by-Step Solution

Step 1: Understand What "Flash Together" Means

This happens when all the lamps flash **at the same time** — like a little **light show**!

This is an **LCM problem** — we need to find the **Least Common Multiple** of 20, 45, and 120.

Step 2: Find LCM

LCM = 360 as shown on the right side with ladder division method.

Step 3: Convert 1 Hour to Seconds

1 hour = 60 minutes = **60 × 60 = 3600 seconds**

Step 4: Find How Many Times They Flash Together

They flash together every **360 seconds.**

Now divide total seconds in 1 hour by 360:

3600 ÷ 360 = 10

5	20	45	120
4	4	9	24
3	1	9	6
	1	3	2

LCM = 5 x 4 x 3 x 3 x 2 = 360

The Key Detail:

First Flash Counts!

The first flash is at 0 seconds (exactly at 12:00 noon).

Final Answer: They will flash together 11 times in 1 hour. (1 + 10)

💡 Tip: **First flash counts and then we have 10 more in an hour.**

Watch Video **Prime Factorization Extended Applications** **SCAN ME**

Real Life Application of GCF

<table>
<tr>
<td>

Application 1 – Arranging Tiles

A rectangular floor measures 360 cm by 240 cm. You want to tile it with the largest possible square tiles so there's no cutting. What should be the side length of each tile? How many tiles are needed?

Solution:

Step 1 – Find the GCF of 360 and 240
- Prime factors of 360 = $2^3 \times 3^2 \times 5$
- Prime factors of 240 = $2^4 \times 3 \times 5$
- GCF = $2^3 \times 3 \times 5$ = 120 cm

Step 2 – Tile side length = 120 cm

Step 3 – Find number of tiles
- Floor area = 360 × 240 = 86,400 cm^2
- Tile area = 120 × 120 = 14,400 cm^2
- Number of tiles = 86,400 ÷ 14,400 = 6 tiles

Answer:

Largest tile size = 120 cm × 120 cm, Number of tiles = 6

</td>
<td>

Application 2 – Dividing Ropes Equally

Two ropes are 48 meters and 60 meters long. You want to cut them into equal-length pieces with no leftover rope. What is the maximum possible length of each piece?

Solution:

Step 1 – Find the GCF of 48 and 60
- Prime factors of 48 = $2^4 \times 3$
- Prime factors of 60 = $2^2 \times 3 \times 5$
- GCF = $2^2 \times 3$ = 12 m

Step 2 – Cut into pieces of 12 m each

Step 3 – Find total pieces
- Rope 1: 48 ÷ 12 = 4 pieces
- Rope 2: 60 ÷ 12 = 5 pieces

Answer:

Maximum piece length = 12 m, Total pieces = 9 pieces

</td>
</tr>
</table>

GCF Practice Questions

Q1	The side lengths of a rectangle are 84 cm and 126 cm. Find the largest square tile that can fit perfectly inside without cutting. 42 cm
Q2	A school is organizing sports teams. There are 96 boys and 72 girls. The school wants to form the largest possible number of identical teams so that each team has the same number of boys and the same number of girls. How many teams can be formed, and how many boys and girls will be in each team? 24 teams, each with 4 boys and 3 girls
Q3	In a bag there are blue, green, and red marbles. There are four times as many blue marbles as green. The ratio of blue to red marbles is 3:5. Find the minimum possible number of blue marbles in the bag. 12

LCM Applications – Practice Questions

Q1	Three traffic lights flash at intervals of 30, 45, and 60 seconds. If they all flash together at 9:00 AM, when will they next flash together?
Q2	Buses A, B, and C arrive at a station every 10, 20, and 25 minutes respectively. If they all arrive together at 6:00 AM, when will they next arrive together?
Q3	Two bells ring every 12 and 18 seconds. If they ring together now, how many times will they ring together in 3 minutes?
Q4	Three kids take turns on a slide every 6, 8, and 12 minutes. How often will they all slide at the same time?
Q5	Two rotating lights blink every 9 and 15 seconds. How many times do they blink together in one hour?
Q6	Machines A and B beep every 25 and 30 minutes respectively. How often do they beep together in 5 hours?
Q7	Trains stop at a station every 18, 24, and 30 minutes. What is the shortest time after which they all stop together?
Q8	Sam does jump jacks every 40 seconds, Alex every 50 seconds. When will they both jump at the same time again if they start now?
Q9	Water sprinklers A and B spray every 14 and 21 minutes. How many times do they spray together in 2 hours?
Q10	School bells ring every 20 and 32 minutes. How long until they ring together again?
Q11	The number 18 is a factor of 378. How can you use this information to determine LCM of 18 and 378? Explain your logic.
Q12	N is a positive integer which when divided by 5, 7, and 11 gives remainder 4, 6, and 10 respectively. Find possible value of N.

Answers: LCM Applications – Practice Questions

Q1	9:03 AM	Q2	7:40 AM	Q3	5 times	Q4	Every 24 minutes
Q5.	80 times	Q6	2 times	Q7	Every 6 hours	Q8	After 200 seconds
Q9	2 times	Q10	Every 160 minutes	Q12	384 Hint: 1 less than LCM		

Quiz Rational Numbers and Operations

Q1	Write $1\frac{1}{9}$ as improper fraction.
Q2	Write $\frac{12}{64}$ in simplified form
Q3	Write $\frac{13}{7}$ as mixed number
Q4	Explain why $\frac{5}{24}$ will result into non-terminating decimal.
Q5	Write three fractions between $\frac{3}{5}$ and $\frac{4}{5}$.
Q6	Arrange the fractions in increasing order: $\frac{5}{8}, \frac{6}{11}, \frac{7}{12}$
Q7	Convert the terminating decimal number 1.02 as a rational number.
Q8	Convert the repeating decimal number $1.\overline{217}$ as a rational number.
Q9	Is $\sqrt{4}$ a rational number? Explain your answer.
Q10	Write five unit-fractions of the form $\frac{1}{N}$ with finite decimals.
Q11	Provide an example of a number which is not a rational number. Explain why that is not a rational number?
Q12	Evaluate: $1+\frac{7}{18}+\frac{5}{12}-\frac{2}{15}$ $1\frac{121}{180}$
Q13	$\frac{7}{8}=\frac{1}{A}+\frac{1}{B}$ Find the value of A and B. Fraction of the form $\frac{1}{n}, n \neq 0$ is a unit fraction.
Q14	$\frac{49}{30}=\frac{a}{5}+\frac{b}{6}+\frac{c}{d}$ Use digits 1, 2, 3, and 4 to find the solution of the above equation.
Q15	If $x<y$, then $x<\frac{x+y}{2}<y$. Use this fact to prove that between any two rational numbers there are infinitely many rational numbers.
Q16	Two traffic lights change at intervals of 40 seconds and 60 seconds. If they turn green together at 8:00 a.m., how many times will they turn green together in one hour?
Q17	Show that 6 weeks is equal to $1 \cdot 2 \cdot 3 \cdot 4 \cdot 5 \cdot 6 \cdot 7 \cdot 8 \cdot 9 \cdot 10$ seconds

Ratio and Proportion

What is a Ratio?

A **ratio** is a way of comparing two quantities by showing how many times one value contains or is contained within the other.
It is expressed using **division** or with a **colon (:)**.

SCAN ME

Ratio of $a \; to \; b = \frac{a}{b} = a\!:b$

Examples of Ratios in Real Life

Classroom Example If there are **12 boys** and **8 girls** in a class, the ratio of boys to girls is: 12:8=3:2	*Cooking Example* To make orange juice, mix **2 parts of concentrate** with **5 parts of water**. The ratio of concentrate to water is **2:5**.
Maps and Scale On a map, **1 cm** represents **5 km**. The ratio of map distance to actual distance is: 1:500,000	*Finance* If a company's profit is **$40,000** and expenses are **$80,000**, the ratio of profit to expenses is: 40,000:80,000=1:2
Speed If a car travels **150 km in 3 hours**, the ratio of distance to time is: 150:3=50:1 This means the speed is **50 km per hour**.	*Sports* In a basketball game, a player scores **18 points out of the team's total 54 points**. The ratio of the player's score to the team's score is: 18:54=1:3 This means the player contributed **one-third of the total points**.

A ratio gives us a **simple, proportional relationship** that can be applied in mathematics, science, business, and daily life.

What is a Proportion or Directly Proportion?

A **proportion** is an equation that says two ratios are equal, $\frac{a}{b} = \frac{c}{d} = k$.

$\frac{4}{6} = \frac{12}{18}$ represents proportions as they are both equal to a constant, $\frac{2}{3}$.

Cross multiplication is a quick check for proportions $4 \times 18 = 72; 6 \times 12 = 72$

Illustrated Examples of Ratio and Proportions

Example 1: Basic Representation

If there are 3 apples and 2 oranges in a basket, the ratio of apples to oranges is:
Apples : Oranges = 3 : 2 or 3/2

Example 2: Equivalent Ratio

If a recipe uses 4 cups of flour and 6 cups of milk, the ratio is 4/6 = 2/3.
This means for every 2 cups of flour, you use 3 cups of milk.

Example 3: Classroom Ratio

In a class of 24 students, there are 15 boys. Ratio of boys to girls.
Solution: Girls = 24 − 15 = 9. Ratio of boys to girls = 15 : 9 = 5 : 3.

Example 4: Cooking Recipe

A juice recipe requires concentrate and water in the ratio 3:7.
If you have 210 mL of water, how much concentrate is needed?
Solution: 7 parts ↔ 210 → 1 part = 30. Concentrate = 3 × 30 = 90 mL.

Example 5: Map Scale

On a map, the scale is 1:50,000.
If two cities are 6 cm apart, what is the actual distance?
Solution: 6 × 50,000 = 300,000 cm = 3 km.

Example 6: Speed and Distance

A car travels 180 km in 3 hours. Find distance travelled in an hour.
Solution: Speed is Ratio of distance : time = 180:3 = 60:1. Speed = 60 km/h.

Example 7: Finance

A company spends $90,000 on salaries and $30,000 on rent.
Solution: Ratio of salaries to rent = 90,000 : 30,000 = 3 : 1.

Example 8: Sharing Money

Ali and Ben share $1200 in ratio 2:3. How much money do they get individually?
Solution: Total parts = 5 → 1 part = 240. Ali = 480, Ben = 720.

Communication and Thinking

How many different fraction equations can be written based on the proportion
$a : b = c : d$ 8

Direct Proportion

Two quantities are said to be **directly proportional** if an **increase in one quantity causes a proportional increase in the other**, and a **decrease in one causes a proportional decrease in the other**.

We write this as: $y \propto x \Rightarrow y = kx \Rightarrow \frac{y}{x} = \boldsymbol{k}$

SCAN ME

✨ **Key Idea**

- In **direct proportion**, the **ratio remains constant**.
- Example: If one doubles, the other doubles. If one is halved, the other is halved.

Wages and Hours Worked

If a worker earns $15 per hour:

- 2 hours → $30
- 4 hours → $60

Cost and Quantity

If apples cost $2 each:

- 5 apples → $10
- 8 apples → $16

Distance and Time (Constant Speed)

If a car moves at 60 km/h:

- In 1 hour → 60 km
- In 2 hours → 120 km

Example 1: Linear Quantities

If x and y are directly proportional and $x = 4$ when $y = 10$, then what is x when $y = 15$?

Solution 1:

Since x and y are directly proportion, the ratio $\frac{x}{y}$ is constant.

$$\frac{x}{y} = \frac{x}{15} = \frac{4}{10}$$

$$x = \frac{4}{10} \times 15 = \mathbf{6}$$

Example 2: Non-Linear Quantities

If x and y^2 are directly proportion and $x = 4$ when $y = 3$, then what is x when $y = 5$?

Solution 2:

Since x and y^2 are directly proportion, the ratio $\frac{x}{y^2}$ is constant.

$$\frac{x}{y^2} = \frac{x}{5^2} = \frac{4}{3^2}$$

$$x = \frac{4}{3^2} \times 5^2 = \mathbf{\frac{100}{9}}$$

Check Understanding

Which of the following do not represent direct proportion relation.

a.	$\frac{x}{y} = 10$	b.	$xy = 10$	c.	$\frac{x}{y} = 10$	d.	$10y = x$

Directly Proportion: Real Life Examples

Example 3:

A car uses 12 L of gas in 80 km.

How far will it go in 9 L.

$$\frac{x\ km}{9\ L} = \frac{80\ km}{12\ L}$$

$$x = \frac{80\ km}{12\ L} \times 9L = \mathbf{60\ km}$$

Example 3 Extended:

Visual Representation

Graph on the right displays distance (km) against fuel consumption (L)

Explain:

a. Why the graph is linear?
b. Why it passes through origin?
c. Will that be the case with all the directly proportion applications?

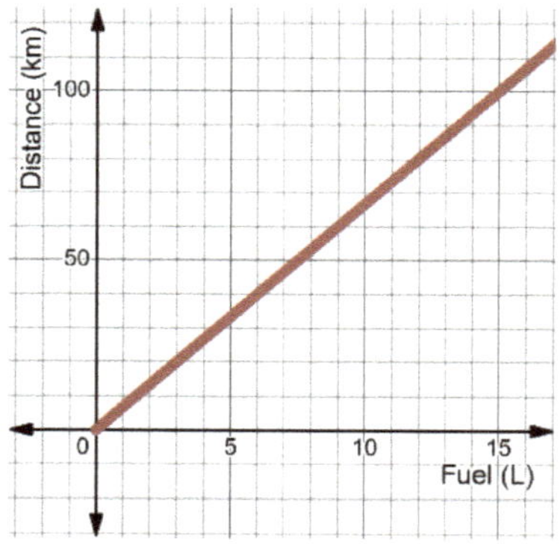

Suggestions:

Distance is proportional to the fuel consumed.

$$Distance \propto Fuel$$

$$Distance = constant \times Fuel$$

$$If\ Fuel = 0, Distance = 0$$

Slope of the line is the rate of **consumption** $= \frac{\mathbf{80\ km}}{\mathbf{12\ L}} = \frac{\mathbf{20\ km}}{\mathbf{3\ L}}$

Increase in one result in proportional increase in the other quantity

On a graph, direct proportion is always a straight line through the origin, $y = kx$

Practice Question: Hint: Unit conversion

If a copper wire of length 4 m costs $12, find the cost of a wire of length 50 cm.

- **Directly proportional** means "moving together in the same ratio."
- One goes up, the other goes up *in exact proportion*.
- One goes down; the other goes down *in exact proportion*.

Conversion of units

Ratio of Study to Play time is 5:2. Find Play time in minutes if Study time is 2 hours.

48 min.

Practice Questions on Ratios and Proportions

Q1	Cut a 21 cm wire in the ratio of 1:2. Find length of longer wire.
Q2.	Ratio of boys to children in a school is 5:8. Find the ratio of boys to girls in the school.
Q3.	If $x: 3 = 20: 12$, find x
Q4.	If $x: 3 = 27: x$, find a possible value of x
Q5	A recipe uses the ratio 3:2 for flour to sugar. If you use 6 cups of flour, how much sugar do you need?
Q6	A map uses a scale ratio of 1:1000. If two cities are 5 cm apart on the map, how far are they in real life?
Q7	The ratio of boys to girls in a class is 5:4. If there are 20 boys, how many girls are there? 16
Q8	The school choir has members in the ratio 2:3:5 for altos, tenors, and sopranos. If there are 50 sopranos, how many total members are there?
Q9	A fertilizer contains nitrogen, phosphorous and potassium in the ratio of 8:7:5. Calculate the mass of potassium in a bag of 50 kg fertilizer.
Q10	If a : b = 2 : 3 and b : c = 4 : 5, find the ratio of a : c. 8:15
Q11	If a : b = 3 : 4 and b : c = 5 : 7, find the ratio of a : c. 15:28
Q12	If A : B = 3 : 4 and B : C = 2 : 5, and A + B + C = 51 then find B. 12
Q13	Ratio of persons who like cars to vans is 5:3 in my workplace. If 32 more persons like car than van then how many persons are in my office?
Q14	Describe differences between directly proportional and inversely proportion.

TIPS

Age of father and son is 50 and 30 years now. After how many years will ratio of father and son's age be 3:2? 10

Watch Video **Ratios of Three** **SCAN ME**

Inversely Proportional

Two quantities are said to be inversely proportional $\left(y \propto \frac{1}{x}\right)$ when an **increase in one** causes a proportional **decrease in the other**, such that their product remains constant.

$$x \times y = k$$

Example 1: Inversely Proportion

If x and y are inversely proportional and $x = 4$ when $y = 10$, then find x when $y = 15$?

$x \times y = k \Rightarrow x \times 15 = 4 \times 10$

$x = \frac{4 \times 10}{15} = \frac{8}{3}$

x decreases as y increases

Example 2:

Speed and Time (Constant Distance)

- If a car travels 120 km:
  - At 60 km/h → takes 2 hours
  - At 40 km/h → takes 3 hours
- Speed × Time = 120 (constant).

Example 3:

Workers and Days (Fixed Work)

- 6 workers build a wall in 10 days.
- 12 workers will finish in 5 days.
- Workers × Days = 60 (constant).

Example 4:

Length and Width of a Rectangle (Fixed Area)

A rectangle of area 100 cm²:

- If length = 20 cm → width = 5 cm
- If length = 25 cm → width = 4 cm
- Length × Width = 100 (constant).

Example 5:

Pressure and Volume of a Gas (Boyle's Law in Physics)

- For a fixed amount of gas at constant temperature:
  - If volume is doubled, pressure is halved.
  - If volume is reduced, pressure increases.
- Pressure × Volume = constant.

Explore and provide few examples of quantities that are inversely proportional.

Extended Example 2: Analyze the Graph

Speed and Time

(Distance of 120 km)

Graph is provided on the right.

The graph is a hyperbola that never touches the coordinate axes.

Take some points from the graph and verify the inverse proportionality rule, product being constant.

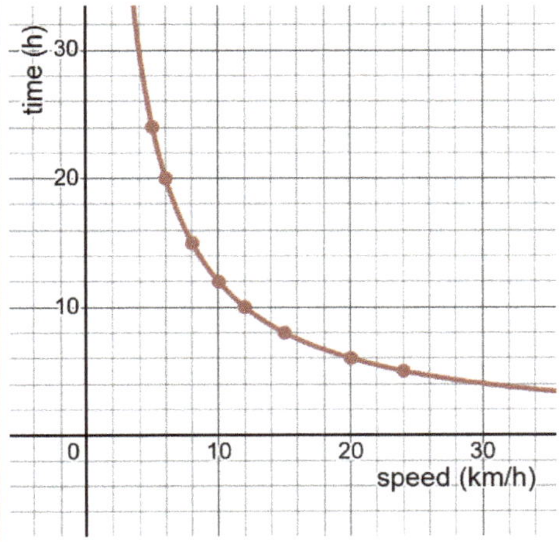

Explore Inversely Proportion

A ship has sufficient food to supply 800 passengers for 3 weeks. How long would the food last for 600 passengers? 4 weeks

Answers

Q1	14 cm	Q2	5:3	Q3	5	Q4	9	Q5	4 cups
Q6	5000 cm	Q7	16	Q8	100	Q9	12.5 kg	Q10	8:15
Q11	15:28	Q12	12	Q13	128				

Extended Solution: Ratio with Three Quantities

Q10	If a : b = 2 : 3 and b : c = 4 : 5, find the ratio of a : c. We need to make b same in both ratios to relate all three a : b = 2 : 3 = 8 : 12 b : c = 4 : 5 = 12 : 15 a : c = 8 : 15

TRY: If $\frac{2}{3}P = \frac{3}{4}Q = \frac{3}{2}R$ find the ratio of $P: Q: R$ $[9: 8: 4]$

Rate and Unitary Method of Problem Solving

What is a Rate?

A **rate** is a special kind of ratio that compares two different quantities with different units. It tells us how much of one quantity corresponds to a unit of another.

Speed = 60 km/hour (means 60 kilometers in 1 hour)

Cost = \$15 per kg (means \$15 for 1 kg)

What is the Unitary Method?

The Unitary Method is a powerful technique in problem-solving. It involves first finding the value of a single unit, and then using it to find the value of any required number of units.

Step 1: Find the value of 1 unit
Step 2: Multiply or divide to find the value of the desired quantity

Example 1: Cost of Dozens

If 12 dozen bananas cost \$24, find the cost of 5 dozen bananas.

Step 1: Cost of 1 dozen = \$24 ÷ 12 = \$2

Step 2: Cost of 5 dozen = \$2 × 5 = \$10

Example 2: Rate with Unit Conversion

A hummingbird flaps its wings 400 times in 8 seconds. How many times will it flap its wings in a minute?

Note: 1 minute is 60 seconds.

Step 1: flapping wings per second = 400 ÷ 8 = 50 times

Step 2: flapping wings in 1 minute (60 seconds) = 50 × 60 = 3000 times

Example 3: Unitary Method with Fractions

If $\frac{3}{7}$ of an amount of money is \$162, then find double the amount.

Step 1: Full Amount $= 162 \div \frac{3}{7} = 162 \times \frac{7}{3} = \$378$

Step 2: Double Amount $= 378 \times 2 = \$756$

Example 4: Tank Capacity in Proportion

If $\frac{2}{5}$ of the tank holds 1500 L of water, how many litters will be in the $\frac{1}{6}$ of tank.

Step 1: Total Capacity of Tank $= 1500 \div \frac{2}{5} = 1500 \times \frac{5}{2} = 3750\ L$

Step 2: $\frac{1}{6}$ of tank $= 3750 \times \frac{1}{6} = 625\ L$

Distance, Speed, and Time: Multilayer Fractions in Real Life

Overview

Every journey tells a story of how **far**, **how fast**, and **how long** – and that story is written by the simple yet powerful relationship:

Distance = Speed × Time.

SCAN ME

These problems help us connect real-life motion with mathematical reasoning.

Connection between distance – speed – time

$d = s \times t$	$s = \frac{d}{t}$	$t = \frac{d}{s}$

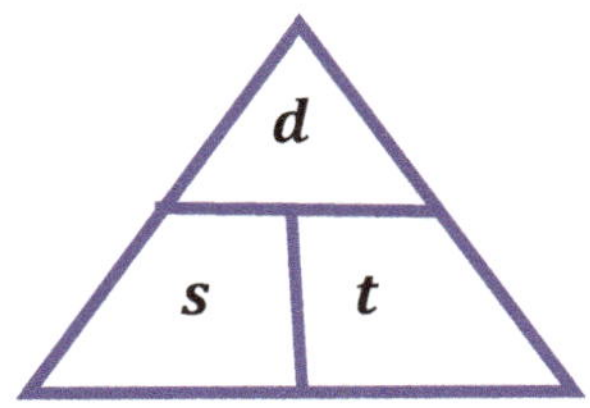

Examples: Understand Distance – Time – Speed Relation

a.	A cyclist travels at a constant speed of 12 km/h for 3 hours. **Total distance travelled,** $d = s \times t$ ⇒12 km/h × 3 hours = 36 km
b.	Find time required to cover a distance of **120 km** at a speed of **40 km/h**. **Time required,** $t = \frac{d}{s} \Rightarrow \quad t = 120\ km \div 40\frac{km}{h} = 3h$

Basic Distance Speed Time Problems

Q1	A car travels **180 km** in **3 hours** at a constant speed. (a) What is its speed (distance travelled in a unit period of time)? (b) How long will it take to travel **300 km** at the same speed?
Q2	Find distance covered in **2 hours** at an average speed of **45 km/h**.
Q3	Find time required to cover a distance of **40 km** at a speed of **50 km/h**.
	A1: 60 km/h, 5 h; A2: 90 km; A3: 48 minutes

🚶 Walking to and from a Destination – Same Path, Different Speeds

In many real-life situations, we travel the **same route** at **different speeds** — walking to school, going to the market, or returning from a class. Although the **distance remains the same**, the **time changes** depending on the speed. These problems often require careful reasoning about **average speed** or **time difference**.
Let's explore how to form and solve equations when speeds or times vary on the same route.

Example: Same Distance Different Speeds

Q1: Walk to School and Return

It takes me a total of **5 hours** if I walk to the **school at 2 km/h** and **return at 3 km/h**. How far is the school from my home?

Strategy: Time = Distance ÷ Speed

Let the distance be d km

Speed to school: 2 km/h

Return speed: 3 km/h

Time to school $= \frac{d}{2}$

Time to return $= \frac{d}{3}$

Solve using LCM:

LCM of 2 and 3 is 6.

Multiply both sides by 6

Simplify to find $\boldsymbol{d}$

Total time is 5 hours: $\frac{d}{2} + \frac{d}{3} = 5$

$$\frac{3d + 2d}{6} = 5$$

$$5d = 5(6) \Longrightarrow \boldsymbol{d = 6}$$

School is 6 km from home

[You can visualize this with a number line showing 6 km one way and two arrows representing different speeds and times.]

Here is representation and quick check of the solution.

🟢 → 6 km @ 2 km/h → 🕒 3 hours

🔴 ← 6 km @ 3 km/h ← 🕑 2 hours

Total Time = 🕔 5 hours

💡 Why can't we use the simple average of 2 km/h and 3 km/h to find the distance?

Challenge Examples: Consider Relative Time and Fractions

🚙 Q2: Two Cars Between City A and B

Two cars travel from city A to B at **30 km/h** and **36 km/h** respectively. One takes **3 hours less** than the other.

- **Find the distance**
- **Find the actual time.**

Strategy: Use Time = Distance ÷ Speed

Let the distance be d km

Time for car 1 $= \frac{d}{30}$

Time for car 2 $= \frac{d}{36}$

Time difference is 3 hours: $\frac{d}{30} - \frac{d}{36} = 3$

$$\frac{6d - 5d}{180} = 3$$

$d = 3(180) \Rightarrow d = 540$ km

City A → City B **distance 540 km**

Find the actual time.

🕒 **Actual time taken by each car:**

🚙 Car 1: speed 30 km/h

Time = 540/30 = **18 hrs**

🚙 Car 2: speed 36 km/h

Time = 540/36 = **15 hrs**

🕒 Check: Time difference = 3 hrs

Q3: Fractional Speeds — Slow and Steady

A person walks to a town at $\frac{3}{4}$ km/h and returns at $\frac{4}{3}$ km/h. The total trip takes $7\frac{1}{2}$ hours. Find the distance to the town.

Strategy: Time = Distance ÷ Speed

Let the distance be d km

Time to town $= \frac{d}{3/4} = \frac{4d}{3}$

Time to return $= \frac{d}{4/3} = \frac{3d}{4}$

Total time

$$\frac{4d}{3} + \frac{3d}{4} = 7\frac{1}{2}$$

$$\frac{16d}{12} + \frac{9d}{12} = \frac{15}{2}$$

$$\frac{25d}{12} = \frac{15}{2}$$

$$d = \frac{15}{2} \times \frac{12}{25} = \frac{18}{5} = 3.6\ km$$

distance to the town = 3.6 km

Summary Strategy Tip:

These problems help reinforce:

- Fractional Reasoning with Units
- LCM for Denominators
- Multi-step Algebraic Thinking
- Meaningful Application of Formulas

✍ **Practice:**

It takes Maria 6 hours to go to school at 2.5 km/h and return at 3.75 km/h. What is the distance to the school?

7.5 km

Practice Questions Unitary Method and Distance

Q1	If $\frac{4}{9}$ of a distance is 60 km, what is half the total distance? 67.5 km
Q2	If two-third of a project is completed in 16 days, how many days are needed to complete the full project? 24 days
Q3	A girl goes to her piano class at x km/h and returns at y km/h. Return takes 30 minutes longer. Write an equation in terms of distance d, x, and y.
Q4	A person walks at $\frac{2}{3}$ km/h and returns at $\frac{3}{2}$ km/h. The round trip takes 7 hours. Find the one-way distance. 3 km
Q5	A plane goes to city Z at $\frac{x}{2}$ km/h. Return takes 2 hours more. Express the distance d in terms of x. $d = \frac{2x^2}{x-\frac{2x}{2}} = 4x$
Q6	A biker goes 15 km at 3 km/h and returns at 5 km/h. What is the total time for the round trip? $\frac{15}{3} + \frac{15}{5} = 8$
Q7	Sumit travelled 15 km on the bus and 72 km on a train. If the train was 18 km/h faster than the bus and the total trip took 2 hours, then what was the average speed of the train. $48\ km/h$

Understanding Percent

What is Percent?

Percent means **"per hundred."**
It is a way to express a number as a part of 100.

The symbol for percent is %.

So, **25%** means **25 out of 100**.

Mathematical Definition:

$$\boldsymbol{Percent = \frac{Part}{Whole} \times 100}$$

This makes it easy to compare quantities of different sizes.

Illustrated Example

Marks obtained:

18 out of 20

$$Percent\ Marks$$

$$= \frac{18}{20} \times 100 = 90\%$$

Examples: Percent Foundation

1 Convert a fraction to percent:	**2 Convert a decimal to percent:**
$\frac{1}{2} = \frac{1}{2} \times 100 = 50\%$	$0.75 = 0.75 \times 100 = 75\%$
3 Convert a percent to a fraction:	**4 Find Percent of a Number:**
$40\% = \frac{40}{100} = \frac{2}{5}$	$20\%\ of\ 80 = \frac{20}{100} \times 80 = 16$

Example 5: Percent Discount is $= \frac{Final - Initial}{Initial} \times 100$

A $860 TV is sold at a discounted price of $550.40. Find the percent discount on the TV. *Note: Percent discount is always over the original price*

$$Percent\ Discount = \frac{860 - 550}{860} \times 100 = 36.05 \approx 36\%$$

The percent discount on the TV is **36%.**

Guided Practice: Percent Change Concept

Q1	Find the cost of $60 shirt after 15% discount. Hint: You pay 85% of the value. Ans: 60×0.85 =$51
Q2	A jacket is sold for $72 after a 10% discount. What was the price before discount? Hint: 72 is 90% of original price. Ans: $72 \div 0.90$ =$80
Q3	After price increase of 10% a house costs $3740,000. What was the price before the increase? Hint: 374,000 is 110% of original price. $340,000

Percent Quiz:

Q1	Convert the percent to fractions					
	a.	20%	b.	65%	c.	5%
Q2	Convert fractions to percent					
	a.	$\frac{1}{4}$	b.	$\frac{2}{5}$	c.	$\frac{3}{8}$

<table>
<tr><td>Q3</td><td>Convert percent to decimals
<table><tr><td>a.</td><td>23%</td><td>b.</td><td>5%</td><td>c.</td><td>110%</td></tr></table></td></tr>
<tr><td>Q4</td><td>What percent of 80 is 24? 30</td></tr>
<tr><td>Q5</td><td>16 is 40% of what number? 40</td></tr>
<tr><td>Q6</td><td>Convert the marks obtained in percent
<table><tr><td>a.</td><td>$27\ out\ of\ 40$</td><td>b.</td><td>$40.5\ out\ of\ 45$</td><td>c.</td><td>$4\frac{1}{2}\ out\ of\ 25$</td></tr></table></td></tr>
<tr><td>Q7</td><td>A shirt costing $60 is on 30% discount. Find its sale price. 42</td></tr>
<tr><td>Q8.</td><td>Cost of computer is $1200 plus 15% tax. What is the net price of the computer? $1380</td></tr>
<tr><td>Q9.</td><td>$75 shirt is being sold for 40% discount over already existing 30% discount. Determine the net price of the shirt after 13% tax on discounted amount. $35.60</td></tr>
<tr><td>Q10.</td><td>Two clocks are on sale. The models have the same features. One sell for $65.50 less 15% and the other sells for $72.50 less 20%. Which is the better buy? $65.50 less 15%</td></tr>
<tr><td>Q11.</td><td>In special sale normal price of a book is reduced by 30%. The sale price is $28.0. Find the normal price of the book. $40.00</td></tr>
</table>

Illustrated Solution of Selected Percent Problems

<table>
<tr><td>Q9.</td><td>$75 shirt is being sold for 40% discount over already existing 30% discount. Determine the net price of the shirt after 13% tax on discounted amount.

Understand situation

First the discount of 30% is applied. That means a payment of 70% of $75

Then a discount of 40% is applied. That means payment of 60%

Then tax on the discounted price is applied. Add 13% to this price

Here is short solution. Students should break it to steps with explanation.

$Discounted\ Net\ Price\ before\ tax = 75 \times 0.70 \times 0.60 = 31.50$

$Final\ after\ Tax = 31.50 \times 0.13 + 31.50 = 35.60$

Net price is $35.60</td></tr>
</table>

Thinking and Application

Samuel wants to buy 35 T-shirts. Each T-shirt costs $11.80

a.	Samuel does the calculations and estimates the cost of $420. Explain why the actual cost will be less than $420.	
b.	There is a special offer: Find the actual cost of buying 35 T-shirts Using the special offer.	T-shirt $11.80 each Buy 30 or more T-shirts Get 10% off the total cost.

Percent Increase and Percent Decrease

What is Percent Increase?

It tells us how much something has grown compared to its original value.

$$Percent\ Increase = \frac{Increase}{Original\ Value} \times 100$$

Example 6:

A book price increased from $40 to $50.

Increase = 50−40 =10

$$Percent\ Increase = \frac{10}{40} \times 100 = \mathbf{25\%}$$

What is Percent Decrease?

It tells us how much something has reduced compared to its original value.

$$Percent\ Decrease = \frac{Decrease}{Original\ Value} \times 100$$

Example 7:

A jacket's price dropped from $80 to $64.

Decrease = 80−64=16

$$Percent\ Decrease = \frac{16}{80} \times 100 = \mathbf{20\%}$$

Challenge Problem

Area of an object is proportional to the square of radius, and its volume is proportional to the cube of the radius. Find the percentage increase in the area and volume of the object if the radius is increased by 20%.

Area: 44%, Volume: 72.8%

The utility services rate has decreased by 20% and its consumption is increased by 20%. What is the net change in the expenditure of utility services? 4% decrease

Unitary Method with Percent

Example 8: Find Value for 1% and then use it to find the other percent

0.2% of population in a town is suffering from COVID 19 and 0.3% of the population has high fever. If 1260 persons have high fever, find the number of persons suffering from COVID 19.

Solution 8:

0.3% of the population suffering from high fever is 1260 persons

1% of the population is $\frac{1260}{0.3} = 4200$

0.2% of the population will be $4200 \times 0.2 = 840$

Number of persons suffering from COVID 19 is 840

Multiple Percent Problems

Sometimes percent change happens more than once, like discounts on discounts or repeated growth. Let us explore such cases.

Example 9: Double Discount Problem

An item is marked at $100. First, a 20% discount, then another 10% discount on the reduced price.

1st discount:

100−20% of 100=100−20=80

2nd discount:

10% of 80=8 ⇒ 80−8=72

Final price = $72, not $70 (which would be if we added 20% and 10%).

Example 10: Compound Increase

A population grows by 10% each year. Starting at 1,000:

After 1 year:

1,000+10% of 1,000 ⇒ 1,000+100=1,100 at the end of the first year

After 2 years:

1,100+10% of 1,100 ⇒ 1,100+110=1,210 at the end of second year

💡 **Tip:** Use multiplication for fast growth:

$$Final = Initial \times \left(1 + \frac{Rate}{100}\right)^n$$

$Rate < 0$ *for decay or discount*

$Rate > 0$ *for growth or increase*

n is the number of times applied

Extended Practice

In Example 10, find the population after 5 years if the growth rate is same.

Word Problems on Multiple Percent

Challenge 1: Price Drop Puzzle

A bicycle originally cost $240. In the first sale, it was discounted by 25%. A week later, another 10% was taken off the new price.
What is the final price of the bicycle? What is the total percent discount?

Challenge 2: Population Growth

A small town has a population of 8,000. It increases by 5% every year.
What will be the population after 2 years?

Challenge 3: Which Discount Is Better?

A store offers two discounts on a laptop:
Option A: 30% off once.
Option B: 20% off, then another 15% off the new price.

Which option gives the bigger discount? Show your calculations.

Challenge 4: Price Reversal

The price of a book increased by 20% to reach $60.
What was the original price before the increase

Challenge 5: Triple Trouble

An item is marked at $150.
Year 1: Price increased by 10%. Year 2: Price decreased by 20%.
Year 3: Price increased again by 5%.

What is the final price after 3 years?

Challenge 6: Reverse Percent Problem

After a 15% discount, a jacket costs $85.
What was the original price before the discount?

Challenge 7: Investment Challenge

A student invests $200 in a savings account. It grows by 6% every year.
How much will the student have after 3 years**?** (Round to the nearest cent.)

Challenge 8: Two Items – One Deal

A shirt is marked at $40 and jeans at $60.
The store offers 25% off on shirts and 15% off on jeans.
What is the total bill if a student buys both?
How much total money is saved?

MORE...

Answers

Q1	$162, 32.5%	Q2	8820	Q3	B: $68	Q4	$50
Q5	$138.60	Q6	$100	Q7	238.20	Q8	$19

Practice Problems: Multi-Step Thinking

Example 11: Find 20% of a number whose 30% is 18.	**Step by Step Solution**

Example 11: Find 20% of a number whose 30% is 18.

Step 1: Find the number whose 30% is 18

Let the number be x.

Then 30% of x is 18

$0.30 \times x = 18$

$x = \frac{18}{0.30} = 60$

So, the number is 60.

It is easier to work with decimal equivalent

$\mathbf{30\% = \frac{30}{100}}$

$\mathbf{= 0.30}$

Step 2: Now find 20% of 60

$$\frac{20}{100} \times 60 = \mathbf{12}$$

30% of 12 is 18

Example 12: Friends do not wish to spend more than $75 for dinner. If sales tax is 7% and they plan 15% tip, what is the most they can spend for the meal.

Step by Step Solution

Let A be the amount that they can spend for the meal.

After adding 7% tax and then 15% tip (note tip is on A) final payment is of $75

Let us now follow the steps

Amount after sales tax

$$= A + 0.07A = 1.07A$$

Add 15% tip on this amount

$1.07A + 1.07A \times 0.15 = 1.07 \times 1.15A$

This should be equal to $75

$1.07 \times 1.15A = 75$

Solve for Amount A

$$A = \frac{75}{1.07 \times 1.15} = \mathbf{60.95}$$

At the most they can spend $60.95

Percent Change Fast Tricks with Video Support

A powerful strategy to beat the competition in percent change problems is to **work with friendly numbers**. By assuming simple, convenient values and then calculating the change, the steps become faster and clearer, as shown in the examples that follow.

Table below provides easy **friendly numbers** to work with.

Change	Meaning	Initial Value	Final Value
10% increase	$\frac{10}{100} = \frac{1}{10}$	10	11
10% decrease		10	9
20% increase	$\frac{20}{100} = \frac{1}{5}$	5	6
20% decrease		5	4
25% increase	$\frac{25}{100} = \frac{1}{4}$	4	5
25% decrease		4	3

Example 13: If length of the rectangle is increased by 25% and the width is decreased by 20% then the area of the rectangle

a.	Increases	b.	Decreases	c.	no change

Solution 13: If length of the rectangle is increased by 25% and the width is decreased by 20% then the area of the rectangle can be calculated with the **friendly numbers**.

	Length 25% increase	Width 20% decrease	Area
Initial	4	5	20
Final	5	4	20
Friendly numbers indicate no change in the area			

Example 14: If length of the rectangle is increased by 10% and the width is decreased by 10% then the area of the rectangle

a.	Increases	b.	Decreases	c.	no change

Solution 14: If length of the rectangle is increased by 10% and the width is decreased by 10% then the area of the rectangle

	Length 10% increase	Width 10% decrease	Area
Initial	10	10	100
Final	11	9	99
Friendly numbers indicate decrease in the area by 1%			

Watch Video **Percent Change Tricks** SCAN ME

Challenge Problem

P is 3 times Q. If P is increased by 10% and Q is decreased by 30% then find the percent change in the product of P and Q. -23%

SAT Practice Test: Percent Change

The figure represents fish population, in thousands, over the past 5 years. The year over which the percent increase was greatest is:

a.	2000 – 2001
b.	2001 – 2002
c.	2002 – 2003
d.	2004 – 2005

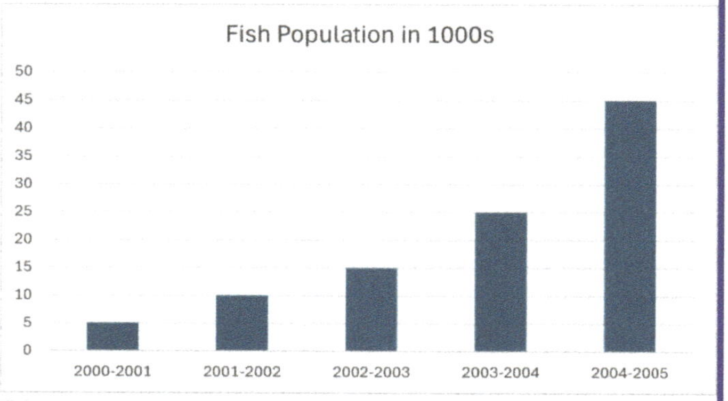

Average of 9 students in Math test for the Grade 8 is 79%. Find the percent mark which the tenth student should get to boost this averake to 80%

a.	80%	b.	88%	c.	89%	d.	91%

a. 2000 – 2001: is incorrect. Correct Answer: c. 89%

Percent Reduction/Increase in Price and Expenditure

Example 15: The price of my favourite coffee increased from \$2.50 to \$3.00. Find the reduction in the consumption so that there is no change in the expenditure.

Solution 15:

Strategy Friendly Number: Assume 100 cups is the consumption to find the reduction in percent.

Old expenditure at \$2.50 each was $2.50 \times 100 = 250$

New consumption for same expenditure of \$250 at \$3.00/cup: $\frac{250}{3} = 83.33$

Reduction in percent will be $100 - 83.33 = 66.67\%$

Practice Problems

Q1	the price of gas increased by 12%. Find the reduction in the consumption so that there is no change in the expenditure. 10.7%
Q2	The duty of imports is reduced by 40%. By how much percent must the consumption increase to boost the revenue by 20%. 100%

In a survey of 200 student's lunch

- 35 students bought soup
- 60 students bought sandwich
- 10 students bought both soup and sandwich

What percent of students bought neither soup nor a sandwich?

Solution

Venn diagram can help visualize the situation

Start from the intersection (both soup and sandwich)

And find how many bought only soup or sandwich.

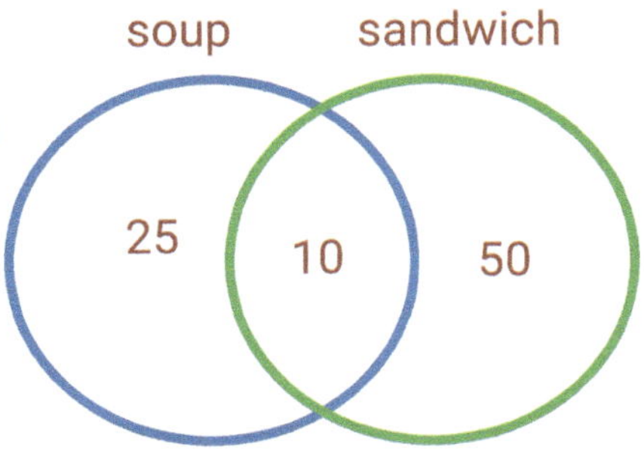

- 10 students bought both soup and sandwich
- 35 − 10 = 25 bought only soup
- 60 − 10 = 50 bought only sandwich
- 10 + 25 + 50 = 85 bought either soup or sandwich or both.

There are total of 200 students

- 200 − 85 = 115 bought neither soup nor sandwich

percent of students bought neither soup nor a sandwich

$$= \frac{students\ who\ bought\ neither\ soup\ nor\ sandwich}{total\ number\ of\ students} \times 100$$

$$= \frac{115}{200} \times 100 = 57.5\%$$

57.5% bought neither soup nor a sandwich

✍ Percent Data Survey

In a school of 300 students:

- 120 students take music
- 100 students take art
- 40 students take both music and art

What percent of students take neither music nor art?

SAT Challenge

If at least 90% students are good in sports, and at least 80% are good in music and at least 70% are good in studies, then find the percentage of students who are at least good in all three.

STEM Project: Design Problems with General Solution

Based of the question we solved regarding Data Survey Application; here is a general problem for you to solve.

In a survey of S students

- M students have Math
- E students have English
- B students have both Math and English

What percent of students have neither Math nor English

a.	$\frac{S-M-E-B}{S} \times 100$	b.	$\frac{S-M-E+B}{S} \times 100$
c.	$\frac{S-M+E+B}{S} \times 100$	d.	$\frac{S-M+E-B}{S} \times 100$

Cost + Markup = Selling Price

- **Cost Price (C):** The amount paid to buy or make a product.
- **Markup:** The extra amount (often a percent of cost) added to earn profit.
- **Selling Price (S):** The price at which the product is sold to customers.

Q1	Markup rate of an item is 40% Find the selling price if the cost price is \$60 Hint: SP = CP + Markup; $60 + 0.4 \times 60 = \$84$
Q2.	An item costing \$46 is being sold for \$69. Find Markup and the Markup rate. Hint: Markup = SP - CP; \$23, 56.5%
Q3	Markup rate of an item is 75%. Find the cost price if the selling price is \$210 \$120

✨ Challenge

An oilwell in Alberta produces 320 barrels in its first day. Its production decreases by 0.1% every year. In how many years will it run dry? There are 42 gallons in one barrel of oil.
13440000

Finance Application: Income Tax

Governments apply a tax rate to income to calculate how much must be paid. In some cases, progressive rates are used, where different portions of income are taxed at different percentages.

Example 1: Flat Rate Tax

Riya earns $50,000 annually.
If the flat income tax rate is **20%**, how much tax does she pay?

Solution 1:
Tax = 20% of 50,000 = $0.20 \times 50000 = 10000$
Riya pays $10,000 in income tax.

Example 2: Tiered Tax Slab

Mark earns $60,000.
The income tax system is:

- No tax on the first $10,000
- 10% on the next $30,000
- 20% on the remaining income

Solution 2:

- First $10,000 → $0
- Next $30,000 → 10% of 30,000 = $0.10 \times 30000 = 3000$
- Remaining $20,000 → 20% of 20,000 = $0.20 \times 20000 = 4000$
  Total Tax = $0 + $3,000 + $4,000 = $7,000

✨ Thinking and Application

Under certain law, the first 50,000 of earnings are subjected to a 25% tax, earnings greater than 50,000 and up to 100,000 are subjected to 35% tax, and earnings greater than 100,000 are taxed at 55%. $T(x)$, is a piecewise function that calculates the tax. Explain the function. Then calculate tax for the earnings of 100,000 and 125,0000.

$$T(x) = \begin{cases} 0.25x, 0 \leq x \leq 50000 \\ 12500 + 0.35(x - 50000), 50000 < x \leq 100000 \\ 30000 + 0.55(x - 100000), \quad 100000 < x \end{cases}$$

Visual Representation: Making a Pie Chart from Percent Data

Pie charts are a great way to represent data in percentage form. Here's an example.

Scenario: Favorite Subjects in a Class of Students

A survey was done in a class asking students about their favorite subject. The results were:

Subject	Math	Science	English	History	Art
Percent	40%	25%	20%	10%	5%

Convert percent data into a pie chart format

How to Make the Pie Chart?

Each percent represents a portion of a **360° circle** (full pie).

Use the formula: **Angle = (Percent ÷ 100) × 360°**

Like: (40 ÷ 100) × 360° = 144°

Subject	Percent	Angle
Math	40%	144°
Science	25%	90°
English	20%	72°
History	10%	36°
Art	5%	18°

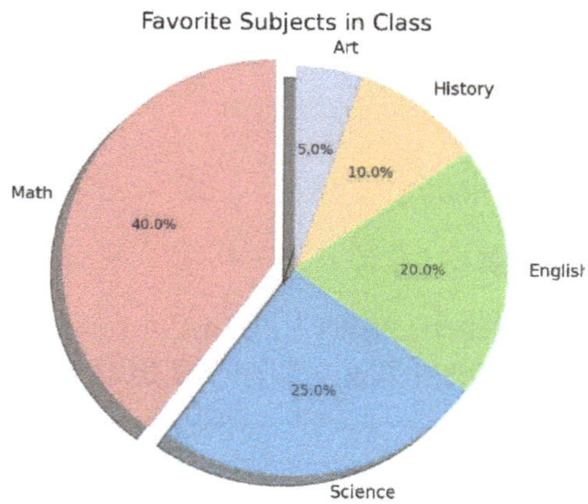

✨ **Quick Tip:**

- Pie charts are great for comparing part-to-whole relationships at a glance. Always ensure the total percent = 100% and total angle = 360°.
- Draw a circle and divided into calculated angles
- Color-code each slice and label with Subject + Percent.

Chapter 1 Test: Foundation Skills

Q1	Evaluate: $3 + 6 \div 2 \times (4 - 1)^2$
Q2	I am a number. Add 13 to me. Divide the sum by 4. You end with 23. What number am I.
Q3	Explain if the number $\frac{231}{192}$ is a terminating decimal or not.
Q4	Find the sum of next three number of the sequence: 1, 2, 3, 5, 8, 13, ...
Q5	Sum of first n natural numbers is given by $n(n+1) \div 2$ Find the sum of all the natural numbers from 20 to 80.
Q6	Express 0.7272... as a rational number
Q7	Find LCM of 12, 16, and 20 using prime factorization
Q8	Simplify: $\frac{\frac{5}{10}}{3} + \frac{1}{3}$
Q9	Divide 84 in the ratio 4:3
Q10	If $\frac{a}{b} = \frac{5}{8}$ and $a = 15$ find b
Q11	If 5 pens cost $7.50, what is cost of 12 pens?
Q12	A machine makes 150 items in 4 h. How many will it make in 10 hours?
Q13	A car travels 216 km in 5 hours. Find its speed in metre per second.
Q14	A number reduced by 20% becomes 36. Find the number
Q15	48 is 60% of what number?
Q16	Length of a rectangle is 7 more than the width. If perimeter of the rectangle is 58 cm find the width.
Q17	the numerator of a fraction is two less than the denominator. If both are increased by one the fraction becomes $\frac{2}{3}$ find the original fraction.

Q18	Computer costs $1449 including 15% tax. Find the tax paid.
Q19	In a tiered government tax system the rates are as follows: • No tax on first $15,000 • 10% on next $35,000 • 20% on the rest If Sam earns $85,000, what is his total tax?
Q20	Create a pie chart from the following information A survey was done in a class asking students about their favorite sport. The results were: **Subject**: Badminton, Soccer, Baseball, Volleyball, others **Percent**: 20%, 25%, 35%, 15%, 5%
Q21	Anita sold 14 cars this week. This is 35% of the cars sold by her team. Calculate the number of cars sold by rest of the members in her team this week.
Q22	Find the percentage increase in the area of rectangle if each side is increased by 20%,
Q23	The price of a laptop decreases by 20%, and later increases again by 20%. Find the net percentage change.
Q24	Arrange in increasing order: $\frac{1}{9}, 0.1, 11\%$
Q25	In a class of 30 students, 17 are studying math, 14 are studying science and 10 are studying both the subjects. Find the number of students studying exactly one subject.

Q20 table:

Subject	Badminton	Soccer	Baseball	Volleyball	others
Percent	20%	25%	35%	15%	5%

Answers

Q1	30	Q2	19	Q3	terminating	Q4	110
Q5	3050	Q6	8/11	Q7	240	Q8	11/6
Q9	48 and 36	Q10	24	Q11	18	Q12	250
Q13	12 m/s	Q14	45	Q15	80	Q16	11 cm
Q17	3/5	Q18	$189	Q19	$10,500	Q21	26
Q22	44%	Q23	4% decrease	Q24	$0.1, 11\%, \frac{1}{9}$	Q25	11

Multiple Choice Test

Q1 A school is planning a car wash to raise \$600.

- There will be 8 teams.
- Each team will wash 2 cars per hour.
- Each car wash will last 5.5 hour.
- Each team will take two 15-minutes break.

How much should the school charge per car to raise \$600?

a.	\$ 9.50	b.	\$ 8.50
c.	\$ 7.50	d.	\$ 6.50

Q2 A ball is dropped from a height of 25 m. The ball's height, H, in metres, after n bounces is represented by the equation below.

$H = 25\left(\frac{1}{2}\right)^n$

What is the height of the ball after 4 bounces?

a.	$\frac{25}{16}\ m$	b.	$\frac{25}{8}\ m$
c.	$\frac{25}{4}\ m$	d.	$\frac{25}{2}\ m$

Q3 The maximum number of tickets that can be sold for school play is 350. The total profit earned, P, can be determined using the equation $P = 4.50n - 1080$, where n is the total number of tickets sold.
Which of the following statement is true?

a.	The maximum profit is 1080
b.	The maximum profit is 1570
c.	the total profit is \$0 when 240 tickets are sold.
d.	the total profit is \$0 when 350 tickets are sold.

Q4 The line $x - 2y = b$ is perpendicular to the line $Ax + 2y - 5 = 0$. Find value of A.

a.	1	b.	2
c.	3	d.	4

Q5 The line with slope $\frac{3}{4}$ passes through the point $(0, c)$ and $(c + 2, 0)$. Find value of c.

a.	6	b.	-6
c.	$\frac{6}{7}$	d.	$-\frac{6}{7}$

Q6 The line $3x - 2y = b$ is perpendicular to the line that pass through the origin. If the two lines intersect at $(a + 1, 2a)$, find the value of a.

a.	$\frac{1}{2}$	b.	$\frac{1}{4}$
c.	$-\frac{1}{4}$	d.	$-\frac{1}{2}$

Q7 AB is the line segment with $A(4, 0)$ and $B(0, 6)$. Mid-point of this segment is $(2, 3)$. Which of the following point is also same distance from A and B.

a.	$\left(1, \frac{2}{3}\right)$	b.	$\left(1, \frac{5}{3}\right)$
c.	$\left(1, \frac{7}{3}\right)$	d.	$\left(1, \frac{3}{5}\right)$

	Q1: c.	Q2: a.	Q3: c.	Q4: d.	Q5: d.	Q6: c.	Q7: c.	

🎯 Bonus Question

In a tiered system:

- No tax on first $15,000
- 10% on next $25,000
- 20% on the rest

If Daniel earns $55,000, what is his **total tax?**

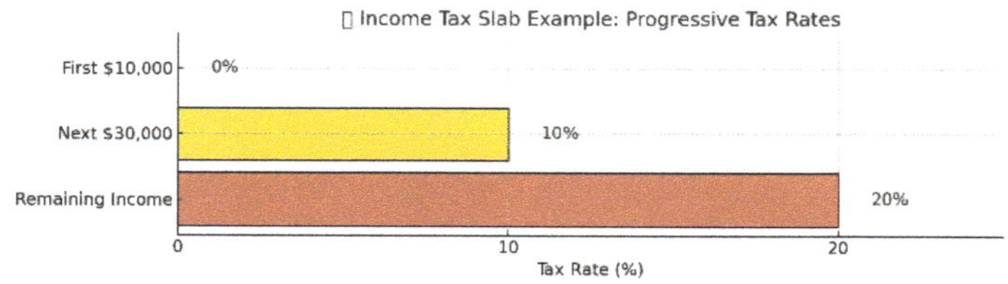

Student Spotlight: Vanessa from the Philippines

Vanessa first reached out in Grade 9 after learning quietly from my YouTube videos. By Grade 12, she joined my classes while preparing for university.

Vanessa - BS Electrical Engineering, Philippines

From watching math videos in Grade 9 to mentoring her sister today — an early start paved the way for STEM success.

Today, she is entering her third year of Electrical Engineering. Her journey from early curiosity to confident pursuit of STEM reminds us why strong foundations matter.

"Good day sir! I hope you still remember me, Vanessa from the Philippines. I am now an incoming 3rd year student taking BS EE... I just wanted to thank you for your videos. They helped me since Grade 9, and now they are also helping my sister prepare for her college entrance exams."

Vanessa's story reflects the heart of this book: to guide learners at every stage, and to show how one spark can light the way for many.

Chapter 2: Introduction to Algebra

Algebra can be thought of as a "language" that connect English (or any language) and mathematics. Phrases can be translated into algebraic expressions and sentences can be turned into algebraic equations. Variables, numbers, and various operations can be used to express a real-life problem in mathematical form. Let us explore further with examples.

SCAN ME

An expression is made up of one or more terms separated by + or − signs. Each term may include a coeficient (numerical part) and a variable (letter part).

Example 1: Express Phrases to Expression:

Use x and y as a variable to translate each phrase into an expression.

Ten less than a number	$x - 10$
Half of a number	$\frac{x}{2}$
Add 3 to twice a number	$2x + 3$
Seven less than thrice a number	$3x - 7$
Three times the sum of number and 7	$3(x + 7)$
Half the sum of number and 4	$\frac{x + 4}{2}$
Quotient of sum and difference of two numbers	$\frac{x + y}{x - y}$
Sum of number and twice the difference of another number and 6	$x + 2(y - 6)$
Two times the sum of thrice a number and the difference between 8 and another number	$2(3x + (8 - y))$
The square of sum of two numbers divided by their product	$\frac{(x + y)^2}{xy}$

Example 2: Common Phrases in Word Problem

Write expressions for the given situation:

Phrases in Word Problems	**Algebraic Expression**
Write an expression for total amount of money if there are n nickels, d dimes and q quarters	$0.05n + 0.10d + 0.25q$
Length of a rectangle is $3\ cm$ more than the width, w. Write an expression for the area of rectangle.	$(w + 3)w$
An arena charges \$15 for children under 18 years, \$25 for others. Write an expression for the total revenue from tickets sold if c children's tickets and d other tickets are sold.	$15c + 25d$
In a soccer league, teams receive 5 points for a win, 3 points for a tie and 1 point for loss. Write an algebraic expression to represent team's total points.	Let w be wins, t be ties and l be losses. $5w + 3t + 1l$
Many times, you need to define the variables with let statement as done above	
Sum of consecutive even numbers	$n + (n + 2)$
Perimeter of a rectangle with l and w as two sides.	$2(l + w)$
Cost of buying 5 sets of 4 pens and 3 pencils together, if each pen costs p and each pencil costs q	$5(4p + 3q)$
The area of a rectangle whose length is doubled, and width is increased by 5, if the original length is l and width is w	$2l(w + 5)$
The product of the sum of a number and 5, and the difference of a number and 5	$(n + 5)(n - 5)$
The square of the difference between four times a number and 9	$(4x - 9)^2$
The square of the sum of twice a number and 7, minus the square of the difference between the number and 4	$(2x + 7)^2 - (x - 4)^2$

Explore Algebraic Expressions

Algebra is the branch of mathematics that deals with variables, numbers and various operations on them. Normally, to solve problems, we need to define variables, relate verbal expressions, form equations, write restrictions and solve.

Example 3: Write algebraic expression in words

$\frac{1}{2}x - 5$	Five less than half the number
$x(x+1)$	Product of consecutive integers
$2(x^3+1)$	Two times sum of number to the third power and 1
$2(x+y^2)^2$	Twice the square of the sum of a number and square of another number
$0.25q + 0.10d$	Amount in cents for q quarters and d dimes

✍ **Practice 1: Write algebraic expressions in words**

a.	Seven less than square of a number
b.	Four-thirds of r cube times π
c.	sum of two number divided by their difference
d.	Quotient of p and q increased by one
e.	Length of a square is increased by one. Write expression for the change in area of the square.
f.	Peter has d dimes and n nickels. Write an expression for the amount of money he has.

✍ **Practice 2: Express phrases to expressions**

a.	$x - 2$	b.	$\frac{x}{2} + x$	c.	$x^2 + 1$
d.	$3x - 2$	e.	$x(x+2)$	f.	$3(x^2 - 1)$

Unit Conversion — Bridging the Real World

Mathematics connects directly to science and engineering through **units**.
Converting between units allows us to compare, calculate, and interpret results accurately — whether measuring speed, area, force, or time.
Before solving applied problems, let's review how to move confidently between different measurement systems.

Refined Method of Working with Units

When solving real-world problems, **work through the units — not just the numbers**. Each conversion factor is a **fraction equal to 1**, arranged so that **units cancel step by step**, leaving only the desired unit in your final answer.

A car travels 72 km/h. Convert this to m/s. $1\ km = 1000\ m, 1\ h = 3600\ s$

$$72\frac{km}{h} \times \frac{1000\ m}{1\ km} \times \frac{1\ h}{3600\ s} = 20\ m/s$$

Car speed is $20\ m/s$

A car travels at 60 mi/h. Find its speed in km/h. $1\ mi = 1.609\ km$

$$60\frac{mi}{h} \times \frac{1.609\ km}{1\ mi} = 96.54\ km/h$$

Car speed is $96.54\ km/h$

Apples cost \$4 for every 3 lb. How much will 5 kg of apples cost? $1kg = 2.205\ lb$

$$\frac{\$4}{3\ lb} \times \frac{2.205\ lb}{1\ kg} \times 5\ kg = \$14.7$$

5 kg of apples will cost $\$14.7$

🚙 **Applied Unit Conversion & Cost Problem**

Gasoline costs: \$3.60 per US gallon. **Fuel consumption**: 8 L per 100 km

Find the cost of gasoline for 100 km trip. $1\ US\ gallon = 3.785\ L$

$$\frac{\$3.60}{gallon} \times \frac{1\ gallon}{3.785\ L} \times 8\ L = \$7.61$$

It will approximately cost = \$7.61 for 100 km trip.

💡 **STEM Insight**

This dimensional (unit) method is a universal tool — used by scientists, engineers, and economists alike. It ensures that every step is logically consistent and error-free, even in complex, multi-step conversions.

When you treat units like algebraic quantities, you can cancel, rearrange, or multiply them — a powerful skill for solving physics, chemistry, and applied math problems.

✍ Practice Problems: Refined Method of Working with Unit Conversion

Q1	Cost of 20 ft long rope is \$4. Find the cost of 5 m rope. 1 ft = 0.3048 m \$3.28
Q2	A car travels at 60 mi/h. How many minutes will it take to travel 100 km. 1 mi = 1.609 km 62 min
Q3	Apples cost \$4 for every 3 kg. How much will 8 lb of apples cost? 1 kg = 2.205 lb \$4.84
Q4	Gasoline costs \$3.75 per US gallon. Fuel consumption: 9 L per 100 km. Find the cost of gasoline for a 100 km trip. 1 US gallon = 3.785 L \$8.92
Q5	The outdoor temperature is 86 °F. Find it in °C. °C = (5/9)(°F - 32) 30 °C
Q6	A rectangular field measures **120 ft × 80 ft**. The land price is **\$45 per m²**. Find the total cost of the land. 1 ft = 0.3048 m \$40,163
Q7	Find the capacity of the container, $2m \times 3m \times 5m$ in Liters. $1\ L = 1000\ cm^3$ 30,000 Litres
Q8	Convert 5 m^3 to cm^3 $5 \times 10^6 cm^3$

Algebraic Equations in Problem Solving

Algebraic equations are the mathematical statements that relate expressions.

Equations can have one solution, multiple solutions or no solution.

Example 4: Write algebraic equations for the given statements.

Let $n, x, and\ y$ be the numbers

Statement	Equation
Sum of three consecutive integers is 66. Consecutive: one after other We can also consider before and after Observe that the centre number is the average 66/3 = 22. The numbers are 21, 22, 23	$n + (n+1) + (n+2) = 66$ Alternate solution $(n-1) + n + (n+1) = 66$
When a number is doubled and then subtracted from 7, the result is 33. Observe doubled and **then** subtracted **from** 7	$7 - 2n = 33$ $2n - 7 = 33$ is incorrect
Angles of a triangle are in the ratio of 1:2:3. Note: Sum of angles is $180°$	$x + 2x + 3x = 180$
The cube less the square of a natural number is one more than twice the number.	$x^3 - x^2 = 2x + 1$
Value of two-digit number is increased by 36 when the numbers are swapped. Let the number be xy. Value is $10x + y$ When swapped it will be yx. Value is $10y + x$	$10x + y + 36 = 10y + x$ It has multiple solutions
Square of a number plus 4 is zero	$x^2 + 4 = 0$ No solution for Real numbers
A right triangle with hypotenuse of 10 have smaller side lengths with difference of 2. Relate length of each side of the triangle. Smaller side lengths are: x and $x + 2$	$x^2 + (x+2)^2 = 10^2$ Three sides of right triangle are linked with Pythagorean Theorem
When you add 8 to a number and then divide by 9, you get the same result as when you subtract 9 and then divide by 8.	$\frac{x+8}{9} = \frac{x-9}{8}$

Simplifying and Evaluating Expressions

Terms in an expression are separated by addition or subtraction operators. Like terms are with same variable raised to the same exponent. These like terms can be combined to simplify expressions. Write final expression in decreasing power of terms.

Example 5: Collect like terms and then simplify expressions

a.	$a - 3 + 3a - 5 + 2a$ $= 6a - 8$ Combine the like terms and simplify	b.	$2x^2 - 3x + 1 - x^2 + 2x$ $= 2x^2 - x^2 - 3x + 2x + 1$ $= x^2 - x + 1$
c.	$3 - (3x^2 - 5x + 4)$ $= 3 - 3x^2 + 5x - 4$ $= -3x^2 + 5x - 1$	d.	$(5x^2 - 3x - 1) - (3x^2 - 5x)$ $= 5x^2 - 3x - 1 - 3x^2 + 5x$ $= 5x^2 - 3x^2 + 5x - 3x - 1$ $= 2x^2 + 2x - 1$
e.	$(x + 2) - (2 + x)$ $= x + 2 - 2 - x = 0$	f.	$(3x - 1) - (2x - 5) - x$ $= 3x - 1 - 2x + 5 - x = 4$

Guided Example 5: Substitute and Evaluate Expressions

First two solved. Students are encouraged to provide solution for the rest.

Evaluate each expression when $\boldsymbol{x = 2}$ and $\boldsymbol{y = -3}$			
a.	$2x + 3y + 6$ $= 2(2) + 3(-3) + 6$ $= 4 - 9 + 6$ $= -5 + 6 = 1$ **Note:** Place negative values in brackets	b	$\frac{1}{x} + \frac{1}{y} + 1$ Find LCM to add fractions $= \frac{1}{2} + \frac{1}{(-3)} + 1 = \frac{1}{2} - \frac{1}{3} + 1$ $= \frac{3}{6} - \frac{2}{6} + \frac{6}{6} = \frac{11}{6} = 1\frac{5}{6}$

Guided Practice 6: Substitute and Evaluate Formula		
Round answer to 2 decimal places.		
a.	$T = 2\pi\sqrt{\frac{l}{g}}$	Evaluate time-period T of a pendulum of length $l = 0.75\ m$. Given gravitational acceleration $g = 9.81 \frac{m}{s^2}$ and $\pi = 3.14$ $1.74s$
b.	$a = \frac{v^2 - u^2}{2s}$	Find acceleration a when the speed of a car changes from $u = 8\ m/s$ to $v = 12\ m/s$ over distance $s = 3.2\ m$ $12.50\ m/s^2$
c.	$V = \frac{4}{3}\pi r^3$	Find volume of a sphere of radius $r = 10\ cm$ $4188.8\ cm^3$

Practice: Simplify and Evaluate each expression when $x = -2$ and $y = 3$

a.	$3x - 2y$	b.	$\frac{1}{2}x - 3y - 4$
c.	$xy - (x + y)$	d.	$\frac{y}{x} + \frac{x}{y}$
e.	$\frac{1}{x} + \frac{1}{y} - x^2y$	f.	$\frac{1}{4}x + \frac{2}{3}y + 1$
g.	$(2x + 1) - (3x - 5)$	h.	$(2x + 1) - (3x - 5)$
i.	$(2x + 1) + 2(3x - 5)$	j.	$5x^3 + 3x^2 - (3x^2 - 5x)$
k.	$3(2x - 1) - 2(3x - 1)$	l.	$(6x^2 - 2x - 1) - 2(x^2 - 2x)$
m.	$\frac{1}{x} + \frac{1}{y} - \left(\frac{x - y}{xy}\right)$	n.	$\left(x + \frac{1}{2}\right) - \left(3x - \frac{5}{4}\right)$
o.	$x \div y - (2xy)^2 - xy^2$	p.	$x \div y - (2xy)^3 \div y$
q.	$(5x^3 - 3y^2 - 1) - (3y^2 - 5x)$		$(2x + 1)^3 - (3y - 5)^2$

Distributive Property: Working with Brackets

The term outside the bracket multiplies each term inside the bracket to expand an expression. This is **distributive law**.

<table>
<tr><td colspan="2">$a(b+c) = a \cdot b + a \cdot c = ab + ac$
$a(b-c+d) = a \cdot b - a \cdot c + a \cdot d = ab - ac + ad$</td></tr>
<tr><td>Example 1
$x(2+y) = 2x + xy$</td><td>Example 2
$9(20-1) = 180 - 9 = 171$</td></tr>
<tr><td>Example 3
$2x(3-x) = 6x - 2x^2$</td><td>Example 4
$-3x(x+y-2) = -3x^2 - 3y + 6x$</td></tr>
<tr><td colspan="2">Like terms are the same variables raised to same power. They can be combined.</td></tr>
<tr><td>Example 5
$3x - 2 + 2y - x + y - 5$
$= 3x - x + 2y + y - 2 - 5 = 2x + 3y - 7$</td><td>Example 6
$3x^2 - 2x^2 + -x + 4x - 5$
$= x^2 + 3x - 5$</td></tr>
<tr><td>Example 7
$2(x-1) + 3x + 4$
$= 2x - 2 + 3x + 4 = 5x + 2$</td><td>Example 8
$3x + 1 - 2x(3-x)$
$= 3x + 1 - 6x + 2x^2 = 2x^2 - 3x + 1$</td></tr>
<tr><td colspan="2">Example 9
$3x(x+y) - (x^2 + xy - 3) = 3x^2 + 3xy - x^2 - xy + 3 = 2x^2 + 2xy + 3$</td></tr>
</table>

✍ Practice Problems: Simplify following expressions

a.	$5(a+2b-c)$	b.	$x(2x+7) - 2x$
c.	$2(1-x) + 3x - 5$	d.	$2x(x-2) + 3x^2 - x - 5$
e.	$3(2x-1) + 2x - 4$	f.	$5(2-x) - 2(x-3)$
g.	$(x-y)(x^2+xy+y^2)$	h.	$(x+y)(x^2-xy+y^2)$

Distributive property helps in calculations: $\mathbf{21^2 = 21(20+1) = 420 + 21 = 441}$

Distributive Property: Two Brackets

Example 10: Simplify Using Distributive Property

- $(a+b)(c+d) = a(c+d) + b(c+d) = ac + ad + bc + bd$
- Combine like terms if possible

a.	$(3+y)(x-3)$ $=3(x-3)+y(x-3)$ $=3x-9+xy-3y$	b.	$(a^2+1)(a-5)$ $=a^2(a-5)+(a-5)$ $=a^3-5a^2+a-5$
c.	$(x-2)(x+3)$ $=x^2+3x-2x-6$ $=x^2+x-6$	d.	$(x+5)(x-5)$ $=x^2-5x+5x-25$ $=x^2-25$
✍ **Practice Problems**			
g.	$(x-1)(x+3)$	h.	$2ab(a+2b-3c)$
i.	$(x+3)(x-3)$	j.	$(2x-3)(2x+1)$
k.	$(2x-2)(2x-2)$	l.	$(x+1)(x+1)$
m.	$(x-3)^2$	n.	$(2x-3)^2$
o.	$(x-1)^2$	p.	$(5-x)^2$
q.	$(2x+3y)^2$	r.	$(2x+7)(2x-7)$
s.	$(x+y)^2-(x-y)^2$	t.	$(x+y)^2+(x-y)^2$

Challenge Question: Square of Trinomial

Prove that $(a+b+c)^2 = a^2+b^2+c^2+2ab+2bc+2ca$

Solution with practice questions at the end of this section

Solution for Perfect Squares.

m.	$(x-3)^2=(x-3)(x-3)$ $=x^2-3x-3x+9$ $=x^2-6x+9$	n.	$(2x-3)^2=(2x-3)(2x-3)$ $=4x^2-6x-6x+9$ $=4x^2-12x+9$

Student Activity: Discovering the Distributive Property with Area Models

Objective:

Students will:
- Understand the distributive property of multiplication over addition.
- Use area models to visualize and apply the property.
- Solve real-world problems using the distributive method.

Concept Introduction

Distributive Property Definition: $a(b + c) = ab + ac$
This means multiplying a number across a sum is the same as multiplying it with each term and adding the results.

Task 1: Draw and Discover

Instructions:
1. Draw a rectangle of width **a** and split the length into **b** and **c**.
2. Label the two sections:
   - Area 1 = $a \times b$
   - Area 2 = $a \times c$
3. Shade each area in different colors.
4. Calculate the total area using both methods:
   - Direct: $a \times (b + c)$
   - Separate: $ab + ac$

Example:
Let $a = 4$, $b = 3$, $c = 5$
- Full rectangle: $4 \times (3 + 5) = 32$
- Split areas: $4 \times 3 = 12$ and $4 \times 5 = 20$
→ Total = $12 + 20 = 32$

Think & Reflect Prompt

"Can we always break up a number into parts and multiply separately? What does this help us with in real life?"

Task 2: Real-Life Shopping Example

You're buying 8 T-shirts. One store says:
- Red T-shirts = \$8 each
- Blue T-shirts = \$12 each
You want 4 of each.
Use the distributive property to find total cost:
$4 \times (8 + 12) = ?$ or $4 \times 8 + 4 \times 12 = ?$

Discussion Questions

1. Why is it helpful to split numbers when multiplying?
2. How can area models help with mental math?
3. How is this connected to algebra?

Answer: $4 \times 20 = 80$ OR $4 \times 8 + 4 \times 12 = 32 + 48 =$ **80**

Task 3: Write an Expression for the given Area Model

x^2	x	x
x	1	1

a^2	ab
ab	b^2

Tips:

Use colored tiles or paper cutouts for positive and negative numbers. Visuals help students **see** how multiplication spreads over addition.

Area is always positive. However, we can have colour code that can represent negative area also. Explore the concept and draw area model for

$$(x+1)(x-3)$$

Solution Challenge Question

Prove that

$(a+b+c)^2 = a^2 + b^2 + c^2 + 2ab + 2bc + 2ca$

Solution with Area Model: Add all areas to get the result

a^2	ab	ac
ab	b^2	bc
ac	bc	c^2

Rectangle Model is very effective approach to show the solution

Consider a square with the side lengths of $a+b+c$

Area of the square is $(a+b+c)^2$

$$= a^2 + b^2 + c^2 + 2ab + 2bc + 2ca$$

Expand: $(a-b+c)^2$

Application of Trinomial Square

a.	Expand and simplify: $(2a - b + 3)^2$
b.	If $a + b + c = 8, a^2 + b^2 + c^2 = 18$ then find the value of $ab + bc + ca$
c.	If $a + b + c = 3, a^2 + b^2 + c^2 = 4$ then find the value of abc
d.	If $a + b + c = 2, and\ \frac{1}{a} + \frac{1}{b} + \frac{1}{c} = 0$ then find $a^2 + b^2 + c^2$ Trinomial Square SCAN ME

Solution

b.	If $a + b + c = 8, a^2 + b^2 + c^2 = 18$ then find the value of $ab + bc + ca$ $(a + b + c)^2 = a^2 + b^2 + c^2 + 2ab + 2bc + 2ca$ $(a + b + c)^2 = a^2 + b^2 + c^2 + 2(ab + bc + ca)$ $(8)^2 = 18 + 2(ab + bc + ca)$ $2(ab + bc + ca) = 64 - 18 = 46$ $ab + bc + ca = \frac{46}{2} = 23$

Challenge Higher Algebra: New Strategy

If $x^2 + 2 = 2x$ then find value of $x^4 - x^3 + x^2$

Strategy: Times x each time to get the following relations

$x^2 + 2 = 2x$

$x^3 + 2x = 2x^2$

$x^4 + 2x^2 = 2x^3$

Add and rearrange

$x^2 + 2 + x^3 + 2x + x^4 + 2x^2 = 2x + 2x^2 + 2x^3$

$x^4 - x^3 + x^2 = -2$

Squares and Square-Roots

Squares of Integers and Perfect Squares

- Integers are whole numbers that can be negative, positive, or zero. Negative integers are additive inverse of natural numbers.
- Square of a number is a number multiplied with itself
- Square of a number is always non-negative.
- A perfect square is a number that can be written as product of two equal integers. $\{0, 1, 4, 9, 16, \dots\}$ are set of perfect squares.
- Square-root of a perfect square is a non-negative integer.

Any number n multiplied by itself results into square of that number.

Square of a number is written as n^2. And read as "n-square", not n 2.

For product or multiplication, we will use three different ways:

- $n \times n$: multiplication sign is used between numbers
- $n \cdot n$: dot in between is not a decimal but sign of multiplication
- $(n)(n)$: no sign in between brackets or variables signify multiplication

Example 7: Let us calculate square of integers using different signs.

$2 \times 2 = 4$	$25 \times 25 = 625$	$3 \cdot 3 = 9$	$0 \cdot 0 = 0$
$(7)(7) = 49$	$(-6)(-6) = 36$	$(-9)^2 = 81$	$20^2 = 400$

Note: Square of all integers are non-negative numbers only

Practice questions to find squares of given numbers.

$5 \times 5 =$	$12 \times 12 =$	$10 \cdot 10 =$	$21 \cdot 21 =$
$(17)(17) =$	$(-10)^2 =$	$(-8)(-8) =$	$35^2 =$
$1^2 =$	$0^2 =$	$-100^2 =$	$45^2 =$

How many perfect squares are there up to 1000? Include assumption.

Squares of Integers: Investigate Patterns and Explore

Example 8: Explore and Learn More

Is $-3^2 = 9$?

-3^2 is not equal to 9. It is equal to -9.

$-3^2 = -3 \times 3 = -9$

Note that the number 3 is only squared not the minus sign, as shown above.

$(-3)^2 = 9$. In this case it is equivalent to $(-3)(-3)$

Pythagorean Triples: Explore and Learn More

Complete the following with a perfect square as shown in the first example.

$3^2 + 4^2 = 5^2$

$5^2 + 12^2 =$

$7^2 + 24^2 =$

The set of three non-negative integers, $a, b, and\ c$, that satisfy the rule $a^2 + b^2 = c^2$ are called **Pythagorean Triples**.

Extended Pattern: Explore and Learn More

Find value of p, q, r in the following pattern.

$3^2 + 4^2 + 12^2 = 13^2$

$5^2 + 6^2 + 30^2 = p^2$

$6^2 + 7^2 + q^2 = r^2$

Extend the pattern to write three more equations relating squares

Sum of Numbers and Sum of Cubes are related.

Sum of first n numbers is given by the formula: $\frac{n(n+1)}{2}$

Sum of first n cubes is given by the formula: $\left[\frac{n(n+1)}{2}\right]^2$

Verify the result for $n = 3$

$$1^3 + 2^3 + 3^3 = (1 + 2 + 3)^2$$

Now, Verify the result for $n = 5\ and\ 10$

Sum of cubes of first n numbers is square of the sum of n numbers

✍ **Practice:** Find the sum of cubes of first 15 numbers.

Squares of Real Numbers

- Real numbers are any numbers that can be represented on a number line. The include **rational** and **irrational numbers**.
- Rational numbers can be written in the form $\frac{p}{q}, q \neq 0$ while irrational numbers cannot be written in this form.
- Any two rational numbers are packed with irrational number on number line.
- Terminating decimal and repeating decimal numbers are rational numbers while non-repeating non-terminating decimals are irrational numbers.
- Square root of numbers other than perfect squares are irrational numbers.
- More examples of irrational numbers are $\sqrt{2}, \pi, e,$ etc.
- Squares of real numbers are always positive.

Example 1: Calculate square of following real numbers

$1.2^2 = 1.44$	$0.6^2 = 0.36$	$(-0.2)^2 = 0.04$	$0.01^2 = 0.0001$
$\sqrt{5}^2 = 5$	$\left(\frac{-1}{2}\right)^2 = \frac{1}{4}$	$-\left(\frac{3}{4}\right)^2 = -\frac{9}{16}$	$\left(\frac{\sqrt{3}}{2}\right)^2 = \frac{3}{4}$

Note: Square of all real numbers is also non-negative number

Practice to find squares of real numbers

0.1^2	-0.9^2	$\sqrt{1.1}^2$	1.2^2
$\left(\frac{\sqrt{5}}{3}\right)^2$	$\left(\frac{2}{5}\right)^2$	$\left(\frac{-1}{10}\right)^2$	$\left(\frac{6}{\sqrt{2}}\right)^2$

Compare Squares of Numbers

Let $a = 0.2,\ b = 0.8,$ and $c = -1.2$
Arrange the following in ascending order (least to greatest)

$$a^2, \quad c^2, \quad (a-c)^2, \quad (b-a)^2, \quad b^2$$

Application

Tiling of a square floor with area of $13\ m^2$ is to be completed with square tiles of length $0.25\ m$. Find the number of tiles required without overlapping or cutting.

208

Concept of Square-Root in the Domain of Real Numbers

Understanding Square Roots: Two Ways to Think About It

1 As an Inverse Operation to Squaring

The **square root** is like "undoing" a square. Square is always positive.

$y^2 = x \ then \ y = \pm\sqrt{x}, x \geq 0$

This means **two numbers** can have the same square: one **positive** and one **negative**.

For example:

- $(5)^2 = 25$
- $(-5)^2 = 25$

So, both 5 and -5 are square roots of 25.
We write this as:

$y = \pm\sqrt{25} = \pm 5$

! Important:

- This only works when x is **positive**.
- We **can't find** the square root of a **negative number** in real numbers.

2 As an Independent Operator

Sometimes, we use the square root symbol like a **machine** that always gives just the **positive result**.

That is:

$\sqrt{x} = y$

means **only the positive number** whose square is x.

So,

- $\sqrt{25} = 5$ (only the **positive** root)
- We call this the **principal square root**.

Note:

Diagonal of a square is square-root two times the side length of the square.

✨ **Can square of a number be less than the number itself?**

✨ **Can square-root of a number be greater than the number itself?**

Square Roots Extended from Real Numbers to Complex Numbers

Square-root of n, $\sqrt{n}$, is an Irrational number if n is not a perfect square. n is called **radicand** and $\sqrt{\ }$ is radical or square-root sign. $\sqrt{36} = 6$, a rational number since 36 is a perfect square. $\sqrt{35}$ is not a rational number.

Square-root of a negative number is a **complex number**. $\sqrt{-1} = i$, where i is an **imaginary number**.

Example: Write Square Root of Numbers Using $\sqrt{-1} = i$ or $i^2 = -1$

a.	$\sqrt{9} = 3$	b.	$\sqrt{-16} = -4i$	c.	$\sqrt{-0} = 0$
d.	$\sqrt{\frac{25}{4}} = \frac{5}{2}$	e.	$\sqrt{-8^2} = \sqrt{-64} = 8i$	f.	$\sqrt{(-6)^2} = 6$
Important Considerations					
a..	$\sqrt{(-9)(-25)}$ $= \sqrt{+225} =$ $= 15$	b.	$\sqrt{(-9)} \times \sqrt{(-25)}$ $= 3i \times 5i$ $= 15i^2 = -15$	c.	$\sqrt{x^2} = \lvert x \rvert$ Always positive $\sqrt{(-2)^2} = 2 \neq -2$

Reflect your learnings from the above Example.

Watch Video **How to Calculate Square Roots** **SCAN ME**

Estimate Square-Roots

We can estimate the approximate value of square-root by finding the two consecutive perfect square nearest to them.

Example 1: Estimate and calculate value of irrational numbers

$\sqrt{2}$	$\sqrt{20}$
$\sqrt{2}$ should be between 1 and 2 as $\sqrt{1}$ and $\sqrt{4}$ are the closest real numbers. So, it can be 1.5 Since, it is closer to $\sqrt{1}$ we can estimate as 1.4 Calculated $\sqrt{2} = 1.414213562$	$\sqrt{20}$ should be between 4 and 5 as $\sqrt{16}$ and $\sqrt{25}$ are the closest real numbers. So, it can be 4.5 Since, it is almost midway we can estimate as 4.5 Calculated $\sqrt{20} = 4.472135955$

Note: The above values are non-terminating and non-repeating decimals. Therefore, they cannot be written as rational numbers. These are irrational numbers.

Basic Operations with Radicals

We will consider radical operations in the domain of Real Numbers if not specified.

Example 2: Simplify radicals and write them as mixed radicals

Identify perfect square factors and then simplify radicals as **mixed radical of the form $a\sqrt{b}$** where a and b are integers.

Write $\sqrt{32}$ and $\sqrt{75}$ as mixed radical.

$\sqrt{32} = \sqrt{16 \times 2} = \sqrt{16} \times \sqrt{2} = 4\sqrt{2}$	$\sqrt{75} = \sqrt{25 \times 3} = \sqrt{25} \times \sqrt{3} = 5\sqrt{3}$

Example 3: Write as an entire radical

Entire radical is of the form $\sqrt{n}$. **It is a radical with coefficient 1.**

Write $a\sqrt{b}$, $3\sqrt{2}$, and $-5\sqrt{2}$ as entire radical.

$a\sqrt{b} = \sqrt{a^2 b}$	$3\sqrt{2} = \sqrt{3^2 \cdot 2} = \sqrt{18}$	$-5\sqrt{2} = -\sqrt{25 \times 2} = -\sqrt{50}$

Practice 3: Write following as mixed radicals

a.	$\sqrt{9 \times 3}$	b.	$\sqrt{4 \times 5}$	c.	$-\sqrt{72 \times 27}$	d.	$\sqrt{98}$

Practice 4: Write following as an entire radicals

a.	$3\sqrt{7}$	b.	$-2\sqrt{5}$	c.	$-3\sqrt{27}$	d.	$5\sqrt{3}$

Example 3: Add / Subtract Radicals

Write each radical in mixed form and then combine like terms.

$\sqrt{8} + \sqrt{32} - \sqrt{2}$	$2\sqrt{48} - 5\sqrt{27}$
$\sqrt{8} + \sqrt{32} = \sqrt{4 \cdot 2} + \sqrt{16 \cdot 2} - \sqrt{2}$	$2\sqrt{48} - 5\sqrt{27} = 2\sqrt{16 \cdot 3} + 5\sqrt{9 \cdot 3}$
$= 2\sqrt{2} + 4\sqrt{2} - \sqrt{2} = 5\sqrt{2}$	$= 2 \cdot 4\sqrt{3} - 5 \cdot 3\sqrt{3} = 8\sqrt{3} - 15\sqrt{3} = -7\sqrt{3}$

Practice 4: Add / Subtract Radicals

a.	$3\sqrt{2} - \sqrt{2}$	b.	$\sqrt{18} + \sqrt{50} - 3\sqrt{2}$

Example 4: Multiply

$$a\sqrt{b} \times c\sqrt{d} = ac\sqrt{bd}$$

- Multiply numbers outside the square root and keep them outside
- multiply the numbers inside the square root and keep them inside
- combine and simplify
- simplify radical if possible

$2\sqrt{2} \times 4\sqrt{2}$ $= 8\sqrt{4} = 8(2) = 16$	$3\sqrt{6} \times 2\sqrt{3}$ $= 6\sqrt{18} = 6 \times 3\sqrt{2} = 18\sqrt{2}$

Practice 5: Estimate and the check square-root values

a.	$\sqrt{2} \times 3\sqrt{6}$	b.	$3\sqrt{2} \times 2\sqrt{2}$	c.	$2\sqrt{12} \times 5\sqrt{3}$	d.	$2\sqrt{21} \times \sqrt{15}$

Note on Radical Solutions

Solutions involving radicals should **always be expressed in radical form** unless instructed otherwise. This is because radical form represents the **exact value**, while decimal approximations may lose precision.

Example 5: Division

$$a\sqrt{b} \div c\sqrt{d} = \frac{a\sqrt{b}}{c\sqrt{d}} = \frac{a}{c}\sqrt{\frac{b}{d}}$$

- Divide and simplify numbers outside the square root and keep them outside
- Divide and simplify the numbers inside the square root and keep them inside
- combine and simplify
- simplify radical if possible

$8\sqrt{2} \div 6\sqrt{50} = \dfrac{8\sqrt{2}}{6\sqrt{50}}$ $\dfrac{8\sqrt{2}}{6\sqrt{50}} = \dfrac{4}{3}\sqrt{\dfrac{2}{50}} = \dfrac{4}{3}\sqrt{\dfrac{1}{25}}$	$\sqrt{39} \div 5\sqrt{3} = \dfrac{\sqrt{39}}{5\sqrt{3}}$ $\dfrac{\sqrt{39}}{5\sqrt{3}} = \dfrac{1}{5}\sqrt{\dfrac{39}{3}} = \dfrac{1}{5}\sqrt{13}$

✍ Practice 6: Divide and Simplify Expressions

a.	$6\sqrt{3} \div 2\sqrt{12}$	b.	$8\sqrt{18} \div 10\sqrt{8}$

Example 6: Distributive Property with Radicals

The Distributive Property allows you to multiply a term outside the parentheses with each term inside the parentheses — and this rule also applies when working with radical expressions.

$$a\sqrt{b}(c\sqrt{d} + e\sqrt{f}) = ac\sqrt{bd} + ae\sqrt{bf}$$

$\mathbf{\sqrt{3}(5 + 2\sqrt{5})}$	$\mathbf{2\sqrt{3}(4\sqrt{2} - \sqrt{3})}$
$= \sqrt{3} \times 5 + \sqrt{3} \times 2\sqrt{5}$	$= 2\sqrt{3} \times 4\sqrt{2} - 2\sqrt{3} \times \sqrt{3}$
$= 5\sqrt{3} + 2\sqrt{15}$	$= 8\sqrt{6} - 2\sqrt{9} = 8\sqrt{6} - 2 \times 3 = 8\sqrt{6} - 6$

Communication and Thinking

$n = a\sqrt{b}, b \geq 0$, for what values of a and b will n be a rational number?

Hint: Perfect squares

Example 7: Multiply Binomial Radicals and Simplify

$$(a\sqrt{b} + c\sqrt{d})(e\sqrt{f} + g\sqrt{h}) = a\sqrt{b}(e\sqrt{f} + g\sqrt{h}) + c\sqrt{d}(e\sqrt{f} + g\sqrt{h})$$

Simplify $(2 - \sqrt{3})(3 + \sqrt{12})$ by first expanding with distributive property and the combining the like terms.

$(2 - \sqrt{3})(3 + \sqrt{12})$

$= 2 \cdot 3 + 2 \cdot \sqrt{12} - \sqrt{3} \cdot 3 - \sqrt{3}\sqrt{12}$

$= 6 + 2 \cdot \sqrt{4 \cdot 3} - 3\sqrt{3} - \sqrt{36}$

$= 6 + 2 \cdot 2\sqrt{3} - 3\sqrt{3} - 6$

$= 4\sqrt{3} - 3\sqrt{3}$

$= \mathbf{\sqrt{3}}$

Explore

$a + \sqrt{b}$ and $a - \sqrt{b}$ are conjugates.

Show that the conjugate sum and conjugate product are both integers if a and b are integers.

Practice 7: Distributive Property

a.	$5\sqrt{2}(2+3\sqrt{5})$	b.	$7\sqrt{3}(2\sqrt{15}+2\sqrt{3})$

Practice 8: Product of Conjugates

a.	$(\sqrt{10}-3)(\sqrt{10}+3)$	b.	$(4-2\sqrt{3})(4+2\sqrt{3})$

Practice 9: Extended Application to Simplify

a.	$(3\sqrt{2}-2\sqrt{5})^2 \div [(3\sqrt{2}-2\sqrt{5})(3\sqrt{2}+2\sqrt{5})]$
b.	$2(-2\sqrt{2}+3\sqrt{5})(-2\sqrt{2}-3\sqrt{5})$

Describe how will you mark the exact length of $\sqrt{2}$ on a number line.

Hint: Diagonal of a square.

Rationalization of Irrational Numbers

Irrational numbers are normally not written in the denominator.

Rationalization is the technique to eliminate radicals from the denominator

Concept: $(a+b)(a-b)=a^2-b^2$

Example 5: Rationalize the irrational numbers $\sqrt{a}\times\sqrt{a}=a$

a.	$\frac{1}{\sqrt{2}}$ Multiply both numerator and denominator by $\sqrt{2}$ $\frac{1}{\sqrt{2}}\times\frac{\sqrt{2}}{\sqrt{2}}=\times\frac{\sqrt{2}}{2}$	b.	$\frac{2}{\sqrt{3}}$ Multiply both numerator and denominator by $\sqrt{3}$ $\frac{1}{\sqrt{3}}\times\frac{\sqrt{3}}{\sqrt{3}}=\frac{\sqrt{3}}{3}$

Example 6: Binomial denominator: Multiply with conjugate

Conjugate of $\sqrt{a}+b$ is $\sqrt{a}-b$

c.	Work	Explanation
	$\dfrac{2}{\sqrt{5}-1}$	Multiply and divide by conjugate
	$\dfrac{2}{\sqrt{5}-1}\times\dfrac{\sqrt{5}+1}{\sqrt{5}+1}$	$\sqrt{5}+1$ is conjugate of $\sqrt{5}-1$
		$(\sqrt{a}-b)(\sqrt{a}+b)=a-b^2$
	$=\dfrac{2(\sqrt{5}+1)}{5-1}$	Strategy: do not apply distributive property but cancel 2 from numerator and denominator.
	$=\dfrac{2(\sqrt{5}+1)}{4}=\dfrac{\sqrt{5}+1}{2}$	

Guided Practice 10: Rationalize the following expressions

a.	$\dfrac{2}{2-\sqrt{3}}$	b.	$\dfrac{2+\sqrt{3}}{2-\sqrt{3}}$	c.	$\dfrac{1}{\sqrt{x+1}+\sqrt{x}}$	d.	$\sqrt{x+1}+\sqrt{x}$

Watch Video **Rationalization of Denominator** **SCAN ME**

Rationalization Expressions Challenge

Rationalization of expressions is important for solving problems.

Example 6: Rationalize Expression

$$\sqrt{x+1}-\sqrt{x}$$

Multiply both numerator and denominator by the conjugate $\sqrt{x+1}+\sqrt{x}$

$$\sqrt{x+1}-\sqrt{x}\times\frac{\sqrt{x+1}+\sqrt{x}}{\sqrt{x+1}+\sqrt{x}}=\frac{(x+1)-x}{\sqrt{x+1}+\sqrt{x}}=\frac{1}{\sqrt{x+1}+\sqrt{x}}$$

Note: Interesting to note how this expression is equivalent to conjugate reciprocal.

Application: Write an expression equivalent to

$$\sqrt{8}-\sqrt{7}$$

Radicals Challenge

💡 Use calculator to check if the following three inequalities are always true.

Take $x = \{0, 1, 9, 99\}$

$2\sqrt{x+1} > \sqrt{x+1} + \sqrt{x}$	$\frac{1}{2\sqrt{x+1}} < \frac{1}{\sqrt{x+1}+\sqrt{x}}$	$\frac{1}{2\sqrt{x+1}} < \sqrt{x+1} - \sqrt{x}$

Evaluate Expression without Calculator: $\sqrt{6+\sqrt{11}} + \sqrt{6-\sqrt{11}}$

Find the larger number without using calculator, between A and B.

$A = \sqrt{3\sqrt{2\sqrt{5}}}$ and $B = \sqrt{2\sqrt{5\sqrt{3}}}$ Hint: Square repeated times

Evaluate Exact Value of Nested Square Roots Infinity Series

$$x = \sqrt{6 - \sqrt{6 - \sqrt{6 + \sqrt{6 \ldots}}}}$$

Hint: Square, rearrange, and then solve the quadratic equation

Challenge Equation: Solve $x = \sqrt{1 - \sqrt{1 - x}}$

Verify that it has following 4 solutions: $x = 0, 1, \frac{-1+\sqrt{5}}{2}, \frac{-1-\sqrt{5}}{2}$

Area of Triangle: Heron's Formula

The area of a triangle with side lengths of $a, b,$ and c can be calculated by **Heron's formula:**

$A = \sqrt{s(s-a)(s-b)(s-c)}$, where $s = \frac{a+b+c}{2}$.

Practice: Calculate the area of triangle with side lengths of 4 cm, 5 cm, and 7 cm.

$$s = \frac{a+b+c}{2} = \frac{4+5+7}{2} = 8$$

$$A = \sqrt{s(s-a)(s-b)(s-c)} = \sqrt{8(8-4)(8-5)(8-7)} = \mathbf{4\sqrt{6}\ cm^2}$$

✍ **Practice:**

Calculate the area of triangle with side lengths of 2 cm, 5 cm, and 6 cm.

Radicals Challenge Solution

Evaluate Expression

$$\sqrt{6+\sqrt{11}}+\sqrt{6-\sqrt{11}}$$

Formulas to Recall

$$\boldsymbol{a(b \pm c) = a \cdot b \pm a \cdot c}$$

$$\boldsymbol{(a \pm b)^2 = a^2 \pm 2ab + b^2}$$

$$\boldsymbol{(a+b)(a-b) = a^2 - b^2}$$

Solution

$$x = \sqrt{6+\sqrt{11}}+\sqrt{6-\sqrt{11}}$$

$$x^2 = \left(\sqrt{6+\sqrt{11}}+\sqrt{6-\sqrt{11}}\right)^2$$

$$x^2 = \left(\sqrt{6+\sqrt{11}}\right)^2 + 2\sqrt{6+\sqrt{11}} \times \sqrt{6-\sqrt{11}} + \left(\sqrt{6-\sqrt{11}}\right)^2$$

$$x^2 = 6+\sqrt{11}+2\sqrt{6^2-\sqrt{11}^2}+6-\sqrt{11}$$

$$x^2 = 12 + 2\sqrt{25} = 12 + 2(5) = 22$$

$$\boldsymbol{x = \sqrt{22}}$$

Exact Value: This is Non-Negative Value

$$x = \sqrt{\boldsymbol{6 - \sqrt{6 - \sqrt{6 + \sqrt{6} \ldots}}}}$$

$$x^2 = 6 - \sqrt{\boldsymbol{6 - \sqrt{6 - \sqrt{6 + \sqrt{6} \ldots}}}}$$

$$x^2 = 6 - x$$

$$x^2 + x - 6 = 0$$

$$(x-2)(x+3) = 0$$

$$x = 2$$

Challenge: Show algebraically that those following equations are true.

a.	$\sqrt{5+2\sqrt{6}} = \sqrt{2} + \sqrt{3}$	b.	$\sqrt{13-2\sqrt{42}} = \sqrt{7} - \sqrt{6}$
c.	$\sqrt{(a+b)+2\sqrt{ab}} = \sqrt{a} + \sqrt{b}$	d.	$\sqrt{(a+b)-2\sqrt{ab}} = \sqrt{a} - \sqrt{b}$

STEM Project: Rationalization Inequality (Solution in Part 7)

Show that

$$\frac{1}{2\sqrt{n+1}} < \sqrt{n+1} - \sqrt{n}, n\epsilon Z, n > 0\ (n\ is\ positive\ integers)$$

Steps #1: Rationalize to show that

$$\sqrt{n+1} - \sqrt{n} = \frac{1}{\sqrt{n+1} + \sqrt{n}}$$

Step #2: Explain why

$$\frac{1}{2\sqrt{n+1}} < \frac{1}{\sqrt{n+1} + \sqrt{n}}$$

Hence Prove

$$\frac{1}{2\sqrt{n+1}} < \sqrt{n+1} - \sqrt{n}$$

Extended Learning: Exploring Square Roots

Claim:

The square root of the product of two real numbers is always equal to the product of their individual square roots.

$$\sqrt{a \times b} = \sqrt{a} \times \sqrt{b}$$

❗ **Task:** Is this always true for **all real numbers**?

Myth Busted

Claim:

The square root of a number is always less than or equal to the square of the number.

$$\sqrt{x} \leq x^2$$

❗ **Task:**

Is this always true for **all real numbers**?

Can you find **counterexamples** that make this claim **false**?

Hint: Try numbers less than 1

Laws of Exponents or Indices

Exponents represent repeated multiplication of the same number.

- Product $3 \cdot 3 \cdot 3 \cdot 3$ is written as 3^4
- Product $2 \cdot 2 \cdot 2 \cdot 2 \cdot 2 \cdot 2$ is written as 2^6

In general, for a number a raised to the power of n can be written as a^n, which means multiply a by itself n time.

$a^n = a \cdot a \cdot a \ldots \cdot a \; (n \; times)$

The entire expression, a^n is called power; the number, a, in the bottom is the **base** and the number, n, on the top is called exponent.

Laws of Exponents

The **laws of exponent** used to simplify operations involving exponents are:

1. **Product Rule**: $a^m \cdot a^n = a^{m+n}$
2. **Quotient Rule**: $\frac{a^m}{a^n} = a^{m-n}$
3. **Power Rule**: $(a^m)^n = a^{mn}$
4. **Zero Exponent**: $a^0 = 1$
5. **Identity Exponent Rule**: $a^1 = a$
6. **Negative Exponent**: $a^{-n} = \frac{1}{a^n}, a \neq 0$
7. **Power of a Product**: $(ab)^n = a^n \cdot b^n$
8. **Power of Quotient**: $\left(\frac{a}{b}\right)^n = \frac{a^n}{b^n}, b \neq 0$

SCAN ME

Find exact value of N

$$2^N = \left(2^{2^{10}}\right)^2$$

Misconceptions

Before we explore further, let us look into **some misconceptions**

- Sum of Exponents Misconception: $(a+b)^m \neq a^m + b^m$
- Difference of Exponents Misconception: $(a-b)^m \neq a^m - b^m$
- Addition of Exponents: $a^m + a^n \neq a^{m+n}$
- Parentheses Rule or Negation vs. Exponents: $(-a)^n \neq -a^n$

Note: Exponents rules apply to multiplication and division and not to addition and subtraction.

Practice 1: Simplify and write as a^n

a.	$a^2 \cdot a^3$	b.	$(a^3 \div a^{-2})^4$	c.	$\left(\frac{a^2}{a^{-3}} \times a^4\right)^5$

Laws of Exponents or Indices: Simple Approach

Example 1: Simple example to illustrate the rules

Product Rule: $2^4 \cdot 2^6 = 2^{4+6} = 2^{10}$

$(2 \cdot 2 \cdot 2 \cdot 2) \cdot (2 \cdot 2 \cdot 2 \cdot 2 \cdot 2 \cdot 2) = 2^{10}$

Quotient Rule: $\frac{3^6}{3^4} = 3^{6-4} = 3^2$

$$\frac{3 \cdot 3 \cdot 3 \cdot 3 \cdot 3 \cdot 3}{3 \cdot 3 \cdot 3 \cdot 3} = 3 \cdot 3 = 3^2$$

Power Rule: $(2^3)^2 = 2^6$

$(2^3)^2 = (2 \cdot 2 \cdot 2)^2 = (2 \cdot 2 \cdot 2) \cdot (2 \cdot 2 \cdot 2) = 2^6$

Zero Exponent Rule: $7^0 = 1$

$\frac{7^3}{7^3} = 1$ a number divided by itself is always 1

$\frac{7^3}{7^3} = 7^{3-3} = 7^0$ from Quotient Rule

From above two statements, it is clear that $7^0 = 1$

Identity Exponent Rule: $9^1 = 9$

Number 9 multiplied by itself just once is the number itself

Negative Exponent Rule: $b^{-1} = \frac{1}{b}, b \neq 0$

$\frac{b^3}{b^4} = b^{3-4} = b^{-1}$ also $\frac{b^3}{b^4} = \frac{b \cdot b \cdot b}{b \cdot b \cdot b \cdot b} = \frac{1}{b}$

Hence $b^{-1} = \frac{1}{b}$

Indeterminates: The mathematical expressions whose value cannot be determined. Examples: $\mathbf{\frac{0}{0}}, \frac{\infty}{\infty}, \infty - \infty, 0^0, 1^\infty$ and 0^∞ are indeterminates

Important to Understand

There is a difference between "indeterminate" and "undefined" values.

- An **indeterminate** expression does not have a fixed value but can be approached or resolved to a meaningful real value depending on the context $\left(\text{e.g.}, \frac{0}{0}\right)$
- An **undefined** expression has no meaning in mathematics and cannot be assigned any real value $\left(\text{e.g.}, \frac{1}{0}\right)$

Laws of Exponents Illustrated Examples

Example 2: Introduction to Zero and Negative Exponents

Identity – Zero – Negative Exponent Relation

Consider the following pattern

$3^4 = 81$	$3^3 = 27$	$3^2 = 9$	$3^1 = 3$	$3^0 = 1$	$3^{-1} = \frac{1}{3}$	$3^{-2} = \frac{1}{3^2}$

As the exponent of 3^n decreases by one the number is divided by 3.

Clearly you can see $3^1 = 3, 3^0 = 1,$ and $3^{-2} = \frac{1}{3^2}$

Note:

- $0^{-1} = \frac{1}{0}$ It is not defined.
- 0^0 is indeterminate.
- $0^n = 0, n > 0$

- $1^n = 1$, for any real number n
- $a^{-n} = \frac{1}{a^n}$
- $\frac{1}{a^{-n}} = a^n$

Power of Product Rule: $(2 \cdot 5)^3 = 2^3 \cdot 5^3$

$(2 \cdot 5)^3 = (10)^3 = 1000$ also $2^3 \cdot 5^3 = 8 \times 125 = 1000$

Power of Quotient Rule: $\left(\frac{6}{3}\right)^4 = \frac{6^4}{3^4}$

$\left(\frac{6}{3}\right)^4 = 2^4 = 16$ also $\frac{6^4}{3^4} = \frac{6 \cdot 6 \cdot 6 \cdot 6}{3 \cdot 3 \cdot 3 \cdot 3} = 2 \cdot 2 \cdot 2 \cdot 2 = 2^4 = 16$

Negative Exponents Rule: $\left(\frac{16}{8}\right)^{-4} = (2)^{-4}$

$\left(\frac{1}{2}\right)^4 = \frac{1}{2^4} = \frac{1}{16}$

SCAN ME

✍ Practice 2: Evaluate each of the following

a.	$\left(\frac{12}{4}\right)^{-3}$	b.	$(2^2 \cdot 5^2)^4$	c.	$\left(\frac{2}{3}\right)^{-3}$

Misconceptions explained with simple examples

Example 3: Misconceptions Overview

Sum of Exponents Misconception: $(a+b)^m \neq a^m + b^m$ $(2+3)^2 \neq 2^2 + 3^2$ $(5)^2 \neq 4+9$ $25 \neq 13$ Note: It is true for $\boldsymbol{m = 1}$ but not for other values	**Difference of Exponents Misconception:** $(a-b)^m \neq a^m - b^m$ $(3-2)^2 \neq 3^2 - 2^2$ $(1)^2 \neq 9-4$ $1 \neq 5$
Addition of Exponents: $a^m + a^n \neq a^{m+n}$ $3^2 + 3^3 \neq 3^{2+3}$ $9 + 27 \neq 3^5$ $36 \neq 24$	**Negative Exponent:** $a^{-n} \neq -a^n$ $2^{-1} = \frac{1}{2} \neq -2$ Exponent of any positive number is always positive

Negation with Parentheses Rule

Parentheses Rule or Negation vs. Exponents: $(-a)^n \neq -a^n$

The difference between $(-a)^n$ and $-a^n$ is due to order of operation.

Expression under the bracket will be evaluated first.

- $(-a)^n$ means that the entire quantity $(-a)$ is raised to the power of n
- $-a^n$ means negation of a^n

Note:

- $(-a)^n = a^n$ for even powers of n
- $(-a)^n = -a^n$ for odd powers of n

✍ Practice 3: Evaluate each of the following

a.	$(-3)^3$	b.	$(-3)^{-4}$	c.	$(-3)^0$
d.	-10^6	e.	$(-1)^{99}$	f.	$-2^3 - 3^2$

Example 4: Evaluate Following Using Laws of Exponents

a.	$(2^2+3^3)^2$ $=(4+27)^2=(31)^2=961$	b.	$(2^2\times 3^3)^0=1$ Note: $\boldsymbol{a^0=1}$
c.	$2^3\div 2^{-7}$ $=2^{3+7}=2^{10}=1024$	d.	$\dfrac{3^3\times 3^{-5}}{3^2}=3^{3-5-2}=3^{-4}=\dfrac{1}{81}$

Guided Practice 4: Evaluate Expressions

g.	$\dfrac{5^3\times 5^7}{5^{-8}}=$	h.	$2^{-3}\times 2^5=$
i.	$2^{-3}\times 5^7\div 5^4$	j.	$(a^2\times b^{-3}\div c)^0=$

Example 5: Determine the exponent that makes the equation true

a.	$3^x=81$	b.	$2^a=32$
c.	$10^{-c}=10000$	d.	$(-1)^p=-1$

Example 6: Evaluate each of the following for r = 2.5. Round to tenths.

a.	$A=\pi(r)^2$	b.	$V=\dfrac{4}{3}\pi(r)^3$

Explore and Learn: Student Activity

Important identities with exponents if p, q, n are positive integers where $p>q>0$.

a.	$-(pq)^n=-(p)^n\cdot(q)^n$	b.	$(p^2+q^2)^2\neq(p)^4+(q)^4$ $(p^2+q^2)^2=(p)^4+2p^2q^2+(q)^4$
	$(p+q)^2>p^2+q^2$ $(p+q)^2=p^2+2pq+q^2$	d.	$(p-q)^2<p^2-q^2$ $(p-q)^2<p^2-2pq+q^2$
e.	$(p)^{2n}=(-p)^{2n}$ $(-1)^{2n}=+1$	f.	$(-p)^{2n+1}=-(p)^{2n+1}$ $(-1)^{2n+1}=-1$

Example 7: Express Numbers as Power of Given Number

Express as Power of 2

Two questions are solved step by step. Provide complete solution for others.

a.	$(2^3)^4 \times 64 \div 2^5$ $= 2^{12} \times 2^6 \div 2^5$ $64 = 2^6$ $= 2^{12+6-5} = 2^{13}$	b.	$4^3 \times 2^4 \times 16^2$ $= (2^2)^3 \times 2^4 \times (2^4)^2$ $16 = 2^4$ $= 2^6 \times 2^4 \times 2^8 = 2^{18}$ $4 = 2^2$
c.	$(2^{-2})^{-3} \times 16 \div 2^3 = 2^7$	d.	$4^2 \div 2^{-4} \times 2 = 2^{4+4+1} = 2^9$
e.	$2^{25} + 2^{25} + 2^{25} + 2^{25}$	f.	$(2^{-2})^6 \div 64 \div 2^5$

Guided Practice 5: Express as Power of 3

Express each of the following number as power of 3. **Practice for Students**

a.	$(3^3)^{-4} \times 81 \div 3^5$	b.	$9^3 \times 27^4 \times 81^2$
c.	$3 \cdot 3^{20} + 9 \cdot 3^{19} + 3^{21}$	d.	$45 \times 25^{-2} \div 5(15)^2$

Practice 6: Evaluate each of the following

a.	$\left(\frac{-3^2}{2}\right)^3$	b.	$\left(\frac{5^2 - 5^0}{-2^2 + 3^2}\right)^{-2}$	c.	$\frac{3^{-3} + 3^0}{3^0 + 3^{-2}}$

Challenge Problems:

Q1	Solve Equation: $x^y = y^x, x \neq y$	Q2	Find the value of x^3 if $4x^5 = 16^2$
Q3a	$A = \sqrt{\frac{b}{c^3}}$, $b = 4.8 \times 10^{-5}$, $c = 1.2 \times 10^{-4}$ Calculate A correct to 3 significant figures.	Q3b Q3c	If b is increased by 10% and c is decreased by 5% find A. Now calculate percentage increase in the value of A.

Rational Indices or Exponents: A Powerful Notation

What Are Fractional or Rational Exponents?

A fractional exponent is a way to represent both powers and roots using a single notation. Instead of writing roots with the radical symbol (√), we can express them as powers with fractional exponents.

$$a^{\frac{m}{n}} = \sqrt[n]{a^m} = \left(\sqrt[n]{a}\right)^m; \; a^{-\frac{m}{n}} = \frac{1}{\sqrt[n]{a^m}}$$

💡 **"The denominator of the exponent is the root, and the numerator is the power."**

Notation	Meaning	Description	Illustrated Example
$a^{\frac{1}{2}}$	$\sqrt{a}$	Square root of a	$9^{\frac{1}{2}} = \sqrt{9} = 3$
$a^{\frac{1}{3}}$	$\sqrt[3]{a}$	Cube root of a	$125^{\frac{1}{3}} = \sqrt[3]{125} = 5$
$a^{\frac{3}{2}}$	$\left(\sqrt{a}\right)^3$	Cube of square root of a	$4^{\frac{3}{2}} = \left(\sqrt{4}\right)^3 = 2^3 = 8$
$a^{-\frac{2}{3}}$	$\frac{1}{\left(\sqrt[3]{a}\right)^2}$	Reciprocal of $a^{\frac{2}{3}}$	$27^{-\frac{2}{3}} = \frac{1}{\left(\sqrt[3]{27}\right)^2} = \frac{1}{(3)^2} = \frac{1}{9}$

Example 8: Write as Single Power of 2

a.	$\sqrt{2} = 2^{\frac{1}{2}}$	b.	$\sqrt{2^3} = 2^{\frac{3}{2}}$	c.	$\sqrt[3]{2} = 2^{\frac{1}{3}}$	d.	$\frac{1}{\sqrt[3]{2}} = 2^{-\frac{1}{3}}$

Example 9: Write as Single Power of 3

a.	$\sqrt{3} = 3^{\frac{1}{2}}$	b.	$\sqrt{27} = 3^{\frac{3}{2}}$	c.	$\sqrt[3]{9} = 3^{\frac{2}{3}}$	d.	$\frac{1}{\sqrt[3]{81}} = 3^{-\frac{4}{3}}$

Watch Video **Rational Exponents Practice Examples** **SCAN ME**

Example 10: Write in Simplest Rational Form

a.	$27^{\frac{1}{3}} = \sqrt[3]{27} = \sqrt[3]{3 \cdot 3 \cdot 3} = 3$	b.	$81^{-\frac{1}{4}} = (3^4)^{-\frac{1}{4}} = 3^{-1} = \frac{1}{3}$
c.	$8^{-\frac{2}{3}} = \frac{1}{\left(\sqrt[3]{8}\right)^2} = \frac{1}{(2)^2} = \frac{1}{4}$	d.	$32^{-\frac{3}{5}} = \frac{1}{\left(\sqrt[5]{32}\right)^3} = \frac{1}{(2)^3} = \frac{1}{8}$

Example 11: Understand Rational Exponents for Decimal Numbers

a.	$\sqrt{0.01} = \sqrt{\frac{1}{100}} = \frac{1}{10} = 0.1$	b.	$\sqrt[3]{0.064} = \sqrt[3]{\frac{64}{1000}} = \frac{4}{10} = 0.4$
c.	$0.2^3 = \left(\frac{2}{10}\right)^3 = \left(\frac{1}{5}\right)^3 = \frac{1}{125}$	d.	$0.25^2 = \left(\frac{25}{100}\right)^2 = \left(\frac{1}{4}\right)^2 = \frac{1}{16}$

Example 12: Simplify Using Laws of Exponents

a.	$(16x^4)^{\frac{1}{2}} = (16)^{\frac{1}{2}}(x^4)^{\frac{1}{2}} = 4x^2$	b.	$\sqrt[3]{27x^6} \div \left(x\sqrt{81}\right) = \frac{3x^2}{9x} = \frac{x}{3}$
c.	$\frac{\sqrt{a} \cdot \sqrt[3]{a}}{\left(\sqrt{a}\right)^3} = a^{\frac{1}{2}+\frac{1}{3}-\frac{3}{2}} = a^{\frac{3+2-9}{6}} = a^{-\frac{4}{6}} = a^{-\frac{2}{3}} = \frac{1}{a^{\frac{2}{3}}} = \frac{1}{\left(\sqrt[3]{a}\right)^2}$		
d.	$\frac{\left(\sqrt{27}\right)^{4x-2}}{3^{3x-1} \cdot 9^{4-3x}} = \frac{\left(3^{\frac{3}{2}}\right)^{4x-2}}{3^{3x-1} \cdot (3^2)^{4-3x}} = \frac{\left(3^{\frac{3}{2}}\right)^{2(2x-1)}}{3^{3x-1+8-6x}} = \frac{3^{6x-3}}{3^{3x-1+8-6x}} = \frac{3^{6x-3}}{3^{-3x-7}} = 3^{9x+4}$		

Challenge Problems: Simplify Each Expression. Write as Mixed Radical

a.	$\sqrt{\frac{a^5b^2c}{a^2c^2}} = \sqrt{\frac{a^3b^2}{c}} = a\lvert b\rvert\sqrt{\frac{a}{c}}$	*Note*: $\sqrt{b^2} = \lvert b\rvert$
b.	$\frac{x^2 - x + \sqrt{x}}{\sqrt{x}} = \frac{x^2}{\sqrt{x}} - \frac{x}{\sqrt{x}} + \frac{\sqrt{x}}{\sqrt{x}} = x^{\frac{3}{2}} - x^{\frac{1}{2}} + 1 = x\sqrt{x} - \sqrt{x} + 1$	

Practice Test: Exponents

Q1.	Write in radical form and then evaluate.
	a. $8^{\frac{1}{3}}$ — b. $25^{\frac{3}{2}}$ — c. $36^{-\frac{3}{2}}$ — d. $256^{-\frac{3}{4}}$
Q2.	Write in exponent form and then evaluate.
	a. $\sqrt[3]{64}$ — b. $\sqrt[3]{10^6}$ — c. $\sqrt[3]{0.027}$ — d. $\sqrt{9^3}$
Q3	Express as single power.
	a. $p^{\frac{5}{3}} \times p^{\frac{2}{5}} \div p^{\frac{7}{5}}$ — c. $\sqrt[3]{\sqrt[4]{x^8}}$
Q4.	Simplify using laws of exponents
	a. $(9x^4y)^{\frac{1}{2}} \times (84x^6y)^{\frac{1}{2}}$ — c. $(16x^7y^{-3})^{\frac{1}{2}} \div (8^3x^{-3}y)^{\frac{1}{2}}$
Q5.	Expand and simplify: $\left(\sqrt{x} - \frac{1}{\sqrt{x}}\right)^2$
Q6.	Simplify using laws of exponents
	a. $\sqrt{\frac{a^2b^3c^4}{b^2c^9}}$ — c. $\frac{\sqrt[3]{x^2} - \sqrt[3]{x} + 2x}{\sqrt{x}}$
Q7.	Evaluate $\sqrt{\frac{x^3-4x}{2x^4}}$ for $x = -2$
Q8.	$a(p)^{2n} = (-3p)^{2n}$. Find the value of a

Challenge Question

Show that $x^{-\frac{1}{2}}\left(3\sqrt{x} + 2\sqrt{x^3} - 5x^{-\frac{1}{2}}\right) = 2x + 3 - \frac{5}{x}$

Extended Skills: Solve Exponential Equations: 3 Methods

We will discuss three methods to solve exponential equations

Method 1: Make the bases the same, if possible

Method 2: Take logarithms (if bases can't be made the same)

Method 3: Use substitution if the equation has multiple terms

Example 9: Solve Equation by making the same base

a.	Solve: $2^x = 32$ $2^x = 32 = 2^5$ $x = 5$	b.	Solve: $3^{-x} = 81$ $3^{-x} = 3^4$ $x = -4$

Example 10: Solve Equation by making the same base

Here we have additional steps to change and relate base on both sides

a.	Solve: $9^x = 27$ $9^x = 3^3$ $(3^2)^x = 3^3$ $3^{2x} = 3^3$ $2x = 3$ $x = \frac{3}{2}$	b.	Solve: $8^x = 128$ $8^x = 2^7$ $(2^3)^x = 2^7$ $2^{3x} = 2^7$ $3x = 7$ $x = \frac{7}{3}$

Guided Practice 9: Solve following equations

a.	$3^x = 243$	b.	$4^x = 32$
c.	$b^4 = 625$	d.	$8^3 = 2^{-x}$
e.	$27^2 = 9^x$	f.	$1 = 7^x$
g.	$4^3 = 8^x$	h.	$16^x = 8^3$
i.	$27^{-3} = 9^x$	j.	$x^3 = 0$

Method 2: Take logarithms (if bases can't be made the same)

Example 10: Solve Equation by taking logarithms

$3^x - 20 = 0$

$3^x = 20$

The solution is between 2 and 3 since

- $3^2 = 9$
- $3^3 = 27.$

And **20 is closer to 27 than to 9**, so x is a number **closer to 3**.

For Accurate Results: Use of Logarithms

To solve this type of exponential equation — where the **base cannot easily be made the same** — we use **logarithms**, a function available on most scientific calculators.

Early Introduction to Logarithms

Take logarithm on both sides (base 10 or natural log):

$\log 3^x = \log 20$

Using the **Power Rule of Logarithms**:

$x \log 3 = \log 20$

$x = \dfrac{\log 20}{\log 3}$ As expected, x is closer to 3

$x \approx 2.73$

SCAN ME

Practice 10: Solve the Equation

a.	$2^x = 10$	b.	$7^x(5^x - 4) = 0$

Logarithms:
Logarithms are a powerful tool for solving exponential equations — though they go beyond the core scope of this book, they open the door to advanced algebra, data science, and engineering. Get ready to explore more in our next journey through Advanced Functions!

Method 3: Use substitution if the equation has multiple terms

Example 11: Solve Equation by substitution

$2^{2x} - 5(2^x) + 6 = 0$

This is an exponential equation in quadratic form

Substitute: $y = 2^x$

$y^2 - 5(y) + 6 = 0$

$(y - 3)(y - 2) = 0$

There are two solutions for each factor to be zero

$\boldsymbol{y - 3 = 0}$ **and** $\boldsymbol{y - 2 = 0}$

Note:
Factoring using product and sum method will be discussed in Chapter 4.

$\boldsymbol{y - 3 = 0}$

$y = 3$

$2^x = 3$

The solution is between 1 and 2 since

- $2^1 = 2$
- $2^2 = 4.$

For Accurate Results: Use of Logarithms

Take logarithms on both sides

$\log 2^x = \log 3$

$x \log 2^x = \log 3$

$$\mathrm{x} = \frac{\log 3}{\log 2}$$

$x \approx 1.58$

$\boldsymbol{y - 2 = 0}$

$y = 2$

$2^x = 2$

$x = 1$

✍ Practice

Solve equation $3^{2x} - 5(3^x) + 6 = 0$

Property of Logarithm

$$\boldsymbol{\log a^x = x \log a}$$

Solution of the equation $\boldsymbol{2^{2x} - 5(2^x) + 6 = 0}$ **is** $\boldsymbol{x = 1}$ **and** $\boldsymbol{x \approx 1.58}$

✍ Practice 11: Solve the Equation

a.	$2^{2x} + 3(2^x) - 10 = 0$	b.	$7^{2x} - 12(7^x) + 35 = 0$

Real Life Applications for Exponents

Example 12: Doubling Bacteria Growth

A single bacterium doubles every hour. How many bacteria will there be after 6 hours?

Solution Strategy:

Visualize with table of values how the bacteria number doubles every hour.

Time (hours)	start	1	2	3	4	5	6
# Bacteria	1	2	4	8	16	32	64
Exponent Form		2^1	2^2	2^3	2^4	2^5	2^6

Answer: After 6 hours, there will be **2^6 or 64 bacteria**.

Alternate Solution

Use the formula:

Total Bacteria $= 1 \times 2^n$ where n number of hours.

Total $= 1 \times 2^6 = 64$

Answer: After 6 hours, there will be 64 bacteria.

Double Life Formula

$N(t) = N_0(2)^{\frac{t}{d}}$

$\boldsymbol{N(t) = N_0(2)^{\frac{t}{d}}}$

$\boldsymbol{N(t)}$: number at time $\boldsymbol{t}$

$\boldsymbol{N_0}$: Initial number at start

$\boldsymbol{d}$: doubling time

Challenge Problems

The population of bacteria can multiply 10 times in 20 hours. If this exponential growth is maintained, then when will the population be double of what it was after 20 hours.

Hint: $N(t) = N_0(10)^{\frac{t}{20}}$

a.	22 h	b.	24 h	c.	26 h	d.	30 h

Watch Video **Exponential Applications** SCAN ME

Example 13: Halving Medicine Dose Application in Medicine

Problem:

A medicine's effect halves every 4 hours. If a patient starts with 160 mg in their bloodstream, how much will be left after 12 hours?

Solution Strategy:

Each 4-hour period cuts the dose in half.

Time in hours	Start	4 h	8 h	12 h
Medicine left	160 mg	80 mg	40 mg	20 mg

Exponents approach

Half after every 4 hours. In 12 h it will half its mass three times.

In 12 hours it will be $160 \times \frac{1}{2} \times \frac{1}{2} \times \frac{1}{2} = 160\left(\frac{1}{2}\right)^3 = 20$

Answer: 20 mg of medicine remains after 12 hours

Reflection:

How the formula is derived for the final amount left, A_f, from initial amount, A_i, amount after t hours if the half time is h

$$A_f = A_i\left(\frac{1}{2}\right)^{\frac{t}{h}}$$

.

Property of Logarithm

$$\log a^x = x \log a$$

$$A_f = A_i\left(\frac{1}{2}\right)^{\frac{t}{h}}$$

A_f: final amount at time t

A_i: Initial amount at start

h: half time

✍ Practice 12: Application of Half-Life in Archaeology

a.	The half-life of Carbon-14 is **5730 years**. A fossil is found to contain only **25% of its original Carbon-14**. Estimate the age of the fossil. 11460 years

Simple and Compound Interest – Exponential Applications

Where money grows with time — linearly or exponentially.

Interest is the extra amount earned or paid when money is invested or borrowed. There are two main types:

- Simple Interest – grows linearly.
- Compound Interest – grows exponentially.

SCAN ME

◆ Simple Interest (Linear Growth)

Interest is calculated only on the original amount (principal).

Formula:

I = (P × R × T) / 100

A = P + I

Where:

- P = Principal (initial amount)
- R = Annual Interest Rate (in %)
- T = Time in years
- A = Amount after interest

Example 1:

You invest $1,000 at 5% simple interest for 3 years. Find the amount due after 3 years.

Solution:

I = (1000 × 5 × 3)/100 = 150

Total Amount, A = P + I = 1000 + 150 = $1,150

Example 2:

A loan of $2,400 is borrowed at 6% simple interest for 4 years. Find the total payable amount after 4 years.

Solution:

I = (2400 × 6 × 4)/100 = 576

Total Payable = $2,976

Example 3:

You deposit $750 at 8% simple interest. It earns $180. How long was it invested?

Solution:

180 = (750 × 8 × T)/100

$T = \frac{180 \times 100}{750 \times 8} = 3.$ **T = 3 years**

✍ Practice Question: Simple Interest

Q1	$1,200 is invested at a 6% simple interest rate for 5 years. Find total amount. $1,560
Q2	A student borrows $3,000 for 3 years. If the total interest paid is $540, what was the annual interest rate? R = 6%
Q3	You deposit money at 7% simple interest. After 4 years, the total amount becomes $1,456. What was the original principal? P = $1,120

Annual Compound Interest (Exponential Growth)

Annual Compound Interest

Interest is calculated on the principal + previously earned interest.

- $A = P(1 + \mathrm{r})^t$
- P = Principal (initial amount)
- R = Annual Interest Rate (in %)
- r = Decimal Value of Annual Rate of Interest = $\frac{R}{100}$
- T = Time in years
- A = Amount after interest
- Compounded Interest, I = A – P

Example 1:
You invest $1,000 at 5% compound interest for 3 years. Find amount and compounded interest after 3 years.

Solution:

$A = P(1 + r)^t$

$A = 1000(1 + 0.05)^3 = 1157.63$

Amount will be $1157.63
Compound Interest = 1157.63 – 1000
= **$157.63**

Practice Question

You invest $2,000 at 6% for 4 years. Find amount and compounded interest after 4 years.

A = 2525.06; Compound Interest = $525.06

Compound n Timer Per Year

Interest is calculated on the principal + previously earned interest.

- $A = P\left(1 + \frac{\mathrm{r}}{n}\right)^{nt}$
- P = Principal (initial amount)
- R = Annual Interest Rate (in %)
- r = Decimal Value of Annual Rate of Interest = $\frac{R}{100}$
- T = Time in years
- A = Amount after interest
- Compounded Interest, I = A – P
- n Number of Compounding times Per Year
- $n = 4$ for quarterly; $n = 2$ for semiannually, $n = 12$ for monthly

Solve for time:
Logarithms can help to find time required in problems.

$\log x^{nt} = nt \log x$

Examples – Compounded Semiannually

Example 2:

You invest \$1,000 at 5% interest compound semiannually for 3 years. Find amount and compounded interest after 3 years.

Solution:

$$A = P\left(1 + \frac{r}{n}\right)^{nt} = 1000\left(1 + \frac{0.05}{2}\right)^{2(3)} = 1159.69$$

$$A = 1000(1 + 0.05)^3 = 1159.69$$

Amount will be \$1159.69

Compound Interest = 1159.69 − 1000 = **\$159.69**

Practice Question

When interest is **compounded n times per year**, the total time and rate per period must be adjusted:

- r is the **annual interest rate**, but if interest is applied more than once a year, we divide it by n to get the rate **per compounding period**:
- t is the number of **years**, but with n compounding periods per year, the total number of periods becomes nt

These adjusted values are used in the formula:

$$A = P\left(1 + \frac{r}{n}\right)^{nt}$$

Find compound interest earned in the following cases

Q1. Annual Compounding

You invest \$1,000 at an interest rate of 8% per year, compounded annually for 3 years.

CI ≈ \$259.71

Q2. Semi-Annual Compounding

A principal of \$2,000 is invested at 6% per annum, compounded semi-annually, for 2 years.

CI ≈ \$251.07

Q3. Quarterly Compounding

You deposit \$1,500 at 4% annual interest, compounded quarterly, for 3 years.

CI ≈ \$190.20

Q4. Monthly Compounding
A savings account pays 5% interest annually, compounded monthly. If you deposit \$3,000 for 2 years, how much will you have at the end?

CI ≈ \$314.70

Q5. Daily Compounding (360-day year)
You invest \$5,000 at 3.65% annual interest, compounded daily, for 1 year

CI ≈ \$185.50

Q6. Continuous Compounding ($A = Pe^{rt}, e\ is\ Euler's\ Number \approx 2.71828$)
You invest \$1,500 at 5.5% annual interest, compounded continuously, for 4 years

CI ≈ \$1869.12

✍️ Practice Questions: Exponents and Applications

Q1	Simplify: $\left(\frac{ab^3}{a^2b}\right)^2$	$\frac{b^4}{a^2}$
Q2	Evaluate: $2^3 \times (2^2)^3 \div 2^5$	2^4
Q3	Write as single power of 2: 4×8^2	2^8
Q4	Express with positive power: $\left(\frac{2^{-3}a}{a^{-2}b^5}\right)^2$	$\frac{a^6}{2^6b^{10}}$
Q5	Solve the equation: $27^x = 81$	$4/3$
Q6.	Solve the equation: $3^{2x+5} + 9 = 10$	$3^{2x+5} = 10 - 9 = 1 = 3^0 \Longrightarrow 2x + 5 = 0 \Longrightarrow x = -5/2$
Q7	Solve the equation: $2^{2x} - 12(2^x) + 32 = 0$	$(2^x - 8)(2^x - 4) = 0 \Longrightarrow 2^x = 8\ or\ 2^x = 4; x = 3\ or\ 2$
Q8.	Solve the equation: $3^{2x+3} + 9^x = 28$	
Q9.	A ball is dropped from a height of 100 meters. Each time it bounces, it reaches half of the previous height. What is the height after 4 bounces?	$6.25\ m$

Q10	The population of bacteria doubles every hour. If the initial population is 500, what will it be after 5 hours? 16,000
Q11	In how many years will the amount invested at 8% per annum simple interest will be doubled. 12.5 years
Q12	Sam invested \$1240. After 4 years this amount was \$1612. Find the annual simple interest rate at which the money was invested 7.5%
Q13	Sketch a graph to show the growth of \$1000 over period of 10 years, if the amount is invested at 8% per annum (Simple Interest).
Q14	How many years would it take for the amount to double if some amount is invested at the rate of 6% compounded quarterly? 11.55 years

Extended Problem: Use Graphing Calculator to Solve

A company releases a new product. The number of **adopters** grows exponentially as: $A(t) = 1000(1 - e^{-0.2t}), t\ in\ months$

At the same time, **machines producing the product** wear out exponentially:

$M(t) = 500(e^{-0.05t})$

After how many months will the number of adopters equal the number of products produced by the machines? 2.84 months

Unit of Inquiry – Extend Your Thinking

Interest is how money grows over time — a real-world example of arithmetic and exponential patterns.

Think about: How does compound interest differ from simple interest in long-term savings or loans? What is significance of Euler's Number?

Try this: Compare two growth models — one using simple interest, another using compound interest at the same rate — over 5, 10, and 20 years.

Reflect: What does this teach you about the power of time, rate, and compounding in financial decisions?

Continuous Compounding

Continuous compounding means that interest is added at every instant of time, rather than at fixed intervals. Here the interest never pauses — it keeps growing every moment. This creates smooth exponential growth modeled by the formula $A = Pe^{rt}$.

$\boldsymbol{A = Pe^{rt}}$ $i: interest\ per\ annum\ in\ decimals$ $t: time\ in\ years$ $e\ is\ Euler's\ Number \approx 2.71828$	Natural logarithms, $\ln x$, is inverse function to be used as $\mathbf{\ln e^{rt} = rt}$

Example 1: Doubling Period

How long will it take to double your principal at an annual interest rate of 8% compounded continuously?

$A = Pe^{rt}$ $A = 2P, r = 0.08$ $2P = Pe^{0.08t}$ $2 = e^{0.08t}$	Take natural logarithm $\ln 2 = \ln e^{0.08t}$ $\ln 2 = 0.08t$ $t = \frac{\ln 2}{0.08} = 8.66$	It will take 8.66 years (8 years 8 months) to double the amount with annual interest rate of 8% compounded continuously. **Property of Logarithm** $\mathbf{\ln e^x = x}$

Example 2: Financial Growth

Find the amount after 5 years if $1000 is invested at 8% annual interest compounded continuously.

$$A = Pe^{rt} = 1000\, e^{0.08(5)} = 1000\, e^{0.4} = \$1491.82$$

Investment will be worth $1491.82 after 5 years.

Example 3: Bacteria Population

A population of bacteria starts with **500 cells** and grows continuously at a rate of **6% per hour**. How many bacteria will there be after **10 hours**?

Number of bacteria are growing continuously.

$$\boldsymbol{A = Pe^{rt}} = 500\boldsymbol{e^{0.06(10)}} = 911$$

There will be 911 bacteria after 10 hours

Chapter 1: Test

Answer the following questions and explain your strategy.

Q1. Evaluate: $7 + 3 \times 4 - \sqrt{16}$ 15

Q2. Average of numbers p and q is given by formula $\frac{p+q}{2}$.
Calculate average of $\frac{2}{3}$ and $\frac{3}{4}$ $\frac{17}{24}$

Q3. Which of the following numbers have terminating decimal?

a.	$\frac{4}{225}$	b.	$\frac{6}{225}$	c.	$\frac{9}{225}$	d.	$\frac{10}{225}$

Q4. Find three fractions between $\frac{2}{7}$ and $\frac{3}{7}$.

Q5. Repeating decimal to rational number: Write $1.\overline{23}$ in the form of $\frac{p}{q}$

Q6. Unit conversion: $1\ m^2 = a\ cm^2$. Find the value of a. 10,000

Q7. Anita sold 14 cars this week. This is 35% of the cars sold by her team. Calculate the number of cars sold by rest of the members this week.

26

Q8. Point, $P = 2^{20}$ on a number line. Identify the point closest to 2^{19}

Q9. $\frac{10^{-\frac{1}{2.5}}}{10^{-\frac{6}{2.5}}} = 10^n$

Hint: $n = -\frac{1}{2.5} - \left(-\frac{6}{2.5}\right)$

Find the value of n.

Q10. $\frac{2+\sqrt{3}}{2-\sqrt{3}} = a + b\sqrt{3}$

Hint: ***rationalize*** $a = 7, b = 4$

Find the value of a and b.

Q11 Try values 1 to 10 in the expression $n^2 + n + 11$. What do you conclude?

Chapter 2 Test: Introduction to Algebra

Q1	Write algebraic expression for: half the square of sum of a number and twice another number.
Q2	Simplify expression: $2(3x-2)-3(1+2x)$
Q3	Solve equation: $\frac{1}{x+1}+\frac{1}{x-1}=\frac{4}{3}$
Q4	How many solution(s) are there for the equation $x^2=4$
Q5	Explain why $\sqrt{a \times b} \neq \sqrt{a} \times \sqrt{b}$
Q6	Provide an example of a number whose square-root is greater than the original number.
Q7	$\sqrt{392}=a\sqrt{b}$. Find $a+b$.
Q8	$\frac{1}{2-\sqrt{3}}=\sqrt{a}+b$. Find a^2b
Q9	Write 23.4 million in scientific notation.
Q10	True or False: $(a-b)^2=a^2-b^2$, if $a>b>0$
Q11	$2^{12}+2^{12}+2^{12}+2^{12}=2^n$. Find the value of n.
Q12	$8^x=4^{10}$. Find the value of x
Q13	The population of bacteria culture is cut in half by an antibiotic every 30 minutes. If the entire bacteria culture is present at 10 am, what fraction of the bacteria culture will be left at 2:30 pm?
Q14	Anil borrowed \$1000 at 12% simple interest for 7 months. Find the total amount due after 7 months.
Q15	Find the future value of an investment of \$1000 at 5% interest after 6 years if the interest is compounded quarterly. Formula: $A=P\left(1+\frac{r}{n}\right)^{nt}$
Q16	In how much time will the amount double in invested at 8% compounded monthly?

Answers Chapter 2 Test

Q1	$\frac{1}{2}(x+2y)^2$	Q2	-7	Q3	$-8, -\frac{1}{3}$	Q4	2
Q5	Both negative	Q6	$\frac{1}{4}$	Q7	16	Q8	18
Q9	2.34×10^7	Q10	False	Q11	14	Q12	$\frac{20}{3}$
Q13	$\frac{1}{512}$	Q14	$1070	Q15	$1347.35	Q16	8.7 years

Real-Life STEM Application: Planning a Solar Panel Installation

Optimizing Solar Energy — A Real-World Math Design Task

Project Scenario:

You have been hired by a local community center to **design a solar panel system** for their rectangular rooftop, measuring **20 meters by 12 meters**. Each **solar panel is 1.6 meters by 1 meter**. Your goal is to **maximize energy generation** while working within real-world constraints.

Each panel produces **320 watts/hour** and receives **5 hours** of effective sunlight per day.

Your Task:

- Determine the maximum number of solar panels that can be installed.
- Estimate the total daily energy output in kilowatt-hours (kWh).
- Visualize your design using diagrams and appropriate mathematical representations.
- Justify your layout and calculations and discuss real-world considerations like space for maintenance or structural limitations.

This project transforms core mathematics into real-world design decisions, cultivating essential habits of clarity, creativity, and justification. To further strengthen the STEM mindset, we also provide additional suggestions and extensions later.

Suggestions provided in Unit 7: STEM Section

Extended Skills: Investigating Patterns in Power Expressions

We will develop a general formula to get the solution of related question in few seconds. This is an amazing trick to beat the competition.

Challenge Problem with General Solution

If $x+\frac{1}{x}=n$ Find General Solution of $x^3+\frac{1}{x^3}$	If $x+\frac{1}{x}=4$ Find $x^3+\frac{1}{x^3}$	If $x+\frac{1}{x}=-1$ Find x^3	If $x+\frac{1}{x}=1$ Find $x^9+\frac{1}{x^9}$

If $x+\frac{1}{x}=n$ Find $x^3+\frac{1}{x^3}$ Note: Formula: $(a+b)^3=a^3+3a^2b+3ab^2+b^3$

$$x+\frac{1}{x}=n$$

$$\left(x+\frac{1}{x}\right)^3=n^3$$

$$x^3+3x^2\frac{1}{x}+3x\frac{1}{x^2}+\frac{1}{x^3}=n^3$$

$$x^3+3x+3\frac{1}{x}+\frac{1}{x^3}=n^3$$

$$x^3+\frac{1}{x^3}+3x+3\frac{1}{x}=n^3$$

$$x^3+\frac{1}{x^3}+3\left(x+\frac{1}{x}\right)=n^3$$

$$x^3+\frac{1}{x^3}+3(n)=n^3$$

$$x^3+\frac{1}{x^3}=n^3-3n$$

Advanced Algebra Reciprocal SCAN ME

Formula for Sum of Cubes

$$(a+b)^3=(a+b)(a+b)^2$$

$$(a+b)^3=(a+b)(a^2+2ab+b^2)=a(a^2+2ab+b^2)+b(a^2+2ab+b^2)$$

$$(a+b)^3=a^3+2a^2b+ab^2+ba^2+2ab^2+b^3$$

$$(a+b)^3=a^3+3a^2b+3ab^2+b^3$$

Substitute $b=-b$ to prove $(a-b)^3=a^3-3a^2b+3ab^2-b^3$

Practice: Expand and simplify $(x-2)^3$

🔑 Why This Type of Question Matters

1. **Pattern Recognition & Substitution**
   - Students must see that powers of $x + \frac{1}{x}$ expand into a pattern:
     $\left(x + \frac{1}{x}\right)^2 = x^2 + \frac{1}{x^2} + 2.$
     $\left(x + \frac{1}{x}\right)^3 = x^3 + \frac{1}{x^3} + 3\left(x + \frac{1}{x}\right).$
   - This shows how algebra builds layers through **structured transformations**.
2. **Efficient Computation**
   - Directly finding $x^3 + \frac{1}{x^3}$ without knowing x saves time—a skill vital for exams like SAT, IB, or Olympiads.
3. **Bridging Concepts**
   - Links quadratic relationships, identities, and higher powers.
   - Serves as a foundation for **polynomial theory, factorization, and complex numbers** later.
4. **STEM Relevance**
   - Simplifying expressions is key in engineering and physics models (e.g., resonance equations, electrical circuits).
   - Encourages thinking in **systems and substitutions**, not just raw calculation.
5. **Habit Formation**
   - Reinforces three habits you emphasize in your teaching:
     ✅ *Clarity* (track substitutions carefully)
     ✅ *Creativity* (spot algebraic patterns)
     ✅ *Justification* (show why the shortcut works).

🌍 Real-World Analogy

Think of this as **compression in coding**: instead of writing long repeated lines, we create a formula to jump directly to the answer. Algebra teaches the same efficiency—using patterns to "compress" calculations.

STEM Exploration: Timeless Puzzle

Suppose you want to arrange **Anil, Ben, Clark, and Dave** in order from **youngest to eldest**.

- Each has a different age.
- They will not reveal their exact ages, but they can answer questions like:
  "Anil, are you older than Ben?"

Question:

What is the least number of such yes/no questions required to determine the complete order of their ages? Provide logical steps and explanation.

Guided Exploration with Video Solution

Define Variables:

Let their ages be A, B, C, D.

- A question like "Anil older than Ben?" translates to $A > B$.
- Slowly, we build a chain of inequalities.

Think Logically:

- With 4 people, there are $4! = 24$ possible orders.
- Each question gives only 2 possible answers (yes/no).
- Since $2^4 = 16 < 24$, 4 questions cannot distinguish all possibilities.
- Therefore, at least **5 questions** are needed in the worst case.

Visual Representation:

Draw arrows for each comparison. For example:

$A \rightarrow B$ means "A is older than B."

Watch Video **Timeless Age Puzzle** **SCAN ME**

Algebra begins where arithmetic ends. It's the art of seeing beyond the obvious, of turning puzzles into possibilities.

Chapter 3: Linear Equations and Word Problems

Change of Subject or Isolate Variable

Change of subject or isolating variable is required to rearrange an equation or formula to write the variable on one side of the equation, normally left.

Rearrange equation (or formula) so that indicated variable is along on the left side of the equation. Practice question also provided.

Example 1: Change the subject to x.

$y-3=\frac{2}{3}(x-3)$	cross multiply
$3(y-3)=2(x-3)$	distributive property
$3y-6=2x-6$	bring x to left side
$2x=3y-6-6$	combine like terms
$2x=3y-12$	divide by 2 both sides to isolate x
$x=\frac{3}{2}y-6$	

Example 2: Change the subject to u

$$\frac{1}{f}=\frac{1}{v}+\frac{1}{u}$$

$$\frac{1}{u}=\frac{1}{f}-\frac{1}{v}$$

$$\frac{1}{u}=\frac{v+f}{fv}$$

$$u=\frac{fv}{f+v}$$

Lense formula in optics

Change the subject to a.

$v=u+at$

Change the subject to T

$PV=RT$

Change the subject to W.

$P=2L+2W$

Change the subject to u.

$s=ut+\frac{1}{2}at^2$

Change the subject to F.

$C=\frac{5}{9}(F-32)$

Change the subject to a

$\frac{a+b}{a-b}=\frac{2}{3}$

Hint: Cross multiply and simplify

Multi-Step Linear Equations with Fractions

Solving equation is to find value(s) that make the equation true.

Basic skills to solve equations that will be applied extensively.

Example 3: Cross Multiply and Simplify for Proportions

a.		b.	
	$\frac{x}{9}=\frac{5}{6}$		$\frac{x+1}{2}=\frac{x-2}{3}$
	$6x = 5(9)$ — Cross multiply		$3(x+1) = 2(x-2)$
	$x = \frac{5(9)}{6}$ — simplify		$3x + 3 = 2x - 4$
	$x = \frac{5(3)}{2} = \frac{15}{2}$		$3x - 2x = -4 - 3$
			$x = -7$

Example 4: Isolate Variables to Left and Simplify

a.		b.	
	$2x + 6 = 30$		$2x + 6 = 13 - 5x$
	$2x = 30 - 6$		$2x + 5x = 13 - 6$
	$2x = 14$		$7x = 7$
	$x = 7$		$x = 1$

Example 5: Zero Principle: Product is zero if each factor is zero

a.		b.	
	$(x+3)(2x-5) = 0$		$2x(3x+4) = 0$
	Equate factors to zero		Equate factors to zero
	$x + 3 = 0$ *or* $2x - 5 = 0$		$x = 0$ *or* $3x + 4 = 0$
	$x = -3$ *or* $x = \frac{5}{2}$		$x = 0$ *or* $x = -\frac{4}{3}$

General Solution: Find the value of variable that makes the equation true.

$$ax + b = 0 \Rightarrow ax = -b \Rightarrow x = -b/a$$

Proportions

Short solution to one part of questions have been provided. Solve others to practice.

Example 6: Solve Equations

Solve each proportion. Provide exact solution.

a.	$\frac{4}{5}=\frac{12}{x}$	b.	$\frac{x}{10}=\frac{24}{15}$	c.	$\frac{6}{5}=\frac{9}{x}$ $\mathbf{6x = 45}$ $\boldsymbol{x = \frac{45}{6} = \frac{15}{2}}$
d.	$\frac{3}{x}=\frac{18}{27}$	e.	$\frac{2}{21}=\frac{x}{14}$		

Illustrated Example with Distributive Property

$$\frac{2}{3}(x-1)=(x+1)\frac{1}{2}$$

Strategy: Multiply by LCM of 6 on both sides to get-rid-of fractions

$$6\times\frac{2}{3}(x-1)=(x+1)\frac{1}{2}\times 6$$

$$4(x-1)=(x+1)\times 3$$

Apply distributive property and simplify

$$4x-4=3x+3$$

$$4x-3x=4+3$$

$$x=7$$

check the solution by substituting $x = 7$ in the given equation.

$$LHS: \frac{2}{3}(7-1)=\frac{2}{3}(6)=4$$

$$RHS: (7+1)\frac{1}{2}=(8)\frac{1}{2}=4$$

Left hand side equals the right hand side and so the solution is correct.

Application – Distributive Property in Real Life

A school orders **15 math kits**, each containing **(3 pencils + 2 notebooks)**.
If each pencil costs **$1.50** and each notebook costs **$2.00**, find the total cost.

Linear Equations One Variable

Short solution to one part of questions have been provided. Explore and learn.

Example 7: Solve Equations

Solve each equation and explain how many solutions it has — none, one, or infinitely many

a.	$2x + 3 = 11$	b.	$2(x - 4) = 2x + 5$	c.	$\frac{3x-1}{5} = \frac{x}{2}$
d.	$\frac{x}{2} + \frac{x}{3} = 10$	e.	$\frac{2x}{3} + 5 = 4x$		$\mathbf{2(3x - 1) = 5x}$
f.	$3(x - 2) = 3x - 6$	g.	$\frac{2x-3}{3} = \frac{x}{2}$		$\mathbf{6x - 2 = 5x}$ $\mathbf{6x - 5x = 2}$ $\mathbf{x = 2.0}$
h.	$\frac{x+1}{3} = \frac{x-1}{3}$	i.	$\frac{2-x}{3} = \frac{x+1}{2}$		

Number of Possible Solutions

Explore and learn from the following examples.

$2(4 - x) + 2x = 8$ $8 - 2x + 2x = 8$ $8 = 8$ **Always True. Infinite Solutions**	$2(1 + 2x) = 4x - 2$ $2 + 4x = 4x - 2$ $4x - 4x = -2 - 2$ $0 = -4$ **Never True. No Solution**
$2(x + 3) = x + 11$ $2x + 6 = x + 11$ $2x - x = 11 - 6$ $x = 5$ **Unique Solution**	**Note:** A **linear equation** usually has **one unique solution**. However, as these examples show, some equations may have **no solution** or **infinitely many solutions**, depending on their form.

$$a(x + 3) = 6b - 2x$$

💡 Find value of a and b for the equation to have one, many or no solution.

Functions and Function Notation

Mathematics is not just about numbers — it is about relationships.
When we perform operations such as addition or multiplication, we are transforming one number into another. Functions extend this idea further: they provide a **rule** that assigns each input to a single, predictable output.

The phrase *"every input lead to exactly one output"* is the heart of the definition of a function. It is also what draws the **fine line of difference** between a general relation (which may connect one input to many outputs) and a true function (which always gives one and only one output).

- **"My mother is a great cook"** is a *function* — because for every child, there is exactly one mother. Each input (child) is connected to a unique output (mother).
- **"My son is a good painter"** is *not a function* — because one mother can have more than one child. A single input (mother) may lead to multiple outputs (sons).

👉 Both statements describe **relations**, but only the first satisfies the strict rule of a function: *one input → exactly one output.*

Relation vs. Function

The table below shows examples of relations that are **functions** and those that are **not functions**, with reasons:

Relation	**Function or Not**	**Reason**
Student and their Student ID	✅ Function	student has a unique ID number.
A student and their subjects	❌ Not a Function	student may take many subjects
$\{(1,1),(0,0),(2,4),(3,9)\}$	✅ Function	Every input $\boldsymbol{x}$ gives exactly one y
$\{(1,1),(0,0),(4,2),(4,-2)\}$	❌ Not a Function	4 is linked with 2 and also -2
$x^2 + y = 1$	✅ Function	$y = 1 - x^2$ Every input $\boldsymbol{x}$ gives one y
$x^2 + y^2 = 1$	❌ Not a Function	$y = \pm\sqrt{1 - x^2}$ two values of y for x

The Vertical Line Test

So far, we have used words, tables, sets, equations, and real-life examples to separate functions from general relations. But how do we recognize a function when looking at its **graph**?

Here is where the **Vertical Line Test** comes in.

What is the Vertical Line Test?

- Imagine sliding a vertical line (parallel to the y-axis) across the graph.
- If the vertical line touches the graph at more than one point at the same time, then the relation is not a function.
- If it touches at most one point everywhere, then the graph represents a function.

Examples with Graphs

1. **Graph of $y = x^2$ (a parabola):**
   - Every vertical line touches the curve only once.
   - ✅ Therefore, it is a function.

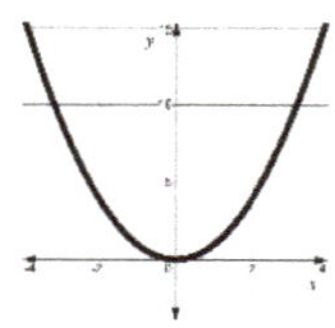

2. **Graph of a circle $x^2 + y^2 = 25$:**
   - A vertical line through $x = 3$ cuts the circle at two points (one above, one below).
   - ❌ Therefore, it is *not* a function.

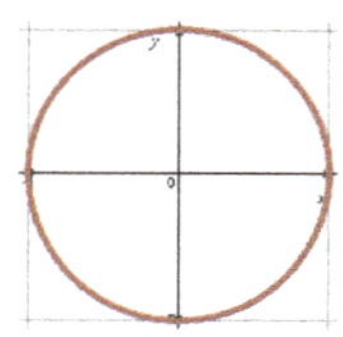

Example with Mapping Diagram

Diagram A

- Inputs: {1, 2, 3}
- Outputs: {5, 7, 9}
- Mappings:
  - $1 \to 5$
  - $2 \to 7$
  - $3 \to 9$

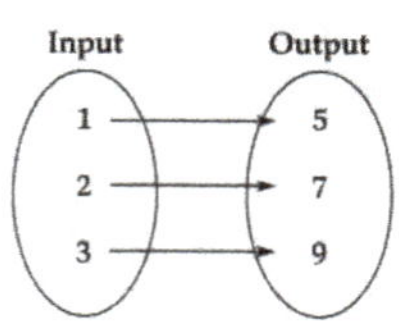

✅ **Conclusion:** Each input is connected to exactly one output. Therefore, Diagram A represents a **function**.

Diagram B

- Inputs: {1, 2, 3}
- Outputs: {5, 7, 9}
- Mappings:
  - $1 \to 5$
  - $2 \to 5$ and $2 \to 9$
  - $3 \to 9$

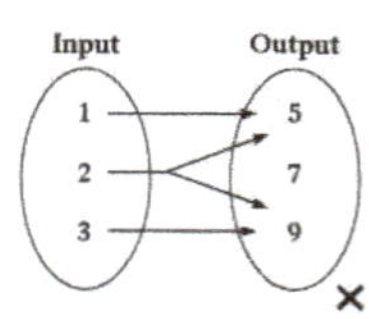

❌ **Conclusion:** Input **2** connects to two outputs (5 and 9). This breaks the function rule. Therefore, Diagram B is **not a function**, but it is still a **relation**.

TIP: Explain why $y = 5$ is a function but $x = 5$ is not a function?

Function Notation

So far, we have described functions using words, tables, and diagrams. Mathematicians also use a special notation that makes functions easier to write and work with. Function notation replaces long verbal descriptions with a compact mathematical expression. Instead of saying "the value of the function at input x," we simply write $f(x)$. This notation tells us three things at once: the name of the function (f), the input (x), and the output ($f(x)$).

- The letter f names the function (it could also be $g, h, a, b, etc.$).
- The parentheses (x) show the input.
- The whole expression $f(x)$ is read as "$f\ of\ x$" and represents the **output** of the function when the input is x. Output now has a name, $f(x)$.

For example:

If $f(x) = 2x + 3$, then:

- $f(1) = 2(1) + 3 = 2 + 3 = 5$
- $f(2) = 2(2) + 3 = 4 + 3 = 7$

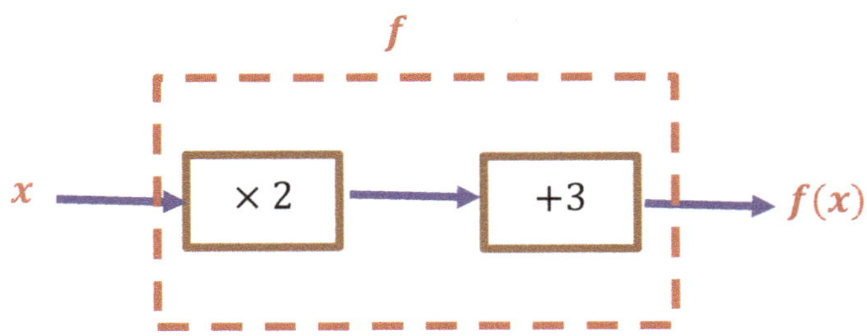

Here, the input values are 1 and 4, and the corresponding outputs are 5 and 11.

Function vs. Equation

- The equation $y = 2x + 3$can also be written as $f(x) = 2x + 3$
- In this case, y and $f(x)$ mean the same thing: they both represent the output.
- Function notation simply tells us **which rule is being used**.

Multiple Functions

We often deal with more than one function at the same time. Using names like $f(x), g(x), h(x)$ helps us keep track.

Example:

- Let $f(x) = x^2$ and $g(x) = 2x + 1$
- Then: $f(3) = 3^2 = 9$ and $g(3) = 2(3) + 1 = 7$

Both use the same input, but different rules give different outputs.

Why Use Function Notation?

Using function notation helps us:

- Label different functions clearly.
- Improves communication when dealing with many relations
- Understand the relationship between input and output.
- Apply real-world meanings (like C(x) for cost when renting x movies).
- Express and visualize complex composition of functions
- Make calculations visual and meaningful as we will further see in the book

Watch Video **Functions and Function Notation** SCAN ME

Example 1: Find Output for Given Input

Consider the function $f(x) = 2x + 3$. Evaluate	
a.	$f(-1)$ $\boldsymbol{f(-1)}$ ***means output when input*** $\boldsymbol{x = -1}$ $f(-1) = 2(-1) + 3 = -2 + 3 = 1$
b.	$f(2)$ $f(2) = 2(2) + 3 = 4 + 3 = 7$
c.	$f(2) - f(-1)$ Since $\boldsymbol{f(2) = 7}$ ***and*** $\boldsymbol{f(-1) = 1}$, as calculated above $f(2) - f(-1) = 7 - 1 = 6$

Example 2: Evaluate Transformed Function

For the function $f(x) = 3x - 2$, determine expression for the following	
a.	$f(a)$ $f(a) = 3(a) - 2 = 3a - 2$
b.	$f(2a - 1)$ $f(2a - 1) = 3(2a - 1) - 2 = 6a - 3 - 2 = 6a - 5$
c.	$f(a^2) - f(a + 1)$ $f(a^2) - f(a + 1) = 3a^2 - 2 - (3(a + 1) - 2) = 3a^2 - 2 - 3a - 3 + 2$ $= 3a^2 - 3a - 3$
d.	$\boldsymbol{2f(a^2 - 1) + 3}$

Example 3: Reverse Calculation to find Input for Given Output

If $g(x) = 5 - x$, determine the input x when output $g(x) = 7$	
$g(x) = 5 - x$ $7 = 5 - x$ $x = 5 - 7 = -2$	For output of 7, input will be $-\mathbf{2}$

Domain and Range of a Function

A **function** assigns exactly **one output** for each valid **input**.

- The **domain** is the set of all possible **input values** (usually x) that the function can accept. Example: Domain of $y = \frac{1}{x}$ is all real values except $x = 0$.
- The **range** is the set of all possible **output values** (usually $f(x)$) the function can produce. Example: Range of $y = \sqrt{x}$ is all real values greater than or equal to zero.

Notation for Domain and Range

(a) Set-Builder Notation

- Describes the set of values using a condition.
- Example: Domain of $y = \sqrt{x}$ is $\{x \in \mathbb{R} | x \geq 0\}$

  This means "the set of all real numbers x such that $x \geq 0$"

(b) Interval Notation

- Uses brackets and parentheses to describe continuous intervals.
- ( ) → endpoint not included
- [ ] → endpoint included
- Inequality → $a < x \leq b$ where a is not included but b is included.

Examples:

- $(-\infty, \infty)$ → all real numbers.
- $[0, \infty)$ → all real numbers greater than or equal to 0.
- $(-\infty, 5]$ → all real numbers less than or equal to 5.
- $-5 \leq x \leq 7$ → all real between and including -5 and 7.

Example 4: Domain and Range from Graph

- Explain why the graph represents a function.
- Write the coordinate points that represents a function $f(x)$ on the graph.
- List the domain and range of function.

A graph represents a function if and only if no vertical line intersects the graph at more than one point.

Given graph is a function since each value of the independent variable corresponds to only one value of the dependent variable. It passes vertical line test.

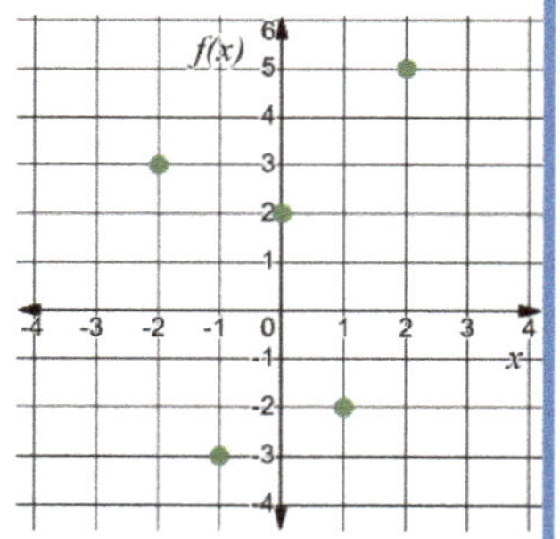

Function $f(x)$ from the graph

$f(x) = \{(-2, 3), (-1, -3), (0, 2), (1, -2), (2, 5)\}$

Domain: $\{-2, -1, 0, 1, 2\}$ **Range**: $\{-3, -2, 2, 3, 5\}$

Note: Function with coordinate points is written in curly brackets

Practice :

Plot given function on coordinate plane and write its domain and range.

$$f(x) = \{(-2,2), (-1, 3), (0, 0), (1, 2), (2, 0)\}$$

Example 5: Real Life Application with Function Notation

Write domain and range from the graph of given functions in set notation and in words:

a.

Domain: $\{x \in \mathbb{R}\}$

Set of all real numbers

Range:

$\{y \in \mathbb{R} | y \leq 4\}$

Set of all real numbers less than and equal to 4.

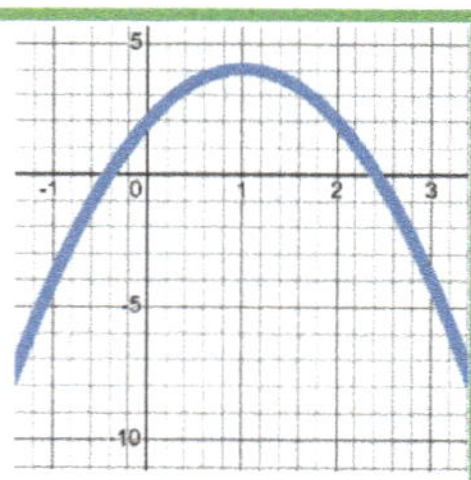

b.

Domain: $\{x \in \mathbb{R} | x \leq 2\}$

Set of all real numbers less than equal to 2

Range: $\{y \in \mathbb{R} | y \leq 0\}$

Set of all real numbers less than and equal to 0.

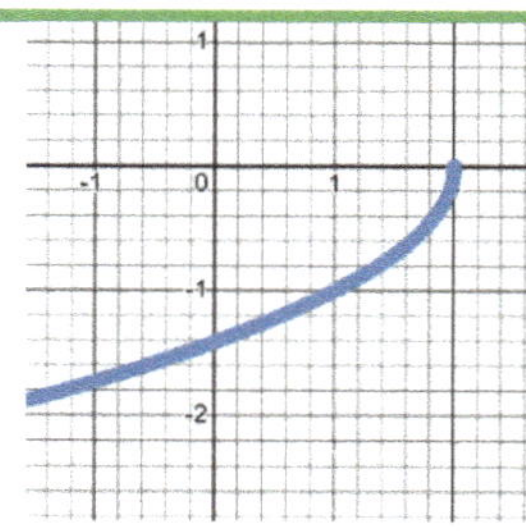

TIPS: Write a function whore domain is all real numbers and range is just zero.

Function Operations

Function operations involve working with functions by adding, subtracting, multiplying, or dividing them, and also by evaluating a function for a given input. They may also include finding the input when the output is known.

Example 8: Find Output for Given Inputs

Consider the function $f(x) = 2x + 3$. Evaluate $f(-1), f(2), and\ f(2) - f(1)$	
$f(-1)$	$f(-1) = 2(-1) + 3 = -2 + 3 = 1$
$f(2)$	$f(2) = 2(2) + 3 = 4 + 3 = 7$
$f(2) - f(-1)$	$f(2) - f(-1) = 7 - 1 = 6$

Example 6: Reverse Calculation to find Input for Given Output

If $g(x) = 5 - x$, determine the input x when output $g(x) = 7$	
$g(x) = 5 - x$ $7 = 5 - x$ $x = 5 - 7 = -2$	Substitute 7 for $\boldsymbol{g(x)}$, rearrange and find input $\boldsymbol{x}$. $\boldsymbol{x = -2}$

Example 7: Transformation of Function. Related Graph Shown.

For the function $f(x) = \sqrt{x - 2}$, determine	
a.	$h(x) = 2f(x) = 2\sqrt{x - 2}$ Vertical stretch by a factor of 2
b.	$g(x) = f(2x) = \sqrt{2x - 2}$ Horizontally compressed by factor of 1/2
c.	$k(x) = f(x) - 2 = \sqrt{x - 2} - 2$ Vertically translated 2 units down

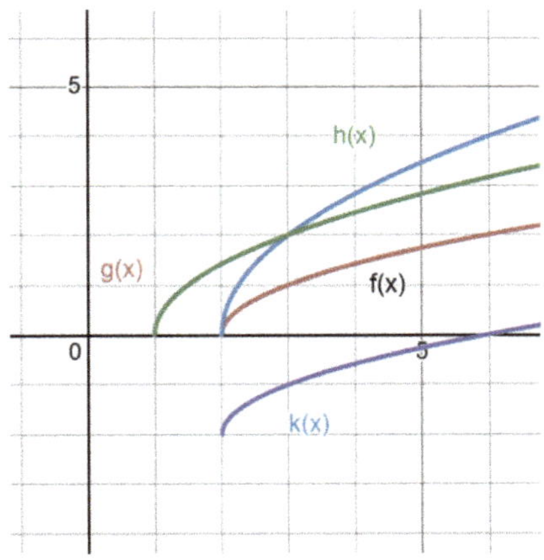

Combining Functions

Sometimes, we don't just use one function at a time. We can **combine functions** to build new ones. The functions are combined on common domain.

1. Adding and Subtracting Functions

If $f(x)$and $g(x)$ are two functions: $(f+g)(x) = f(x) + g(x)$; $(f-g)(x) = f(x) - g(x)$

Example 8: Combination of Functions: Add/Subtract

If $f(x) = x^2 + 1, g(x) = x - 1$ then	
a.	$(f+g)(x) = f(x) + g(x) = (x^2+1) + (x-1) = x^2 + x$
b.	$(f-g)(x) = f(x) - g(x) = (x^2+1) - (x-1) = x^2 - x + 2$
c.	$(f-g)(2) = x^2 - x + 2 = (2)^2 - (2) + 2 = 4$

2. Multiplying and Dividing Functions

$(f \cdot g)(x) = f(x) \cdot g(x)$; $\left(\frac{f}{g}\right)(x) = \frac{f(x)}{g(x)},\ g(x) \neq 0$

Example 9: Combination of Functions: Division

If $f(x) = x^2 - 1, g(x) = x + 1$ then
$\left(\frac{f}{g}\right)(x) = \frac{f(x)}{g(x)} = \frac{x^2-1}{x+1} = \frac{(x+1)(x-1)}{x+1} = x + 1, x \neq -1$

Composition of Functions

One of the most powerful ways to combine functions is **composition**:

Notation: $(f \circ g)(x) = f(g(x))$

- This means we apply $g(x)$ first, then apply $f(x)$ to the result.
- The domain of $f(g(x))$ is the subset of the domain of g. It is the set of values of in the domain of g for which $g(x)$ is in the domain of f.
- Note that $f(g(x))$ is not always equal to $g(f(x))$

Example 10: Composition of Functions

If $f(x) = x^2 - 1, g(x) = x + 1$ then	
a.	$(f \circ g)(x) = f(g(x)) = f(x+1) = (x+1)^2 - 1$
b.	$(f \circ g)(2) = f(g(2)) = f(2+1) = f(3) = (3)^2 - 1 = 8$
c.	$(g \circ f)(x) = g(f(x)) = g(x^2 - 1) = x^2 - 1 + 1 = x^2$

Example 11: Domain of Composite Functions

If $f(x) = x^2 - 1, g(x) = \sqrt{x}$ then find domain of $(f \circ g)(x)$

$(f \circ g)(x) = f(\sqrt{x}) = \sqrt{x}^2 - 1 = x - 1$

Domain of $(f \circ g)(x) = [0, \infty)$

Note the domain is restricted since the inside function is restricted to $x \geq 0$

Practice

Q1	Is $y^2 - x = 1$ a function? Explain					
Q2	$f(x) = \{(-2,2), (-1,3), (0,0), (1,2), (2,0)\}$ $g(x) = \{(3,0), (2,-1), (1,3), (0,-2)\}$					
	a.	$(f+g)(x)$	b.	$(f \circ g)(2)$	c.	$(g \circ f)(2)$
Q3	If $f(x) = x^2, g(x) = -2x + 5$ then evaluate the following					
	a.	$(f \circ g)(3)$	b.	$(g \circ f)(3)$	c.	$g(x) = f(x)$
Q4.	If $f(x) = x^2$ find an expression for $h(x) = f(x+3)$ Sketch graph of $f(x)$ and $h(x)$. Explain how the two graphs are related.					

Watch Video **Composition Domain** SCAN ME

Example 12: Read and Understand the Graph of the Given Function

- Write domain and range of the function
- Find the value of $f(4)$
- When is the function decreasing
- When is $f(x)$ positive
- Find x when $f(x) = 1$

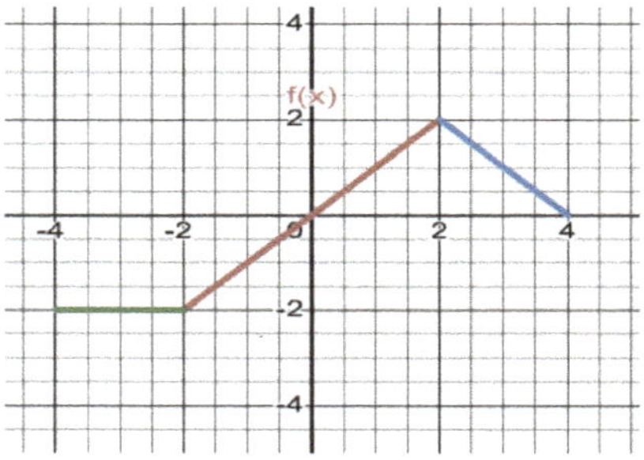

Solution 12:

Domain: $[-4, 4]$	Range: $[-2, 2]$	$f(4) = 0$
Decreasing: $(2, 4)$	Positive: $(0, 4)$	$f(x) = 1\ at\ x = 1, 3$

Example 13: Real Life Application with Function Notation

Water is leaking at the rate of 5 L/h from a tank with capacity of 100 L. Relate

a.	Let $W(t)$ be the water remaining in the tank at any instance of time t. Write a relation to represent the situation for the first 20 hours. $W(t) = 100 - 5t$
b.	Explain why the relation is a function. Relation is a function since at any instance of time a particular amount of water will remain in the tank.
c.	Evaluate: $W(0), W(10), W(20)$ $W(0) = 100 - 5(0) = 100;\ \ W(10) = 100 - 5(10) = 50;$ $W(20) = 100 - 5(20) = 100 - 100 = 0$
c.	Find domain of the function Domain: $0 \le t \le 20$
d.	Find range of the function Range: $0 \le W(t) \le 100$

Watch Video **Domain and Range of Functions** SCAN ME

Functions: Practice Test

Q1	Which of the following relation is not a function
	a. $x^2 - y = 1$ b. $x - y^2 = 1$ c. $xy = 1$
Q2	Explain transformation $f(x) = -4(x+3)^2 + 1$ from $y = x^2$
Q3	$f(x) = 2x - 3$. Evaluate $f(2) - f(1)$
Q4	Consider the function $f(x) = (x-2)^2 - 1$. Determine the following values
	a. $f(3-2)$ b. $f(2)$ c. $f(2) - f(-2)$
Q5.	For the function $f(x) = -2x - 3$, determine the input x when $f(x)$ is 5
Q6.	For function $f(x) = (x-3)^2 - 4$ determine the inputs x when $f(x) = 0$
Q7	If $f(x) = x^2 + 1, g(x) = x + 1$ then evaluate the following
	a. $(g+f)(2)$ $(g \circ f)(-2)$ $(f \circ g)(a)$
Q8	$f(x) = x^2 + 1$. Write simplified expression for $\frac{f(x+h)-f(x)}{h}$
Q9	Write domain for the functions:
	a. $f(x) = \dfrac{1}{x^2 + x}$ $\{x \in \mathbb{R}\}$ b. $f(x) = \dfrac{\sqrt{x-5}}{\sqrt{x^2-16}}$ $\{x \in \mathbb{R} \mid x \neq \pm 4\}$
Q10	Write domain and range for each of the following relations. Identify the function. **Note**: Functions pass vertical line test.
	a. 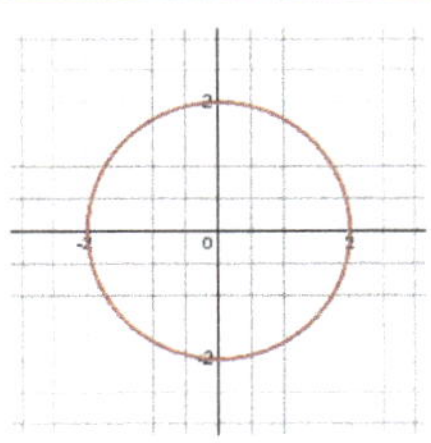 b. 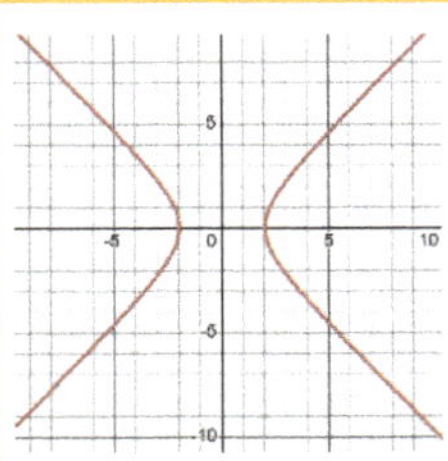 c. 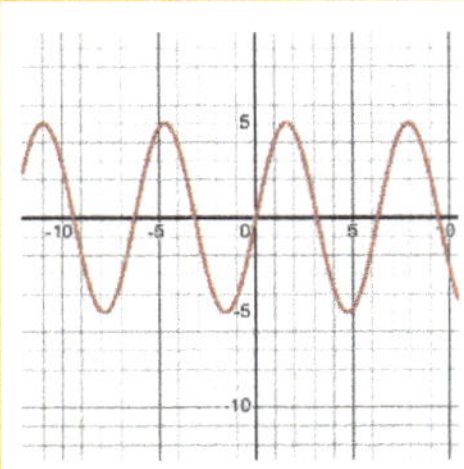

Linear Functions and Applications

Linear functions are not just algebraic tools—they're powerful models for solving real-world problems involving cost, planning, and decision-making. By analyzing their graphs, we can **interpolate** and **extrapolate** to estimate unknown values, predict future trends, and make informed decisions. Understanding how to write, interpret, and solve linear equations strengthens logical reasoning and problem-solving skills essential for success in higher-level mathematics and everyday life.

Example 1: Video Rental Cost

A video rental store charges a **one-time membership fee** of \$12 and \$3 for **each movie rented**.

Your Tasks:

Define a function that models the total cost.

Calculate the cost of renting 6 movies.

Determine the maximum number of movies that can be rented with \$56.

Solution

Let m be the number of movies rented, and C(m) be the total cost as a function of m. Membership, \$12, is the fixed cost and \$3 is variable cost of renting each movie.

$C(m) = 12 + 3m$ — total cost is sum of fixed and variable cost

$C(6) = 12 + 3(6)$ — substitute $\boldsymbol{m} = \mathbf{6}$ for renting 6 movies

$C(6) = 12 + 18 = 30$

Total cost of renting 6 movies will be \$30

$56 = 12 + 3m$ — $\boldsymbol{C}(\boldsymbol{m}) = \mathbf{56}$ Given total cost

$56 - 12 = 3m$ — Rearrange equation to solve for $\boldsymbol{m}$

$44 = 3m$ — Note: $\boldsymbol{m}$ is a whole number

$m = \frac{44}{3} = 14.66$ — Caution: Do not round the answer.

Maximum number of movies that can be rented for \$56 are 14.

Extended Example 1 for Students to Explore and Connect

Visual Insight

Represent this situation on a **scatter plot for graph** of $C(m) = 12 + 3m$

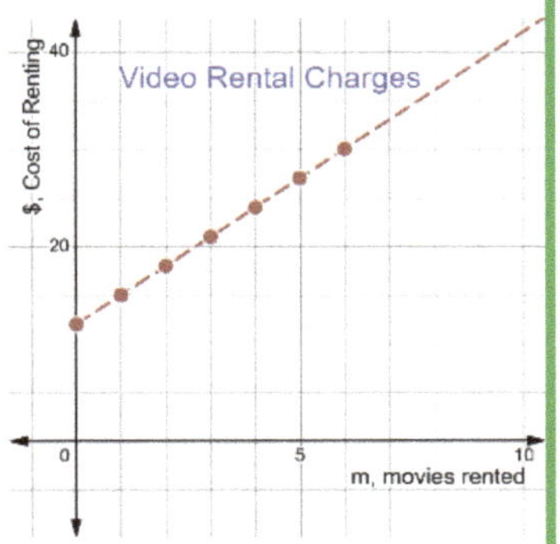

- Plot points for different movie counts (e.g., 0 to 15).
- Identify the **slope** (rate of change), **y-intercept**, and the meaning of each in context.

Use the graph to explore:

- When is the cost exactly $45?
- How fast does the cost increase per rental?
- What kind of function does the graph represent?

Let's Investigate:

a. Why is a **scatter plot** preferred over a **solid line** in this situation?
(Hint: Think about whether renting half a movie makes sense!)

b. What does the **y-intercept** represent in this context?
(What is the cost when zero movies are rented?)

c. Find the **rate of change** (slope) using any two points from the plot.
(Use the formula: rate of change = change in cost ÷ change in number of movies)

d. Based on your observations, what conclusions can you draw?

- Describe how this is a **linear relation**.
- Explain the connection between the **real-life situation** and the **equation of a line**.

Critical Thinking

Write domain and range for the function $C(m)$ if the budget is $56.

A **linear function** is the simplest way to model real-life situations where there is:

- a **fixed starting cost, b,** (base amount, membership fee, or setup charge), and
- a **constant rate per unit, m** (per km, per hour, per item, etc.).

The general form is: $\boldsymbol{C(x) = mx + b}$

Extension Task: Create Your Own Linear Scenario

Now that you understand how linear functions can represent real-life situations like video rentals...

Can you think of other situations where there is a fixed starting cost and a constant rate per unit?

Some examples to get you started:

- Renting a **banquet hall** with a flat booking fee and a charge per hour
- Paying for a **gym membership** with a one-time registration fee and a monthly fee
- Hiring a **bike or car** with a fixed deposit and a rate per kilometer or hour
- Provide examples to math the graph shown below.

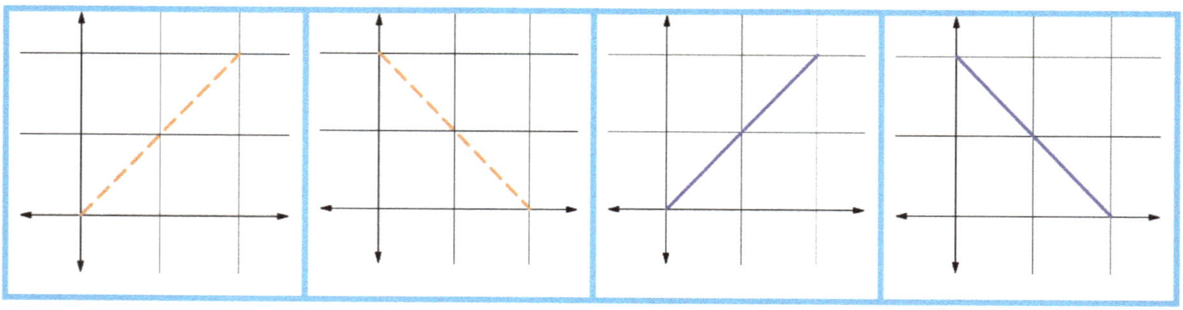

Your Turn:

Choose one example and answer the following:

1. What is the **fixed cost** (y-intercept)?
2. What is the **rate per unit** (slope)?
3. Write a **linear equation** to represent the total cost.
4. **Restrict function** defining maximum capacity or limit.
5. Plot a **few points** (e.g., for 1, 2, 3, 4 units) and create a graph.
6. Write **domain** and **range** for the real-life situation.
7. Reflect: How does your example show the same kind of linear pattern?

Example 2: Water Leaking from a Tank — A Continuous Linear Relation

A 20-liter water tank has a small leak. Water drips out steadily at a rate of 0.5 liters per minute.

Investigate:

1. Write a **linear equation** to model the volume of water remaining in the tank after *t* minutes.

   Let V be the volume in liters and t be time in minutes.
   Equation: $V(t) = 20 - 0.5t$

2. What is the **initial value** (y-intercept)?

   20 liters — that's the full tank at time $t = 0$.

3. What does the **slope** or rate of change represent?

   A **decrease of 0.5 liters per minute** — the water is leaking out steadily.

4. When will the tank be empty?

   Set $V(t) = 0$
   $0 = 20 - 0.5t$

   $0.5t = 20$

   $t = \frac{20}{0.5} = 40$

   In 40 minutes, tank will be empty

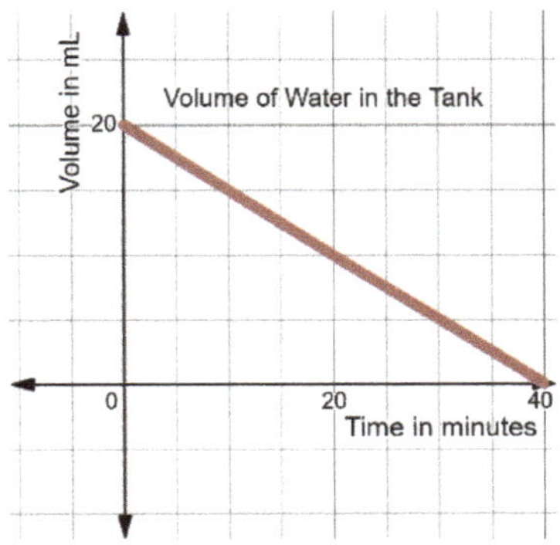

Graphing:

- Since time and water volume **change steadily**, the graph is a **straight line**.
- All values **in between** (e.g., 12.5 minutes or 7.2 L) **make sense**.
- Plotting these forms a **continuous linear relation**, unlike the **scatter plot of discrete movie rentals**.

Reflection:

- This model shows how linear functions help predict real-world situations where **quantities decrease (or increase) at a steady rate**.
- It also demonstrates the difference between **discrete** (counted, like movies rented) and **continuous** (measured, like water volume) variables.

From Real-World Applications to Linear Equations

In the previous examples — whether it's **video rentals** with fixed membership fees or **a leaking water tank** — we discovered something remarkable:

- All these situations follow a **predictable, steady pattern**.
- When plotted on a graph, the pattern forms a **straight line**.

This tells us that they represent a **linear relationship**.

What Do These Linear Applications Have in Common?

Let's highlight the key features we observed:

- The **graph is a straight line**
- The **y-intercept** represents a **fixed starting value** (like an upfront cost or initial volume)
- The **slope** (also called the **rate** or **gradient**) tells us **how fast or slow** the value changes

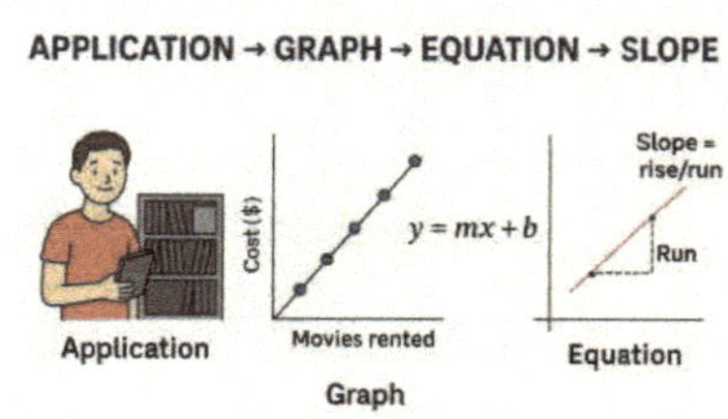

Let's Define a Linear Relation:

> A linear relation is a mathematical relationship between two variables that forms a straight line when graphed. It can be written in the form:
>
> $y = mx + b$
>
> Where:
>
> - y: the dependent variable (e.g., total cost, volume left)
> - x: the independent variable (e.g., number of items, time)
> - m: the **slope** or **rate of change**
> - b: the **y-intercept** (starting value or fixed cost)

Formula for converting a temperature in degree Celsius to degree Fahrenheit is

$$F(C) = \frac{9}{5}C + 32$$

Q1	Is the given function linear? Verify after plotting a graph.
Q2	Find the $F(25)$ and interpret its meaning.
Q3	For what value of C is $F(C) = 95$

Practice Quiz 1

Question 1: Gym Membership Plan

A gym charges a one-time registration fee of **$25**, and **$15 per month** for ongoing membership.

Tasks:
a. Write a linear equation to represent the **total cost (C)** for **m months** of gym membership.
b. What will be the total cost for 6 months?
c. Plot a graph for 0 to 6 months.
d. Identify the **y-intercept** and **slope**. Explain what they represent.

Question 2: Taxi Fare

A taxi service charges **$4.50 as base fare** and **$2 per km** travelled.

Tasks:
a. Write a linear equation to represent **total fare (F)** for **k kilometers**.
b. What is the total fare for a 10 km trip?
c. Draw a graph to represent the relationship between distance and cost.
d. Identify slope and intercept from the equation and graph, and interpret their meanings.

Question 3: Non-Linear Function

Provide three reasons to explain that the function $h(t) = -5t^2 + 1$ is non-linear.

Answer Key

Q1	$C = 15m + 25$	Q2	$F = 2k + 4.5$

Unit of Inquiry – Extend Your Thinking

Linear functions describe constant change — but what happens when change isn't constant?
Think about: How can we use a straight line to *approximate* a curve near a specific point?
Try this: Take a nonlinear graph (like $y = x^2$) and draw the tangent line at one point. Compare how close the tangent stays to the curve near that point.
Reflect: How does this idea of "local straightness" connect algebra to future STEM studies

Linear Applications Practice Problems

Q1	Sum of three consecutive numbers is 261. Find the lowest number. [86]
Q2	Anil and Ben travel the distance D at the speed of 40 km/h and 50 km/h respectively. If the time difference of their journey is 15 minutes, find the distance D. [50 km]
Q3	Terry have $4.50 consisting of nickels, dimes, and quarters. He has twice as many dimes as nickels and four times as many dimes as quarters. How many dimes does he have? [24]
Q4	Every year a person is paid $500 more than the previous year. If he receives $22200 over five years, what was he paid in the second year? [$3940]
Q5	Students sell T-shirts to raise money for the library. The cost of the T-shirt includes design cost of $800, plus $4 per T-shirt. They sell each T-shirt for $20. How many T-shirts should be sold to get profit of at least $10,000. [675]
Q6	Andrew drove 500 km from Windsor to Peterborough in 5.5 h. He drove part of the way at 100 km/h and the rest at 80 km/k. How far did he drive at each speed? [300 km at 100 km/h and 200 km at 80 km/h]
Q7	Ritu is six years younger than John. Eighteen years ago, she was two-thirds his age. How old is she now? $[J = 42, R = 36]$
Q8	Father is 3 times as old as his son. Five years ago, he was 4 times as old as his son. How old is each person now? $[S = 15, F = 45]$
Q9	A rectangle has perimeter of 120 cm. Three times the width is 5 more than double the length. Determine the area of the rectangle. $[x = 31.25, y = 28.75, Area = 898.44]$
Q10	Peter's daily earnings are $90 plus 8% commission on his sales. What should be the minimum sales for him to earn $250 on a day. [2000]
Q11	An isosceles triangle has 2 equal sides. The third side is 4 cm shorter than double of the length of one of the equal sides. The perimeter of triangle is 84 cm. Find dimensions of the triangle. [22, 22, 40]
Q12	Peter invested $8000. He invested part at 5% per year and the remainder at 6% per year. If the total interest for the year was $457.50, how much did he invest at each rate? [2250 at 5% and 5750 at 6%]
Q13	Sam needs to make 35 L of 38% sulphuric acid solution. There are 30% acid and 50% acid solutions available. How much litres of each solution must be mixed to make 38% solution? [21 L of 30% and 14 L of 50%]

Q14	A class trip is being planned. For one option each student will pay \$630. This includes two meals daily and accommodation for 9 day trip. The other option offers three meals a day and a accommodation of 9-days for \$720. What is the cost per meal? [\$10]
Q15	Akain follows a cycle path every morning along the river. If be bikes 3 km/h faster, he would take 40 minutes less. If he travelled 2 km/h slower he would have taken 40 minutes more. How many kilometres is the cycling path? [40 km]
Q16	7 Adult and 4 children ticket cost \$41.5. 8 Adult and 6 children ticket cost \$51.0. Find the cost of 5 Adult and 3 children tickets. [\$30]

Solution of Guided Practice Application Problems

Q1	$x + x + 1 + x + 2 = 261 \Rightarrow x = 86$		
Q2	$\frac{D}{40} - \frac{D}{50} = \frac{1}{4} \Rightarrow D = 50\ km$		
Q3	$5x + 10(2x) + 25\left(\frac{x}{2}\right) = 450 \Rightarrow x = 12; dimes = 2x = 24$		
Q4	$5x + 500 + 1000 + 1500 + 2000 = 5x + 5000 = 22200$ $\Rightarrow x = 3440 \Rightarrow 2nd\ year = 3440 + 500 = \$3940$		
Q5	$20x - (800 + 4x) \geq 10000 \Rightarrow x \geq 675$		
Q6	$\frac{x}{100} + \frac{500 - x}{80} = 5.5 \Rightarrow x = 300 \frac{km}{h} at \frac{100km}{h}, 200\ km\ at\ 80 \frac{km}{h}$		
Q7	$R = J - 6; J - 6 - 18 = \frac{2}{3(J - 18)} \Rightarrow J = 42, R = 36$		
Q8	$3S - 5 = 4(S - 5) \Rightarrow S = 15, F = 45$		
Q9	$2x + 2y = 120\ and\ 3y = 2x + 5 \Rightarrow x = 31.25, y = 28.75, Area = 898.44$		
Q10	$90 + 0.08x = 250 \Rightarrow x = 2000$		
Q11	$x + x + (2x - 4) = 84 \Rightarrow x = 22. Sides: 22,22,40$		
Q12	$0.05x + 0.06(8000 - x) = 457.5 \Rightarrow x = 2250\ at\ 5\%; 5750\ at\ 6\%$		
Q13	$0.3x + 0.5(35 - x) = 0.38 \times 35 \Rightarrow x = 21L$ (30% solution); 14L(35% solution)		
Q14	$\$90\ for\ 9\ extra\ meals \Rightarrow \$10\ per\ meal$	Q15	$d = 40\ km$
Q16	$7A + 4C = 41.5, 8A + 6C = 51.0 \Rightarrow A + 2c = 9.5 \Rightarrow 5A + 3C = 30$		

Two Variable Problem Solved in Single Variable

The difference between two numbers is 56. The ratio of these numbers is 7:3. Find the sum of the numbers.

Solution

The numbers are in the ratio of $7:3$

Then the numbers are $7x$ and $3x$

Critical thought process and an important strategy.

$7x - 3x = 56$

now write equations to solve

$4x = 56$

$x = \frac{56}{4} = 14$

substitute the value of x to get the required numbers.

The numbers are

$7(14) = 98$ and $3(14) = 42$

check solution

98 and 42 the required numbers

Challenge Problem

A car averages 12 L/100 km in city driving and 8 L/100 km on the highway. In a week of mixed driving, the car used 48 L of fuel and travelled 500 km. Determine the distance travelled in highway driving.

Strategy and Short Solution

	Fuel L/km	Distance km	Consumption L
City	0.12	x	$0.12x$
highway	0.08	y	$0.08y$

$x + y = 500$

$0.12x + 0.08y = 48$

Solve equations

Answer: 300 km

UOI – Beyond the Test: Mathematics for Innovation

Mathematics sparks invention and problem-solving. **Think about:** How can one problem lead to many solutions? **Try this:** Apply two concepts from the book to a real-world challenge. **Reflect:** How did mathematical thinking help you create or innovate?

Guided Practice Application Problems

<table>
<tr><td>Q1</td><td>A man is 24 years older than his son. Ten years ago, he was five times as old as his son. Find present age, x, of his son.

Hint

	Man	Son
Present age	$x + 24$	x
Ten years ago, age relation	$x + 24 - 10$	$x - 10$
Man was 5 times his son	$x + 24 - 10 = 5(x - 10)$	

Above table helps to relate age

[16 years]</td></tr>
<tr><td>Q2</td><td>The perimeter of a triangle is 57. The first side is twice the length of the second. The third side is seven more than the second. Find length of each side of the triangle.

Hint: Let x be the length of the second side.

$2x + x + (x + 7) = 57$ Solve the equation to find all sides.

[12.5, 25, 19.5]</td></tr>
<tr><td>Q3</td><td>The sum of two-digit number and the number obtained by reversing the order of its digit is 165. If the digits differ by 3, find the number.

Hint: $10x + y\ is\ the\ value\ of\ two - digit\ number\ xy$ [96]</td></tr>
<tr><td>Q4</td><td>The larger of the two numbers is 4 more than the smaller. Twice the larger added to smaller equals 86. Find the numbers.

Hint: Let the numbers be x and $x + 4$

$2(x + 4) + x = 86$ [26 and 30]</td></tr>
<tr><td>Q5</td><td>Amita has $200 less invested at 9% than she does at 6.5%. If the annual return from the two investments is the same, how much is invested at each rate? [$520 at 9%; $720 at 6.5%]</td></tr>
</table>

Q1	I walk at an average rate of x km/h to the office and return at the speed of y km/h. If the return time is more by t hours, find the distance to the office.
Q2	A student goes to school at the rate of 2.5 km/h and reaches 16 minutes late. If he travels at the speed of 3 km/h, he is 10 minutes early. What is the distance to the school. [Shortcut method $t = 16\,min = \frac{4}{15}h, x = 3, y = 2.5$.
Q3	If I go from A to B at 4 km/h and return from B to A at 6 km/h, what is my average speed?
Q4	Akain follows a cycle path every morning along the river. If be bikes 3 km/h faster, he would take 40 minutes less. If he travelled 2 km/h slower, he would have taken 40 minutes more. How many kilometres is the cycling path? 40 km
Q5	Two trains are approaching each other along parallel tracks. One train travel at 100 km/h and goes twice as fast as the other. If their distance apart is 225 km, in how many hours will they pass each other. 1.5
Q6	To travel 816 km, an Amtrak train takes 9 hours more than the VIA Rail. If however, the speed of the Amtrak is doubled, it takes 4 hours less than VIA. What is the speed of Amtrak train?
Q7	A journey of 600 km takes 30 minutes longer when the speed of the airplane is decreased by 200 km/h. Find the normal speed of the plane. 600 km/h
Q8	Alan and Eros went out of town at the same time and at the same speed to the next town, that is 18 km away. After they went for 2 hours together, they split up. Alan kept going at the same speed, and Eros stopped for an hour. After the break, Eron went at 1 km/h hogher compared to the previous speed and got to the town at the same time as Alan. Find Alan's speed. 3 km/h
Q9	Amy drives a distance of 50 miles along a highway from A to C. The first half of the journey was at an average speed of 50 miles/hour and the later was at an average speed of 60 miles/hour. Find the difference in time, to the nearest minute, between the two parts of the journey. 5 min Watch Video **D – S – T Problems** **SCAN ME**

Strategy for Quick Solution to the Model Questions Above

Q1

Walk to Office and Return with Different Average Speed

I walk at an average rate of x **km/h** to the office and return at y **km/h**. If the return time is t **hours** more than the time taken to reach, how far is the office from home?

Strategy: General Solution

Let the distance to office be d

Time to office $= \frac{d}{x}$ Time to return $= \frac{d}{y}$ Difference: $\frac{d}{y} - \frac{d}{x} = t$

$$\frac{xd - yd}{xy} = t$$

Note: we are assuming $x > y$

Take LCM of the denominators

$$\frac{d(x-y)}{xy} = t$$

Common factor d and simplify:

$$d = \frac{txy}{x-y}$$

$$\boldsymbol{d = \frac{txy}{x-y}}$$

Distance to the office is $d = \frac{txy}{x-y}$

Q2

General Solution

Average speed both ways:

$$x = 3\ km/h, y = 2.5 km/h$$

Time difference for same distance:

$$t = 16\ min = \frac{4}{15} hours$$

$$\boldsymbol{d = \frac{txy}{x-y}}$$

$$d = \frac{\frac{4}{15}(3)(2.5)}{3 - 2.5} = 4$$

distance to the school is 4 km.

Q3

Strategy: Friendly Numbers

Given average speed both ways:

$$x = 4\frac{km}{h}, y = 6\frac{km}{h}$$

Let distance be 24 km

24 is the friendly number. You could work with LCM 12 also.

Time A to B: $\frac{24}{4} = 6\ h$

Time B to A: $\frac{24}{6} = 4\ h$

Average speed $= \frac{Total\ distance}{total\ time}$

$$= \frac{24 + 24}{6 + 4} = \frac{48}{10} = 4.8\frac{km}{h}$$

Average speed is $4.8\frac{km}{h}$

💡 **Innovative Solution Strategy**

Q6	To travel **816 km**, an Amtrak train takes **9 hours more** than the VIA Rail. If however, the speed of the Amtrak is doubled, it takes 4 hours less than VIA. What is the speed of Amtrak train?
	Let the time taken by Amtrak train be t_A and that by VIA Rail be t_V $t_A - t_V = 9$ — Amtrak takes 9 hours more than VIA Rail $\frac{1}{2}t_A - t_V = -4$ — Doubled speed implies half time and 4 less hours (-4) Subtract 2nd eqn. from 1st — Solve t_A by elimination $\frac{1}{2}t_A = 9 + 4 = 13$ — speed = distance/time $t_A = 2(13) = 26$ Speed of Amtrak train = distance/time **= 816 km/26 h= 31.4 km/h**

Student Spotlight: Sruti Pandey

During her IB journey, Sruti found mathematics intimidating—until she began learning with Anil Kumar. His patient and structured teaching made even the most challenging topics, like calculus and statistics, clear and approachable.

With his support, Sruti's confidence and grades grew, and the skills she developed played an important role in her successful university applications.

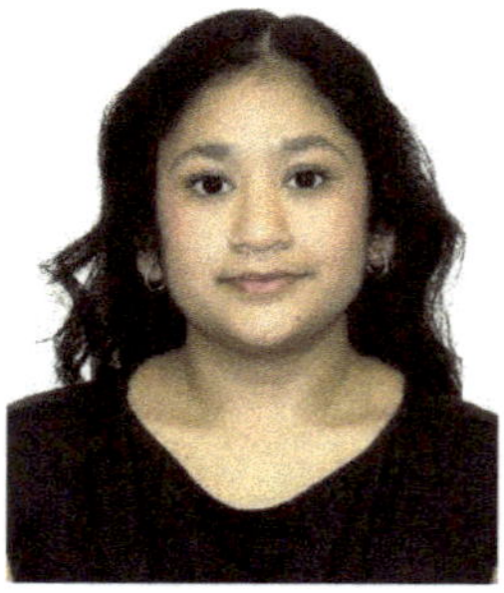

Today, Sruti is pursuing Medicine at the **Royal College of Surgeons in Ireland**, carrying forward the problem-solving mindset she built through math.

— ***Sruti Pandey***

Lines: Understanding the Slope from Two Points

Lines have constant slope or gradient. How do we find it?

If we pick **any two points** on the line, $A(x_1, y_1)$, $B(x_2, y_2)$, then

$y_1 = mx_1 + b$ and $y_2 = mx_2 + b$

Then the **change in y** over the **change in x** is:

$$\frac{y_2 - y_1}{x_2 - x_1} = \frac{(mx_2 + b) - (mx_1 + b)}{x_2 - x_1} = \frac{mx_2 - mx_1}{x_2 - x_1} = \frac{m(x_2 - x_1)}{x_2 - x_1} = m$$

So, the **slope m** can be calculated **from the graph** by taking **rise over run**:

$$slope, m = \frac{y_2 - y_1}{x_2 - x_1}$$

The slope represents the rate of change, connecting mathematics to speed, growth, cost, profit, or any linear relationship.

Let's Practice!

- Use your video rental graph and pick two points to find the slope.
- In the water tank example, try plotting a line and calculating slope using two points.
- Now write the equation in the form $y = mx + b$ for each of the above cases.
- Write equation of the two lines given in the graph below.

a.

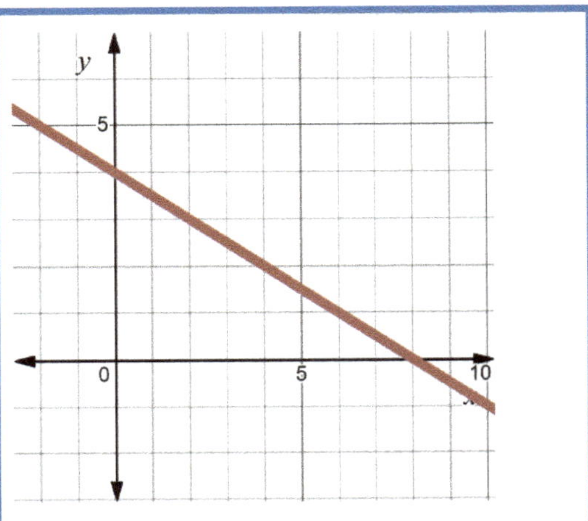

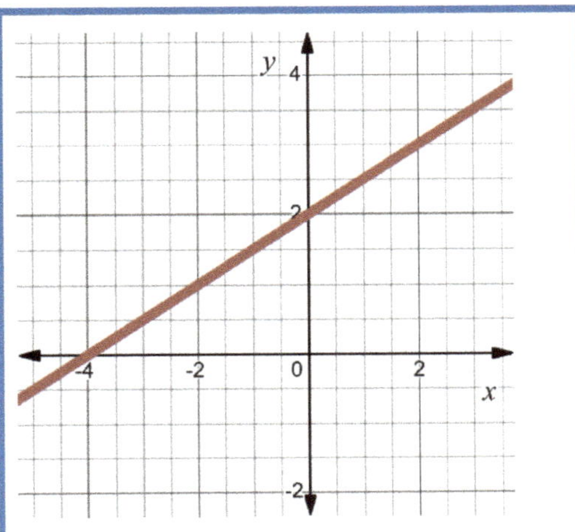

Understanding First Finite Differences

In many real-life situations, we work with data tables showing how one quantity changes with another — like time and distance, or number of items and total cost.

What Are Finite Differences?

Finite differences are a way of studying how a sequence of numbers of changes as the input increases step by step.

Case I: Equal x-Intervals

First differences are the changes in the consecutive y-values when the x-values increase by a constant amount.

- When the first differences are constant, the relation is linear.
- If the first differences are not constant, the relation is non-linear.

Example Table: Distance (km) versus Time (hours)

c $(hours)$	y (km)	$First\ Finite\ Difference$
1	10	$20 - 10 = 10$
2	20	$30 - 20 = 10$
3	30	$40 - 30 = 10$
4	40	

Since the first differences are all 10, this is a linear relation as shown in the graph.

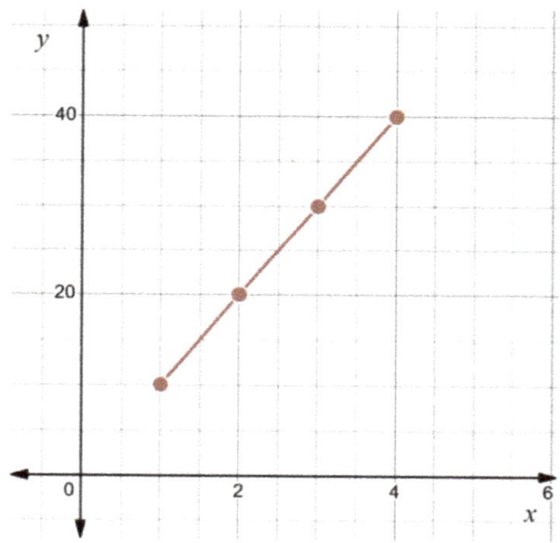

💡 Can you also find first finite difference if the x intervals are not equal?

✍ **Practice Quiz 2**

Use first difference to confirm that the given function is linear.

x	-2	-1	1	2	3
$f(x)$	-7	-4	2	5	8

Case II: How to Find First Finite Difference with Unequal x-Intervals

Step-by-step Strategy:

1. Create a table with x-values and corresponding y-values.
2. Find the change in x (Δx) between consecutive x-values.
3. Find the change in y (Δy) between corresponding y-values.
4. Calculate the rate of change for each interval:

Example

x	y	Δx	Δy	$\frac{\Delta y}{\Delta x}$
1	4	$3 - 1 = 2$	$10 - 4 = 6$	$\frac{6}{2} = \mathbf{3}$
3	10	$6 - 3 = 3$	$19 - 10 = 9$	$\frac{9}{3} = \mathbf{3}$
6	19	$10 - 6 = 4$	$31 - 19 = 12$	$\frac{12}{4} = \mathbf{3}$
10	31			

Since the **rate of change is constant (3)**, the relation is **linear** even though x is not increasing by 1

Summary:

For unequal intervals, **first difference** becomes **rate of change**:

$$first\ Difference = \frac{\Delta y}{\Delta x}$$

If all are equal, the relation is linear.

💡 $\mathbf{2x + y = 6}$ is equation of a straight line.

Show your work to prove that the first difference is constant.

Different Forms of Equation of Lines with Derivation

Did You Know?

The equation of line isn't always written in the same way! Depending on what information is given, you can use different forms to describe the same linear relationship:

Point – Slope Form: $y - y_1 = m(x - x_1)$ where m is slope with given point (x_1, y_1)

Slope – Intercept Form: $y = mx + b$ where m is slope and b is y-intercept.

Standard Form: $ax + by + c = 0$ general form for many algebraic operations

Intercept Form: $\frac{x}{a} + \frac{y}{b} = 1$ where a *and* b are x- and y-intercepts respectively.

Point – Slope Form

Derive Point-Slope Form from Slope Definition

Given a point (x_1, y_1) on a line with slope m. Let a general point be (x, y)

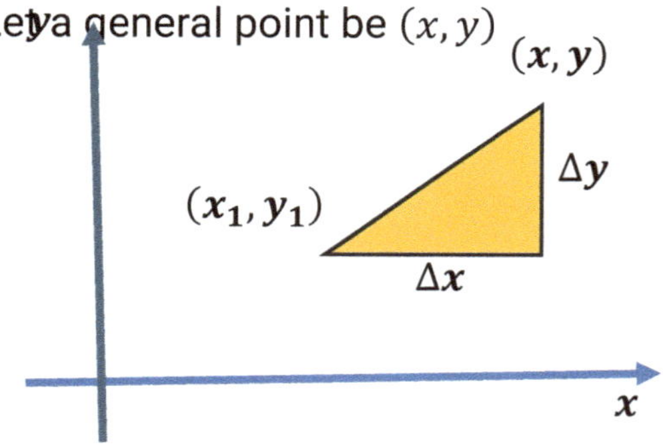

$$m = \frac{\Delta y}{\Delta x} = \frac{y - y_1}{x - x_1}$$

Cross multiple to get the equation

$m(x - x_1) = y - y_1$

rearrange

$\boldsymbol{y - y_1 = m(x - x_1)}$

This is Point – Slope Form of Equation for a Line

Example 1: Equation of Line in Point Slope Form

Write equation of line passing through the point $P(-2, 3)$ with slope of $-\frac{2}{3}$.

Equation of Line: $y - y_1 = m(x - x_1)$, where $m = -\frac{2}{3}, x_1 = -2, y_1 = 3$

$y - 3 = -\frac{2}{3}(x + 2)$ is the equation of line in point-slope form.

Slope – Intercept Form:

Let us now consider a special case when the point on the line is the $y-intercept, (0, b)$

Point – Slope Form of Equation is $\boldsymbol{y - y_1 = m(x - x_1)}$

Substituting the point $(0, b)$, we get: $\boldsymbol{y - b = m(x - 0)}$

Rearranging, we get: $\boldsymbol{y = mx + b}$

💡 **The Truth About $y = mx + b$**

Most students think the **equation of a line** is always

$\boldsymbol{y = mx + b}$

But in reality, this is just a **special case** of the more general **point–slope form**:

$$\boldsymbol{y - y_1 = m(x - x_1)}$$

- The general form works for **any point** $(\boldsymbol{x_1, y_1})$on the line.
- The slope–intercept form happens only when we choose the **specific point** $(\boldsymbol{0, b})$ the $y - intercept$.
- So, $\boldsymbol{y = mx + b}$ is simply the **point–slope form anchored at the y-axis.**

Quick Recall: Key Line Properties

X-Intercept • The point where the line crosses the x-axis. • At this point: $\boldsymbol{y = 0}$. • Solve for x when $\boldsymbol{y = 0}$. **Y-Intercept** • The point where the line crosses the y-axis. • At this point: $\boldsymbol{x = 0}$. • Solve for $\boldsymbol{y}$ when $\boldsymbol{x = 0}$.	**Parallel Lines** • Lines that never intersect. • Have the same slope $(\boldsymbol{m_1 = m_2})$ but different intercepts. **Perpendicular Lines** • Lines that meet at a 90° angle. • Slopes are negative reciprocals: $\boldsymbol{m_1 = -\frac{1}{m_2} or\ m_1 \cdot m_2 = -1}$

Example 3: Visualize Solution

a..	Find slope and a point on the line: $y - 2 = 3(x + 4)$ Slope 3, point $(2, -4)$
b.	Find slope and y-intercept of the line: $y = -2x + 5$ Slope -2, y-intercept 5
c.	Find equation of the line with slope of 5 and y-intercept of -2 $y = 5x - 2$
d.	Find equation of line passing through the points $(2, 0)$ and $(0, -6)$ $m = \frac{6-0}{0-2} = -3, y - intercept\ of - 6$, Equation: $y = -3x - 6$
e.	Find equation of line passing through the points $(2, 5)$ and $(-1, 11)$. $m = \frac{11-5}{-1-2} = \frac{6}{-3} = -2$ Equation: $y - 2 = -2(x - 5)$ $y - 2 = -2x + 10 \Rightarrow y = -2x + 12$
f.	Find equation of line passing through the points $(3, 9)$ and the origin. $m = \frac{9-0}{3-0} = \frac{9}{3} = 3$, Equation of line $y = 3x$

Example 4: Explore Slope-Intercept Form for Line L, $y = \frac{5}{2}x - 6$

The point $(6, p)$ lies on the line L. Find p.

$y = \frac{5}{2}(6) - 6 = 5(3) - 6 = 15 - 6 = 9$

$p = 9$

Find $x - intercept$ of the line L.

Substitute $y = 0$, solve for x

$0 = \frac{5}{2}x - 6$

$\frac{5}{2}x = 6$

$x = \frac{2}{5} \times 6 = \frac{12}{5}$

Write equation of a line passing through $(10, 6)$ and parallel to the given line L.

Parallel line will have slope $m = \frac{5}{2}$

Equation will be $\mathbf{y = \frac{5}{2}x + b}$

Find $b\ using\ point\ (10, 6)$

$6 = \frac{5}{2}(10) + b$

$6 = 25 + b$

$b = 6 - 25 = -19$

equation of parallel line will be

$y = \frac{5}{2}x - 19$

Standard Form

The standard form of the line is $ax + by + c = 0$, where $a, b,$ and c are real numbers and a and b are not both zero. Normally a, the coefficient of x, is a positive integer.

Note: Standard form is also written as $ax + by = c$. This is easier to work with.

This form helps identify **x-intercept** and **y-intercept** easily and is widely used in coordinate geometry and algebraic applications.

For $ax + by + c = 0$ slope $m = -a/b, x - intercept = -c/a, y - intercept = -c/b$

Example 5: Convert Equations to Standard Form

a.	$y - 3 = \frac{2}{5}(x + 2)$ $5(y - 3) = 2(x + 2)$ $5y - 15 = 2x + 4$ $2x - 5y + 15 + 4 = 0$ $2x - 5y + 19 = 0$	b.	$y = \frac{3}{2}x - 5$ $2y = 3x - 5$ $3x - 2y - 5 = 0$ **Note:** In the standard form the coefficient of x is a positive number.

Example 6: Find Slope and Intercepts from Standard Form

Given: Equation of line as $2x + 3y - 24 = 0$. Rearrange to $2x + 3y = 24$

$x - intercept: y = 0$	$y - intercept: x = 0$	$Rearrang\ to\ Slope\ Intercept\ Form$
$2x + 3(0) = 24$	$2(0) + 3y = 24$	$2x + 3y = 24$
$2x = 24$	$3y = 24$	$3y = -2x + 24$
$x = 12$	$y = 8$	$y = -\frac{2}{3}x + 8$

$x - intercept = 12, y - intercept = 8, slope, m = -\frac{2}{3}$

Challenge Example: Verify Equivalent Forms

Verify $2x + 3y - 24 = 0$ can also be represented by $\frac{x}{12} + \frac{y}{8} = 1$

💡 This is the intercept form

$$\frac{x}{x - intercept} + \frac{y}{y - intercept} = 1$$

How to Graph Lines

Introduction: Why Graph Lines?

Algebraic equations describe relationships between variables.
Graphing these equations helps us *see* the relationship. A linear equation, when plotted on the coordinate plane, always produces a straight line.

This visual approach allows us to quickly identify slope, intercepts, and patterns in real-world data, such as cost vs. items purchased, distance vs. time, or velocity in physics.

Methods of Graphing a Line

(a) Table of Values

One of the simplest ways to graph a line is by choosing values of x, calculating the corresponding y, and then plotting the points.

Example 1: Graph $y = 2x + 1$ using table of values

Choose $x - values$: $-2, -1, 0, 1, 2$ and compute $y - values$

x	-2	-1	0	1	2
$y = 2x + 1$	-3	-1	1	3	5

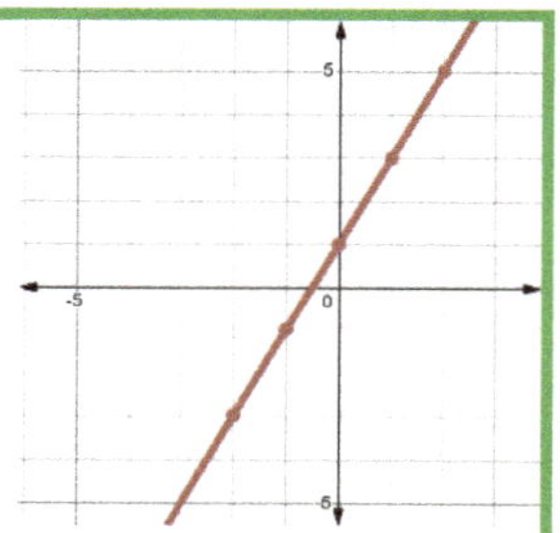

Plot the points and then connect with a solid line as shown.

(b) Use Intercepts

Line can be formed by finding where it crosses axes.

Example 2: Graph $3x + 2y = 6$ by finding the intercepts

$\mathbf{3x + 2y = 6}$

- $x - intercept$ where $y = 0 \Rightarrow 3x + 2(0) = 6 \Rightarrow 3x = 6 \Rightarrow x = 2$
- $y - intercept$ where $x = 0 \Rightarrow 3(0) + 2y = 6 \Rightarrow 2y = 6 \Rightarrow y = 3$

Draw a line connecting the intercepts as shown in the figure.

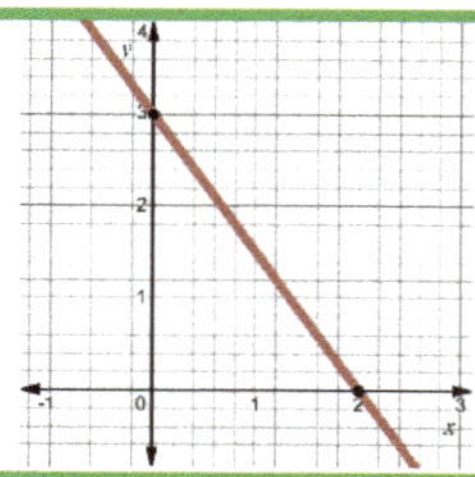

(c) Use Slope – Intercept Form

Directly read the slope and $y - intercept$ from the line to get two points to draw line.

Example 3: Find Points on the line Using Slope and Intercept

Find three points on the given line starting from the $y - intercept$

a.	$y = \frac{1}{2}x + 4$ • $y - intercept$ is $(0, 4)$ • $slope, m = \frac{1}{2} = \frac{change\ in\ y}{change\ in\ x} \Rightarrow$ one up, two right • x increase by 2; y increase by 1 Points: $(0,4), (2,5), (4,6)$.	b.	$y = -\frac{2}{3}x + 3$ • $y - intercept$ is $(0, 3)$ • $slope, m = -\frac{2}{3} = \frac{change\ in\ y}{change\ in\ x} \Rightarrow$ two down, three right • x increase by 3; y decrease by 2 Points: $(0,3), (3,1), (6,-1)$.

Example 4: Graph $y = -\frac{1}{2}x + 3$ by reading the y-intercept and slope

$y = -\frac{1}{2}x + 3$

- $y - intercept$ is $y = 3$
- $slope, m = -\frac{1}{2} = \frac{change\ in\ y}{change\ in\ x} \Rightarrow$ one down, two right
- Start from $y - intercept$ $P(0, 3)$. Next point is $Q(2, 2)$
- Draw line through $P(0, 3)$ and $Q(2, 2)$

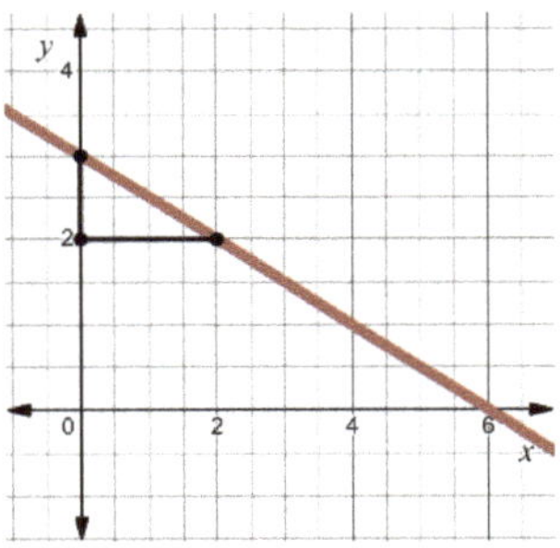

Types of lines

Horizontal Line: Slope is zero, equation is $y = b$. Equation of $x - axis$ is $y = 0$

Vertical Lines: Slope is undefined, equation is $x = a$. Equation of $y - axis$ is $x = 0$

Slant line: $y = mx + b$. Increasing if $m > 0$. Decreasing if $m < 0$

Note: Vertical line is not a function as a value of x is mapped to many $y - values$.

Practice: Draw line with following equations

a.	$y = -3x + 6$	b.	$x - 2y = 8$	c.	$y = \frac{2}{3}x - 4$

Example 5: Equation in Standard Form $ax + by + c = 0$

Write equation of line in standard form that is parallel to the line $y = \frac{2}{3}x - 7$ and pass through the point $P(6, -8)$,

Solution 5

Parallel lines have same slope. Line is parallel to $y = \frac{2}{3}x - 7$ which has slope $m = \frac{2}{3}$

Now, we need to write equation of line through point $P(6, -8)$ with slope $m = \frac{2}{3}$

Point slope form of the linear equation is $y - y_1 = m(x - x_1)$

$y - (-8) = \frac{2}{3}(x - x_1)$

$y + 8 = \frac{2}{3}(x - 6)$

$3(y + 8) = 2(x - 6)$

$3y + 24 = 2x - 12$

$2x - 3y - 12 - 24 = 0$

$\mathbf{2x - 3y - 36 = 0}$

Insight:

The **standard form:** $ax + by + c = 0$ is especially useful for:

- Analyzing **parallel** and **perpendicular** lines
- Finding **x- and y-intercepts** quickly
- Solving systems of equations using elimination

Example 6: Equation in Standard Form $ax + by + c = 0$

Write equation of line in standard form that is perpendicular to the line $y = \frac{1}{2}x + 1$ and pass through the point $P(3, 5)$,

Solution 6

Perpendicular lines have negative-reciprocal slope. Slope of given line is $\frac{1}{2}$

Slope of perpendicular line will be $m = -2$.

equation of line through point $P(3, 5)$ with slope $m = -2$ in Point slope form

$y - y_1 = m(x - x_1)$

$y - (5) = -2(x - 3)$

$y - 5 = -2x + 6$

$\mathbf{2x + y - 11 = 0}$

Challenge: **Refer page 170 for distance formula**

Find equation of a line perpendicular to $y = 2x + 6$ which is one unit away from the origin

Example 7: Intercepts from Standard Form $ax + by + c = 0$

Find $x - intercept$ and $y - intercept$ of the line $3x - 2y + 12 = 0$

Solution 7

$x - intercept$ is the point where the line crosses the $x - axis$. Here $y = 0$.

$y - intercept$ is the point where the line crosses the $y - axis$. Here $x = 0$.

$x - intercept, y = 0$	$y - intercept, x = 0$
$3x - 2y + 12 = 0$	$3x - 2y + 12 = 0$
$3x - 2(0) = -12$	$3(0) - 2y = -12$
$3x = -\frac{12}{3} = -4$	$y = \frac{-12}{-2} = 6$
$x - intercept, x = -4$ and $y - intercept, y = 6$	

Example 8: Convert Equation of Line in Different Forms

Graph of a line with x-intercept of 4 and y-intercept of -2 is shown on the right.

a.	Write the equation of line in intercept form. $\frac{x}{a} + \frac{y}{b} = 1, a = 4, b = -2$ $\frac{x}{4} + \frac{y}{(-2)} = 1 \Rightarrow \frac{x}{4} - \frac{y}{2} = 1$
b.	Convert the above equation in slope-intercept form, $y = mx + b$, to find the slope of the line. $\frac{x}{4} - \frac{y}{2} = 1 \Rightarrow \frac{y}{2} = \frac{x}{4} - 1$ $y = 2 \times \frac{x}{4} - 2 \Rightarrow y = \frac{1}{2}x - 2$ $m = \frac{1}{2}$

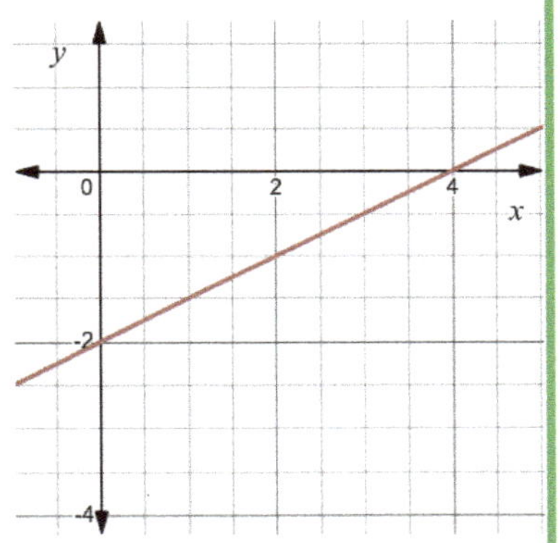

Challenge Questions

Consider three-dimensional space like your room. If two lines do not intersect, does it mean that they are parallel. Hint: Consider planes

The line perpendicular to PQ given by the equation $3x + 2y = 7$. If $P(3,4)$ and $Q(a,b)$ find an expression for b in terms of a. $b = \frac{2}{3}a + 2$

Challenge Example 1: Area Formed with Coordinate Axes

Determine area of triangle formed by the line $3x + 4y = -12$ and the coordinate axes.

Step #1: Find x-intercepts	Step #2: Find y-intercepts	
$x - intercept: y = 0$ $3x + 4(0) = -12$ $x = \frac{12}{3} = 4$	$y - intercept: x = 0$ $3(0) + 4y = -12$ $y = \frac{12}{4} = 3$	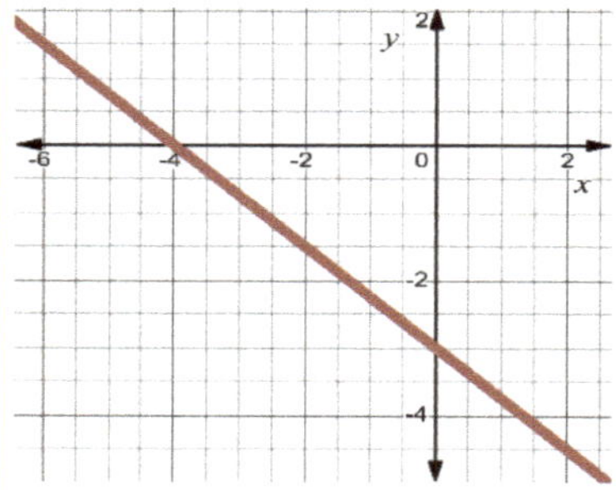

Step #3: Calculate Area of Right Triangle

Graph of the line is shown to visualize the triangle formed.

Area of the triangle, $A = \frac{1}{2} \times base \times height$

$$A = \frac{1}{2} \times base \times height = \frac{1}{2}(4)(3) = 6\ unit^2$$

Practice

Q1	Sketch the line $x - 2y + 4 = 0$ on the coordinate plane. Find area of triangle formed by the line $3x + 4y = -12$ and the coordinate axes
Q2	Sketch the line $y = -\frac{2}{3}x + 4$ on the coordinate plane. Find area of triangle formed by this line and the coordinate axes.
Q3.	The line $y = ax + 5$ intersects the line $y = 4x + b$. Find the value of $a + b$.

Unit of Inquiry – Extend Your Thinking

Every linear equation tells a geometric story — of direction, intersection, and area.
Think about: How does changing the slope or intercept affect the shape and size of the triangle formed with the coordinate axes?
Try this: Sketch several lines with different slopes and intercepts. Observe how their intercepts with the axes change the area of the triangle they form.
Reflect: What patterns emerge between the line's equation and the area enclosed? How does algebra translate into geometry on the coordinate plane?

TIPS Example 2: Choosing the Right Form — Application of Intercept Form

Situation:

A water tank with total capacity C liters is being drained **at a constant rate** of r liters per minute. It takes T **minutes** to empty the tank completely.
Let A be the amount of water in the tank at time t minutes.

How to Model This Situation?

We can model the draining process in **two different ways**, depending on what information is known:

Intercept Form

$$\frac{\mathrm{t}}{T} + \frac{A}{C} = 1$$

- Use this form when the **total drain time** (T) and **tank capacity** (C) are known.
- This relates time passed and amount of water left directly in a proportional form.
- It's especially helpful for interpreting scenarios graphically where **intercepts** are meaningful:

$$\mathrm{t} = 0, \mathrm{A} = \mathrm{C}\ (\text{Tank Full});\ \ \mathrm{t} = \mathrm{T}, \mathrm{A} = 0\ (\text{Tank Empty})$$

Slope-Intercept Form

$$A = rt + C$$

- Use this form when the **draining rate** (r) and **initial amount** (C) are known.
- It clearly shows:
    - The **starting amount** C as the **y-intercept**
    - The **rate of change** (negative slope r) as how fast the tank is losing water

💡 **Insight:**

Both forms describe the **same situation** — but depending on what values you know or what you want to visualize, **one form may be more useful than the other**.

Perpendicular Lines Strategy

Verify that the following lines are perpendicular.

$$Ax + By = C \text{ and } Bx - Ay = D$$

Extended: Derive Intercept Form of Linear Equation from Similar Triangles

Intercept Form of a Line is

$$\frac{x}{a}+\frac{y}{b}=1$$

where:

- $x - intercept\ is\ (a, 0)$ and
- $y - intercept\ is\ (0, b)$

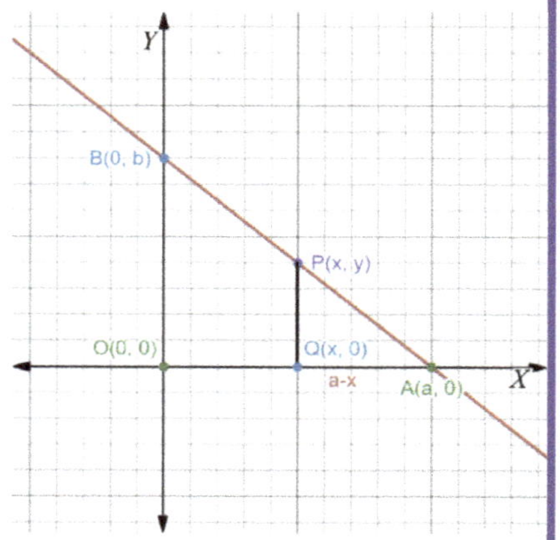

Step-by-Step Derivation Using Similar Triangles

ΔAOB and ΔAPQ are similar as their corresponding angles are equal.

$\Delta AOB \approx \Delta APQ \ \ from\ AA\ Property$

Ratio of the sides are proportional

$$\frac{\mathrm{OA}}{OB}=\frac{\mathrm{QA}}{QP}$$

$$\frac{\mathrm{a}}{b}=\frac{\mathrm{a}-\mathrm{x}}{y}$$

$$\mathrm{ay}=\mathrm{ab}-\mathrm{xb}$$

$$\mathrm{ay}+\mathrm{xb}=\mathrm{ab}$$

$$\frac{x}{a}+\frac{y}{b}=1$$

Divide both sides by $\mathbf{ab}$

Insight:

This form is especially useful when you know the x- and y-intercepts of a line. It's easy to graph and interpret geometrically — making it a perfect connection between algebra and coordinate geometry.

Challenge Question

$p(y-2)=p-x$ is an equation of a line where $p>0$

Q1	Find $x - intercept$ and $y - intercept$ of the line
Q2	Write the equation of the line in intercept form.
Q3	Find slope of the line
Q4	Write the equation in slope-intercept form.

Equation of Line from Graph in 2 Forms (if possible)

Example 1: Find Equation of Line from Graph with Different Strategies Mentioned

a.

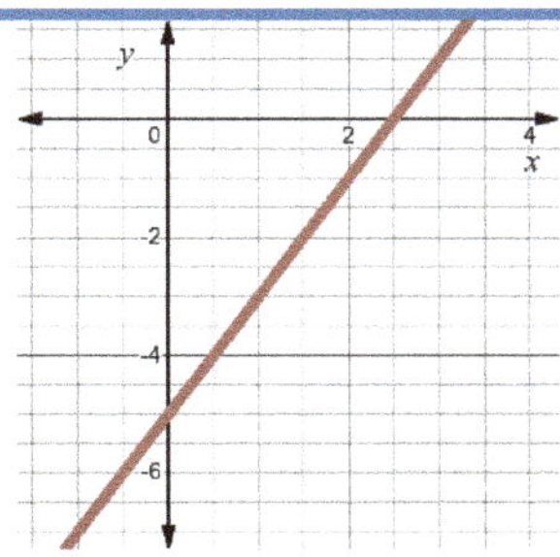

Strategy: Slope and y-intercept

b.

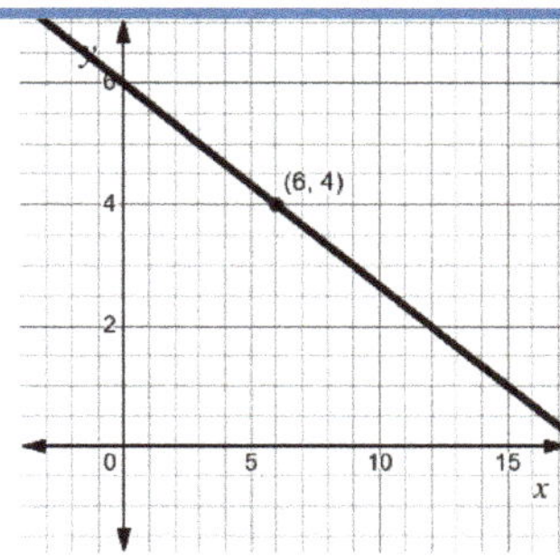

Strategy: Point-Slope Form

c.

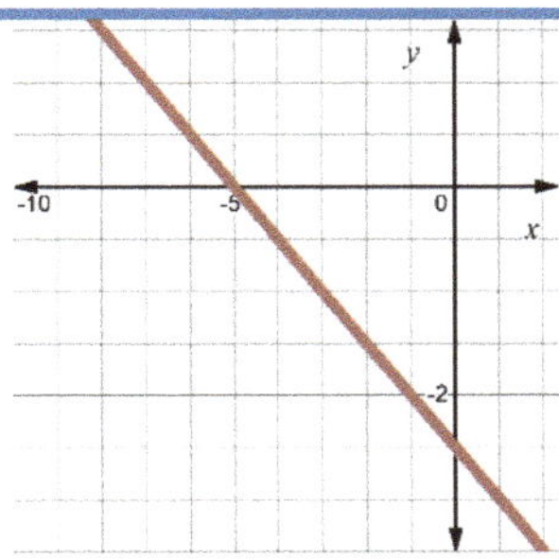

Strategy: Point-Slope Form

d.

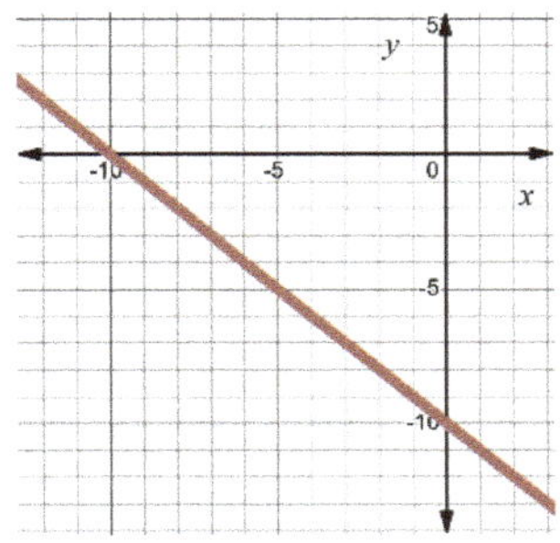

Strategy: Intercept Form

e.

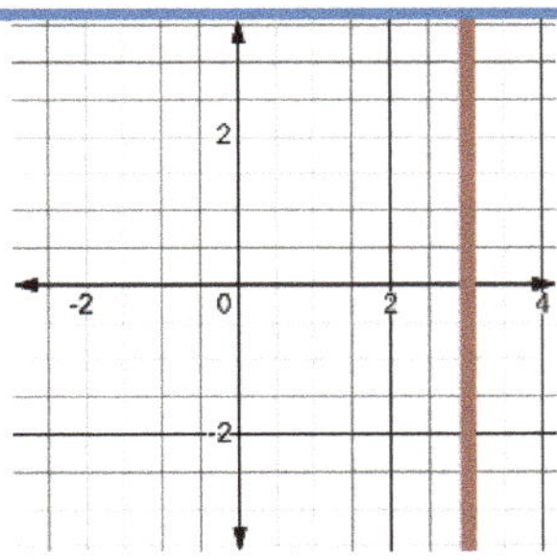

Vertical lines: Same x-value

f.

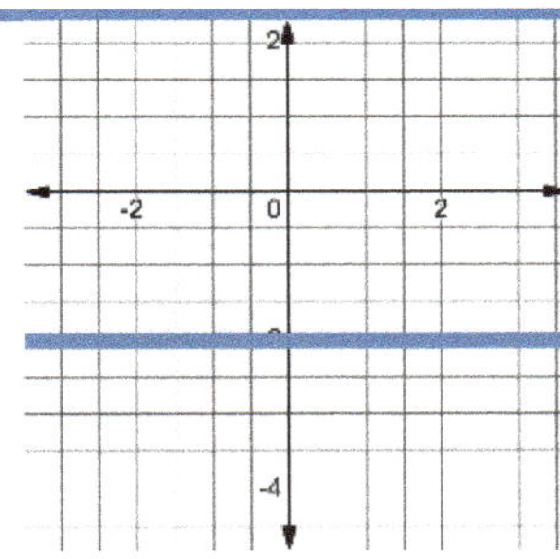

Horizontal lines: Same y-value

Solution 1: Equation of Lines from Graph

a.	$y = 2x - 5$ $2x - y - 5 = 0$	b.	$y - 4 = -\frac{1}{3}(x - 6)$ $x + 3y - 6 = 0$	c.	$y = -\frac{1}{2}x - \frac{5}{2}$ $x + 2y + 5 = 0$
d.	$\frac{x}{10} + \frac{y}{8} = 1$ $4x + 5y - 40 = 0$	e.	$x = 3$	f.	$y = -2$

Extended Application on Linear Equations

It is found that the temperature decreases linearly with the depth of lake. Temperature on top is $24°C$ and it decreases by $1.5°C$ for every $8m$.

a.	Write a model to relate depth, d, with temperature, $T, in\ °C$ $T = 24 - \frac{1.5}{8}d = 24 - \frac{3}{16}d$
b.	Use the model to find the temperature $in\ °C$ at depth of $40\ m$. $T = 24 - \frac{3}{16}(40) = 16.5$
c.	Find the depth in m at which the temperature is $12°C$. $12 = 24 - \frac{3}{16}d \Rightarrow d = 64$

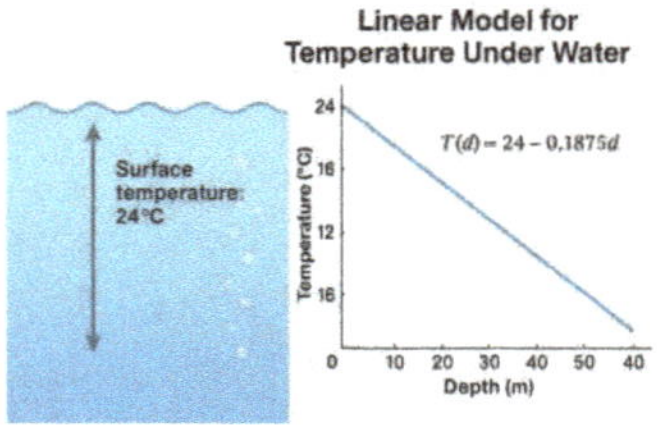

Challenge Problem on Linear Equations

The highest and the lowest marks awarded on an examination were 150 and 75. All the marks must be reduced so that the highest and the lowest marks become 110 and 60. Determine a linear function that will scale the marks as required.

Hint: Let the equation be $y = mx + b$. Then $110 = 150m + b$ and $60 = 75m + b$

Solve the system of equation to get $y = \frac{2}{3}x + 10$

SAT Practice

The points $A(x, -1), B(2, 1), C(4, 5)$ are collinear (on the same line)

Find the value of x.

a.	1	b.	0	c.	-1	d.	-2

Challenge: Directly and Inversely Proportion

d is directly proportional to x^2 and y is inversely proportional to d^2.

Given: $d = 24\ when\ x = 2$ and $y = 4\ when\ d = 10$.

Find the formula for y in terms of x.

Equations of Parallel and Perpendicular Lines

Lines on a coordinate plane can be compared by looking at their slopes.

- Lines that never meet are parallel.
- Lines that meet at right angles (90°) are perpendicular.

Understanding these relationships is very useful in geometry, design, physics, and engineering.

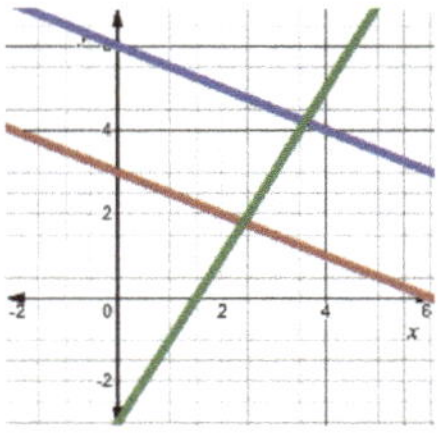

Parallel Lines

Two lines are parallel if they have the same slope but different y-intercepts.

- $y = -\frac{1}{2}x + 3$ is parallel to the line $y = -\frac{1}{2}x$
- Symbol Usage: $L_1 \parallel L_2$ Read as line L_1 is parallel to the line L_2

> 💡 **Why different y-intercept?**
>
> If slope and y-intercept are same, then the line will be coincident (overlapping) intersecting at infinite points. They are effectively the same lines

Perpendicular Lines

Two lines are perpendicular if the product of their slopes is -1.
This means their slopes are negative reciprocals of each other.

- $y = -\frac{1}{2}x + 3$ is perpendicular to the line $y = 2x$
- $y = \frac{2}{3}x - 1$ is perpendicular to the line $y = -\frac{3}{2}x + 2$
- Symbol Usage: $L_1 \perp L_2$ Read as line L_1 is perpendicular to the line L_2

Example 1: Identify Parallel and Perpendicular Line to $y = 4x - 1$

a.	$y = 1 - 4x$	b.	$y = 0.25x - 1$	c.	$y = -0.25x - 1$
d.	$4x = y$	Parallel is d; Perpendicular is c.			

Example 2: Perpendicular Line Passing Through a Point in Standard Form

Write Equation of a line perpendicular to $3x + 4y = 12$ that pass through the point $(2, -3)$

Step #1: Rearrange and find slope

$$3x + 4y = 12$$
$$4y = -3x + 12$$
$$y = -\frac{3}{4}x + \frac{12}{4}$$

Slope of the line: $m = -\frac{3}{4}$

Step #2: Slope of perpendicular line

$$m_{\perp} = \frac{4}{3}$$

Step #3: Equation through $(2, -3)$

Point-Slope Form

$$y - (-3) = \frac{4}{3}(x - 2)$$

Step #4: Simplify to Standard Form

$$3(y + 3) = 4(x - 2)$$
$$3y + 9 = 4x - 8$$
$$4x - 3y - 8 - 9 = 0$$
$$4x - 3y - 17 = 0$$

Example 3: Line Passing Through a Point and Parallel to Another Line

Write Equation of a line passing through $A(2, 4)$ and parallel to $x - 2y = 0$.

Step #1: Rearrange and find slope

$$x - 2y = 0$$
$$2y = x$$
$$y = \frac{1}{2}x$$

Slope of the line: $m = \frac{1}{2}$

Step #2: Slope of parallel line

$$m_{\parallel} = \frac{1}{2}$$

Step #3: Equation through $(2, 4)$

Slope Intercept Form

$$y = \frac{1}{2}x + b$$

Step #4: Find b using point $(2, 4)$

$$4 = \frac{1}{2}(2) + b = 1 + b$$
$$b = 4 - 1 = 3$$

Equation of required line:

$$y = \frac{1}{2}x + 3$$

Tip: Parallel Lines: $a_1x + b_1y = p$ and $a_1x + b_1y = q$ where $p \neq q$

✍ Practice Test: Parallel and Perpendicular Lines

Q1	Are the lines y = 2x + 3 and y = 2x - 5 parallel, perpendicular, or neither? parallel
Q2	Are the lines y = -1/2 x + 1 and y = 2x - 4 parallel, perpendicular, or neither? perpendicular
Q3	Find the slope of a line parallel to $3x - 2y = 0$. $m = 3/2$
Q4	Write the equation of the line parallel to y = 3x - 1 that passes through (2,4). $y = 3x - 2$
Q5	Write the equation of the line perpendicular to y = 2/3 x + 1 that passes through (3,2). y = -3/2x + 13/2
Q6	Determine whether the lines 2x - 3y = 6 and 3x + 2y = 4 are parallel, perpendicular, or neither. perpendicular
Q7	Find the equation of a line perpendicular to the x-axis through point (4,5). $x = 4$
Q8	Find the equation of a line parallel to the y-axis through point (-2,7). $y = 7$
Q9	The line $12x + By = 2$ is parallel to $y = 2 - 3x$. Find the value of B. $B = 4$
Q10	Find equation of a line perpendicular to the line $2x + 3y = 15$ with same y-intercept. $3x - 2y = -10$

SAT Practice

A line $y = mx + b$ has $x - intercept$ at $A(-4, 0)$ and $y - intercept$ at $B(0, b)$. What is the slope, m, if the area of the triangle AOB is $12\ unit^2$?

a.	2	b.	$-\frac{3}{2}$	c.	$\frac{3}{4}$	d.	$-\frac{3}{4}$

💡 Challenge Problem: Equation of Perpendicular Line

$x^2 + y^2 = 625$ is equation of a circle with centre at the origin. Verify that the point $(7, 24)$ lies on the circle. Find the equation of line tangent to the circle at this point.

$7x + 24y = 625$

Perpendicular Lines with Standard Form of Equation

Example 4: Shortcut to Perpendicular Lines

Prove that the lines $ax + by + c = 0$ and $bx - ay + k = 0$ are perpendicular

Find slope for both the line

$$ax + by + c = 0$$

$$by = -ax - c$$

$$y = -\frac{a}{b}x - \frac{c}{b}$$

Slope of the line is $m_1 = -\frac{a}{b}$

Rearrange to find slope for 2nd line

$$bx - ay + k = 0$$

$$ay = bx + k$$

$$y = \frac{b}{a}x + \frac{k}{a}$$

Slope of the line is $m_2 = \frac{b}{a}$

$$m_1 \times m_2 = -\frac{a}{b} \times \frac{b}{a} = -1$$

Product of the slopes is -1, so the lines are perpendicular.

Example 5: Shortcut to Perpendicular Lines Application

Write the equation of a line that passes through the point $P(-1, 0)$ and is perpendicular to $2x - 5y - 8 = 0$.

Equation of line perpendicular to the line

$2x - 5y - 8 = 0$ is

$5x + 2y + k = 0$

It passes through the point $P(-1, 0)$

$5(-1) + 2(0) + k = 0 \Rightarrow k = 5$

The perpendicular line is:

$$5x + 2y + 5 = 0$$

TIP

You need to swap the coefficients and change any one of the signs.

Challenge Problem: Equation of Perpendicular Line

General equation of line perpendicular to $ax + by + c = 0$ and passing through point $P(x_0, y_0)$ is $-bx + ay + (bx_0 - ay_0) = 0$.

Prove the statement.

Linear Systems – Equations Working Together

Introduction

A linear system is a set of two or more linear equations working together. These systems often model real-life situations such as ticket sales, mixture problems, business profit models, or supply and demand. Solving linear systems helps us understand how different conditions or rules interact at the same time.

Types of Solutions

Common point of intersection between all the lines is the solution of the system. Linear system will three conditions for solutions depending on slope and y-intercepts.

Consider the equation of two lines as: $a_1x + b_1y = c_1$ and $a_2x + b_2y = c_2$

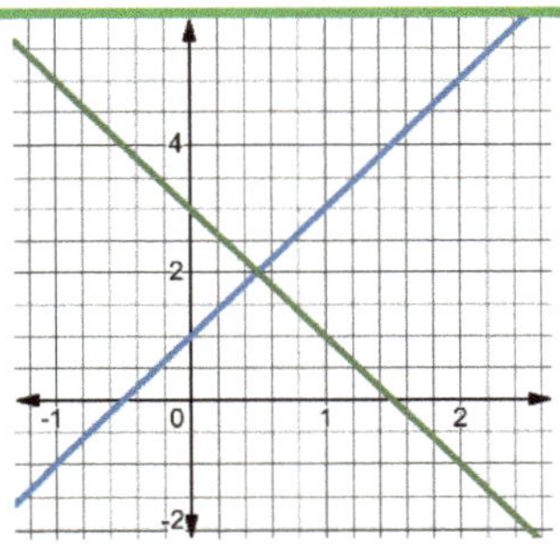

Unique Solution:

The two lines intersect at exactly one point.

This system of equation is called **consistent and independent**.

Condition:

different slopes

$$\frac{a_1}{a_2} \neq \frac{b_1}{b_2}$$

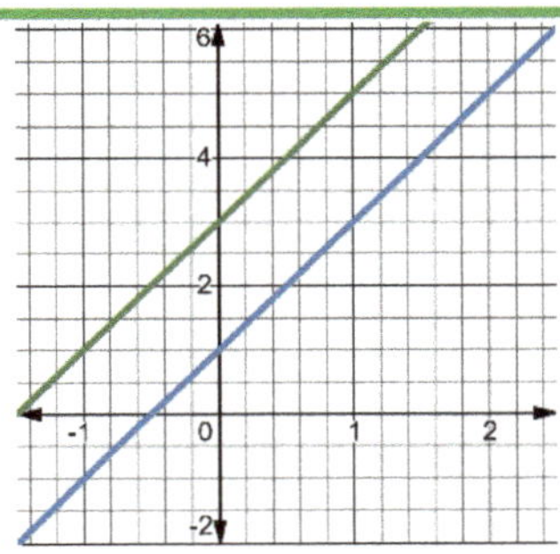

No Solution:

The two lines are parallel and never meet. This system of equation is called **inconsistent and independent**.

Condition:

Same slope

different y-intercept

$$\frac{a_1}{a_2} = \frac{b_1}{b_2} \neq \frac{c_1}{c_2}$$

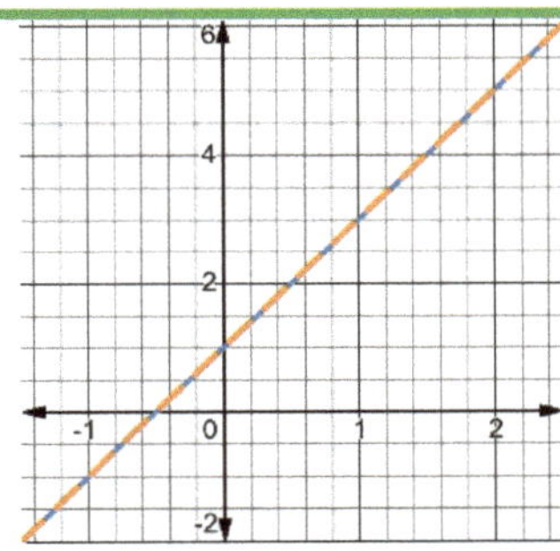

Infinitely Many Solutions:

The two lines are coincident (overlapping). This system of equation is called **consistent and dependent**.

Condition:

Same slope

same y-intercept

$$\frac{a_1}{a_2} = \frac{b_1}{b_2} = \frac{c_1}{c_2}$$

Example 1: Investigate the number of solutions for the given linear system

a.	$y = 2x - 3$ $y = x + 3$	b.	$y = 2x$ $y = 2x - 3$	c.	$y = 2x - 3$ $2x - y = 3$
	One solution different slope $m_1 = 2, m_2 = 1$		**No solution** same slope and different y-intercept These are parallel lines		**Infinite solutions** same slope and same y-intercepts. overlapping lines

Practice 1: Write the number of solutions for the following system of equations

a.	$y = x - 3$ $x - y = 3$	b.	$2x - 3y = 0$ $3y = 2x - 5$	c.	$y = -3$ $x = -3$

Methods to Solve Linear Systems (Two Variables)

1. Graphical Method

Plot both equations on the coordinate plane and identify the point of intersection. This method provides a visual understanding but may lack precision.

Watch Video **Linear Systems** **SCAN ME**

Example 2: Graph and Investigate

Solve the system:

y = 2x + 1

y = -x + 4

Sketch the graph of both the lines on x-y coordinate plane.

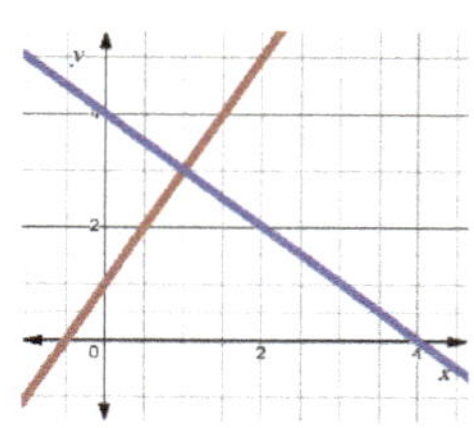

Graphing shows the lines intersect at (1, 3). Thus, the solution is (1, 3).

Practice

Graph the following system of linear equations to find the solution:

$x - y = 6$ and $y = 2x - 10$

2. Substitution Method

Solve one equation for one variable and substitute into the other equation.

Example 3: Direct Substitution

Solve the system:

$y = 2x - 3$ Eqn. 1

$x + 2y = 4$ Eqn. 2

Substitute y from Eqn. 1 to Eqn. 2

$x + 2(2x - 3) = 4$

Solve for x

$x + 4x - 6 = 4$

$5x = 10 \Rightarrow x = 2$

Substitute $x = 2$ in Eqn. 1 to find y

$y = 2(2) - 3 = 1$

Solution is $(2, 1)$

Example 4: Isolate a Variable

Solve the system:

$x + y = 5$ Eqn. 1

$3x + 2y = 4$ Eqn. 2

Isolate x from Eqn. 1

$x = 5 - y$ Eqn. 3

Now substitute x from Eqn. e in Eqn. 2

$3(5 - y) + 2y = 4$

$15 - 3y + 2y = 4$

$-y = 4 - 15 = -11 \Rightarrow y = 11$

Substitute $y = 11$ in Eqn. 3 to find x

$x = 5 - 11 = -6$

Solution is $(-6, 11)$

Note: Isolate any variable which is easier to work with.

Example 5: Application

The larger of the two numbers is 4 more than the smaller. Twice the larger added to the smaller equals 86. Find the two numbers.

Let the numbers be x and y where $x > y$.

Write the system of equations

$x = y + 4$ Eqn. 1

$2x + y = 86$ Eqn. 2

Substitute x from Eqn. 1 to Eqn. 2

$2(y + 4) + y = 86$

Solve for y

$2y + 8 + y = 86$

$3y = 86 - 8 = 78$

$y = \frac{78}{3} = 26$

Substitute $y = 26$ in Eqn 1.

$x = 26 + 4 = 30$

The numbers are 30 and 26

3. Elimination Method

Add or subtract equations to eliminate one variable, then solve for the other.

Example 6: Direct Elimination	**Example 7: Indirect Elimination**
Solve the system $2x + y = 7$ Eqn. 1 $3x - y = 8$ Eqn. 2 Add equations to eliminate y $2x + 3x = 7 + 8$ $5x = 15$ $x = 3$ Substitute in Eqn. 1 to find y $2(3) + y = 7$ $y = 7 - 6 = 1$ Solution: $(3, 1)$	Solve the system $2x + 3y = 1$ Eqn. 1 $x + 2y = 3$ Eqn. 2 Multiply Eqn. 2 by 2 then subtract $2x + 4y = 6$ Eqn. 3 Now Subtract Eqn. 1 from Eqn. 3 $5y - 3y = 6 - 1$ $y = 5$ Substitute $y = 5$ in Eqn. 2 to find x $x + 2(5) = 3$ $x = 3 - 10 = -7$ Solution: $(-7, 5)$
Example 8: Application	**Example 9: Application**
When three time a number is added to four times another, the result is 44. When five times the first is added to two times the second, the result is 50. What are the numbers? Equations $3x + 4y = 44$ Eqn. 1 $5x + 2y = 50$ Eqn. 2 Multiply Eqn. 2 by 2 and then subtract ***Solution left for the students*** The numbers are 8 and 5.	Peter is 9 years older than Sam. Two years hence he will be twice as old as Sam. How old are they now. Let present age be P and S for Peter and Sam respectively $P = S + 9$ Eqn. 1 $P + 2 = 2(S + 2) = 2S + 4$ Eqn. 2 Subtract to eliminate P and solve for S $2 = 2S + 4 - S - 9 \Rightarrow S = 7$ $P = S + 9 = 7 + 9 = 16$ Peter is 16 years and Sam is 7 years old

Practice Problems on Linear Systems

✍ Graph and Solve

Q1	$y = -x + 4$ $y = x - 2$	Q2	$y = -\frac{1}{2}x + 1$ $y = x + 4$	Q3	$y = \frac{3}{2}x + 3$ $x + y = -2$

✍ Solve Using Substitution Method

Q4	$y = x - 9$ $y + x = 3$	Q5	$2x + 3y = 6$ $2x - y = 7$	Q6	$x + 2y = 2$ $x - 2y = 6$

✍ Solve Using Elimination Method

Q7	$2x - 3y = 2$ $3x + 2y = 3$	Q8	$6x + 3y = -6$ $4x - 5y = 38$	Q9	$13x - 5y = 1$ $3x = 7y - 1$

✍ Without Solving Find the Number of Solutions to the System

Q10	$2x - y = 25$ $x - \frac{1}{2}y = 33$	Q11	$4x + 2y = -5$ $-8x - 4y = 10$	Q12	$x = 11$ $y = 7$

✍ Solve Linear System

Q13	$\frac{x}{5} - \frac{3y}{2} = -\frac{9}{5}$ $\frac{x}{3} - 5y = -8$	Q14	$6x - (y - 3) = 25$ $2(5x - 1) + 3y$ $= 16$	Q15	$\frac{5}{x} + \frac{1}{y} = 2$ $\frac{6}{x} - \frac{3}{y} = 1$

Given: $4x + by = 13$ and $ax - 4y + 7 = 0$ with solution of the system as $(1, b)$.

Find the value of a and b.

Challenge

The pair of linear equations $px + 3y = p - 2$ and $12x + py = p$ has no solution. Find the value of p. $p = \pm 6$

Challenge

The pair of linear equations $2x + 3y = 7$ and $(p - 1)x + (p + 1)y = 3p - 1$ has infinitely many solutions. Find the value of p. $p = 5$

Linear System with Three Variables

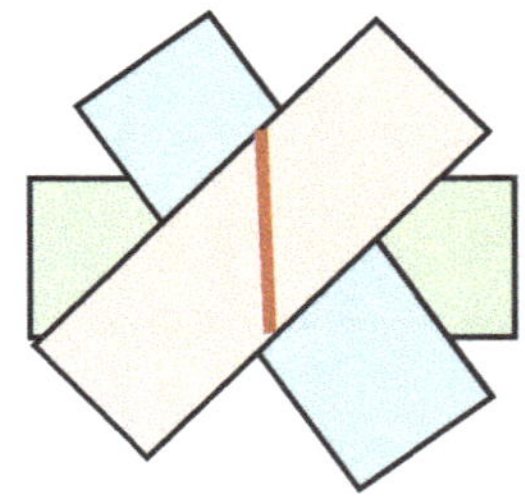

A **linear system of three equations in three variables** is a set of equations where each equation represents a plane in 3D space. The solution is the point(s) where these planes intersect.

A system of 3 equations in 3 variables may have:

1. **Unique Solution** – The three planes intersect at a single point.
2. **Infinitely Many Solutions** – The three planes intersect along a line (or all coincide).
3. **No Solution** – The planes do not intersect at a common point (parallel planes or inconsistent system).

Systematic Elimination Method is preferred for solution of these systems.

- Eliminate one variable at a time.
- Reduce system of three equations into two equations with two variables.
- Solve the 2-variable system, then back-substitute.

Example 10: Solve the system of given three equations.

$x + y + z = 2$ Eqn. 1

$2x + y - z = 2$ Eqn. 2

$x + 3y + 2z = 1$ Eqn. 3

Eliminate x from Eqn. 2 and Eqn. 3

$2 \times Eqn. 1 - Eqn. 2$

$y + 3z = 2$ Eqn. 4

$Eqn. 1 - Eqn. 3$

$-2y - z = 1$ Eqn. 5

Eliminate y from Eqn. 4 and Eqn. 5

$2 \times Eqn. 4 + Eqn. 5$

$5z = 5 \Rightarrow \mathbf{z = 1}$

Sub $z = 1$ in Eqn. 4, find y

$y + 3(1) = 2 \Rightarrow y = \mathbf{2 - 3 = -1}$

Sub $y = -1, z = 1$ in Eqn. 1, find x

$x + (-1) + (1) = 2 \Rightarrow x = 2$

Solution: $x = 2, y = -1, z = 1$

The solution is intersection of the three planes at a point $(2, -2, 1)$

Visualize like corner of a room where the walls meet.

Linear Systems 3 Equations SCAN ME

Example 11: System with Three Equations

Systematically eliminate variables to get the solution

<table>
<tr>
<td>
Solve the system:
$x + y + z = 6$ Eqn. 1
$x - y + z = 2$ Eqn. 2
$2x + y - z = 3$ Eqn. 3
Add Eqn. 3 to Eqn.1 and Eqn. 2
$3x + 2y = 6 + 3 = 9$ Eqn. 4
Add Eqn. 3 to Eqn. 2
$3x = 5 \Rightarrow \boldsymbol{x} = \mathbf{\frac{5}{3}}$
Substitute $x = \frac{5}{3}$ in Eqn. 4
</td>
<td>
$3 \cdot \frac{5}{3} + 2y = 9$
$5 + 2y = 9 \Rightarrow 2y = 9 - 5 = 4$
$\boldsymbol{y} = \mathbf{\frac{4}{2} = 2}$
Substitute $x = \frac{5}{3}, y = 2$ in Eqn. 1
$\frac{5}{3} + 2 + z = 6$
$z = 6 - \frac{5}{3} - 2 = 4 - \frac{5}{3} = \frac{12 - 5}{3} = \frac{7}{3}$
Solution: $\boldsymbol{x} = \mathbf{\frac{5}{3}},\ \boldsymbol{y} = \mathbf{2}, \boldsymbol{z} = \mathbf{\frac{7}{3}}$
</td>
</tr>
</table>

Example 12: System with Two Equations and Three Unknowns

Find the line of intersection of $2x - 2y + 5z = -4$ and $x + 2y - 2z = -8$

When we have three unknowns with two equations the solution will be in terms of a parameter. We will take the value of $z = t$ and then find solution in terms of t.

<table>
<tr>
<td>
Add the two equations
$3x + 3z = -12 \Rightarrow x + z = -4$
$x + z = -4 \Rightarrow x = -z - 4$
Let $z = t$
$x = -t - 4$
Now substitute
$x = -t - 4, z = t$ in $x + 2y - 2z = -8$
</td>
<td>
$-t - 4 + 2y - 2t = -8$
$2y = t + 4 + 2t - 8 = -4 + 3t$
$y = -2 + \frac{3}{2}t$
Now, we get infinite solutions. This is a line for intersection for two planes.
$x = -t - 4, y = -2 + \frac{3}{2}t$ and $z = t$
</td>
</tr>
<tr>
<td colspan="2">Two independent planes will intersect in a line if they are not parallel.</td>
</tr>
</table>

✍ Practice: System of Equations Challenge

Q1	$x + 2y - z = 1$ $2x - y + 3z = 7$ $3x + y + 2z = 9$ *no solution*	Q2	$x + y + z = 6$ $2x + 3y + z = 11$ $-x + y + 2z = 7$ $(1, 2, 3)$
Q3	$x + y + z = 4$ $2x - y + z = 8$ $3x + y - 2z = -1$ $(2, 1, -3)$	Q4	$x + y + z = 1$ $2x + 2y + 2z = 3$ $x - y + z = 2$ Show that the above system has no solution

Challenge Application: Sports Drink Blending (two nutrients + total volume)

Goal: Mix three base solutions (A, B, C) to make 10 L of sports drink with target sugar and salt.	
• Per-liter data A: 40 g sugar, 1 g salt B: 60 g sugar, 2 g salt C: 20 g sugar, 0.5 g salt **Target:** 10 L total, 460 g sugar, 14 g salt. Let x, y, zbe liters of A, B, C.	**Hint: Equations to Solve** $x + y + z = 10$ $40x + 60y + 20z = 460$ $x + 2y + 0.5z = 14$ 3L of A, 5L of B, 2L of C

Challenge Application: Ticket Sales (counts + revenue + relation)

A venue sells Adult (A), Student (S), Child (C) tickets at \$12, \$8, \$5.	
On a particular day: • 300 tickets sold, • \$2560 revenue, • Adults were 20 more than children.	**Hint: Equations to Solve** $A + S + C = 300$ $12A + 8S + 5C = 2560$ $A - C = 20$ A=100, S=120, C=80

Application Questions on Linear Systems

Everyday Contexts	
Q1	A movie ticket for an adult is \$12 and for a child is \$8. If 5 tickets cost \$52, how many adults and children attended?
Q2	A taxi charges \$3 per km for the first 10 km and \$2 per km thereafter. If a person pays \$46 for 20 km, how much did he pay for each part?
Q3	Two pens and three pencils cost \$15, and three pens and two pencils cost \$18. Find the cost of each.
Q4	Two numbers add to 25. Their difference is 7. Find the numbers.
Mixture & Work	
Q5	A chemist mixes a 10% alcohol solution with a 40% alcohol solution to make 20 liters of a 25% solution. How much of each should he mix?
Q6	A and B together can finish a job in 12 days. A alone can do it in 20 days. How long will B take?
Q7	15 liters of a solution contains 60% sugar. How much pure water must be added to make it 40% sugar?
Q8	A farmer has some cows and chickens. Total heads = 30, total legs = 74. Find the number of chickens.
Motion & Finance	
Q9	A train leaves a station at 60 km/h. Another leaves 2 hours later at 90 km/h. When will they meet?
Q10	A father is twice as old as his son. In 10 years, the father will be 20 years older than the son. Find present age of son.

Q11	Two numbers differ by 5. Their product is 84. Find them
Q12	A person invests \$5000, part at 6% and part at 8%. Annual income = \$370. How much at higher rate?
STEM & Challenge	
Q13	A plane flies 480 km with the wind in 1 hour, and 360 km against the wind in the same time. Find the speed of the wind.
Q14	Three types of tickets for a concert: \$100, \$50, \$20. Total sales = \$15,000. Number of \$20 tickets = twice the number of \$100 tickets. Total tickets sold = 300. Find tickets sold for \$20.
Q15	Three friends share \$72 in such a way that the first gets twice as much as the second, and the third gets \$8 more than the second. How much does each receive?
Q16	Rent a car is a company in the business of renting vehicles. It has cars, vans, and jeeps. The ratio of Jeeps to Vans is 5:6 and the ratio of Vans to Cars is 3:4. If there are 60 more cars than jeeps, find the total number of vehicles operated by the company.
Q17	Club membership is offering 20% discount for December. Normal charges are \$1200, plus monthly fee of \$120. Express this disclunted cost as a function.

Answers

Q1	3 adults, 2 children	Q2	\$30 + \$16	Q3	4, 2.33
Q4	16, 9	Q5	10 L, 10 L	Q6	30 days
Q7	7.5 L water	Q8	23 chickens).	Q9	6 hrs
Q10	10 years	Q11	12, 7	Q12	\$3500
Q13	w=60 km/h	Q14	150	Q15	\$32, \$16, and \$24
Q16	380	Q17	$A(m) = 960 + 96m$		

Chapter 3 Test: Linear Equations

Q1	$y = \frac{2x-1}{x+2}$. Rearrange the formula to change the subject to x.
Q2	Solve the equation: $\frac{x+7}{6} + \frac{1}{2} = \frac{x-2}{4}$
Q3	If $x = -4$ and $y = 3$ satisfy the equation $3x^2 + ky = 24$ find the value of k
Q4	a linear function is given by $y = 5 - 3x$. If x increases by 4, how much does y change?
Q5	8 men and 12 boys can complete a piece of work in 10 days, while 6 men and 8 boys can finish it in 14 days. Find the time taken by one man alone to complete this work.
Q6	The equation $2x + 3y - 7 = 0$ and $(p-1)x + (p+1)y = (3p-1)$ has infinitely many solutions. Find the value of p.
Q7	Jim is 4 years older than Sara. Five years from now, the sum of their ages will be 100. How old is Sara now?
Q8	One metal alloy is 25% copper, while the other is 50% copper. How much of 25% alloy should be used to make 1500 g of metal alloy that is 40% copper?
Q9	Show that the line $ax + by = c$ is perpendicular to the line $bx - ay = d$.
Q10.	The point $P(1,2)$ lies on the circle $x^2 + y^2 = 5$.Find equation of tangent to the circle at point P.

Answers

Q1	$x = -\frac{2y+1}{y-2}$	Q2	-11	Q3	-8	Q4	-12
Q5	$140\ days$	Q6	5	Q7	$43\ years$	Q8	$600\ g$
Q10	$x + 2y - 5 = 0$						

Distance Between Points

Distance AB between two points $A(x_1, y_1)$ and $B(x_2, y_2)$ is given

$$AB = \sqrt{(x_2 - x_1)^2 + (y_2 - y_1)^2}$$

Example 1: Verify Right Triangle

Verify that the triangle with vertices $\boldsymbol{A(-2,2), B(1,4), C(3,1)}$ is a right triangle.

Right triangle will have two perpendicular sides. Check slope of each line to verify.

$$m = \frac{y_2 - y_1}{x_2 - x_1}$$

$\boldsymbol{A(-2,2), B(1,4)}$

$$m_{AB} = \frac{4-2}{1-(-2)} = \frac{2}{3}$$

$\boldsymbol{B(1,4), C(3,1)}$

$$m_{BC} = \frac{1-4}{3-1} = \frac{-3}{2}$$

Since

$$m_{AB} \times m_{BC} = \frac{2}{3} \times \left(\frac{-3}{2}\right) = -1$$

Sides AB and BC of this triangle are perpendicular. Therefore, the triangle is a right triangle.

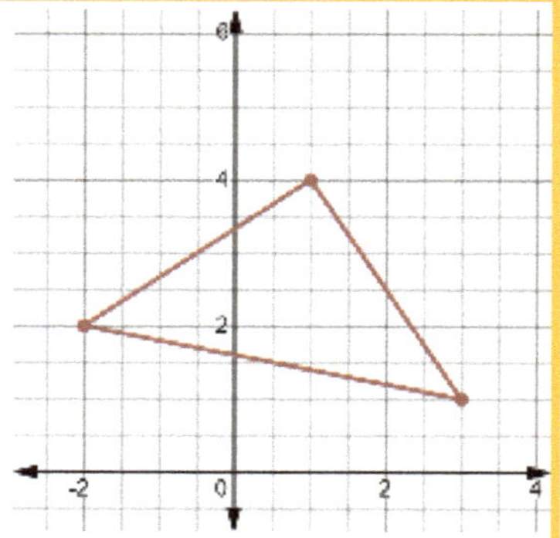

Extended: Sides of right triangle are related with Pythagorean Theorem

$AB^2 = (1+2)^2 + (4-2)^2 = 9 + 4 = 13$

$AC^2 = (3+2)^2 + (1-2)^2 = 25 + 1 = 26$

$BC^2 = (3-1)^2 + (1-4)^2 = 4 + 9 = 13$

$26 = 13 + 13 \Rightarrow AC^2 = AB^2 + BC^2$

Triangle ABC is a right triangle

Practice: Verify, in two different ways, that the triangle with vertices $A(7,7), B(4,-8), C(-6,-6)$ is a right triangle.

STEM Project: Derive the Distance Formula on Coordinate Plane

Hint: Distance can be derived using Pythagorean Theorem in a right triangle with lengths of the sides as $a, b,$ and c.

$$c^2 = a^2 + b^2 \Rightarrow c = \sqrt{a^2 + b^2}$$

Extended: Mirror Image of Points

Example 2: Mirror Image Across Horizontal Line, $y = p$

Step 1: Understand Reflection Rule

When reflecting across $y = p$, the x-coordinate remains unchanged. The y-coordinate flips symmetrically about the line $y = p$

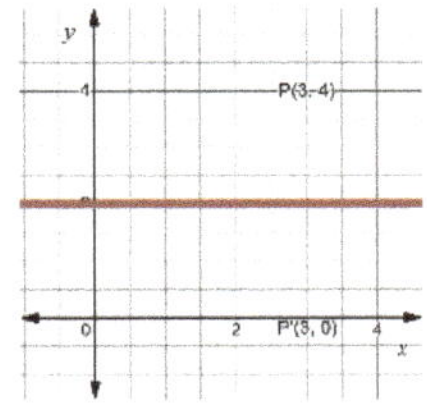

Step 2: Distance from the line

The vertical distance of y from the line is: $d = y - p$

Step 3: Find the reflected y-coordinate

If original point is above the line, its image will be the same distance below the line.

New y' is: $y' = p - d = p - (y - p) = 2p - y$

General Formula

For a point (x, y), its reflection across the line $y = p$ is: $(x', y') = (x, 2p - y)$

Check: Reflection of the point $P(3,4)$ on $y = 2$ is $P'(3,0)$

Watch Video **Mirror Image Concept** **SCAN ME**

Explore

a.	Show that the coordinates of a point $P(x, y)$ reflected on vertical line $x = q$, is $(x', y') = (2q - x, y)$
b.	Find the coordinates of a point $P(3, 4)$ reflected on vertical line $x = 0$.
c.	Find the coordinates of a point $P(3, 4)$ reflected on vertical line $x = -1$.
d.	Show that the coordinates of a point $P(x, y)$ reflected on vertical line $y = x$, is $(x', y') = (y, x)$. Hint: Reflection on both axis.

Challenge Problem

Determine the coordinates of the image of the point $P(5, 1)$ which is reflected in the mirror placed along the straight line $y = x + 1$.

$P'(0, 6)$

Example 3: Reflection of a Point on a Line

Determine the coordinates $P'(m,n)$ of the image of the point $P(5,4)$ which is reflected in the mirror placed along the straight line $y = x + 1$.

Given: Point P(5, 4) and mirror line y = x + 1.

Step 1 — Find the Foot of the Perpendicular

Let **F(h, k)** be the foot of the perpendicular drawn from P to the mirror line. The line y = x + 1 has slope 1, therefore any line perpendicular to it has slope −1. Equation of the perpendicular line through P(5, 4) is:

$y - 4 = -1(x - 5) \rightarrow y = -x + 9$

Now, find the intersection of this perpendicular with the mirror line y = x + 1.

$x + 1 = -x + 9 \rightarrow 2x = 8 \rightarrow x = 4 \rightarrow y = x + 1 = 5$

Hence, the foot of the perpendicular is **F(4, 5).**

Step 2 — Use the Midpoint Property

The mirror line acts as the perpendicular bisector of the segment joining P and its image P'. Therefore, F(4, 5) is the midpoint of P(5, 4) and P'(m, n).

$F = ((5 + m)/2, (4 + n)/2) = (4, 5)$

Equating components:

$(5 + m)/2 = 4 \rightarrow 5 + m = 8 \rightarrow m = 3$

$(4 + n)/2 = 5 \rightarrow 4 + n = 10 \rightarrow n = 6$

Step 3 — Result

The coordinates of the image of P after reflection in the line y = x + 1 are: **P'(3, 6)**

Challenge Question

Show that the coordinates $P'(q-a, p+a)$ of the image of the point $P(p,q)$ which is reflected in the mirror placed along the straight line $y = x + a$.

This result is **perfect for multiple-choice questions**, because once you recognize the pattern, you can answer instantly — no lengthy calculation needed.

Shortest Distance of a Point from a Line

In coordinate geometry, we often need to calculate how far a point lies from a given line. The shortest distance is always the **perpendicular distance** from the point to the line.

Formula: For a line in general form:

$ax + by + c = 0$ and a point $P(x_1, y_1)$, the shortest distance d is given by:

$$d = \left|\frac{ax_1 + by_1 + c}{\sqrt{a^2 + b^2}}\right|$$

Note: Since distance is always positive, the absolute value is taken above.

Derivation (Step-by-Step Idea)

1. Equation of line: $ax + by + c = 0$.
2. Equation of any line **perpendicular** to it and passing through

$$bx - ay + k = 0$$

   (since perpendicular slope is negative reciprocal).
3. Find k using the given point $P(x_1, y_1)$
4. Solve for the intersection point $Q(x_2, y_2)$ of the two lines.
5. Use the **distance formula** to calculate distance between $P(x_1, y_1)$ and $Q(x_2, y_2)$
6. Simplifying gives the formula above.

Example 3

Find the shortest distance of point $P(3, 4)$ from the line $3x + 4y - 5 = 0$

Solution 3

$$d = \left|\frac{3(3) + 4(4) - 5}{\sqrt{3^2 + 4^2}}\right| = \left|\frac{9 + 16 - 5}{\sqrt{9 + 16}}\right| = \frac{20}{5} = 4$$

Challenge Question

Find the equation of line perpendicular to the line $3x - 4y + 6 = 0$ which is 7 units away from the origin.

Hint: Perpendicular line is $4x + 3y + k = 0$

$$d = 7 = \left|\frac{4(0) + 3(0) + k}{\sqrt{4^2 + 3^2}}\right| = \left|\frac{k}{\sqrt{9 + 16}}\right| = \frac{k}{5} \Rightarrow k = 35$$

$4x + 3y + 35 = 0$

Test: Projection, Mirror Image, and Distance from a Point

Q1	Find the projection of the point $P(3,5)$ onto the x-axis.	$P(3,0)$
Q2	Find the projection of the point $P(3,5)$ onto the line $x + y - 5 = 0$.	$P(2,3)$
Q3	Find the mirror image of $P(1,4)$ across the line $y = x$.	$P(4,1)$
Q4	Find the shortest distance from $P(2,1)$ to the line $3x + 4y = 12$.	$\frac{2}{5}$
Q5	A boat is moving along the line $x + 2y - 10 = 0$. The port is located at $A(4,1)$). Find the shortest distance of the port from the boat's path.	$\frac{4\sqrt{5}}{5}$
Q6	A fishing boat sails along the line $y = x + 3$. The lighthouse (port) is at $P(-2,4)$. Find the perpendicular distance of the port from the sailing route.	$\frac{3\sqrt{2}}{2}$
Q7	A ship moves along the line $2x + 3y = 6$. A port is at $D(7,2)$ Determine the shortest distance from the port to the ship's route.	$\frac{14\sqrt{13}}{13}$
Q8.	Which point is closer to the line $y = 1 - 2x$. Find this distance. $A(-1,2)$ or $B(2,-1)$	A
Q9	Projection of a point $A(6,b)$ on the line $y = 2x - 3$ is $A'(2,a)$. Find $a + b$.	0

Challenge: Section Formula for Segment Division

Point C divides straight line joining points $A(x_1, y_1)$ and $B(x_2, y_2)$ in the ratio of $m{:}n$. Prove that the coordinates of the point C are:

$$\left(\frac{mx_2 + nx_1}{m+n}, \quad \frac{my_2 + ny_1}{m+n}\right)$$

Determine point C that divides straight line joining points $A(-2,1)$ and $B(8,6)$ in the ratio of $1{:}4$.

Challenge Problems

The point $P(4,3)$ lies on a circle $x^2 + y^2 = 25$. Centre of the circle is at the origin O. The tangent to the circle at P has x-intercept at Q and y-intercept at R.

Q1	Find equation of the tangent line at P.
Q2	Find the area of the triangle ROQ

Given the circle $x^2 + y^2 = 20$ with centre at the origin O.

Q3	Verify that the points $A(-4,2)$ and $B(2,4)$ lie on the circle.
Q4	Find the equation of the right bisector of the chord from $A(-4,2)$ to $B(2,4)$.

The three lines $y = x + 1, y = 4 - 2x$, and $y = -2$ form a triangle.

Q5	Find the coordinates of the triangle ABC formed by the intersection of the given lines.
Q6	Find the area of the triangle ABC.

Challenge

Q7. Choose servings x, y, z of **Oatmeal, Milk, Nuts** to hit exact targets.

- Per serving:
  Oatmeal: 150 calories, 5 g protein, 3 g fat
  Milk: 100 calories, 8 g protein, 2 g fat
  Nuts: 200 calories, 6 g protein, 18 g fat

Target: 450 calories, 19 g protein, 23 g fat.

Explore Grandi's Series

Grandi's Series is $1 - 1 + 1 - 1 + 1 - 1 + 1 - 1 + \cdots$

What can be the sum of this infinite series?

Show that by grouping differently it can be 0, 1 or 1/2

Advanced Algebraic Skills: Investigate Patterns

Special Identities: Componendo and Dividendo

Extremely useful technique to simplify calculations involving ratios and proportions.

Rule 1

If $\frac{a}{b} = \frac{c}{d}$ then $\frac{a+b}{b} = \frac{c+d}{d}$

Verify:

$\frac{3}{4} = \frac{6}{8} \Rightarrow \frac{3+4}{4} = \frac{6+8}{8}$

Proof: Add 1 on both sides

$$\frac{a}{b} + 1 = \frac{c}{d} + 1$$

$$\frac{a+b}{b} = \frac{c+d}{d}$$

Rule 2

If $\frac{a}{b} = \frac{c}{d}$ then $\frac{a-b}{b} = \frac{c-d}{d}$

Verify:

$\frac{3}{4} = \frac{6}{8} \Rightarrow \frac{3-4}{4} = \frac{6-8}{8}$

Proof: Subtract 1 from both sides

$$\frac{a}{b} - 1 = \frac{c}{d} - 1$$

$$\frac{a+b}{b} = \frac{c+d}{d}$$

Rule 3

If $\frac{a}{b} = \frac{c}{d}$ then $\frac{a+b}{a-b} = \frac{c+d}{c-d}$

Verify:

$\frac{3}{4} = \frac{6}{8} \Rightarrow \frac{3+4}{3-4} = \frac{6+8}{6-8}$

Proof: Divide one by the other

$$\frac{(a+b)/b}{(a-b)/b} = \frac{(c+d)/d}{(c-d)/d}$$

$$\Rightarrow \frac{a+b}{a-b} = \frac{c+d}{c-d}$$

Example 4:

If $\frac{p+q}{p-q} = \frac{7}{5}$ find $\frac{p}{q}$

Apply Rule 3:

$$\frac{p+q+(p-q)}{p+q-(p-q)} = \frac{7+5}{7-5}$$

$$\frac{2p}{2q} = \frac{12}{2}$$

$$\frac{p}{q} = 6$$

Example 5:

If $\frac{\sqrt{p}+\sqrt{q}}{\sqrt{p}-\sqrt{q}} = \frac{4}{3}$ find $\frac{p}{q}$

$$\frac{\sqrt{p}+\sqrt{q}+(\sqrt{p}-\sqrt{q})}{\sqrt{p}+\sqrt{q}-(\sqrt{p}-\sqrt{q})} = \frac{4+3}{4-3}$$

$$\frac{2\sqrt{p}}{2\sqrt{q}} = 7 \Rightarrow \frac{\sqrt{p}}{\sqrt{q}} = 7$$

$$\frac{p}{q} = 49$$

Challenge Problem

If $x = \frac{\sqrt{3}}{2}$ then find the value of $\frac{\sqrt{1+x}+\sqrt{1-x}}{\sqrt{1+x}-\sqrt{1-x}}$

Guided Practice

Example 19: If $x = \frac{\sqrt{3}}{2}$ show that $\frac{1+x}{1-x} = (2+\sqrt{3})^2$

$$x = \frac{\sqrt{3}}{2} \Rightarrow \frac{1}{x} = \frac{2}{\sqrt{3}}$$

Since we need $\frac{1+x}{1-x}$ take the reciprocal

$$\frac{1+x}{1-x} = \frac{2+\sqrt{3}}{2-\sqrt{3}}$$

Apply Rule 3

Rationalize denominator

$$\frac{1+x}{1-x} = \frac{2+\sqrt{3}}{2-\sqrt{3}} \times \frac{2+\sqrt{3}}{2+\sqrt{3}}$$

Difference of squares in the denominator

$$\frac{1+x}{1-x} = \frac{(2+\sqrt{3})^2}{4-3} = (2+\sqrt{3})^2$$

Example 20:

If $\frac{1+x}{1-x} = (2+\sqrt{3})^2$ show that $\frac{\sqrt{1-x}+\sqrt{1-x}}{\sqrt{1-x}-\sqrt{1-x}} = \sqrt{3}$

$$\frac{\mathbf{1+x}}{\mathbf{1-x}} = (2+\sqrt{3})^2$$

$$\frac{\sqrt{1+x}}{\sqrt{1-x}} = (2+\sqrt{3})$$

$$\frac{\sqrt{1-x}+\sqrt{1-x}}{\sqrt{1-x}-\sqrt{1-x}} = \frac{(2+\sqrt{3})+1}{(2+\sqrt{3})-1} = \frac{3+\sqrt{3}}{1+\sqrt{3}} = \frac{\sqrt{3}(\sqrt{3}+1)}{1+\sqrt{3}} = \sqrt{3}$$

Next Challenge: Find the Larger Number

$$\sqrt{3\sqrt{2\sqrt{5}}} \quad or \quad \sqrt{2\sqrt{5\sqrt{3}}}$$

Try before checking the solution on the next page

Next Challenge: Find the Larger Number

$\sqrt{3\sqrt{2\sqrt{5}}}$ *or* $\sqrt{2\sqrt{5\sqrt{3}}}$

Strategy: Square three times to get rid of the radical. Then evaluate the simplified term

	$\sqrt{3\sqrt{2\sqrt{5}}}$	$\sqrt{2\sqrt{5\sqrt{3}}}$
Square first time	$3\sqrt{2\sqrt{5}}$	$2\sqrt{5\sqrt{3}}$
Square second time	$3^2 \cdot 2\sqrt{5}$	$2^2 \cdot 5\sqrt{3}$
Again square them	$3^4 \cdot 2^2 \cdot 5$	$2^4 \cdot 5^2 \cdot 3$
Simplify divide by common factors	$3^3 \cdot (3 \cdot 2^2 \cdot 5)$	$2^2 \cdot 5 \cdot (3 \cdot 2^2 \cdot 5)$
Evaluate the remaining part	$3^3 = \mathbf{27}$	$2^2 \cdot 5 = \mathbf{20}$
Final result	$\sqrt{3\sqrt{2\sqrt{5}}} > \sqrt{2\sqrt{5\sqrt{3}}}$	

Student Spotlight: Alex Strapko

Midway through Grade 11, Alex was barely passing IB Math despite his efforts. After discovering Anil Kumar's YouTube channel, he began tutoring sessions and learned new ways to study and approach problem-solving.

By the end of Grade 11, Alex achieved one of the highest marks in his class on the final exam and finished with an 80% average. In his senior year, his grade rose to 94%, and he earned a Math Award. His teacher wrote on his report card:

"In my twenty-seven years of teaching, it has only happened a handful of times that a student has raised their grade so substantially from Grade 11. Alex has done it!"

Alex credits this dramatic improvement to Anil Kumar's guidance and support.

— Alex Strapko, Toronto, Ontario

Absolute Value

Absolute value of a number is its distance from zero on a number line, regardless of direction. It is always positive. Absolute value of a number n is written as $|n|$.

Example: Explore Absolute Value (Modulus Value)

Q1 Evaluate expressions

a.	$|-2| = |2| = 2$	b.	$|3-7| = |-4| = 4$
c.	$|-34| + |34|$ $= 34 + 34 = 68$	d.	$|-1| + |-1|$ $= 1 + 1 = 2$

Q2 Understand Common Mistakes

	Incorrect Statement	Correct Statement
a.	$|x| > 0$	$|x| \geq 0$
b.	$|x+y| = |x| + |y|$	$|x+y| \leq |x| + |y|$
c.	$\sqrt{x^2} = x$	$\sqrt{x^2} = |x|$
d.	If $|x-2| \geq 0$ then $x \geq 2$	If $|x-2| \geq 0$ then $x \geq 2, x \leq -2$
e.	If $|x-2| \leq 0$ then x has no solution	If $|x-2| \leq 0$ then If $x = 2$ is the only solution

Q3 Find the solution of the equation $|x| = 8$

Solution: $x = \pm 8$

Q4 Solve: $|x+3| = 7$

Solution: $x + 3 = \pm 8 \Rightarrow x = 3 \pm 8 \Rightarrow x = -5, 11$

Q5 Find the solution of the equation $|x| \leq -1$

Solution: No solution as absolute value is non-negative.

Q6 Solution of the equation $|x| > -1$

Solution: All real numbers as $|x|$ is always ≥ 0

Solve Linear and Absolute Value Inequalities

Linear inequalities involve $>, <, \geq, \leq$, symbols instead of equal to sign. Steps to solve inequalities are similar to those of equations. The solution of inequality is a range of possible values, not specific value(s) as in equation. These soltions can be represented on a number line or written as an interval.

Example 1: Solve the following inequalities

Q1	$2x+4<12$ $2x+4-4<12-4$ $2x<8$ $x<4$ **Note:** Zero is a solution. It is an easy check for the result.	Q2	$\frac{2x+3}{3}\geq 7$ — Is zero a solution? No $2x+3\geq 7\cdot 3$ $2x\geq 21-3$ $2x\geq 18$ $x\geq 9$ Solution confirms that zero is not the solution.
Q3	Three-fourths of a number decreased by two is at least thirty-seven. $\frac{3}{4}x-2\geq 37$ — zero is not a solution $\frac{3}{4}x\geq 37+2$ $\frac{3}{4}x\geq 39$ $x\geq 39\cdot\frac{4}{3}$ $x\geq 52$	Q4	15 is no more than three times sum of twice a number and seven. $15\leq 3(2x+7)$ — zero is a solution $\frac{15}{3}\leq 2x+7$ $5\leq 2x+7$ $5-7\leq 2x$ $-1\leq 2x$ $-1\leq x$ $-1\leq x$ has zero in solution set. So, it is correct

TIP: **Zero check is a fast way to check the solution**

Understand Common Phrases Related with Inequalities

x is at least p	**x is at most q**	**x is between p and q**
x is no less than p $x\geq a$	x is no greater than q $x\leq a$	$p<x<q$ **x is between p and q (inclusive)** $p\leq x\leq q$

Graph Linear Inequalities with Two Variable

Example 2: Draw a sketch and shade the area represented

$1 < x < 3$	$-1 \leq y \leq 3$	$x + y \geq 4$
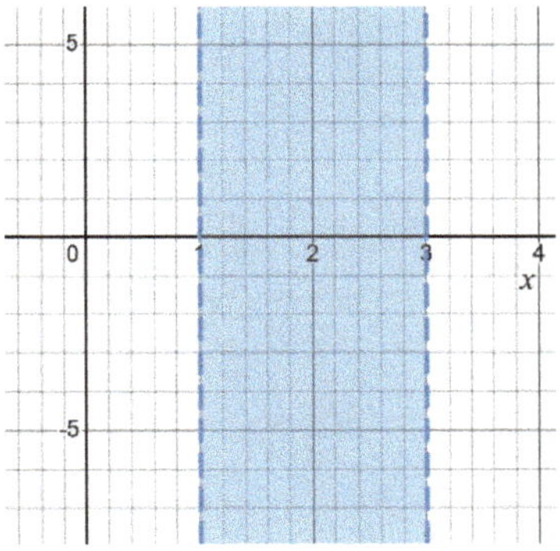	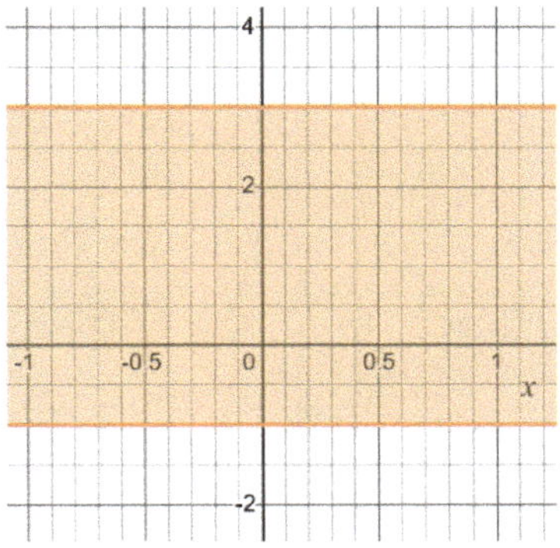	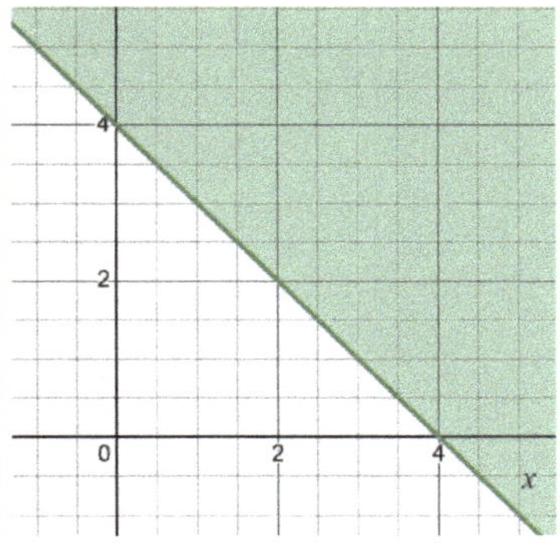
Vertical broken lines $x = 1$ and $x = 3$ indicate that the points on the line are not included.	Horizontal solid lines $y = 3$ and $y = -1$ indicates that the points on the line are included.	Area above the solid line is shaded to indicate that the line is included.

Example 3: Area defined by the Inequality

Draw the inequality to indicate area between the lines $f(x) = -2x - 2$, $g(x) = 2x + 1$, $x = 0$ and $x = 3$. This can be defined as: $f(x) \leq g(x), 0 \leq x \leq 3$

Trapezoid is the area shown. Students are encouraged to graph and verify the points of intersection.	$f(x) \leq g(x), 0 \leq x \leq 3$
$f(3) = -2(3) - 2 = -8$ $g(3) = 2(3) + 1 = 7$ Area of Trapezoid: $\frac{(a+b)h}{2}$ $a = 1 - (-2) = 3, b = 7 - (-8) = 15, h = 3$ $A = \dfrac{(3 + 15)3}{2} = 27$ Area of the region between lines is $27\ unit^2$	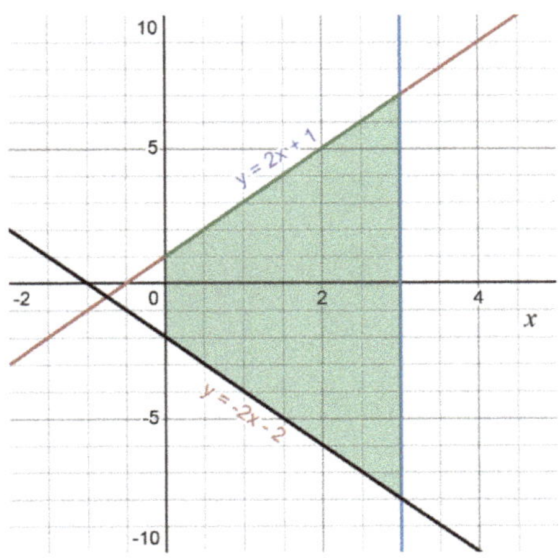

Example 4: Application Based on Inequality

a. Jean owns a flower shop. She needs at least 60 bunches of flowers. She buys x bunches of daffodils and y bunches of tulips for her shop.

Express this information as an inequality: $x + y \geq 60$

b. Jean pays 40 cents for each bunch of daffodils and 80 cents for each bunch of tulip. Jean does not want to pay more than $\$40$. Express this information as an inequality.

$$0.40x + 0.80y \leq 40 \Rightarrow x + 2y \leq 100$$

c. Jean is going to buy at least 20 bunches of tulips. $y \geq 20$

On a grid, show by shading, the region which satisfies all the above conditions.

Find the points of intersections. (They represent probable solution)

d. Jean will make profit of 20 cents on each bunch of tulip and 12 cents on each bunch of daffodils. If she sells all the bunches then what can be her maximum profit.

Graph equations

$x + y \geq 60$

$x + 2y \leq 100$

$y \geq 20$

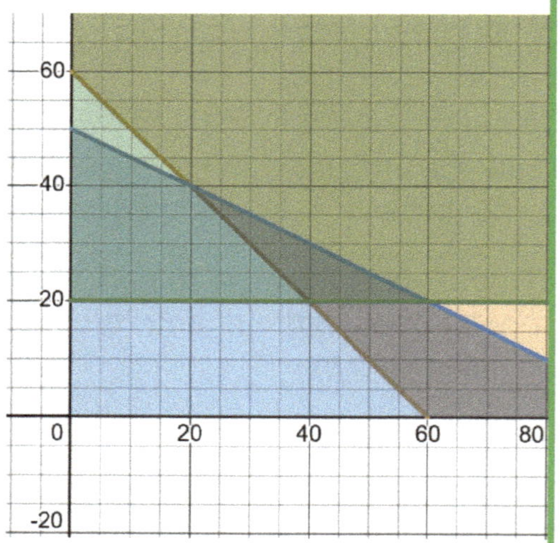

Check the critical points (the points of intersection from graph) and calculate the profit for each.

The points of intersection are:

$(60, 20), (20, 40)$

Profit for each critical point:

$(60, 20)$: $60 \times 0.12 + 20 \times 0.20 = 11.20$

$(20, 40)$: $20 \times 0.12 + 40 \times 0.20 = 10.40$

Maximum profit is $\$11.20$ by selling 60 daffodils and 20 tulips.

Note: Points of intersection are the critical numbers to be investigated.

Watch Video **Inequalities Concept** **SCAN ME**

Double Inequalities and Absolute Value Inequalities

Double inequalities combine two inequalities to describe a range of values

<table>
<tr><td>Q5</td><td>Linear Double Inequality
$-1 \leq -x + 9 \leq 13$
$-1 - 9 \leq -x + 9 - 9 \leq 13 - 9$
$-10 \leq -x \leq 4$
$10 \geq x \geq -4$

KEY:
Multiplication by a negative, reverses the sign also in any inequality</td></tr>
<tr><td>Q6</td><td>Double Inequality with Absolute Value
$|2x - 3| < 11$
$-11 < 2x - 3 < 11$
$-11 + 3 < 2x - 3 + 3 < 11 + 3$
$-8 < 2x < 14$
$-4 < x < 7$

KEY:
Absolute value less than inequality is within an interval</td></tr>
<tr><td>Q7</td><td>Double Split Inequality with Absolute Value
$|x + 4| > 5$
<table><tr><td>$x + 4 < -5$
$x < -5 - 4$
$x < -9$</td><td>$x + 4 > 5$
$x > 5 - 4$
$x > 1$</td></tr></table>KEY: Absolute value greater than inequality is beyond an interval</td></tr>
</table>

✍ **Practice: Solve Inequalities. Check if zeros is pert of solution.**

a.	$\lvert x\rvert < 2$	b.	$\lvert x - 4\rvert < 20$	c.	$\lvert 2x\rvert \geq 10$
d.	$\lvert x + 3\rvert \geq 6$	e.	$-4 < x + 5 < 12$	f.	$\lvert 3 - x\rvert \geq 20$

Quadratic Inequalities with Absolute Value Concept

Less Than Inequality with Quadratic Relations (within zeros)

$$\sqrt{x^2} = |x|$$

Q8

$x^2 - 4 < 0$

$x^2 < 4$

$\sqrt{x^2} < \sqrt{4}$

$|x| < 2$

$-2 < x < 2$

KEY:

Taking square-roots both sides

$\sqrt{x^2} = |x|$

$|x| < a \Longrightarrow -a < x < a$

Within an interval

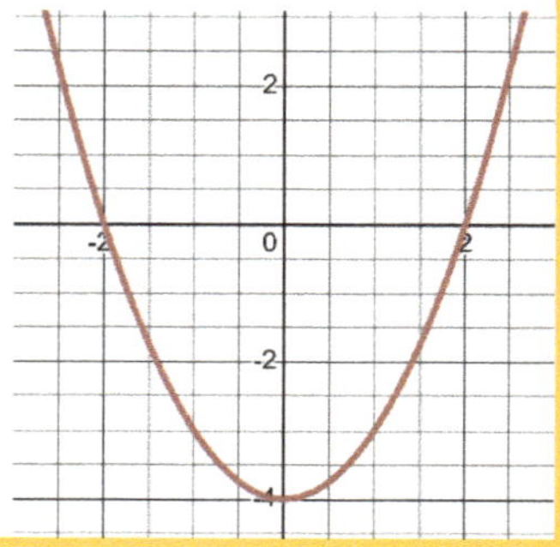

Q9

$(x + 2)^2 - 9 < 0$

$(x + 2)^2 < 9$

$\sqrt{(x + 2)^2} < \sqrt{9}$

$|x + 2| < 3$

$-3 < x + 2 < 3$

$-3 - 2 < x + 2 - 2 < 3 - 2$

$-5 < x < 1$

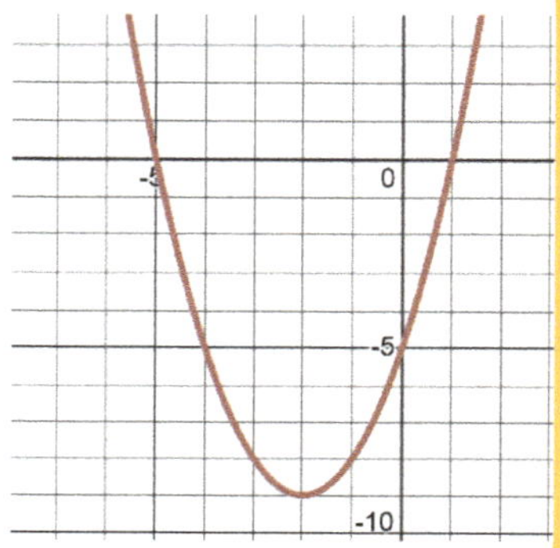

💡

$$x^2 < a \Longrightarrow |x| < \sqrt{a}$$

Advantage of the above approach is to avoid the interval tables and get the solution directly with algebra concept.

✍ Practice: Solve Inequalities. (less than means within the interval between zeros)

a.	$x^2 - 9 < 0$	b.	$(x - 2)^2 < 4$	c.	$\frac{1}{2}(x + 2)^2 \le 8$
Challenge Problem					
d.	$\lvert x - 2\rvert \le \lvert 2x - 1\rvert$	e.	$x^2 - 7\lvert x\rvert - 8 = 0$	f.	$\lvert 4 - x\rvert \le \lvert 3 - 2x\rvert$

Greater Than Inequality with Quadratic Relations (Away from zeros)

$$x^2 > a \Longrightarrow |x| > \sqrt{a}$$

Q10

$x^2 - 1 > 0$

$x^2 > 1$

$\sqrt{x^2} > \sqrt{1}$

$|x| > 1$

$x < -1$ or $x > 1$

KEY:

Taking square-roots both sides

$\sqrt{x^2} = |x|$

$|x| > a \Longrightarrow x < -a$ or $x > a$

beyond the interval

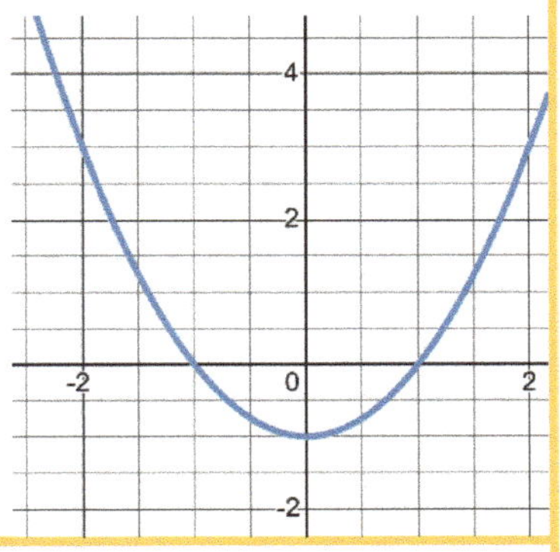

Q11

$(x-2)^2 - 4 > 0$

$(x-2)^2 > 4$

Take square-root both sides

$\sqrt{(x-2)^2} > \sqrt{4}$

$|x-2| > 2$

$x - 2 < -2$ or $x - 2 > 2$

$x < -2 + 2$ or $x - 2 > 2 + 2$

$x < 0$ or $x > 4$

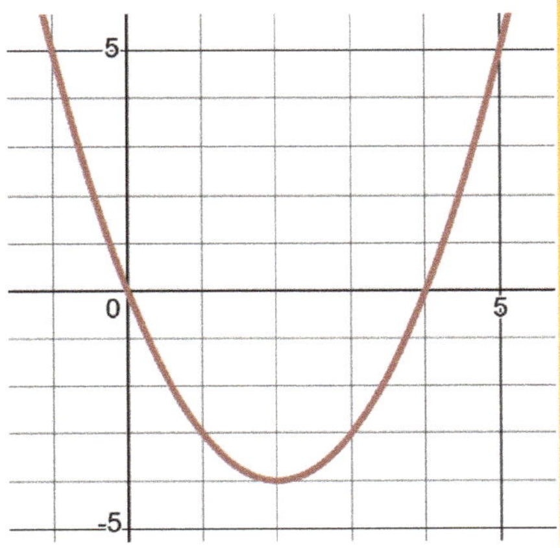

💡 **Greater Than Inequality means away from the interval between zeros**

✍ Practice: Solve Inequalities (away from zeros)

a.	$x^2 - 5^2 > 0$	**b.**	$(x-2)^2 \geq 0$	**c.**	$\frac{1}{2}(x-2)^2 \geq 2$
d.	**Communication and Thinking** $a(x-2)^2 + b > 0$. Write the condition for the given inequality to be always true.				

Double Inequality and Equivalent Absolute Value Inequality

Absolute value inequalities often express **symmetric constraints** around a center point. A double inequality can sometimes be rewritten as an absolute value inequality to better reveal the symmetry and simplify graphing or solving.

Write following double inequalities as absolute value inequality

<table>
<tr>
<td>

Case I: Symmetric with less than or equal to

$$-5 \leq x \leq 5$$

This represents all values of x within 5 units from 0. Since the inequality is symmetric about the origin (midpoint = 0), we can write:

$$|x| \leq 5$$

</td>
<td>

Case II: Symmetric with greater than

$$x < -3 \cup x > 3$$

This represents values of x **outside** the interval [−3, 3]. Again, it's symmetric about the origin.

We write:

$$|x| > 3$$

</td>
</tr>
<tr>
<td>

Case III: Not Symmetric with less than

$$-5 \leq x \leq 9$$

This interval is **not symmetric** about the origin. But we can still express it using an absolute value centered around its **midpoint**.

Step 1: Find midpoint of the interval:

$$\frac{-5+9}{2} = \frac{4}{2} = 2$$

Midpoint is at $x = 2$

Step 2: Find the distance from midpoint to an endpoint:

$$9 - 2 = 7$$

Now we can write the inequality

$$|x - 2| \leq 7$$

This means: "x is within 7 units of 2."

</td>
<td>

Case IV: Not Symmetric with greater than

$$x < -10 \cup x > 2$$

This describes values of x **outside** the interval [−10, 2].

$$\frac{-10+2}{2} = -4$$

Midpoint is at $x = -4$

Find the distance from one endpoint:

$$-4 - (-10) = 6$$

Equivalent absolute value inequality:

$$|x| > 6$$

This means: "x is more than 6 units away from -4."

</td>
</tr>
<tr>
<td colspan="2">

Challenge IB HL Math: Write a linear function, $g(x)$ such that it is greater than $f(x) = 4 - x^2$ when $|2x - 3| \geq 2.5$.

</td>
</tr>
</table>

Test Absolute Value and Inequalities

Q1	Solve the following linear inequalities a. $x - 3 > 4$ b. $2 - x < 4$ c. $\frac{2}{3}x - 1 \leq 5$
Q2	Solve the following absolute value inequalities a. $\lvert x \rvert > 4$ b. $\lvert x - 2 \rvert \leq 5$ c. $\lvert 3 - x \rvert \leq 1$
Q3	11 is no more than four times sum of twice a number and three.
Q4	Three-fourths of a number increased by four is at least thirty.
Q5	Solve the following quadratic inequalities a. $x^2 - 1 > 8$ b. $x^2 + 2 \leq 6$ c. $(x - 2)^2 > 1$
Q6	$x - 3 < 4$; x is a positive integer $\{1, 2, 3, 4, 5, 6\}$
Q7	$2x + 5 \leq 11$; x is a positive integer
Q8	$0 < 2x - 4 < 46$; x is a prime number
Q9	State the smallest perfect square x for which $x - 3 > 50$
Q10	A company rents car for \$35 per day plus a flat rate of \$220. Susan has no more than \$400 to pay for car hire. Find the maximum number of days she can hire the car for.
Q11	Sketch graph and shade the following inequalities for $x \geq 0$ and $y \geq 0$. a. $y \leq 3$ *and* $x \leq 4$ b. $x + y > 6$ c. $y < x + 4$
Q12	Sketch graph and shade $\lvert x + 2 \rvert < 10$ and $x \geq -15$
Q13	**Sketch graph and shade $\lvert x \rvert + \lvert y \rvert \leq 1$**

Chapter 4: Quadratic Relations

Factorising: Multiple Variables

Earlier we expanded expressions using **distributive property**. The reverse of this process is called **factorising**. Here we will split the expression into multiplication of simple expressions called **factors**.

Let us again have a close look at the expansion to understand factorization better.

- $a(b \pm c) = a \cdot b \pm a \cdot c$
- $(a + b)(c + d) = a \cdot c + a \cdot d + b \cdot c + b \cdot d$
- $(a \pm b)^2 = a^2 \pm 2ab + b^2$
- $(a + b)(a - b) = a^2 - b^2$

Explore and find expressions for

- $(a + b)^n$
- $a^n - b^n$

Here is a simple categorization to make it simple.

CASE I: Common Factor

Common factor is reversal of $a(b \pm c) = a \cdot b \pm a \cdot c$

Example 1: Factorise $a \cdot b \pm a \cdot c = a(b \pm c)$

a.	$3x + 3y$ $= 3(x + y)$ 3 is common factor	b.	$2s^2 + 3s$ $= s(2s + 3)$ s is a common factor	c.	$8s^2t + 6st$ $= 2st(4s + 3)$ $2st$ is common now
d.	the common factor can be taken among more than two terms also. $6a^2b + 3ab - 9ab^2 = 3ab(2a + 1 - 3b)$				

Guided Practice: Factorise $a \cdot b \pm a \cdot c = a(b \pm c)$

a.	$10x + 5y$	b.	$21c^2 + 6bc$	c.	$9p^2qr + 6prs$
d.	$2a^2b + 4ab - 6ab^2$	e.	$3xy + 3y - 9$	f.	$9s^3t - 6s^2t$

CASE II: Group Factor with 4 Terms

Group factor is reversal of $(a+b)(c+d) = ac + ad + bc + bd$

Example 2: Factorise $ac + ad + bc + bd = (a+b)(c+d)$

<table>
<tr><td>a.</td><td colspan="3">$ap - aq + bp - bq$
$= a(p-q) + b(p-q)$
$= (p-q)(a+b)$
$(p-q)$ and $(a+b)$ are the two factors.</td><td>Divide into groups with common elements.
• $\boldsymbol{a}$ is common to the first pair and $\boldsymbol{b}$ is common to second pair.
Now take the factor common to both
• $\boldsymbol{p-q}$ is common to both</td></tr>
<tr><td>b.</td><td>$x^3 + 3x^2 + 6 + 2x$
$= x^2(x+3) + 2(x+3)$
$= (x+3)(x^2+2)$</td><td>c.</td><td colspan="2">$4abc - 16ab - c + 4$
$= 4ab(c-4) - (c-4)$
$= (c-4)(4ab-1)$</td></tr>
</table>

Extra Practice: Factorise

a.	$pr + qr - ps - qs$	b.	$x^3 - x^2 - x + 1$
c.	$x^3 + 3x^2 + 6 + 2x$	d.	$y(x-1) - z(1-x)$
e.	$ax^2 + bx^2 + ay + by$	f.	$6xy + 2xz + 3yb + bz$

CASE III: Quadratic Trinomial Factorising: Product Sum Strategy

Factoring a quadratic expression is the reverse process of expansion. For example:

$(x+p)(x+q) = x^2 + px + qx + pq = x^2 + (p+q)x + pq$

This shows that any quadratic expression in standard form: $x^2 + bx + c$

can be factored as $(x+p)(x+q)$ If and only if

- $b = p + q \ (sum)$
- $c = pq \ (product)$

So, factoring $x^2 + bx + c$ is essentially finding two integers that satisfy both **the sum and the product conditions.**

Example 3: Factorise $x^2 + bx + c$

<table>
<tr><td>a.</td><td colspan="3">$x^2 + 5x + 6$
$= (x+3)(x+2)$

Find two numbers $\boldsymbol{p}$ $\boldsymbol{and}$ $\boldsymbol{q}$ so that
• $\boldsymbol{pq = 6}$ and $\boldsymbol{p + q = 5}$
Here the numbers are $\boldsymbol{3}$ $\boldsymbol{and}$ $\boldsymbol{2}$</td></tr>
<tr><td>b.</td><td>$x^2 - x - 6$
$= (x-3)(x+2)$
$pq = -6, \quad p + q = -1$
$p = -3, q = 2$</td><td>c.</td><td>$x^2 - 4x - 12$
$= (x-6)(x+2)$
$pq = -12, \quad p + q = -4$
$p = -6, q = 2$</td></tr>
<tr><td>d.</td><td>$x^2 - 8x + 15$
$= (x-3)(x-5)$
$pq = 15, \quad p + q = -8$
$p = -3, q = -5$</td><td>e.</td><td>$x^2 + 3x - 108$
$= (x-9)(x+12)$
$pq = -108, \quad p + q = 3$
$p = -9, q = 12$</td></tr>
<tr><td colspan="4">Given $f(x) = x^2 - x - 5$. If $f(n) = 7$, find the value(s) of n. $4, -3$</td></tr>
</table>

CASE IV: Factoring Trinomial $ax^2 + bx + c$

Trinomial where the leading coefficient is not one. It can be factored if

- $b = p + q\ (sum)$ and $ac = pq\ (product)$

Note: Case III is a special case where $a = 1$

Example 4: Factorise $ax^2 + bx + c$

<table>
<tr><td>a.</td><td colspan="3">$2x^2 + 7x + 6$
$= 2x^2 + 4x + 3x + 6$
$= 2x(x+2) + 3(x+2)$
$= (x+2)(2x+3)$

$\boldsymbol{pq = 12\ (2 \times 6)\ and\ p + q = 7}$
Here the numbers are 4 and 3
Split $\boldsymbol{7x}$ as $\boldsymbol{4x + 3x}$</td></tr>
<tr><td>b.</td><td>$3x^2 - 5x - 2$
$= 3x^2 - 6x + x - 2$
$= 3x(x-2) + (x-2)$
$= (x-2)(3x+1)$
$pq = -6, \quad p + q = -5$
$p = -6, q = 1$</td><td>c.</td><td>$6x^2 + 7xy - 3y^2$
$= 6x^2 + 9xy - 2xy - 3y^2$
$= 3x(2x+3y) - y(2x+3y)$
$= (2x+3y)(3x-y)$
$pq = -18y^2, \quad p + q = 7y$
$p = 9y, q = -2y$</td></tr>
</table>

Guided Practice: Factorise

a.	$x^2 - 3x - 28$	b.	$x^2 - x - 20$
c.	$6x^2 - 27x + 30$	d.	$4x^2 - 23x + 15$

CASE V: Perfect Square

Perfect square is a special case of Case III where the two numbers $p \text{ and } q$ are equal.

$(x + p)^2 = (x + p)(x + p) = x^2 + px + px + pq = x^2 + 2px + p^2$

Example 4: Factorise $x^2 + bx + c$

a.	$x^2 + 6x + 9$ $= (x + 3)(x + 3)$	Find two numbers $\boldsymbol{p \text{ and } q}$ so that $\boldsymbol{pq = 9}$ and $\boldsymbol{p + q = 6}$; $\mathbf{p = q = 3}$	
b.	$x^2 - 10x + 25$ $= (x - 5)(x - 5) = (x - 5)^2$ $pq = 25, \quad p + q = -10$ $p = -5, q = -5$	c.	$x^2 - 14x + 49y^2$ $= (x - 7y)(x - 7y) = (x - 7y)^2$ $pq = 49y^2, \quad p + q = -14y$ $p = -7y, q = -7y$

Guided Practice: Factorise

a.	$x^2 - 4x + 4$	b.	$x^2 + 20x + 100$
c.	$x^2 - 8x + 16$	d.	$x^2 - 6xy + 9y^2$

Challenge Problem

Find the value of k so that the trinomial $x^2 + kx + 36$ is a perfect square.

Solution

$(x \pm y)^2 = x^2 \pm 2xy + y^2$

$k = \pm 2y$

$x^2 + kx + 36 = x^2 + kx + 6^2$

This can be a perfect square if and only if $k = \pm 12$

Check $\boldsymbol{(x \pm 6)^2 = x^2 \pm 12x + 36}$

CASE VI: Difference of Squares

Product of sum and difference of same binomial is difference of squares.

$(x-y)(x+y) = x^2 - y^2$

Example 5: Factorise $x^2 - y^2$

a.	$x^2 - 9$ $= (x-3)(x+3)$	b.	$4x^2 - 25$ $= (2x-5)(2x+5)$	c.	$x^2 - \frac{1}{4}$ $= \left(x - \frac{1}{2}\right)\left(x + \frac{1}{2}\right)$

Guided Practice: Factorize

a.	$x^2 - y^2$	b.	$x^2 - 81$	c.	$x^2 - \frac{1}{9}$
d.	$16x^2 - 49$	e.	$4x^2 - \frac{1}{36}$	f.	$100x^2 - 1$

CASE VII: Combination of Above Cases

Factoring is an important skill, and we can have many combinations.

Here are few examples to explore.

a.	$2x^3 - 6x^2 - 20x$ $= 2x(x^2 - 3x - 10)$ $= 2x(x+2)(x-5)$ Group factor and then product-sum for trinomial	b.	$x^2 - 6xy + 9y^2 - a^2$ $= (x-3y)^2 - a^2$ $= (x-3y-a)(x-3y+a)$ Perfect square and then difference of squares

Guided Practice: Factorise

a.	$3(x-1)^2 - 12$	b.	$x^2 + 4x + 4 - 100y^2$
c.	$x^2 - 10x + 25 - 16y^2$	d.	$x^2 - 6x + 9 - 9y^2$

CASE VIII: Factor Sum and Difference of Cubes

Verify that following is true. Then use the formula to factor

Difference of Cubes	$a^3 - b^3 = (a-b)(a^2+ab+b^2)$
Sum of Cubes	$a^3 + b^3 = (a+b)(a^2-ab+b^2)$

Example 6

$a^3 + 3^3 = (a+3)(a^2-3a+9)$	$x^3 - 8$ $x^3 - 2^3 = (x-2)(x^2+2x+4)$
$8x^3 - 27$ $= (2x)^3 - 3^3 = (2x-3)(4x^2+6x+9)$	$64x^3 + 1$ $= (4x)^3 + 1 = (4x+1)(16x^2-4x+1)$

CASE IX: Factor Sum and Difference of Cubes Combinations

Variation of Results in Factoring. First option is correct

$a^6 - b^6 = (a^3)^2 - (b^3)^2 = (a^3-b^3)(a^3+b^3) = (a-b)(a^2+ab+b^2)(a+b)(a^2-ab+b^2)$
$a^6 - b^6 = (a^2)^3 - (b^2)^3 = (a^2-b^2)(a^4+a^2b^2+b^4) = (a+b)(a-b)(a^4+a^2b^2+b^4)$
$a^6 - b^6 = (a-b)(a^5+a^4b+a^3b^2+a^2b^3+ab^4+b^5)$

Challenge Problem: Verify the following generalized factorizing formula

$a^n - b^n = (a-1)(a^{n-1}+a^{n-2}b+a^{n-3}b^2+\cdots+a^2b^{n-3}+ab^{n-2}+b^{n-1})$
$a^n - 1 = (a-1)(a^{n-1}+a^{n-2}+a^{n-3}+\cdots+a^3+a^2+a+1)$
$a^n + b^n = (a+1)(a^{n-1}-a^{n-2}b+a^{n-3}b^2-\cdots+(-1)^{n-1}b^{n-1}), when\ n\ is\ odd$

Factor Polynomials SCAN ME

Challenge: Factor $x^5 + x + 1$. Solution in Unit 7 page 474

Case X: Trinomials of Degree 4 Where Product Sum Method Does Not Work

Strategy: Complete the Squares to Factor: Note $(a+b)^2 = a^2 + 2ab + b^2$

$x^4 + x^2 + 1$ is un-factorable	Now, apply this strategy to factor
To factor let us complete the square by adding and subtracting x^2	$9x^4 + 8x^2 + 4$
$= x^4 + x^2 + 1 + x^2 - x^2$	$= 9x^4 + 8x^2 + 4 + 4x^2 - 4x^2$
$= x^4 + 2x^2 + 1 - x^2$	$= 9x^4 + 12x^2 + 4 - 4x^2$
$= (x^2+1)^2 - x^2$	$= (3x^2+2)^2 - 4x^2$
Now apply difference of squares	$= (3x^2 + 2 + 2x)(3x^2 + 2 - 2x)$
$= (x^2 + 1 + x)(x^2 + 1 - x)$	$= (3x^2 + 2x + 2)(3x^2 - 2x + 2)$
$= (x^2 + x + 1)(x^2 - x + 1)$	

✍ Cumulative Practice: Factor Fully

a.	$x^2 - 4x - 5$	b.	$x^2 - 3x + 2$	c.	$x^2 - 7x - 8$
d.	$x^2 - 3x - 28$	e.	$2x^2 + 7x + 6$	f.	$21x^2 - 3xy + 7xy - y^2$
g.	$4x^2 - 8x - 60$	h.	$15x^3 - 4x^2 - 4x$	i.	$12x^2 + 17x - 14$
j.	$25x^2 - 70xy + 49y^2$	k.	$18x^3 - 50x$	l.	$6x^2 - 23xy + 15y^2$
m.	$x^8 - 1$	n.	$x^6 - 1$	o.	$8x^3 + 27$
p.	$(x-y)^2 - (x+y)^2$	q.	$3(x-1)^2 - 27$	r.	$x^2 - 4xy + 4y^2 - 9$
Challenge Problems					
s.	$x^4 + 4$	t.	$27x^4 + 24x^2 + 12$	u.	$x^4 - 23x^2 + 49$
v.	Find possible integer values of k for $2x^2 + kx + 40$ to be factorable				
w.	Find the value of k, if $x^3 + 15 = (x+3)(x^2 - 3x + 9) + k$ $\quad k = -12$				
x.	Solve Equation: $3x^3 + x^2 - 5x - 12 = 2x^3 - x^2 + 4x + 6$				

Solve Quadratic Equations by Factoring

Solve equation is to find the value of x when $y = 0$

Steps to Solve

1. **Arrange in standard form:** Move all terms to one side so the equation equals zero: $ax^2 + bx + c = 0$
2. **Factorize:** Write the quadratic as a product of two binomials.
3. **Zero Product Rule:** If $(p)(q) = 0$, then either $p = 0$ or $q = 0$.
4. **Solve for x:** Find the roots of the equation.

Example 7: Solve Quadratic Equations by Factoring

a.	$x^2 - 3x - 28 = 0$ $(x-7)(x+4) = 0$ $x - 7 = 0 \Rightarrow x = 7$ $x + 4 = 0 \Rightarrow x = -4$ Solution is $x = 7$ and $x = -4$	b.	$2x^2 + 7x + 6 = 0$ $2x^2 + 4x + 3x + 6$ $2x(x+2) + 3(x+2) = 0$ $(x+2)(2x+3) = 0$ $x = -2, x = -\frac{3}{2}$
c.	$4x^2 - 125 = 0$ $(2x+5)(2x-5) = 0$ $x = \pm\frac{5}{2}$	d.	$x^2 + 1 = 0$ $x^2 = -1$ No solution in domain of Real numbers

Example 8: Application

One integer is five larger than the other. The product of two numbers is 104. Find the other integer if one of them is x.

$x(x+5) = 104$ $x^2 + 5x - 104 = 0$ $(x+13)(x-8) = 0$	If $x = -13$ other is $-13 + 5 = -8$ If $x = 8$ other is $8 + 5 = 13$

Challenge: Prove the Identity

$$(2+1)(2^2+1)(2^4+1)(2^8+1)\ldots\left(2^{2^{10}}+1\right)+1 = 2^{2048}$$

Hint: Multiply by $(2-1)$

Extended Strategies to Solve Equations

Example 9: Solve Cubic Equations by Factoring

a.	$x^2 + 2x^2 - x - 2 = 0$ $x(x+2) - (x+2) = 0$ $(x+2)(x-1) = 0$ $x + 2 = 0 \Rightarrow x = -2$ $x - 1 = 0 \Rightarrow x = 1$ Solution is $x = -2$ and $x = 1$	b.	$2x^3 + 16 = 0$ $2(x^3 + 8) = 0$ $2(x+2)(x^2 - 2x + 4) = 0$ $x + 2 = 0 \Rightarrow x = -2$ $(x^2 - 2x + 4) = 0\ has\ no\ solution$ Solution is $x = -2$

Example 10: Solve Equations by Rearranging

a.	$2(x-1)^2 - 8 = 0$ $2(x-1)^2 = 8$ $2(x-1)^2 = \frac{8}{2} = 4$ $x - 1 = \pm\sqrt{4} = \pm 2$ $x = 1 \pm 2$ Solution is $x = -1$ and $x = 3$	b.	$3(x+2)^3 - 81 = 0$ $3(x+2)^3 = 81$ $(x+2)^3 = \frac{81}{3} = 27$ $x + 2 = \sqrt[3]{27} = 3$ $x = 3 - 2 = 1$ Solution is $x = 1$

Example 11: Solve Equations by Squaring

a.	$\sqrt{x+2} - 8 = 0$ Isolate Square-root, then square $x + 2 = 8^2 = 64$ $x = 64 - 2 = 62$ Solution is $x = 62$	b.	$\sqrt{2x-3} - 7 = 0$ $2x - 3 = 7^2 = 49$ $2x = 49 + 3 = 52$ $x = \frac{52}{2} = 26$ Solution is $x = 26$

✍ Practice: Solve following equations

a.	$2(x-1)^2 - 32 = 0$	b.	$3(x+2)^3 - 81 = 0$

Rational Expressions

Rational expressions are of the form $\frac{P(x)}{Q(x)}$ where the denominator $Q(x)$ cannot be zero. This leads to restrictions.

Example 1: Find restriction in the following expressions:

a.	$\frac{x}{2x-1}$	b.	$\frac{x-1}{x^2-1}$	c.	$\frac{x}{x^2-2x-3}$
Factor the denominator and find all possible zeros as restrictions					
a.	$2x-1=0$ $2x=-1$ $x=-\frac{1}{2}$ Restriction: $x \neq -\frac{1}{2}$	b.	$x^2-1=0$ $x^2=1$ $x=\pm 1$ Restrictions: $x \neq \pm 1$	c.	$x^2-2x-3=0$ $(x-3)(x+1)=0$ $x=-1,3$ Restrictions: $x \neq -1,3$

Example 2: Take common denominator to add expressions and write restrictions

a.	$\frac{2}{x+1}+\frac{3}{(x-2)(x+1)}=\frac{2(x-2)+3}{(x-2)(x+1)}=\frac{2x-4+3}{(x-2)(x+1)}=\frac{2x-1}{(x-2)(x+1)}$ Restrictions: $x \neq -1, 2$

Example 3: Multiply expressions and write restrictions

a.	$\frac{x-1}{x^2-4}\times\frac{x+2}{x}=\frac{x-1}{(x+2)(x-2)}\times\frac{x+2}{x}=\frac{x-1}{x(x-2)}$ Restrictions should include -2 although it was a common factor that cancelled Restrictions: $x \neq -2, 0, 2$

Example 4: Divide expressions and write restrictions

a.	$\frac{x-1}{x+3} \div \frac{x+2}{x-1} = \frac{x-1}{x+3} \times \frac{x-1}{x+2} = \frac{x^2-2x+1}{(x+3)(x-2)}$ Division is reciprocal of multiplication. Write the equation in product form. **Note:** Restrictions should include zeros of numerator and denominator of the divisor. Restrictions: $x \neq -3, -2, 1$

Practice 1: Write Restrictions, if any

a.	$\frac{x-1}{x^2-x}$	b.	$\frac{2x}{x^2+1}$	c.	$\frac{x}{x^2-7x+6}$

Practice 2: Simplify Rational Expressions and Write Restrictions

a.	$\frac{2}{x^2} - \frac{1}{x} + \frac{3}{x^3}$	b.	$\frac{x}{x+2} - \frac{1}{x+2}$
c.	$\frac{x}{x-3} + \frac{1}{3-x}$	d.	$\frac{x}{x-2} + \frac{1}{x}$
e.	$\frac{1}{x^2-1} - \frac{2}{x^2+1}$	f.	$\frac{2}{x+2} - \frac{x}{x^2+4x+4}$

Practice 3: Simplify Rational Expressions and Write Restrictions

a.	$\frac{x}{x-2} + \frac{1}{x}$	b.	$\frac{2}{x+2} - \frac{x}{x^2+4x+4}$
c.	$\frac{x+1}{3x} \times \frac{2x+2}{5}$	d.	$\frac{x+1}{2x^2-2} \times \frac{4x}{6}$
e.	$\frac{2x}{x-2} \div \frac{6x}{x^2-4}$	f.	$\frac{x+5}{x+3} \div \frac{2x^2+10x}{3-x}$

Rational & Radical Equations: Extraneous Roots

An extraneous root (or *extraneous solution*) is a solution that **appears valid algebraically** during the solving process but **does not satisfy the original equation**.

Example 12: Extraneous Roots as Restriction Cannot be a Solution

$$\frac{x}{x-2}+\frac{2}{3}=\frac{2}{x-2}$$ multiply both sides by $(x-2)$ $$x+2(x-2)=2$$ $$x+2x-4=2$$	$$3x=6 \Rightarrow x=\frac{6}{3}=2$$ $x=2$ is a restriction as that will make the denominator zero. It is an **extraneous root**. Therefore, there is **no solution.**
Note: Always write the restrictions before solving to avoid mistakes	

Example 13: Extraneous Roots as Restriction Cannot be a Solution

$$\frac{1}{2x-2}=\frac{3}{x^2-1}$$ Restriction: $x \neq \pm 1$ $$\frac{1}{2(x-1)}=\frac{3}{(x+1)(x-1)}$$ $$(x+1)(x-1)=6(x-1)$$ $$(x+1)(x-1)-6(x-1)=0$$	$$(x-1)(x+1-6)=0$$ $$(x-1)(x-5)=0$$ $$x=1, x=5$$ Since $x=1$ is a restriction, it is an **extraneous root**. **Only solution to this equation is $x=5$**
Note: Always write the restrictions before solving to avoid mistakes	

Example 14: Extraneous Roots as Restriction Cannot be a Solution

$$\sqrt{x+2}=4-x$$ Square both sides (it may cause error) $$x+2=(4-x)^2$$ $$x+2=16-8x+x^2$$ $$x^2-8x-x+16-2=0$$	$$x^2-9x+14=0$$ $$(x-7)(x-2)=0$$ $$x=7, x=2$$ Check: $\sqrt{2+2}=4-2 \Rightarrow 2=2$ $\sqrt{7+2}=4-7 \Rightarrow 3 \neq -3$
$x=7$ is an **extraneous root**. **Only solution is $x=2$.**	

Test: Solve Equations

Solve Equations

a.	$(3x-2)+1=5$	b.	$3x(x-2)=0$
c.	$x^2-16=0$	d.	$(x+2)(x-4)=0$
Factor and Solve Equations			
e.	$x^2+x-2=0$	f.	$x^2+2x-15=0$
g.	$x^2-7x+12=0$	h.	$10x^2-33x-7=0$
i.	$32x^3-20x^2+3x=0$ $\boldsymbol{x=0,\frac{1}{4},\frac{3}{8}}$	j.	$1-x^2+6x-9=0$ Hint: Write as $\mathbf{1-(x-3)^2=0}$

Solve Radical Equations

k.	$\sqrt{x-1}-2=0$	l.	$\sqrt{2x-1}+3=6$
m.	$\sqrt{x-3}+\sqrt{x}=3$ Hint: Square both sides and then solve. Check for extraneous roots. $\sqrt{x-3}=3-\sqrt{x}\Rightarrow x-3=\left(3-\sqrt{x}\right)^2\Rightarrow x=4$		
Solve Rational Equations. Write Restrictions. Check for Extraneous Roots			
n.	$\frac{3}{3x+2}=\frac{6}{5x}$ $\boldsymbol{x=-3/10}$	o.	$\frac{2x}{x+1}+\frac{4}{x-1}=2$ $\boldsymbol{x=-\frac{3}{2}}$
p.	$\frac{5}{x-2}=\frac{2}{x+1}$ $\boldsymbol{x=-3}$	q.	$\frac{x-5}{1-x}=\frac{3-x}{x-1}$ Extraneous root
r.	Find the value of k that will make the following expression true. $\frac{3}{x-5}-\frac{k}{x+2}=\frac{46-5x}{(x-5)(x+2)}$ $\boldsymbol{k=8}$		

Quadratic Functions

A ball thrown into the air follows a parabolic path that opens downward. Similar curved shapes can be seen in everyday objects like satellite dishes, glass bowls, and reflectors – many of which open upward. These curves can be mathematically modeled using a non-linear equation of the form:

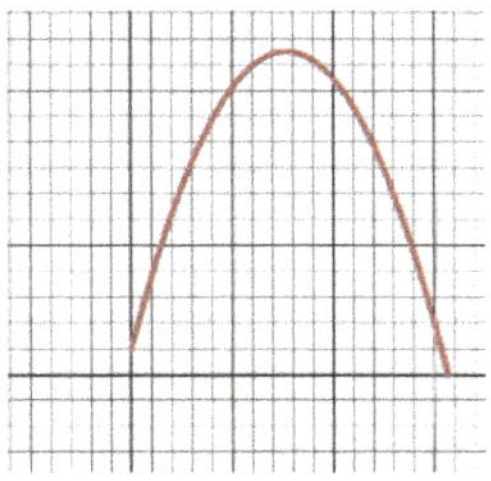

$y = ax^2 + bx + c$, where $a, b,$ and c are the real numbers and $a \neq 0$

This type of equation represents a quadratic function, one of the most important families of functions in mathematics.

Properties of Quadratic Function

- Shape of the Graph: The graph of a quadratic function is a curve called a parabola.
- Direction of Opening: If a > 0, the parabola opens up; if a < 0, it opens downward.
- Axis of Symmetry: A vertical line that divides the parabola into two mirror images, given by $\boldsymbol{x = -\frac{b}{2a}}$.
- Vertex: It is a turning point representing the optimal value (maximum or minimum). Vertex is located on the axis of symmetry.
- Maximum or Minimum: If the parabola opens downward, the y-coordinate of the vertex is a maximum; if it opens upward, it is a minimum.
- Y-Intercept: The point where the graph crosses the y-axis (when x = 0).
- X-Intercepts: The points where the graph crosses or touches the x-axis, found by solving $\boldsymbol{ax^2 + bx + c = 0}$.
- $y = a(x-p)^2 + q$ may not have any $x - intercept$ if $aq > 0$
- First finite difference is in steps $\boldsymbol{a(1, 3, 5, 7, \ldots)}$ from the vertex.
- The second finite difference is constant, but the first difference is not constant. Leading coefficient is half of the constant finite second difference.
- Sum of roots of the quadratic equation is -b/a and the product is c/a.
- End-behaviour: If a > 0 then as $x \to \pm\infty$ $y \to +\infty$; If a < 0 then as $x \to \pm\infty$ $y \to -\infty$
- **4 Forms** of quadratic equations help to investigate and understand them.

Standard Form $y = ax^2 + bx + c$	Factored Form $y = a(x-r)(x-s)$
Vertex Form $y = a(x-p)^2 + q$	Partial Factored Form $y = x(ax - b) + c$

Finite Difference Analysis for Quadratic Functions

For a quadratic function, the **second finite difference is constant**.
Finding the equation from data can be tricky, but here's the concept with a formula.

Let the general function be $f(x) = ax^2 + bx + c$.

Following table provides finite difference analysis for $x = \{1, 2, 3, 4, 5\}$

x	$\boldsymbol{ax^2 + bx + c}$	**1st Difference, Δ_1**	**2nd Difference, Δ_2**
1	$\boldsymbol{a + b + c}$		$2a$
2	$4a + 2b + c$	$\boldsymbol{3a + b}$	$2a$
3	$9a + 3b + c$	$5a + b$	$\boldsymbol{2a}$
4	$16a + 4b + c$	$7a + b$	
5	$25a + 5b + c$	$9a + b$	

Reflections of Analysis

- $\boldsymbol{2a}$ is 2nd Finite Difference
- $\boldsymbol{3a + b}$ is difference of first two terms
- $\boldsymbol{a + b + c}$ is first term

Example 1: Find Explicit Rule for Sequence: 2, 7, 16, 29, 46, ... In Function Notation

x	y	Δ_1	Δ_2	Equation: $\boldsymbol{ax^2 + bx + c}$
1	**2**			2nd finite difference constant: Quadratic
2	7	**5**		$\boldsymbol{2a = 4} \Rightarrow a = 2$
3	16	9	**4**	$\boldsymbol{3a + b = 5} \Rightarrow 3(2) + b = 5 \Rightarrow b = -1$
4	29	13	4	$\boldsymbol{a + b + c = 2} \Rightarrow 2 + (-1) + c = 2 \Rightarrow c = 1$
5	46	17	4	Equation: $\boldsymbol{2x^2 - x + 1}$
$\boldsymbol{f(x) = 2x^2 - x + 1}$				

Example 1: Find the Next Three Terms of 2, 7, 16, 29, 46, ...

Method 1: Use the explicit rule from above: $f(x) = 2x^2 - x + 1$

6th term: $x = 6$	$f(6) = 2(6)^2 - 6 + 1 = 67$
7th term: $x = 7$	$f(7) = 2(7)^2 - 7 + 1 = 67 = 92$
8th term: $x = 8$	$f(8) = 2(8)^2 - 8 + 1 = 67 = 121$

Method 2: Use Reverse Calculations from Finite Difference

1st finite difference is increasing by 4. 21, 25, 29 will be the finite differences Add them to the previous term	2, 7, 16, 29, 46, **67, 92, 121**, ... 5, 9, 13, 17, **21, 25, 29,** $46 + 21 = 67, 67 + 25 = 92, 92 + 29 = 121$
Note: An important strategy is applied to get the next three terms using difference	

Practice 1: Find Explicit Rule for the Sequences and then write Next 3 Terms.

a.	1, 6, 15, 28, 45, ...	b.	4, 12, 24, 40, 60, ...	c.	1, 8, 21, 40, 65, ...

Practice 2: Analyze Sequence for Quadratic Relation

Given sequence P, Q, R, 7, 16, 29, 46, 67, S

a.	Verify that the given sequence is quadratic in nature.
b.	Find the value of P, Q, R, and S
c.	Explore to find a quadratic relation of the form $y = ax^2 + bx + c$

Quadratic Function Standard Form: $y = ax^2 + bx + c$

Example 1: Connect Quadratic Function Graph and Properties

Graph of a quadratic function: $y = x^2 + 4x - 5$ is given below.

Comparing with standard form: $a = 1, b = 4, c = -5$

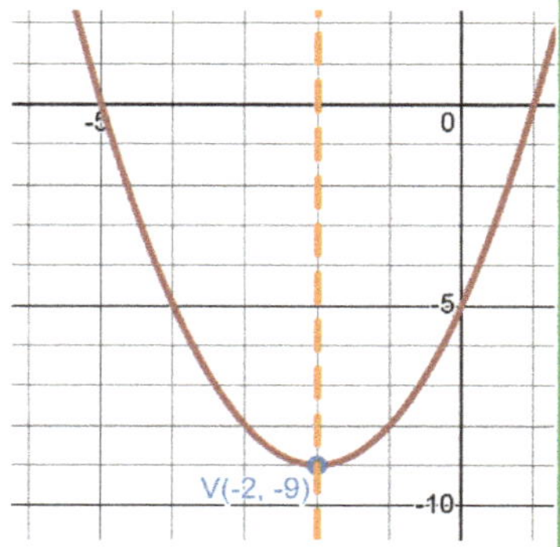

Properties

- Graph is a parabola as shown in the right
- Parabola opens up as $a = 1$ $(a > 0)$
- Axis of symmetry is $x = -2$ $\left(x = -\frac{b}{2a} = -\frac{4}{2(1)} = -2\right)$
- Minimum value is -9
- Turning Point, Vertex is at $(-2, -9)$
- $x - intercept$ at $x = -5$ and $x = 1$
- $y - intercept$ at $y = -5$ $(c = -5)$
- End-Behaviour: What happens when x approaches a large value: $x \to \pm\infty$ $y \to +\infty$
- Graph goes from quadrant II to quadrant I

✍ Practice 1:

Graph of a quadratic relation: $y = -2x^2 + 8x$ is given here.

Write properties of this quadratic relation as mentioned above.

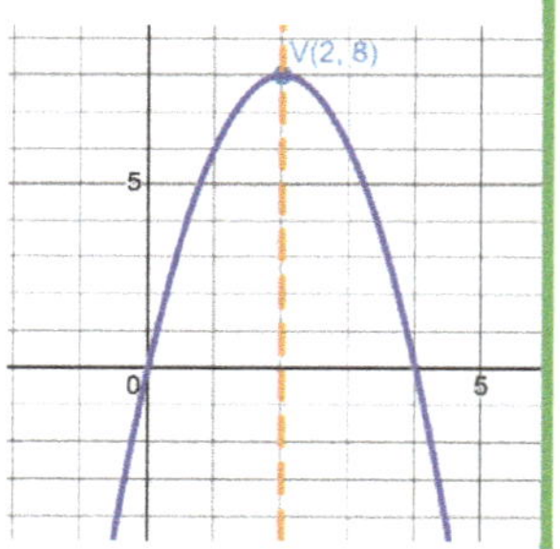

✍ Practice 2:

Draw a parabola that meets all the following characteristics:

- Minimum value of 1 at $x = 3$
- $y - intercept$ at $y = 5$

Write the number of possible $x - intercept$ for the given situation

✍ Practice 3:

Sketch graph of following equations on graphing calculator and then write their characteristics:

a.	$y = x^2 + 6x + 9$	b.	$y = -0.5x^2 + 2x - 5$

Example 2: Relate Quadratic Relation with Graph and Finite Difference

Given: $y = -x^2 - 4x + 5$

Show that the second finite difference is constant by completing the given table.

a. Complete the table below for the relation

x	y	1st Difference	2nd Difference
-6	-7		
-5	0	$0-(-7)=7$	
-4	5	$5-0=5$	$5-7=-2$
-3	8	$8-5=3$	$3-5=-2$
-2	9	$9-8=1$	
-1	8		
0	5		
1			

Here we have explored another important property of quadratic functions. The second finite difference is constant, but the first difference is not constant. Leading coefficient is half of the constant finite second difference.

b. Verify that second finite difference is a constant

c. Find $x - intercept$ from the graph

d. Find $y - intercept$ from the graph

e. Vertex Coordinates

f. Summarise properties of graph.

g. How is it rated to Example 1?

h. Graph and label the parabola with key points.

Quiz: Characteristics of Parabola

Describe the characteristics of parabola $y = 3 + 2x - x^2$.

a.	Write 7 points on the given curve for $x = \{0, 1, 2, 3, 4, 5, 6\}$.
b.	Use the points in part a) and find second difference.
c.	Find the $x - intercepts$ of the parabola
d.	Find the $y - intercepts$ of the parabola
e.	Find coordinates of vertex of the parabola

Projectile Motion

A ball is thrown upward from a height of 4 feet with an initial velocity of 48 feet per second. Its height h (in feet) after t seconds is given by $h = -16t^2 + 48t + 4$

At what time(s) will the ball be at a height of 36 feet?

STEM Exploration: Discovering Triangular Numbers

What Are Triangular Numbers?

Triangular numbers represent dots arranged in the shape of an equilateral triangle. First five terms of triangular numbers are:

1, 3, 6, 10, 15, ...

a.	Extend the pattern to ten terms.
b.	Create a table and justify that this is a quadratic pattern.
c.	$t_n = an^2 + bn + c$ can represent the general nth term for triangular numbers. Find the value of $a, b, and\ c$.
d.	Use the derived formula to find the value of 11th term.
e.	Find the term number for the triangular number 253.

Quadratic Formula and Discriminant Applications

Quadratic equations in standard form is $\boldsymbol{ax^2 + bx + c = 0}$

Solution of a quadratic equation means to find its zeros. In other words, solve the equation $\boldsymbol{ax^2 + bx + c = 0}$

Quadratic equation $ax^2 + bx + c = 0$ can be solved using the formula:

$$x = \frac{-b \pm \sqrt{b^2 - 4ac}}{2a}$$

Solve Quadratic Equation Using Quadratic Formula

Example 1: Solve $x^2 + 3x - 18 = 0$

Solution 1:

$x^2 + 3x - 18 = 0$

$$x = \frac{-b \pm \sqrt{b^2 - 4ac}}{2a}$$

$a = 1, b = 3, c = -18$

Substitute in quadratic formula and find solution

$$x = \frac{-(3) \pm \sqrt{(3)^2 - 4(1)(-18)}}{2(1)}$$

$$x = \frac{-3 \pm \sqrt{9 + 72}}{2} = \frac{-3 \pm \sqrt{81}}{2}$$

$$x = \frac{-3 \pm 9}{2}$$

$$x = \frac{-12}{2} = \boldsymbol{-6} \text{ and } x = \frac{6}{2} = \boldsymbol{3}$$

2 real solutions
$\boldsymbol{b^2 - 4ac > 0}$

Example 2: Solve $4x^2 + 12x + 9 = 0$

Solution 2:

$4x^2 + 12x + 9 = 0$

$$x = \frac{-b \pm \sqrt{b^2 - 4ac}}{2a}$$

$a = 4, b = 12, c = 9$

Substitute in quadratic formula and find solution

$$x = \frac{-b \pm \sqrt{b^2 - 4ac}}{2a}$$

$$x = \frac{-12 \pm \sqrt{(12)^2 - 4(4)(9)}}{2(4)}$$

$$x = \frac{-12 \pm \sqrt{144 - 144}}{8}$$

$$x = \frac{-12}{8} = -\boldsymbol{\frac{3}{2}}$$

NOTE: Only one solution as discriminant, $b^2 - 4ac = \boldsymbol{0}$

Communication

Explain why $y = 2(x - 3)^2 + 1$ does not have any real roots in two different ways.

Solve Quadratic Equation Using Quadratic Formula

Example 3: Solve $2x^2 - 2x - 5 = 0$

Solution 3:

$2x^2 - 2x - 5 = 0$

$$x = \frac{-b \pm \sqrt{b^2 - 4ac}}{2a}$$

$a = 2, b = -2, c = -5$

Substitute in quadratic formula and find solution

$$x = \frac{-b \pm \sqrt{b^2 - 4ac}}{2a}$$

$$x = \frac{-(-2) \pm \sqrt{(-2)^2 - 4(2)(-5)}}{2(2)}$$

$$x = \frac{4 \pm \sqrt{4 + 40}}{4} = \frac{4 \pm \sqrt{44}}{4} = \frac{4 \pm 2\sqrt{11}}{4}$$

$$x = \frac{2 \pm \sqrt{11}}{2}$$

2 conjugate radical roots

Example 4: Solve $2x^2 - x + 3 = 0$

Solution 4:

$2x^2 - x + 3 = 0$

$$x = \frac{-b \pm \sqrt{b^2 - 4ac}}{2a}$$

$a = 2, b = -1, c = 3$

$$x = \frac{-(-1) \pm \sqrt{(-1)^2 - 4(2)(3)}}{2(2)}$$

$$x = \frac{1 \pm \sqrt{1 - 24}}{4}$$

$$x = \frac{1 \pm \sqrt{-23}}{4}$$

No real solution as

$\boldsymbol{b^2 - 4ac < 0}$

However, there are two complext roots

Quadratic equations may have two solutions, one solution or no solution, in real numbers, depending on the value of the **discriminant** $\boldsymbol{D = b^2 - 4ac}$

$b^2 - 4ac > 0$	$b^2 - 4ac = 0$	$b^2 - 4ac < 0$
Two real solutions	One real solution	No real solution Two complex roots

Note: $\boldsymbol{D}$ ***or*** Δ *are commonly used symbols for* **discriminant** $\boldsymbol{b^2 - 4ac}$

Challenge Problem: Exam Practice

The function $f(x) = \frac{x^2-(a-3)x+3a}{x-2}, x \neq 2$

Show that the function will always have at least one real root for any value of a.

Watch Video **Exam Practice Roots Analysis** SCAN ME

✍ Practice Set 4: Find Solution of Equations. Round answer to tenths.

a.	$x^2 + 3x - 18 = 0$ [3.0, -6.0]	b.	$2x^2 - 3x - 7$ [2.8, -1.3]
c.	$4x^2 + 13x = x - 9$ [-1.5]	d.	$-3x^2 + 2x + 4 = 3 - 2x$ [1.5, -0.2]

✍ Practice Set 5: Discriminant Analysis

Discriminant $D\ or\ \Delta = b^2 - 4ac$ for the relation $y = ax^2 + bx + c$

Find the number of distinct real roots of the following quadratic relations by analysing the discriminant.

a.	$y = x^2 - 4x + 4$	b.	$y = 2x^2 - 5x$
c.	$y = x^2 + 2x + 3$	d.	$y = x^2 - 3$

✍ Practice Set 6: Properties of Quadratic Relation

Describe properties of given quadratic relations.

a.	$y = x^2 - 6x + 9$	b.	$y = 3 + 5x - 2x^2$
c.	$y = x^2 - x$	d.	$y = x^2 + 9$

Example 5: Find Range of k for Given Roots

Given the quadratic relation $y = 3x^2 - 6x + k$

Find the value (s) of k for

a.	Two	b.	One	c.	none

Solution 5:

$D = b^2 - 4ac$ helps to find the number of real root.

$D > 0$	$D = 0$	$D < 0$
Two distinct real roots	One real root	No real roots

Find D in the given equation: $y = 3x^2 - 6x + k \Rightarrow a = 3, b = -6, c = k$

$$D = b^2 - 4ac = (-6)^2 - 4(3)(k) = 36 - 12k$$

$D > 0$	$D = 0$	$D < 0$
$36 - 12k > 0$	$36 - 12k = 0$	$36 - 12k < 0$
$36 > 12k$	$36 = 12k$	$36 < 12k$
$\frac{36}{12} > k$	$\frac{36}{12} = k$	$\frac{36}{12} < k$
$\boldsymbol{k < 3}$	$\boldsymbol{k = 3}$	$\boldsymbol{k > 3}$
Two distinct real roots	**One real root**	**No real roots**

Communication and Thinking

Explain why $y = 2(x-3)^2 + 1$ does not have any real roots in two different ways.

Practice Set 7: Discriminant Application for Real Roots

a.	Determine the value of k so that the equation $y = 3x^2 - 2kx + 12$ has two roots.
b.	Determine the values of k so that the equation $y = x^2 - 4kx + k$ has equal roots.
c.	Determine the value(s) of k so that the equation $y = 4kx^2 + 8k + k$ has no real solution.
d.	The equation $y = (k+3)x^2 + 2(k+3) + 4$ has only one real solution. Find the value of k.

Watch Video **Discriminant Analysis** **SCAN ME**

Practice 8:

Table of Values: A rectangular piece of cardboard 30 cm long and 20 cm wide has equal squares of side x cm cut from each corner. It is folded into an open box. What integer value of x gives **maximum volume**?

Volume = $x(30-2x)(20-2x)$

Hint: Make a table of values for integer values of x.

Investigate Quadratic Function: Real World Application

Practice 9: Trajectory of a Ball

A ball is thrown upward at an initial velocity of 15 m/s, from a height of 1.0 m. The height, h, in metres, of the ball above the ground after t seconds can be calculated using the quadratic function $\boldsymbol{h = -4.9t^2 + 15t + 1.0}$. Round your answers to tenths.

a. Find the height when the ball was released.
b. Find height of the ball at t = 2 seconds
c. Calculate the maximum height of the ball.
d. Explain how you can find the time duration for validity of the relation.
e. How long was the ball in the air?
f. When will the ball hit the ground?
g. Describe assumptions for the given situation.
h. Label the graph with your findings.
i. Find the interval when the ball is above 10 m.
j. Describe the characteristic properties for the given equation.

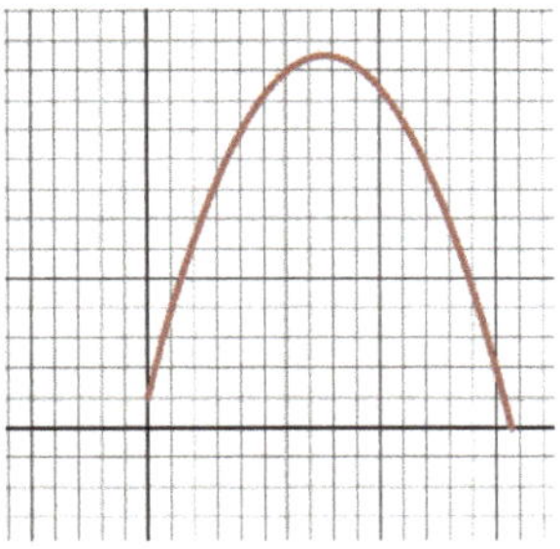

Model Situation: Upward Toss from a Platform, Projectile Motion

A ball follows a **parabolic path** when it is thrown upward against the force of gravity. If it is thrown with an **initial velocity** v_0m/s from an **initial height** $h_0$metres, then its **height** $h(t)$, in metres, after $t$seconds is modeled by the quadratic relation:

$$h(t) = \frac{1}{2}gt^2 + v_0t + h_0$$

where g is the acceleration due to gravity (approximately 9.8 m/s^{2}near Earth's surface). Substituting $g = 9.8$, we get:

$$h(t) = -4.9t^2 + v_0t + h_0$$

STEM Insight

This quadratic model is used in **physics, engineering, and sports science** to predict motion under constant acceleration. Explore and find its applications.

Challenge: Inequalities

$\boldsymbol{n}$ is an **integer** such that $7n + 6 \leq 34$ and $\frac{6n}{5+n^2} > 1$. Find all possible values of $\boldsymbol{n}$.

Solve: $x^2 + 4 \geq |3x + 2| - 7x$ Solution: $(-\infty, -9.36) \cup (-0.64, -0.67) \cup (-0.59, \infty)$

Exam Practice: Multiple Choice

Q1	A ball is thrown upward at an initial velocity of 48 feet per second. The height, h, in feet, of the ball above the ground after t seconds can be modelled by the function $\boldsymbol{h = -12t^2 + 48t}$. **How many seconds after the ball is thrown will it return to the ground**
	a. 2 | b. 3.5 | c. 4 | d. 4.5
Q2	A football is thrown in the air. Its height h in metres above the ground at any instant t seconds after the throw is approximately given by $h = -16t^2 + kt + 1, t \geq 0$. Find the value of k if the maximum height of the ball occurs at 3 seconds after it is launched.
	a. 48 | b. 72 | c. 80 | d. 96
Q3	Find the $y - intercept$ for a quadratic function with axis of symmetry at 1, $x - intercept\ at\ 3$ and maximum value of 32
	a. -3 | b. 32 | c. 16 | d. 30
Q4	The side of a square exceeds the side of another square by 4 cm and the sum of the areas of the two squares is 400. Find the perimeter of the larger square.
	a. 40 | b. 48 | c. 52 | d. 64
Q5	$y = 0.04x^2 - 2x + 513$ is equation of a parabola. Find the distance between the vertex of the parabola and the point $(25, 500)$
	a. 11 | b. 12 | c. 13 | d. 14

Answers

Q1	c.	Q2	d.	Q3	d.	Q4	d.	Q5	b.

Quadratic Equations in Factored Form

Standard form of the equation $y = ax^2 + bx + c$ can be written in factored form by using factoring skills.

- A quadratic relation in factored form is $y = a(x - r)(x - s)$
- r and s are the $x - intercepts$ or zeros of this equation
- Axis of symmetry is the average of the $x - intercepts$.
- $Axis, x = \frac{r+s}{2}$

Write Quadratic Relation in Factored Form from Graph

Example 7: Write Equation from the Graph in the form $y = a(x - r)(x - s)$

Step #1:

Read $x - intercepts$

They are at $x = \mathbf{-4}$, $\text{and } x = \mathbf{2}$

$r = -4, s = 2$

$y = a(x + 4)(x - 2)$

Step#2:

To find the value of a

use $y - intercept, (0, -8)$

$-8 = a(0 + 4)(0 - 2)$

$-8 = -8a$

$a = 1$

Therefore, the equation is: $\boldsymbol{y = (x + 4)(x - 2)}$

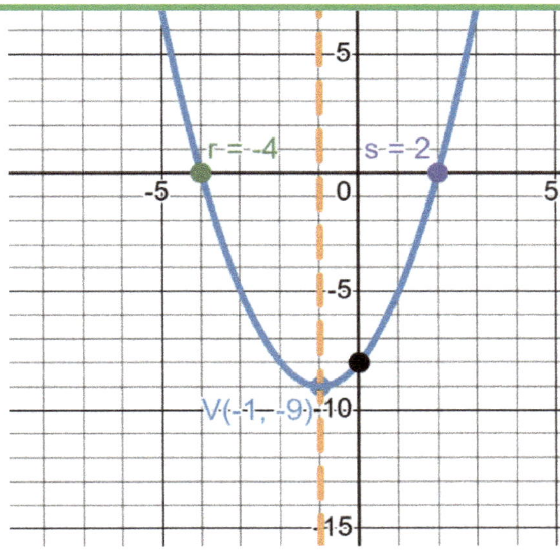

Practice: Write Equation of following graphs.

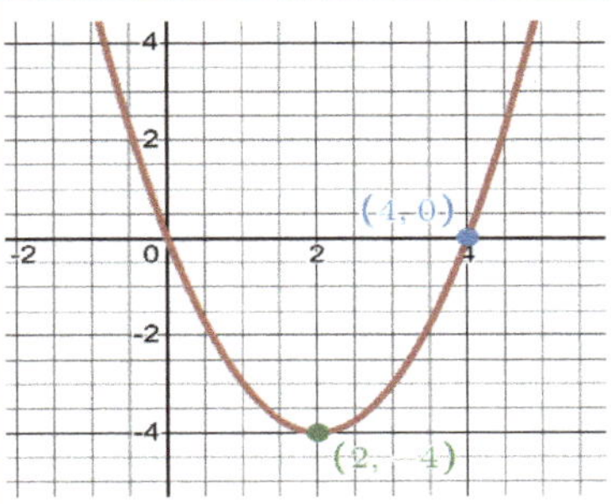

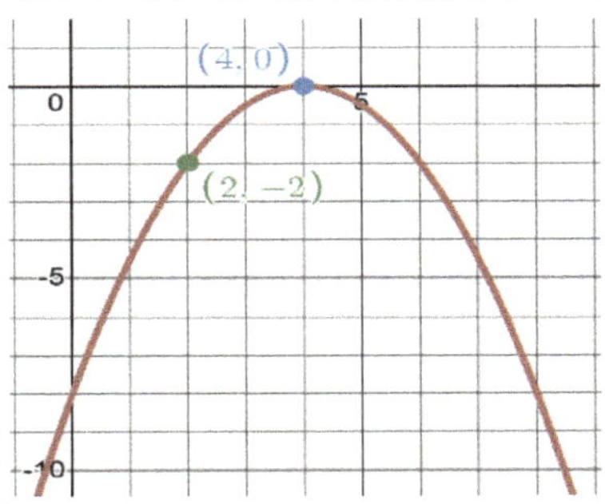

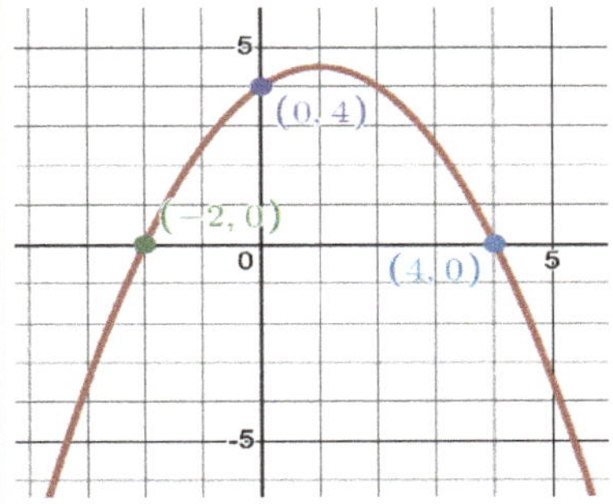

Understand Factored Form: $y = a(x - r)(x - s)$

Example 8: Consider quadratic relation $y = -0.5(x + 5)(x - 3)$

Find key parameters (intercepts, axis of symmetry, optimum value, vertex coordinate) and then graph the parabola.

Write $x - intercepts$

$x - intercepts$ can be directly read from the equation.

$x - intercepts$ are at $x = -5$ and $x = 3$

Find $y - intercepts$

Substitute $x = 0$ to find $y - intercepts$

$y = -0.5(0 + 5)(0 - 3) = -0.5(5)(-3) = 7.5$

$y - intercepts$ is at 7.5

Find $axis\ of\ symmetry$

It is the average value of the roots

$$axis\ of\ symmetry = \frac{(-5) + (3)}{2} = -1$$

Find maximum value

Maximum will be on the axis, $x = -1$

$y = -0.5(-1 + 5)(-1 - 3) = -0.5(4)(-4) = 8$

Write vertex of the parabola

Vertex, $(-1, 8)$

Graph of the quadratic relation: $y = -0.5(x + 5)(x - 3)$

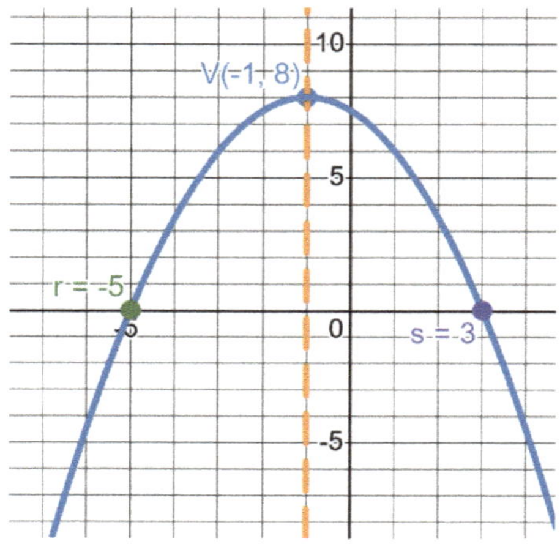

Practice: Sketch graph the given functions after finding all the key parameters.

a.	$y = (x + 4)(x - 2)$	b.	$y = -(2x + 1)(x - 2)$
c.	$y = -\frac{1}{2}(x + 1)(x + 3)$	d.	$y = 2x(4 - x)$

Application:

A ball kicked off the ground goes just over a one-meter fence at a distance of 10 m then drops 5 m further away. Model the situation in the form $h = a(x - r)(x - s)$.

Application Thinking and Communication Problems

Applications on Quadratic Function

Example 9: Area of Plot

A factory is to be built on a lot that measures 80 m by 60 m. A lawn of uniform width, equal to the area of the factory must surround it. How wide is the strip of the lawn, and what are the dimensions of the factory.

Solution 9: Area of Plot

Consider the figure on the right

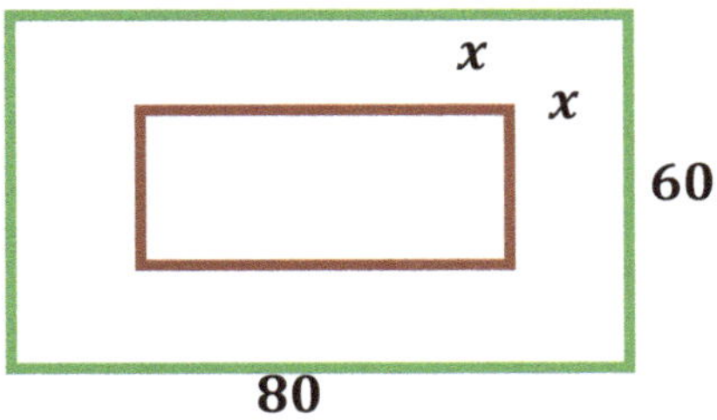

Lot that measures 80 m by 60 m

Let width of the strip be x from each side

Area of the factory = $(80-2x)(60-2x), x<30$

Total area of the lot = $(80)(60)$

Relating area: Area of the lawn = Area of the factor

$(80)(60)-(80-2x)(60-2x)=(80-2x)(60-2x)$

Dividing by 4 each term

$(40)(30)-(40-x)(30-x)=(40-x)(30-x)$

$(40)(30)=2(40-x)(30-x)$

$(20)(30)=(40-x)(30-x)$

$600=1200-40x-30x+x^2$

$x^2-70x+1200-600=0$

$x^2-70x+600=0$

$(x-10)(x-60)=0$

$x=10$

Uniform width of the strip is 10 m

Dimensions of the factor $\mathbf{(80-20)\ by\ (60-20)=60m\ by\ 40\ m}$

Practice Review Problem

Sum of area od two squares is $400\ m^2$ and the difference of their perimeter is $16\ m$. Find the ratio of the side lengths of the two squares.	4/3
If $f(x)=x^2+5x+3k$ and $f(k)=-16$, find $f(3)$.	12
If $f(5a-1)=15a+1$, find an expression for $f(x)$.	$f(x)=3x+4$

Example 10: Rate of Filling the Tank

Two water taps together can fill a tank in 6 hours. The tap of larger diameter takes 9 hours less than the smaller one to fill the tank separately. Find the time in which each tap can fill the tank.

Let the time for smaller tap be x

Time for larger will be $x - 9$

Rate of filling together

$\frac{1}{x} + \frac{1}{x-9} = \frac{1}{6}, x > 9$

Solve by multiplying by LCM

$6(x-9) + 6x = x(x-9)$

$6x - 54 + 6x = x^2 - 9x$

$x^2 - 9x - 12x + 54 = 0$

$x^2 - 21x + 54 = 0$

$(x-3)(x-18) = 0$

$x = 3, x = 18$

$x = 18$ is the only valid solution

time for smaller tap be 18 hours

time for larger tap be 9 hours

Example 11: Contribution for Gift

A group of students sent flowers costing $20 to sick classmate. There were four fewer students contributing than was planned, requiring each of other to give 25 cents more. How many students planned for the gift?

Solution 11: Contribution for Gift

Let the students planned be

contribution by each= $\frac{20}{x-4}$

contribution planned= $\frac{20}{x}$

extra paid $= 0.25 = \frac{1}{4}$

$\frac{20}{x-4} = \frac{20}{x} + 0.25$

$\frac{20}{x-4} = \frac{20}{x} + \frac{1}{4}$

Multiply by LCM

$80x = 80(x-4) + x(x-4)$

$80x = 80x - 320 + x^2 - 4x$

$x^2 + 80x - 320 - 4x - 80x = 0$

$x^2 - 4x - 320 = 0$

$(x-20)(x+16) = 0$

$x = 20$

20 students planned for the gift.

✍ Practice Application Problem

Tina bought some T-shirts for $80. If she had bought 4 more T-shirts fo the same amount, each T-shirt would have costed $1 less. Find the number of T-shirts she bought. 16

Find the point of tangency of the line with the slope equal to 4 to the parabola $y = 0.5 + 2x - 8$ $(2, -2)$

Example 12: Difference Between X-Intercepts and Zeros in the Graph of a Parabola

Understand Zeros of a Function

The graph of a quadratic function is a parabola. Let us examine three distinct cases:

- **Case I:** Two x-intercepts
- **Case II:** One x-intercept
- **Case III:** No x-intercept

Case I	Case II	Case III
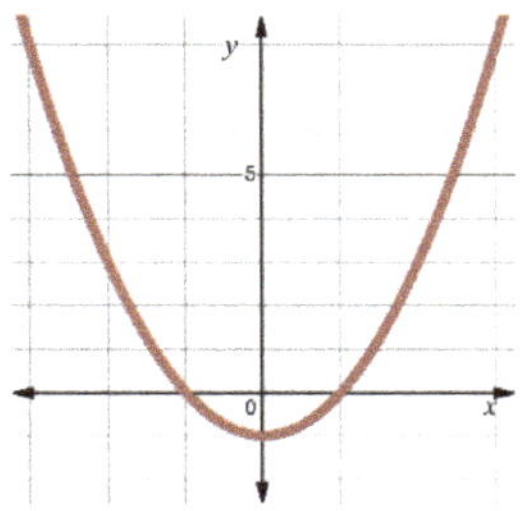	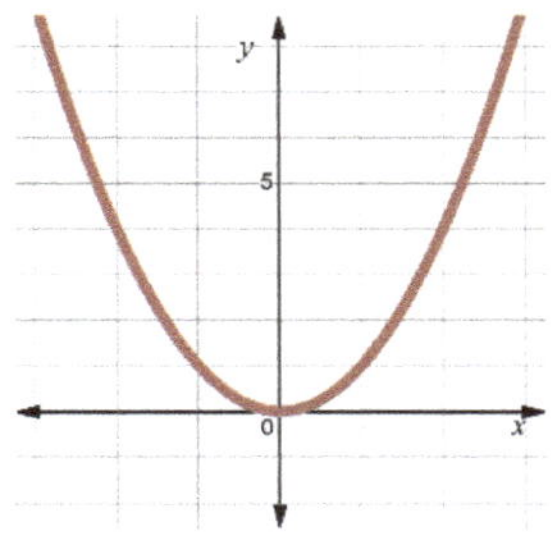	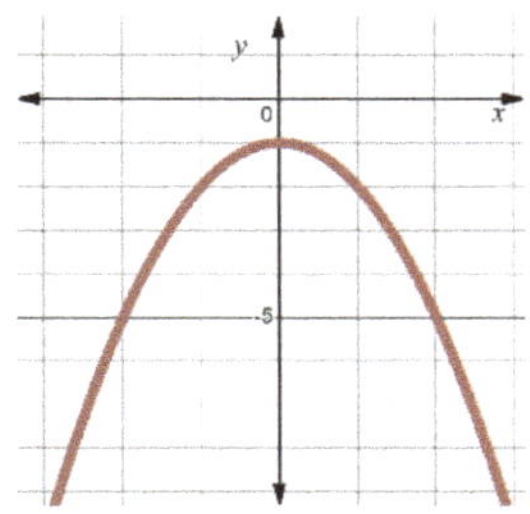

X-intercepts are the points where the graph crosses or touches the x-axis. At these points, the value of the function is zero. For real numbers, these intercepts also represent the *real solutions* of the corresponding quadratic equation.

However, **zeros** (or roots) refer to the values of x that make the function equal to zero—these can be real or complex.

Notice the difference in the curve's behaviour at the x-intercepts:

- In **Case I**, the curve *crosses* the x-axis, like a straight line.
- In **Case II**, the curve *touches* the x-axis and turns back, without crossing.

This difference is due to the **multiplicity** (order) of the zeros:

- If a zero has multiplicity **1**, the graph crosses the x-axis at that point.
- If a zero has multiplicity **2**, the graph only touches the x-axis and turns back.

For a degree-two polynomial, there will always be **two zeros** in total. These may be:

- Two distinct real zeros (Case I)
- A repeated real zero (Case II)
- Two complex (imaginary) zeros with no real intercept (Case III)

Zero decides behaviour at x-intercept and can also be imaginary.

Completing the Square Method with Optimization Applications

This section introduces the **Completing the Square method** not just as an algebraic tool to solve quadratic equations but as a powerful strategy for optimization problems, such as maximum area, minimum cost, and maximum profit. It bridges algebra and real-world problem-solving—an essential leap toward mastering quadratic modeling.

Understanding the Method

Completing the square is a method used to rewrite a quadratic expression in the form:

$$y = a(x-h)^2 + k$$

- This is vertex form of quadratic equation where the vertex is at (h, k).
- If $a > 0$ then the parabola opens and has minimum value of k at $x = h$
- If $a < 0$ then the parabola opens and has maximum value of k at $x = h$
- Range (all possible $y - values$): $y \geq k$ if $a > 0$ or $y \leq k$ if $a < 0$.

Vertex form is useful for directly reading optimum (maximum or minimum) value.

Knowledge and Communication Question

Parabola $y = 2(x-2)^2 - 4$ is shown on the right.

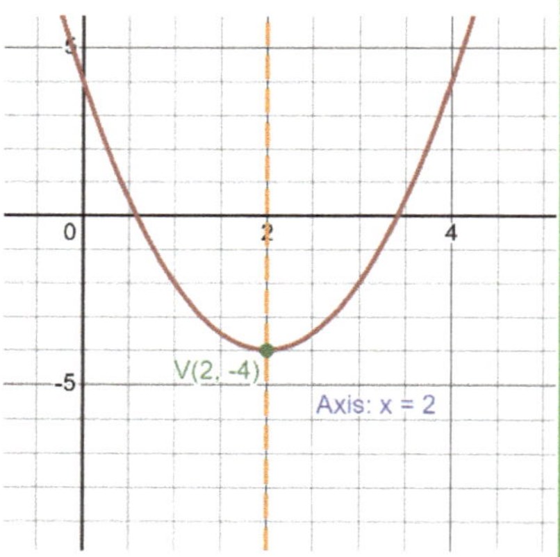

Find

- Vertex
- Axis of symmetry
- $x - intercept$
- $y - intercepts$
- Range of relation
- Explain how the given graph is related with $y = 2x^2$
  Hint: Steps $2(1, 3, 5, 7, \dots)$

Communication and Thinking

Explain the differences between zeros and x-intercepts.

Standard Steps to Complete the Square

Let's complete the square and write the quadratic relation

$y = 2x^2 + 12x + 5$ as $y = a(x-h)^2 + k$ for

Step 1: Group the first two terms and factor.

$y = 2[x^2 + 6x] + 5$

Step 2: Add and subtract $\left(\frac{6}{2}\right)^2$

$y = 2\left[x^2 + 6x + \left(\frac{6}{2}\right)^2 - \left(\frac{6}{2}\right)^2\right] + 5$

Step 3: First three terms form a perfect square

$y = 2[(x+3)^2 - 9] + 5$

Step 4: Open the brackets and simplify

$y = 2(x+3)^2 - 18 + 5$

$\boldsymbol{a = 2, h = -3, k = -13}$

Final Form:

Vertex is $(\mathbf{-3, -13})$

$y = 2(x+3)^2 - 13$

Practice Set 10

Given: $2x^2 - 8x + 1 = a(x-h)^2 + k$

a.	Find the value of $a, h\ and\ k$
b.	Write axis of symmetry
c.	Find the coordinates of the vertex
d.	Find intercepts of the parabola.
e.	Sketch the graph of the quadratic relation

Communication and Thinking

Explain why $y = 2(x-3)^2 + 1$ does not have any real roots in two different ways.

Complete the Square and Derivation of Quadratic Formula

Let's complete the square and write the quadratic relation

$y = ax^2 + bx + c$ as $y = a(x-h)^2 + k$ for:

Step 1: Group the first two terms and factor.

$$y = a\left[x^2 + \frac{b}{a}x\right] + c$$

Step 2: Add and subtract $\left(\frac{b}{2}\right)^2$

$$y = a\left[x^2 + \frac{b}{a}x + \left(\frac{b}{2a}\right)^2 - \left(\frac{b}{2a}\right)^2\right] + c$$

Step 3: First three terms form a perfect square

$$y = a\left[\left(x + \frac{b}{2a}\right)^2 - \frac{b^2}{4a^2}\right] + c$$

Step 4: Open the brackets and simplify

$$y = a\left(x + \frac{b}{2a}\right)^2 - \frac{b^2}{4a} + c$$

$$\Rightarrow y = a\left(x + \frac{b}{2a}\right)^2 - \frac{b^2 - 4ac}{4a}$$

Vertex is

$$\left(-\frac{\boldsymbol{b}}{\boldsymbol{2a}}, -\frac{b^2 - 4ac}{4a}\right)$$

Solve Equation: $\boldsymbol{y = 0}$

$$0 = a\left(x + \frac{b}{2a}\right)^2 - \frac{b^2 - 4ac}{4a} \Rightarrow a\left(x + \frac{b}{2a}\right)^2 = \frac{b^2 - 4ac}{4a} \Rightarrow \left(x + \frac{b}{2a}\right)^2 = \frac{b^2 - 4ac}{4a^2}$$

$$\Rightarrow x + \frac{b}{2a} = \pm\sqrt{\frac{b^2 - 4ac}{4a^2}} \Rightarrow x = -\frac{b}{2a} \pm \sqrt{\frac{b^2 - 4ac}{4a^2}}$$

$$x = \frac{-b \pm \sqrt{b^2 - 4ac}}{2a}$$

Quadratic Formula Mystery Solved

Determine the equation of the parabola, in standard form, with roots as

$$\frac{-9 \pm 3\sqrt{5}}{2}$$

$y = x^2 + 9x + 9$

How to Solve a Quadratic Equation Given in Vertex Form

A quadratic in vertex form looks like: $y = a(x-h)^2 + k$
To solve the equation $a(x-h)^2 + k = 0$ (i.e., find x when y = 0):

Step-by-Step Method

$a(x-h)^2 = -k$	Isolate the squared part
$(x-h)^2 = -k/a$	Divide by a (assuming a ≠ 0):
	Square-root both sides (remember ±)
$x - h = \pm\sqrt{\frac{-k}{a}}$	Solve for x
$x = h \pm \sqrt{\frac{-k}{a}}$	**Interpret the result (real vs complex)** If $-k/a$ > 0 → two distinct real solutions If $-k/a$ = 0 → one real solution (double root) at x = h If $-k/a$ < 0 → no real solutions (complex roots)

Example 1 — Two Real Solutions	Example 2 — One Double Root
Solve: $(x-3)^2 - 16 = 0$	Solve: $5(x+2)^2 = 0$
$(x-3)^2 = 16$	$(x+2)^2 = 0$
$x - 3 = \pm\sqrt{16} = \pm 4$	$x + 2 = 0$
$x = 3 \pm 4 \Rightarrow x = -1 \; and \; x = 7$	$x = -2$

✍ **Practice: Solve the equations using the method of completing the squares:**

a.	$x^2 + 8x + 15 = 0$	b.	$2x^2 + 3x - 20 = 0$
c.	$\frac{1}{2}x^2 - 5x + 8 = 0$	d.	$-5x^2 + 20x = 0$
e.	$x^2 + 2x + 3 = 0$	f.	$x^2 + 10x + 25 = 0$
g.	$2x^2 - 16x = 5$	h.	$x^2 - 3x = 3x - 14$

Answers:

a.	$-5, -3$	b.	$\frac{5}{2}, -4$	c.	$8, 2$	d.	$0, 4$
e.	No real roots	f.	-5	g.	$4 \pm \sqrt{\frac{37}{2}}$	h.	No real roots

Why Completing the Squares Matter in Optimization

Quadratic functions model many real-world situations where something must be maximized or minimized.

Examples:

- Maximizing area with limited fencing
- Minimizing cost for packaging
- Maximizing profit for selling items

The vertex of a parabola gives us the optimal value.

Real-World Optimization Examples

Example 12: Maximum Area Problem

A farmer has 100 meters of fencing to enclose a rectangular field against a barn (no fence on one side). What dimensions give the maximum area?

Let width be x and length be y

Perimeter $= 2x + y = 100 \Longrightarrow y = 100 - 2x$

Area: $\boldsymbol{A = x(100 - 2x)} = -2x^2 + 100x$

Complete the square:

$A = -2[x^2 - 50x]$

$A = -2[x^2 - 50x + 25^2 - 25^2]$

$A = -2[(x - 25)^2 - 625]$

$A = 2(x - 25)^2 - 1250$

Maximum area = 1250 m² when $x = 25$

> **Extended Problem**
>
> Solve the same problem with a partition in the rectangular field.

> **Alternate Solution**
>
> Area: $A = x(100 - 2x)$ axis will be midway between the roots $0 \; and \; 50$
>
> Axis is at $x = \frac{0+50}{2} = 25$ Now find the area at $x = 25$
>
> $A = 25(100 - 2(25)) = 1250$

Example 14: Profit Optimization

A company sells a product for \$50 and sells 100 units. For every \$1 decrease in price, 5 more units are sold. What price maximizes revenue?

- Let $x =$ number of \$1 reductions
- Price $= 50 - x$
- Units sold $= 100 + 5x$
- Revenue = Price × Units Sold

Revenue:

$R = (50 - x)(100 + 5x)$

$R = 5000 + 250x - 100x - 5x^2 = -5x^2 + 150x + 5000$

Complete the square:

$R = -5x^2 + 150x + 5000 = -5[x^2 - 30x] + 5000$

$R = -5[x^2 - 30x + 15^2 - 15^2] + 5000$

$R = -5[(x - 15)^2 - 225] + 5000$

$R = -5(x - 15)^2 + 1125 + 5000 =$

$R - 5(x - 15)^2 + 6125$

Maximum revenue: **\$6125** when $x = 15$
Price: $50 - 15 = 35$
Units sold: $\mathbf{100 + 5(15) = 175}$

Alternate Solution

$R = (50 - x)(100 + 5x)$ axis will be midway between the roots 0 *and* 50

Axis is at $x = \frac{50+(-20)}{2} = 15$ Now find the revenue at $x = 15$

$R = (50 - 15)(100 + 5(15)) = 6125$

Explore

Determine the condition when the following equations will have both common roots.

$\boldsymbol{a_1x^2 + b_1x + c_1}$ and $\boldsymbol{a_2x^2 + b_2x + c_2}$

Partial Factoring: Discover a New Skills

When optimizing quadratic functions, most students are taught to **complete the square**.
But there's a **faster and more intuitive method** — especially when factoring fully is difficult:

Partial Factoring

Instead of factoring the entire quadratic, we factor **only the first two terms** and leave the constant untouched.

A quadratic in **partial factored form** looks like:

$y = ax(x - m) + n$

at $x = 0$ and $x = m, the\ value\ of\ y = n$

This symmetry tells us the **axis of symmetry** is halfway between 0 and m:

This indicated $axis\ of\ symmetry$ is midway between 0 and m or

$axis\ of\ symmetry\ is = \dfrac{0 + m}{2} = \dfrac{m}{2}$

Since the **vertex lies on the axis of symmetry**, substitute $x = \frac{m}{2}$ back into the equation to find the **y-coordinate** of the vertex.

Illustrative Example 15: Equation from Graph

In the given graph $x - intercepts$ and $vertex$ are easy to read. However, you notice that the $y - intercept$ along the horizontal line crosses the graph at $x = 6$. The $y - value$ is 4 all along the horizontal line. Equation in partial factored form can be easily written as $\boldsymbol{y = ax(x - 6) + 4}$

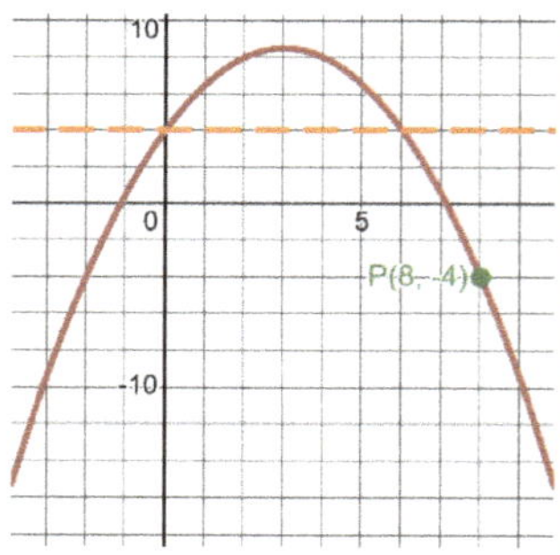

To find the value of a we can use a point $P(8, -4)$

$-4 = 8a(8 - 6) + 4 \Rightarrow -4 = 16ax + 4$

$16a = -8 \Rightarrow a = -\dfrac{8}{16} = -1/2$

$\boldsymbol{y = -\frac{1}{2}x(x - 6) + 4}$

✍ **Student Exercise:** Convert above equation to Standard Form

Optimization Application: Partial Factoring

Example 16:

A flair is fired from the deck of a ship. The height of the flair, h metres, above the water after t seconds is given by the function $h = -4.9t^2 + 98t + 8$. Determine the maximum height of the flair.

Solution 16:

Partial Factoring involves factoring common terms from the first two terms.

$h = -4.9t^2 + 98t + 8$

$h = -4.9t(t - 20) + 8$

for $t = 0$ and $t = 20$ height is 8

axis is midway between these values

axis of symmetry,

$$t = \frac{0 + 20}{2} = 10$$

Maximum will be the vertex which is always on the axis of symmetry.

Substitute: $t = 10$

to calculate the maximum height

$h = -4.9(10)^2 + 98(10) + 8$

$h = -490 + 980 + 8 = 498$

maximum height of the flair is 498 m.

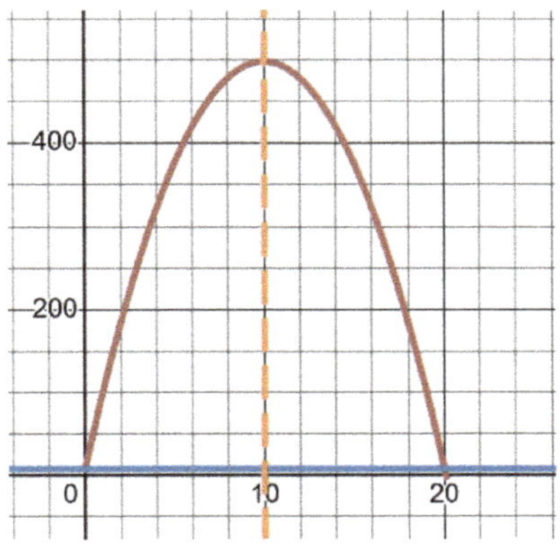

Key Concept

If two x-values give the same y-value, the axis of symmetry lies right in the middle

Practice Problem 11

a.	Find the coordinates of vertex of the equation $y = -5x(x - 16) + 12$
b.	A parabola passes through the points $(0, -5)$ and $(8, -5)$. If the vertex is $(a, 10)$ find the possible values of a.
c.	Find range of $y = -0.5x^2 + 3x - 1$
d.	Find vertex with partial factoring: $h = -5x^2 + 20x + 60$

Example 17: Maximum Profit

Profit, P, of a cosmetic company, in thousands of dollars, is given by the equation $P = -5x^2 + 600x - 1810$ *where* x is the amount spent in advertising in thousands of dollars. Determine the maximum profit the company can make.

Solution 17:

Partial Factoring involves factoring common terms from the first two terms.

$P = -5x^2 + 600x - 1810$

$P = -5x(x - 120x) - 1810$

for $x = 0$ and $x = 120$ profit is 1810

axis is midway between these values

axis $x = \frac{0+120}{2} = 60$

Maximum will be the vertex on the axis of symmetry, $x = 60$

Now calculate the maximum height

$P = -5(60)^2 + 600(60) - 1810$

$P = -5(60)^2 + 600(60) - 1810$

$P = 16190$

Maximum profit of the company can be $16190 million.

Practice Applications

Q1	A football is thrown in the air. Its height h in metres above the ground at any instant t seconds after the throw is approximately given by $$h = -4.9t^2 + 14.7t + 2, t \geq 0$$ Find the maximum height of the ball using three different methods. 13 m
Q2	Find two positive numbers with a product 200 such that the sum of one number and twice the second number is as small as possible. 20, 10
Q3	Nancy owns a hardware store. For every increase of 15 cents in the price of a package of batteries, she estimates that sales will decrease by 20 packages per day. The store normally sells 840 packages a day at $6.00 each. Find maximum revenue $5043
Q4	A farmer wants to create rectangular pen for cows to graze. He has 1500 m of fence and wants to have the pen border on a creek so that the cows will have water and he will have to fence only three sides. What is the maximum area that can be enclosed? 281250 sq. m

Linear and Quadratic Systems

Example 1

A parabola is given by $y = 2x - x^2$.

A line is given by $y = mx + 1$

What value(s) of the **slope** m would make the line **tangent** to the parabola?

Concept:

- A **line is tangent** to a parabola if it **touches it at exactly one point**.
- This happens when the system of equations has **exactly one solution**.
- For a quadratic equation, this means the **discriminant (Δ)** must be **zero**.

Step-by-Step Solution 1:

Equate equations equal to each other

We want points where the **line intersects the parabola**, so set:

$2x - x^2 = mx + 1$

$x^2 + mx - 2x + 1 = 0$

$x^2 + (m - 2)x + 1 = 0$

discriminant (Δ) must be **zero**

$a = 1, b = m - 2, c = 1$

$b^2 - 4ac = 0$

$(m - 2)^2 - 4(1)(1) = 0$

Solve the discriminant equation

$(m - 2)^2 = 4$

$m - 2 = \pm\sqrt{4} = \pm 2$

$m - 2 = 2$	$m - 2 = -2$
$m = 2 + 2 = 4$	$m = -2 + 2 = 0$

For $m = 4$ and $m = 0$ the line will be tangent to the parabola.

✍ Practice Problem 1

The line $y = 5 - kx$ is tangent to the parabola $y = 3x^2 + 4x - 2$. Find the integer value of k.

✍ Practice Problem 2

The line $x + ky - 5 = 0$ is tangent to the circle $x^2 + y^2 = 5$. Find the value of k.

Example 2

The graph of parabola $y = ax^2 + bx + 1, a \neq 0$ pass through point $P(1,2)$. For what values of a does the graph intersect the $x - axis$ at two distinct points.

Concept:

- Equation of $x - axis$ is $y = 0$
- Discriminant of $ax^2 + bx + 1 = 0$ should be greater than zero for two distinct roots.
- Parabola pass through $(1,2)$ will provide equation in one variable.
- Use quadratic formula to find zeros and then apply the concept of away from zeros for greater than inequality.

Step-by-Step Solution 2:

Write equation in one variable	**Strategy**
Substitute $x = 1$ *and* $y = 2$ in $y = ax^2 + bx + 1, P(1,2)$ $2 = a(1)^2 + b(1) + 1$ $2 = a + b + 1$ $b = 2 - 1 - a = 1 - a$	Find solution for the zeros and then get the inequality solution as away from those zeros. (Refer to quadratic inequalities discussed earlier)
substitute $b = 1 - a$ $y = ax^2 + (1 - a)x + 1$ For two distinct $x - intercepts$ **discriminant (Δ)** must be **>0** $a = 1, b = 1 - a, c = 1$ $b^2 - 4ac = 0$ $(1 - a)^2 - 4(a)(1) > 0$ **Solve the discriminant equation** $1 - 2a + a^2 - 4a > 0$ $a^2 - 6a + 1 > 0$	Use quadratic formula to find a and solve for $a^2 - 6a + 1 = 0$ $a = \frac{6 \pm \sqrt{36 - 4(1)(1)}}{2} = \frac{6 \pm \sqrt{32}}{2}$ $a = \frac{6 \pm 4\sqrt{2}}{2} = 3 \pm 2\sqrt{2}$ Now, solve for inequality $(1 - a)^2 - 4(a)(1) > 0$ $a < 3 - 2\sqrt{2}$ or $a > 3 + 2\sqrt{2}$ (away from the zeros)

Practice Problem 3

The graph of parabola $y = ax^2 + bx + 2, a \neq 0$ pass through point $P(1,3)$. For what values of a does the graph intersect the $x - axis$ at two distinct points.

$a < 5 - 2\sqrt{6}$ or $a > 5 + 2\sqrt{6}$

Strategy: Writing the Equation of a Parabola from a Graph

To determine the equation of a parabola from its graph, **choose the right form based on the key points provided**:

Factored Form

Use when the **x-intercepts** and **one additional point** are given.
Ideal for solving when roots (zeros) are known.

Vertex Form

Use when the **vertex** and **one additional point** are given.
Best when the turning point of the parabola is clear.

Partial Factoring

Use when the **y-intercept** and its **image point** (symmetric across the axis) are given, along with **one more point**.
Quick method when full factoring is hard or unnecessary.

Example 1: Equation of Graph

Case I: Factored Form

General Form: $y = a(x - r)(x - s)$

$x - intercepts$ at $x = r, x = s$

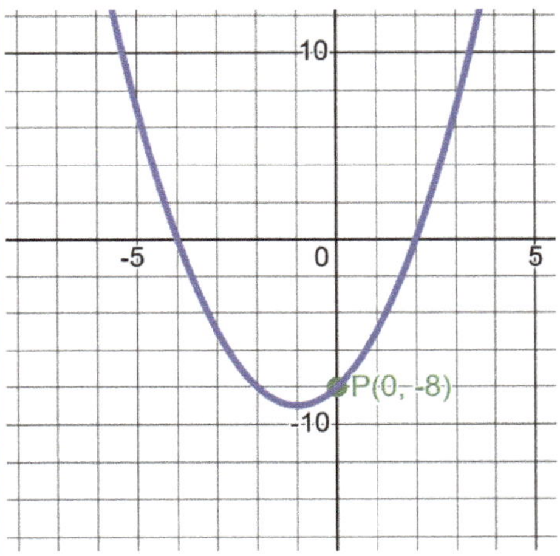

Easy to read $x - intercepts$

$x - intercepts$ at $x = -4$ and 2

Additional point $P(0, -8)$

$y = a(x + 4)(x - 2)$

Use point $P(0, -8)$ to find a

$-8 = a(0 + 4)(0 - 2)$

$-8 = -8a$

$a = 1$

Equation of Parabola

$\boldsymbol{y = (x + 4)(x - 2)}$

Note:

Since the $\boldsymbol{x - intercepts}$ were easy to read from the graph, use factored form.

Case II: Vertex Form

General Form: $y = a(x-p)^2 + q$

$Vertex$ at (p, q)

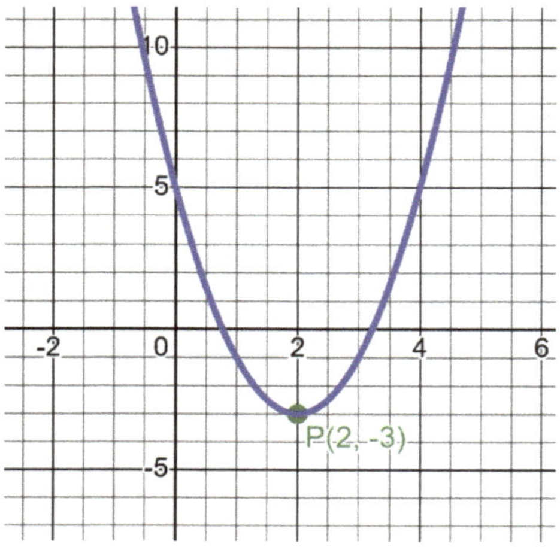

Easy to read: $Vertex$

$Vertex(2, -3)$

Additional point $P(0, 5)$

$y = a(x-2)^2 - 3$

Use point $P(0, 5)$ to find a

$5 = a(0-2)^2 - 3$

$5 = 4a - 3$

$4a = 5 + 3$

$a = \frac{8}{4} = 2$

Equation of Parabola

$$\boldsymbol{y = 2(x-2)^2 - 3}$$

Case III: Vertex Form

General Form: $y = ax(x-m) + n$

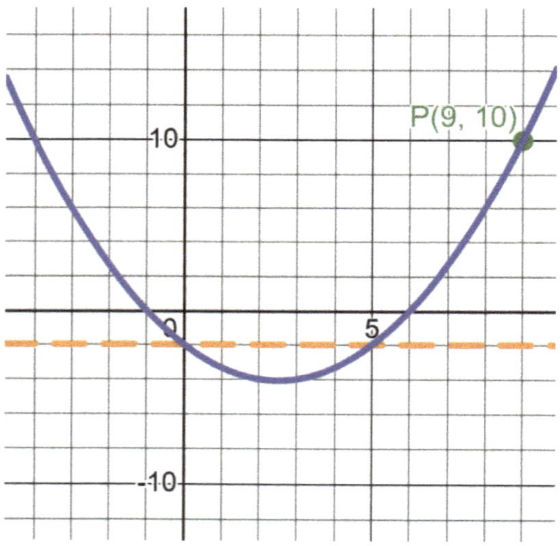

Easy to read $x - intercept$

$y - intercept\ and\ image$: $(0, -1), (5, -1)$

Additional point $P(9, 10)$

$y = ax(x-5) - 1$

Use point $P(9, 10)$ to find a

$10 = 9a(9-5) - 1$

$10 + 1 = 9(4)a$

$\frac{11}{36} = a$

$\boldsymbol{y = \frac{11}{36}x(x-5) - 1}$

Convert this to standard form.

Note: *Vertex coordinates are easy to read from the graph, use Vertex form.*

Communication and Thinking

The graph of $y = 2x^2 + 6x + 1$ is reflected over the line $y = -1$. How many points of intersection will the two graphs have. Justify your answer.

How to Graph a Parabola

Different strategies are used to graph a parabola depending on the given form of the quadratic equation.

Graphing From Vertex Form: $y = a(x-p)^2 + q$

A quadratic in vertex form is written as: $y = a(x-p)^2 + q$

- **Vertex:** The parabola starts at (p, q)
- **Axis of Symmetry:** The graph is symmetric about the vertical line $x = p$.
- **Direction:** If $a > 0$, the parabola opens upward; if $a < 0$, it opens downward.

Step Pattern

From the vertex, the parabola grows outward according to the sequence $\boldsymbol{a\{1, 3, 5, 7, ..\}}$.

- 1st step: move **1 unit right/left** → go up/down by a.
- 2nd step: move **2 units** → go up/down by $3a$.
- 3rd step: move **3 units** → go up/down by $5a$.
- And so on.

This pattern allows us to plot one side of the parabola and then mirror the points across the axis of symmetry.

Coordinates of Points

On right side of the vertex, the coordinates can be generated as:

$(p, q), (p+1, q+a), (p+2, q+a+3a), (p+3, q+a+3a+5a), ...$

Reflect these points across the line $x = p$ to complete the graph.

Example 2: Graph Parabola: $y = 2(x-1)^2 - 8$

$\boldsymbol{y = 2(x-1)^2 - 8}$

Vertex: $(1, -8)$. Axis: $x = 1$. Opens up as $a > 0$

Steps: $\boldsymbol{2}\{1, 3, 5, 7, ..\}$ or $\{2, 6, 10, 14, ..\}$.

$(1, -8)$	$(2, -6)$	$(3, 0)$	$(4, 10)$

reflect the points across the axis

draw the parabola as shown on the right.

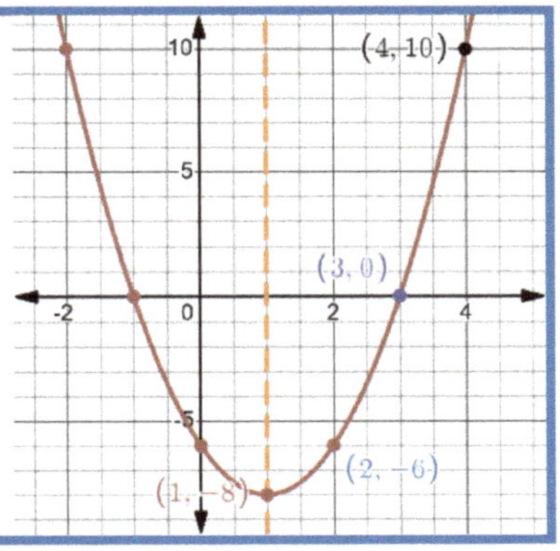

Graphing From Standard Form: $y = ax^2 + bx + c$

A quadratic in standard form is written as: $y = ax^2 + bx + c$

- **Axis of Symmetry:** The line of symmetry is $x = -\frac{b}{2a}$
- **Vertex (Extreme Value):** Substitute $x = -\frac{b}{2a}$ into the equation to find the corresponding y.
  This gives the maximum (if $a < 0$) or minimum (if $a > 0$) point.
- **Step Pattern:** Once the vertex is known, apply the same step pattern as before: $a\{1, 3, 5, 7, ..\}$ to plot additional points.
- **y-intercept:** Found by setting $x = 0$. In this form, the y-intercept is simply c.

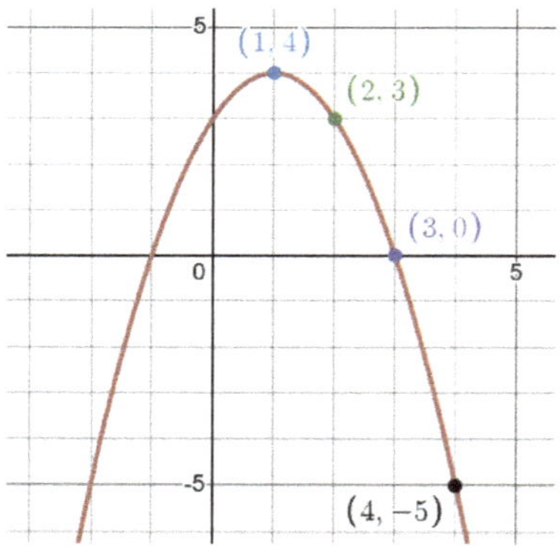

Example 3: Graph $y = -x^2 + 2x + 3$

- Axis of symmetry is $x = -\frac{2}{2(-1)} = 1$
- Extreme value $\mathbf{y = -(1)^2 + 2(1) + 3 = 4}$
- **Vertex:** $(1, 4)$
- **Opens down as** $a < 0$
- **Step Pattern:** $-\{1, 3, 5, 7, ..\} = \{-1, -3, -5, -7, ..\}$
- Points on the right of axis: $(1, 4), (2, 3), (3, 0), (4, -5)$
- Points on the left will be symmetric: $(1, 4), (0, 3), (-1, 0), (-2, -5)$
- $y - intercept, y = 3$

Graphing from Factored Form

A quadratic in factored form is written as: $y = a(x - r)(x - s)$

x-intercepts (roots): The parabola crosses the x-axis at $(r, 0) and\ (s, 0)$.

Axis of Symmetry: The parabola is symmetric about the vertical line through the midpoint of the roots: $x = \frac{r+s}{2}$

Vertex: Substitute the midpoint into the equation to find the vertex.

Direction: If $a > 0$, parabola opens upward; if $a < 0$, it opens downward.

y-intercept: Found by setting $x = 0$.

✍ **Practice Problem:** Graph $y = -2x^2 + 3x + 10$

Example 4: Write coordinate points and graph parabola: $y = (x-2)(x+4)$

Graph of $y = (x-2)(x+4)$

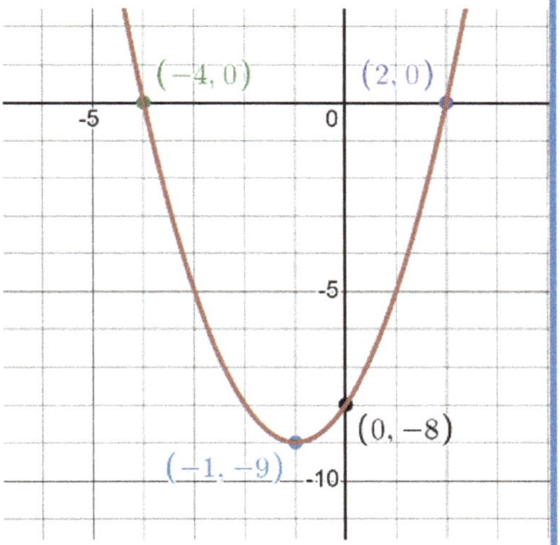

Key Points for the Graph are

x-intercepts (roots): $(2,0)$ *and* $(-4,0)$.

Opens upwards: coefficient of x^2 is positive

Axis of Symmetry: $x = \frac{2+(-4)}{2} = -1$

Minimum: $y = (-1-2)(-1+4) = -9$

Vertex: $(-1,-9)$

Direction: If $a > 0$, parabola opens upward

y-intercept: $y = (0-2)(0+4) = -8$

Draw a smooth curve that opens up passing through the intercepts and turning at the vertex $(-1,-9)$

Practice

Sketch graph of $y = (x-2)(x+4)$.

Show all the key points on the graph.

Tip: Use graphing calculator to check graph details.

Explore: Relate the Graph of Quadratic Functions

Sketch the graphs of the following quadratic relations on the **same coordinate axes**. Use the reference points $A(-1,1), B(0,0), C(1,1),$ *and* $D(2,4)$ from the graph of the **parent function** $y = x^2$.

Trace how these points are transformed in each new graph. Explain the relationship between the original and transformed graphs.

a.	$y = x^2$	Parent quadratic function
b.	$y = -\frac{1}{2}x^2$	Vertically compressed by factor of half and Reflected on the $x - axis$
c.	$y = -\frac{1}{2}(x+1)^2$	Same compression and reflection, plus a horizontal shift **left by 1 unit**.
d.	$y = -\frac{1}{2}(x+1)^2 + 2$	Same compression, reflection, and left shift, followed by a vertical shift **up by 2 units**

Graphing by Transformations

Another powerful method to sketch a parabola is by using the transformation of functions. We begin with the parent function: $y = x^2$. This is the basic parabola with vertex at (0,0), axis of symmetry $x = 0$, and opening upward. Every other quadratic graph can be obtained by applying transformations to this parent parabola.

Types of Transformations

1. Vertical Stretch or Compression:

Multiplying by a: $y = ax^2$.
- If $|a| > 1$, the parabola becomes narrower.
- If $0 < |a| < 1$, the parabola becomes wider.
- If $a < 0$, the parabola reflects downward.

2. Horizontal Shift:

Replacing x by (x - p): $y = (x - p)^2$.
- Shifts the parabola right by p (if $p > 0$) or left by $|p|$ (if $p < 0$).

3. Vertical Shift:

Adding q: $y = x^2 + q$.
- Moves the parabola up by q units if $q > 0$, or down if $q < 0$.

Transformed Image Points

When transformations are applied, the coordinates of key points are updated:
- Parent function points: $(-2,4),(-1,1),(0,0),(1,1),(2,4)$
- After transformation $y = a(x-p)^2 + q$:
  $(x,y), \rightarrow (x+p, ay+q)$

> **Note**: Transformations should be applied in order. The correct order is to take **stretches, compressions and reflections before translations**.

Writing Transformed Equation

Thus, we can take simple points from the parent parabola, apply the transformation rules, and directly plot the new image points.

Graph of $y = x^2$ is vertically stretched by a factor of 3 and translated 2 units left and 3 units down. Write its equation after transformation.

$$y = 3(x+2)^2 - 3$$

Example 5: Describe Transformation with related coordinates points then Graph

Consider function $\mathbf{y = 2(x-1)^2 + 3}$.

Transformation in order for $y = 2(x-1)^2 + 3$ are:

- Vertical stretch by a factor of 2. Translation of 1 unit right and 3 units up.

Steps to sketch the graph with transformation in steps:

- Start with $y = x^2$.
- Apply transformations step by step:
  - Multiply by 2 → vertical stretch (narrower parabola).
  - Shift right by 1 → move points 1 unit right.
  - Shift up by 3 → move points 3 units up.
- Vertex: (1,3).
- Transformed points:
  - From (0,0) → (1,3).
  - From (1,1) → (2,5).
  - From (2,4) → (3,11).
  - Reflect across axis x = 1.
- Draw smooth parabola through these points.

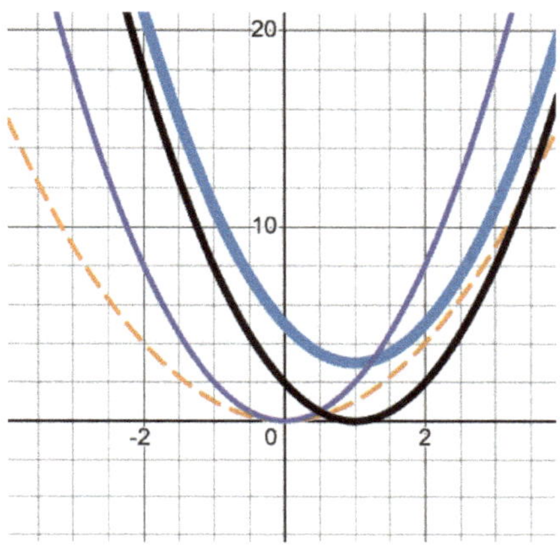

Summary

Table belows show the image of transformed point

After each transformation

x^2 (x, y)	$2x^2$ $(x, 2y)$	$2(x-1)^2$ $(x+1, 2y)$	$2(x-1)^2 + 3$ $(x+1, 2y+3)$
$(-2, 4)$	$(-2, 8)$	$(-1, 8)$	$(-1, 11)$
$(-1, 1)$	$(-1, 2)$	$(0, 2)$	$(0, 5)$
$(0, 0)$	$(0, 0)$	$(1, 0)$	$(1, 3)$
$(1, 1)$	$(1, 2)$	$(2, 2)$	$(2, 5)$
$(2, 4)$	$(2, 8)$	$(3, 8)$	$(3, 11)$

Note:

Transformation is applied in the correct order.

vertical stretches before translations

Q1	Prove that the maximum area of a rectangle with perimeter P will be $\frac{P^2}{16}$.
Q2	Find the range for the function $f(x) = 3x(x+6) - 10$
Q3	A ball is thrown: $h(t) = -5t^2 + 20t + 1$ (meters). When is the maximum height reached and what is it?
Q4	At \$20, a show sells 300 tickets. For each \$1 increase, 15 fewer tickets are sold. Show that the maximum revenue is \$6000.
Q5	You have 120 m of fencing to build a rectangular pen against a river (3 sides). Show that the maximize enclosed area will be $1800\ m^2$.
Q6	A farmer needs six identical pens for the farm animals. He has 1800 m of fencing. If the area enclosed is to be maximized, then what are the dimensions of each pen? $100\ m\ by\ 112.5\ m$
Q7	A store sells 200 compact disc players a week at \$350 each. A market survey indicates that for every \$10 rebate offered to the buyers, the number of sets sold will increase by 20 a week. Find the rebate that will maximize the revenue. $\$125$
Q8	Find the point on $y = 2x$ closest to $(6,0)$. Hint: $distance^2 = (x-6)^2 + (2x)^2$ $(1.2, 2.4)$
Q9	A rectangle's corner lies on the line $y = 12 - 3x$ in the first quadrant. Find maximum possible area if the other vertices are on the axes. Hint: $Area = x(12-3x)$ $12\ square\ units$
Q10	Find the point on the line $y = -x + 4$ that is closest to $(3,5)$. $(1,3)$

Challenge

Diagram shows a square inside a square with sides of 10 cm.

a. Find the area of the inside as a function of x.

b. Find the domain and range of this area.

SCAN ME

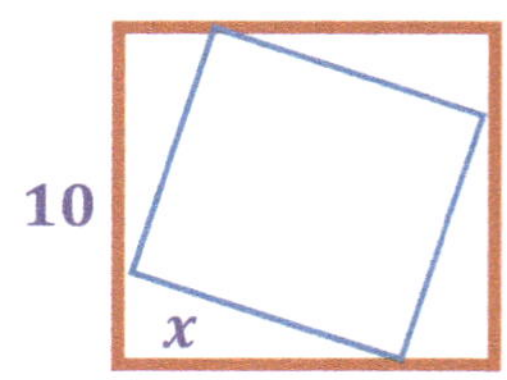

Challenge

Find the relation between x and y such that the point (x, y) is equidistant from the points $(7,1)$ and $(3,5)$.

Hint: Right bisector $y = x - 2$

Linearization of Curves

Turning curve from non-linear to linear of the form $Y = mX + c$

Many relationships in nature are not linear – they curve, bend, or flatten out. But by changing the way we look at data or by transforming variables, we can make a curved graph appear as a straight line. Once a relationship is linearized, it becomes easier to analyze and predict. The slope and intercept can still give valuable information, and it helps students recognize patterns hidden in nonlinear data.

Reciprocal Function $y = \frac{k}{x}$

As x increases y decreases.

y is inversely proportional to x

The graph can be linearized by transformation

$$Y = mX; Y = \frac{1}{y}, X = x, x > 0$$

Graph

$$\frac{1}{y}\ vs\ x, slope\ will\ be \frac{1}{k}$$

Example 1:

$$f(x) = \frac{4}{x}, x > 0$$

Transformed to

$$y = \frac{1}{f(x)} = \frac{1}{4}x$$

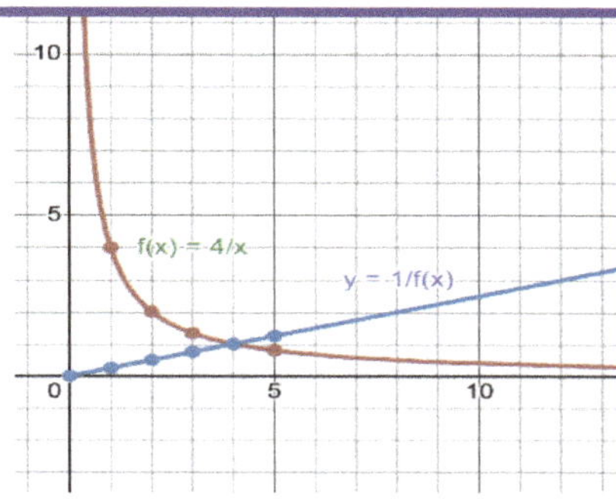

Linear graph with slope of $\frac{1}{4}$

x	1	2	3	4	5
$y = \frac{4}{x}$	4	2	4/3	1	4/5
$\frac{1}{y}$	$\frac{1}{4}$	$\frac{1}{2}$	$\frac{3}{4}$	1	$\frac{5}{4}$

Real Life Connection: Distance – Speed – Time relations

STEM Project: Communication and Thinking

Take an example of the type

$$y = \frac{c}{kx + d}$$

Provide transformation for linearization

Support wth table of values and graph

Hint: Consider reciprocal

$$\frac{1}{y} = \frac{kx + d}{c} = \frac{k}{c}x + \frac{d}{c}$$

Linearization of the Graph:

$$Y = mX + b$$

$$Y = \frac{1}{y}, m = \frac{k}{c}, b = \frac{d}{c}$$

Quadratic Function $y = ax^2$

Example 2: $y = 2x^2$

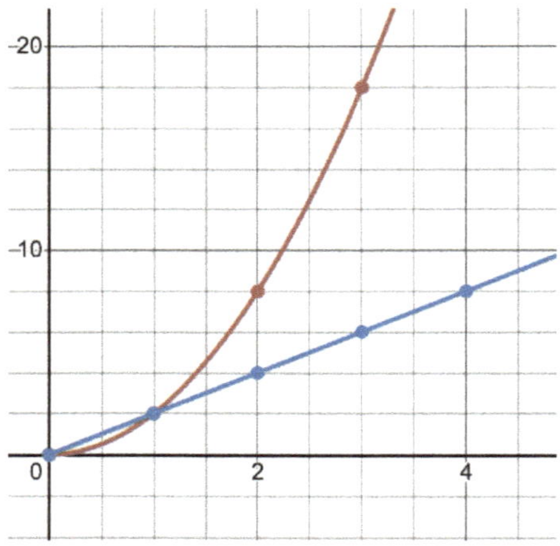

y is proportional to $square\ of\ x$

Transformation: $Y = mX + c$

$$Y = y, x = X, m = a$$

Graph $y\ vs\ x^2$ for Linearization

Graph $y\ vs\ x^2$

x	0	1	2	3	4
$y = 2x^2$	0	2	8	18	32
x^2	0	1	4	9	16

Square-Root Function $y = \sqrt{a}$

Example 3: $y = 2\sqrt{x}$

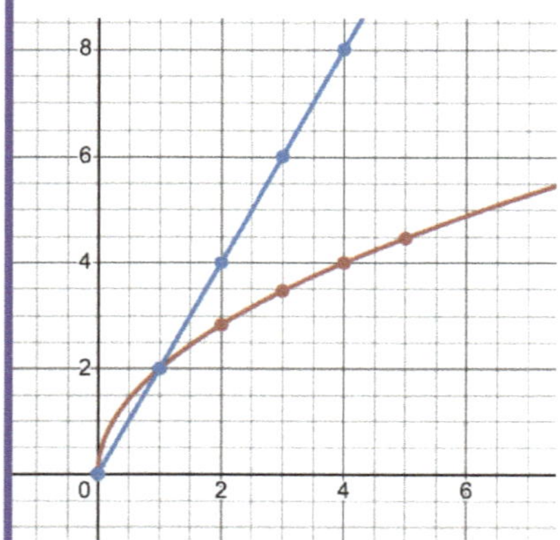

$The\ square\ of\ y$ is proportional to x

Transformation: $Y = mX + c$

$$Y = y^2, x = X, m = a^2$$

Graph $y^2\ vs\ x$ for Linearization

x	0	1	2	3	4
$y = 2\sqrt{x}$	0	2	$2\sqrt{2}$	$2\sqrt{3}$	4
y^2	0	4	8	12	16

Challenge Problem:

The variable x and y are related by the equation $ay = x + bx^2$. When a graph of y/x is plotted against x, the resulting line has a slope of 5 and an intercept on the $\frac{y}{x} - axis$ of 0.5. Calculate the values of a and b. Sketch the graph.

Watch Video **Linearization Playlist** **SCAN ME**

New Equation from Root Transformation of Quadratic Equations

When you know the roots of a quadratic equation, you can build a new equation based on changes to those roots.

Key Properties of Roots: The roots r_1 and r_2 of equation $x^2 + bx + c = 0$

- Sum of roots: $r_1 + r_2 = -b$
- Product of roots: $r_1 r_2 = c$

So, we can write the quadratic as $x^2 - (r_1 + r_2)x + r_1 r_2 = 0$

Example 1:

Find the equation whose roots are each 6 more than the roots of $x^2 + 8x - 1 = 0$

Solution 1:

Step 1: Identify original roots

Let r_1 and r_2 be the roots of the equation $x^2 + 8x - 1 = 0$

- Sum of roots: $r_1 + r_2 = -8$
- Product of roots: $r_1 r_2 = -1$

Step 2: Define new roots

New roots are $r_1 + 6$ and $r_2 + 6$

- Sum of roots: $r_1 + 6 + r_2 + 6 = r_1 + r_2 + 12 = -8 + 12 = 4$
- Product of roots: $(r_1 + 6)(r_2 + 6) = r_1 r_2 + 6r_1 + 6r_2 + 36$
  $= r_1 r_2 + (r_1 + r_2)6 + 36 = -1 + (-8)6 + 36 = -13$

Final Answer:

New quadratic equation:

$x^2 - (new\ sum)x + (new\ product) = 0$

$x^2 - 4x - 13 = 0$

✍ Practice Problems

Q1	Find the equation whose roots are each 5 more than of the roots of $2x^2 - 17x = -2$ $2x^2 - 37x + 137 = 0$
Q1	Find the equation whose roots are each 2 less than of the roots of $(x-1)^2 = 4$ $x^2 - 2x - 3 = 0$

Example 2:

Find the equation whose roots are reciprocal of the roots of $x^2 - 5x + 6 = 0$

Solution 2:

Step 1: Identify original roots

Let r_1 and r_2 be the roots of the equation $x^2 - 5x + 6 = 0$

- Sum of roots: $r_1 + r_2 = 5$
- Product of roots: $r_1 r_2 = 6$

Step 2: Define new roots

New roots are $\frac{1}{r_1}$ and $\frac{1}{r_2}$

- Sum of roots: $\frac{1}{r_1} + \frac{1}{r_2} = \frac{r_2 + r_1}{r_1 r_2} = \frac{5}{6}$
- Product of roots: $\frac{1}{r_1} + \frac{1}{r_2} = \frac{1}{r_1 r_2} = \frac{1}{6}$

Final Answer:

New quadratic equation:

$x^2 - (new\ sum)x + (new\ product) = 0$

$x^2 - \frac{5}{6}x + \frac{1}{6} = 0$

$6x^2 - 5x + 1 = 0$

Practice Problems

1	Find the equation whose roots are reciprocal of the roots of $x^2 + 4x + 11 = 0$ $11x^2 + 4x + 1 = 0$
2	Find the equation whose roots are reciprocal of the roots of $x^2 + 2x - 3 = 0$ $3x^2 - 2x - 1 = 0$
3	Find the equation whose roots are the squares of the reciprocal of the roots of parabola $x^2 + 6x - 2 = 0$ $4x^2 - 40x + 1 = 0$
4.	Find quadratic equation whose roots are the squares of roots of the parabola $4x^2 - 9x - 2 = 0$ $16x^2 - 97x + 4 = 0$

Quadratic Equation from Given Irrational Roots

When you're given two **irrational roots** of a quadratic equation — say r_1 and r_2 — you can use the **factored form** of a quadratic:

$y = a(x - r_1)(x - r_1) = a[x^2 - (r_1 + r_2)x + r_1 r_2]$

This gives the general form of the parabola, where:

- r_1 and $r_2$are the roots
- a is a constant determined using a known point on the parabola

Example 3

Find equation of parabola which pass through the point $(1, -1)$ with roots at $2 \pm \sqrt{3}$

Solution 3

Step 1: Use the roots to write the factored form

$r_1 = 2 + \sqrt{3}$ and $r_2 = 2 - \sqrt{3}$

$y = a[x^2 - (r_1 + r_2)x + r_1 r_2]$

$r_1 + r_2 = (2 + \sqrt{3}) + (2 - \sqrt{3}) = 4$

$r_1 r_2 = (2 + \sqrt{3})(2 - \sqrt{3}) = 4 - 3 = 1$

$y = a[x^2 - 4x + 1]$

Step 2: Use the point (1, −1) to find a

Substitute $x = 1, y = -1$ into the equation $y = a[x^2 - 4x + 1]$

$-1 = a[(1)^2 - 4(1) + 1]$

$-1 = -2a \Rightarrow a = \frac{1}{2}$

Step 3: Write the final equation

$y = \frac{1}{2}[x^2 - 4x + 1] \Longrightarrow y = \frac{1}{2}x^2 - 2x + \frac{1}{2}$

Challenge: Find the sum of the infinite series

$$\sqrt{12 + \sqrt{12 + \sqrt{12 + \sqrt{12 \ldots}}}}$$

NEW: Layered Difference Method for Quadratics

Discovering Quadratic Equations $ax^2 + bx + c$ from Data

The **Layered Difference Method** is a novel and efficient technique to determine the quadratic equation that fits a given sequence of data. This strategy is particularly useful in classroom learning, where students can visually see how quadratic patterns evolve. It builds upon the method of finite differences and introduces a layered approach to separate quadratic and linear components.

💡 **Core Idea:**

Use the **finite second difference** to quickly determine the **leading coefficient**, $\boldsymbol{a}$, then subtract the corresponding quadratic terms to isolate the **linear pattern** – and rebuild the full equation step by step i.e. $\boldsymbol{ax^2 + bx + c - ax^2 = bx + c}$

Step-by-Step Explanation

1. Identify the given sequence and calculate the first and second differences.
2. If the second difference is constant, divide it by 2 to find the coefficient 'a' of the quadratic term ax^2.
3. Subtract the quadratic component from each term of the sequence. This leaves behind a linear sequence.
4. Analyze this new linear sequence to determine the coefficients 'b' and 'c'.
5. Combine to form the complete quadratic function: $y = ax^2 + bx + c$

Instead of memorizing formulas, students **visualize the structure** of a function – turning patterns into precise algebra.

Worked Example

Given Sequence with first 5 terms: 0, 4, 10, 18, 28, ...

Step 1: Calculate differences
First differences: 4, 6, 8, 10
Second differences: 2, 2, 2

Step 2: Since second difference is 2, we divide by 2 → $a = \frac{2}{2} = 1$

Step 3: Subtract x^2 from each term to get the remaining linear sequence
Original: 0, 4, 10, 18, 28
x^2: 1, 4, 9, 16, 25
Subtract: -1, 0, 1, 2, 3,

Step 4: Remaining sequence is linear: -1, 0, 1, 2, 3, → 1, common difference

Linear Equation: $bx + c = x - 2$

Final Equation: Combination of quadratic and linear part

$$y = x^2 + x - 2$$

Why It Works

This method efficiently decomposes a quadratic sequence into its layered components: quadratic and linear. It's accessible for students and reveals the structure behind seemingly complex sequences, making it ideal for deeper understanding and classroom STEM exploration.

Watch Video **Layered Difference Method** **SCAN ME**

Example: Explore and Learn New Strategies

Consider the non-linear number pattern 7, 18, 35, 58, 87, 122, ...

a.	Show that the sequence is quadratic in nature
b.	Write the quadratic equation for the general term t_n.

Solution: Explore and Learn New Strategies

Let us start with finite difference of 7, 18, 35, 58, 87, 122, ...

Given Sequence	7, 18, 35, 58, 87, 122,	
1st difference	11, 17, 23, 29, 35	Quadratic Function, since 2nd finite difference is constant
2nd difference	6, 6, 6, 6	
Leading coefficient is half of finite difference	$t_n = at^2 + bt + c$ $t_n = 3t^2 + bt + c$	$a = \frac{6}{2} = 3$

To find the linear term, $bt + c$, let us subtract $3t^2$ from the terms.

$bt + c = t_n - 3t^2$

Increases by 2 each time: $b = 2. c = 2$

$$bt + c = 2t + 2$$

n	1	2	3	4	5	6
t_n	7	18	35	58	87	122
$3t^2$	3	12	27	48	75	108
$bt + c$	4	6	8	10	12	14

Quadratic Equation is

$$t_n = 3t^2 + 2t + 2$$

✍ Practice 2: Write quadratic equation and the next three terms

a.	0, 3, 10, 21, 36, 55, ...	$t_n = 2t^2 - 3t + 1$
b.	4, 16, 34, 58, 88, 124, ...	$t_n = 3t^2 + 3t - 2$
c.	4, 15, 32, 35, 84, 119...	$t_n = 3t^2 + 2t - 1$

Application Problem

A lidless box is to be created from a square tin by cutting 10 cm square from each corner and bending up the sides for the box. If the volume of the box is $1200\ cm^3$, find the dimensions of the box.

$2\sqrt{30} \times 2\sqrt{30} \times 10$

Challenge Problem

AB is diameter of a circle with point C on the circumference such that its distance from A is 7 less than its distance from B.

- Draw the diagram to illustrate the situation
- Find the perimeter of the triangle ABC if the diameter of the circle is 17 m.
- Find area of the triangle ABC.

Student Spotlight: Akshith Kandivanam

"Learning mathematics with Anil Kumar has been transformative. His structured approach and engaging style made complex concepts intuitive, helping me excel in high school math and strengthen my problem-solving skills in STEM. This book is essential for mastering trigonometry and building a strong foundation for advanced math."

— Akshith, Schulich Leader Scholarship Recipient

Chapter 4 Test: Quadratic Functions

✍ Exam Review 1: Quadratic Relations (With Solutions in Videos)

Q1	Isolate variable t from the equation $s = ut + \frac{1}{2}at^2, t \geq 0$
Q2	Find the minimum value for the parabola $2x^2 - 6x + 1$
Q3	Show that the solution of the equation a$x^2 + bx + c = 0$ is $x = \frac{-b \pm \sqrt{b^2 - 4ac}}{2a}$
Q4	Show that $x^2 - 6x + 10$ is always positive.
Q5	Factor fully $14x^4 - 13x^2 + 9$.
Q6	the product of two consecutive even integers increased by their sum is 98. Find the possible integers.
Q7	Line $y = 10$ intersects the parabola $y = (x - 2)^2 + 1$ at A and B. Find the area of triangle ABC, where C is the vertex of the parabola.
Q8.	The equation $y = 2x^2 + ax + 23$ can also be written as $y = (x - 3)^2 + b$. Find the value of constants a and b.
Q9.	Find the range of values for k for which the equation $kx^2 + 2kx = 8x - k$ has two distinct roots.
Q10.	One of the $x - intercepts$ of the graph of $y = 3x^{\frac{2}{3}} - 5x^{\frac{1}{3}} + k$ is at $x = 1$. Find the other $x - intercept$.
Q11	Determine the value(s) of k in the equation $x^2 - kx + 50 = 0$ so that the root is double the other.
Q12	Equation $y = ax^2 + bx + c$ has roots at $\frac{5 \pm 3\sqrt{2}}{2}$. Find value of $a, b, and\ c$.
Q13.	Describe transformation and sketch the graph of $y = -2(x + 1)^2 + 8$
	Watch Video **Quadratic Exam Review** SCAN ME

Practice 14: Application Problem

Hunter is shooting a duck. The height of the duct is modelled by the relation $h = -5t^2 + 30t + 6$, where h is height in metres t seconds after the bullet is released. The path of the hunter's bullet is modelled by the relation $y = 40t + 1$, with the same units. How long will the bullet take to hit the duck? 0.4 seconds
A playground which measures 60 m by 40 m, is to be doubled in area by extending each side an equal amount. By how much should each side be extended? 10 m
The area of a rectangle is 24 square units, and its perimeter is 20 units. Find the dimensions of the rectangle. 4 by 6
Two numbers differ by 7 and the sum of their squares is 389. Find the numbers. -7, -10 and 7, 10
Solve the systems: $x^2 - y^2 = 12$ and $xy - 8 = 0$ $(-4, -2), (4, 2)$
Solve the systems: $x^2 + y^2 = 13$ and $x - y + 5 = 0$ $(-2, 3), (-3, 2)$
Solve the systems: $y = x^2 - 2x + 1$ and $y = -x^2 + 4x - 3$ $(1, 0), (2, 1)$
Solve equation: $n(n + 3) + 2 = 0$ -1, -2

Exam Review 3: Extended Linear Quadratic Systems Test (Video Support)

Q1	Find the coordinates of point where the line $x + y = 0$ intersects the circle given by equation $x^2 + y^2 = 8$
Q2	Find the value of k for which the line $y = kx - 5$ is tangent to the graph of $y = 3x^2 + 4x - 2$. Hence for this value of k, find the point of intersection.
Q3	Find the point of intersection between the curve $3x^2 - 8x - 2y = 10$ and the line $3x + 2y + 8 = 0$. SCAN ME Watch Video Linear Quadratic Linear System Extended

✍ Exam Review 4: Solve Simultaneous Linear Quadratic Systems (Video Support)

Q1	$\begin{cases} 2x + y = 7 \\ x^2 - xy = 6 \end{cases}$ $\left[\left(-\frac{2}{3}, \frac{25}{3}\right), (3, 1)\right]$	**Q2**	$\begin{cases} \frac{4}{x} + 3y = 14 \\ \frac{3}{x} - 4y = 23 \end{cases}$ $\left[\left(\frac{1}{5}, -2\right)\right]$
Q3	$\begin{cases} \frac{x+y}{xy} = 5, x \neq 0, y \neq 0 \\ \frac{x-y}{xy} = 7, x \neq 0, y \neq 0 \end{cases}$ $\left[\left(-1, \frac{1}{6}\right)\right]$	**Q4**	$\begin{cases} \frac{x+y}{xy} = \frac{12}{35}, x \neq 0, y \neq 0 \\ \frac{x-y}{xy} = \frac{2}{35}, x \neq 0, y \neq 0 \end{cases}$ [(7,5)]
Q5	$\begin{cases} ax + by - a + b = 0 \\ bx - ay - a - b = 0 \end{cases}$ $[(1, -1)]$	**Q6**	$\begin{cases} x^2 + y^2 = 13 \\ y^2 + z^2 = 25 \\ x^2 + z^2 = 20 \end{cases}$ $(\pm 2, \pm 3, \pm 4)$
	Watch Video **Simultaneous Equations** **SCAN ME**		

Mersenne Number with Factoring Connection

A **Mersenne Number** is a special number of the form:

$$M_n = 2^n - 1, n \text{ is positive integer}$$

63 is a Mersenne number: $2^6 - 1 = 64 - 1 = 63$

$63 = 3(21) = (2^2 - 1)(2^4 + 2^2 + 2^0)$

$63 = 7(9) = (2^3 - 1)(2^3 + 1)$

✍ **Practice:** Provide two examples where the numbers can be treated as polynomials and factored as shown above

Factoring Strategy: Case IX

$$a^6 - 1 = (a^2)^3 - (1)^3 = (a^2 - 1)(a^4 + a^2 + 1)$$

$$a^6 - 1 = (a^3)^2 - (1)^2 = (a^3 - 1)(a^3 + 1)$$

✍ Exam Review 5: Parabolic Trajectory Applications (Video Solutions)

Q1	A football is thrown in the air. Its height h in metres above the ground at any instant t seconds after the throw is approximately given by $$h = -4.9t^2 + 14.7t + 2, t \geq 0$$
Q1a	Determine the height of the ball after 2 seconds
Q1b	Determine how long the ball is 8 m above the ground.
Q2	a water balloon is catapulted into the air from top of a platform. The height h in metres of the ballon after t seconds is $$h = -1.25t^2 + 7.5t + 2.5, t \geq 0$$
Q2a	When will the balloon hit the ground?
Q2b	Can the ballon ever reach the height of 15 m? Show your work.
Q2c	Find the maximum height reached by the balloon?
Q3	the height of a football in metres t seconds after it is kicked can be modelled by $$h = -4.9t^2 + 22.54t + 1.1, t \geq 0$$
Q3a	What was the height of the football when punter kicked it?
Q3b	Determine the maximum height of the football to one decimal place.
Q4	a baseball is hit from a height of 1 m. The height h in metres of the ball after t seconds is modelled by $$h = -5t^2 + 9t + 1, t \geq 0$$
Q4a	Determine the maximum height reached by the ball.
Q4b	When will the ball reach at maximum height? Watch Video **Ball Trajectory** **SCAN ME**

STEM Project: Investigate Patterns – Area of Trapezoid in Quadrant I

$$\boldsymbol{Area\ of\ Trapezoid} = \frac{(\boldsymbol{a}+\boldsymbol{b})\boldsymbol{h}}{\boldsymbol{2}}$$

	Area of trapezoid formed between line $y = 2x + 1$ and the coordinates axes bounded by the vertical line $x = n$, where $n = \{1, 2, 3, 4, 5\}$ is given below.

$x = n$	$x = 1$	$x = 2$	$x = 3$	$x = 4$	$x = 5$
$Area$	$\frac{(1+3)(1)}{2} = 2$	$\frac{(1+5)(2)}{2} = 6$	$\frac{(1+7)(3)}{2} = 12$	$\frac{(1+9)(4)}{2} = 20$	$\frac{(1+11)(5)}{2} = 30$

Graph below is given for $x = 1, x = 3,$ and $x = 5$ to visualize the trapezoids.

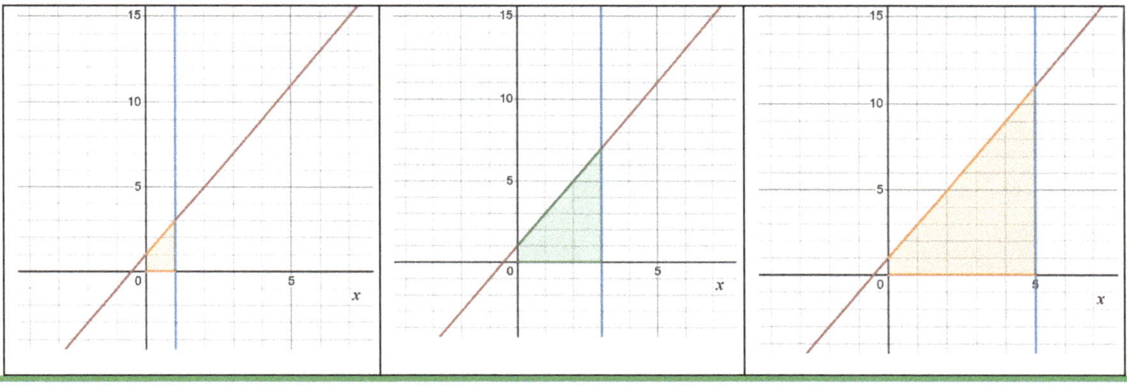

a.	Show that the area of trapezoid is a quadratic relation. Hint Example 8a: Finite Difference
b.	Area of trapezoid, $A(n) = an^2 + bn + c$. Find the value of $a, b,$ and c. **Hint Example 8a: Reflections of Analysis** • $\boldsymbol{2a}$ is 2nd Finite Difference • $\boldsymbol{3a+b}$ is difference of first two terms • $\boldsymbol{a+b+c}$ is first term
c.	Verify that $A(n) = n(n + 1)$
d	Find the area of trapezoid for $n = 10$
e.	For what value of n will the area $A(n) = 182$.

Exploring the Number System

Hierarchy of number systems is provided in which the group below includes all the above. Rational numbers include natural numbers, whole numbers and integers. Integers do not include rational numbers.

All the number systems have infinite numbers.

Symbol with Set of Numbers	Description
$\mathbb{N} = \{1, 2, 3, \ldots\}$	Natural Numbers ($\mathbb{N}$) are counting numbers from 1 onwards.
$\boldsymbol{W} = \{0, 1, 2, 3, \ldots\}$	Whole Numbers ($\boldsymbol{W}$) include natural numbers and zero.
$\mathbb{Z} = \{\ldots, -2, -1, 0, 1, 2, \ldots\}$	Integers ($\mathbb{Z}$)include whole numbers and negative numbers also.
$\mathbb{Q} = \left\{\frac{p}{q}, q \neq 0, p, q \in Z\right\}$ $q \neq 0$ since division by zeros is not valid	Rational Numbers ($\mathbb{Q}$) are numbers that can be written as ratio of integers. $\in$ is symbol for "belongs to" They include all integers also. Repeated decimal numbers are rational numbers as they can be written as $\frac{p}{q}$. Example: $0.333 \ldots = \frac{1}{3}$
$\overline{\mathbb{Q}}$ =Numbers that are not Rational numbers. These cannot be written in the form of $\frac{p}{q}$.	Irrational Numbers ($\overline{\mathbb{Q}}$) are numbers that cannot be expressed as fractions; their decimal form is non-terminating and non-repeating. $\sqrt{2}, e, \pi$ are examples.
$\mathbb{R} = \{\mathbb{Q}\ or\ \overline{\mathbb{Q}}\}$ Square root of negative numbers are not real numbers. They are classified as Complex numbers	Real Numbers ($\mathbb{R}$) include all rational and irrational numbers. Number line represents all real numbers.

Symbol with Set of Numbers	Description
We will restrict to real numbers in this book if not specified otherwise.	
$\mathbb{C} = \{z = a + bi \mid a, b \in R, i = \sqrt{-1}\}$ i is the imaginary number. $z = a + bi$ is a complex number with a as real part and bi as complex part. In a complex plane the x-axis is the axis for real part and y-axis is for complex part.	Square root of negative numbers is not defined in the set of real numbers. Complex numbers ($\mathbb{C}$) include i, which is square root of negative one. It cannot be represented on number line. Imaginary numbers are represented on complex planes.

Nomenclature for Set of Numbers:

$\mathbb{Z} = \{\dots, -2, -1, 0, 1, 2, \dots\}$

$\mathbb{Z}$ indicates the symbol for the number system

$\{\quad\}$ curly brackets enclose complete set of numbers included for that system.

$\{\dots, -2, -1, 0, 1, 2, \dots\}$ These set of numbers are also referred to as the domain.

Domain, in general, is the set of numbers in which we are working.

Symbols of Number System should be written as capital letters only

$\boldsymbol{N}, \boldsymbol{W}, \boldsymbol{Z}, \boldsymbol{Q}, \overline{\boldsymbol{Q}}, \boldsymbol{R}, \boldsymbol{C}$ are acceptable.

Number systems include infinite set of numbers still there are gaps which we have seen in a very simple classification shared above. Let us explore a bit!

Communication and Thinking Test

Q1	**Continuity of Number System** Rational numbers form a set of infinite numbers and can be represented on a number line. How can you show that the rational numbers have discontinuities.
Q2	**Irrational Numbers on Number Line** How to represent $\sqrt{2}$ and other irrational numbers on number line?

Example: Number Systems

<table>
<tr><td>a.</td><td colspan="6">List the set of numbers to which each of the following belong.</td></tr>
<tr><td></td><td>(i)</td><td>0 : $\boldsymbol{W, Z, Q, R, C}$</td><td>(ii)</td><td>$\pi$: $\boldsymbol{\bar{Q}, R, C}$</td><td>(iii)</td><td>$\sqrt{-1}$: $\boldsymbol{C}$</td></tr>
<tr><td>b.</td><td colspan="6">State why the following statements are true.</td></tr>
<tr><td></td><td>a.</td><td colspan="5">Infinity is not a real number.
Infinity does not have specific location on the number line.</td></tr>
<tr><td></td><td>b.</td><td colspan="5">Equation $2x = 3$ has no solution in the domain of integers.
$\mathbf{3/2}$ is not an integer.</td></tr>
<tr><td></td><td>c.</td><td colspan="5">$\sqrt{-1}$ is not a real number.
It is a complex number $\boldsymbol{i}$</td></tr>
<tr><td></td><td>d.</td><td colspan="5">$x^2 + 1 = 0$ has no solution in domain of real numbers.
$\sqrt{-\mathbf{1}}$ is not a real number.</td></tr>
<tr><td></td><td>e.</td><td colspan="5">There are infinite real numbers between any two numbers.
You can keep getting a number half way between them.</td></tr>
</table>

Communication and Thinking Problems

Q1. How can you visualize infinity?

Here is one way to visualize infinity.

Draw a 2 by 1 rectangle. Keep on colouring half of it in different colours. This is an infinite series where the ultimate-coloured region will be 2 unit square.

$1 + \frac{1}{2} + \frac{1}{4} + \frac{1}{8} + \cdots = 2$

Explore and provide more examples to visualize infinity.

Q2. Explain why there are more irrational numbers than rational numbers.

Complex Roots of Quadratic Equation

A quadratic equation of the form $ax^2 + bx + c = 0$ has two zeros (or roots) that can be determined using the quadratic formula: $x = \frac{-b \pm \sqrt{b^2-4ac}}{2a}$

Earlier, on pages 200, we explored how these zeros can be found algebraically. The expression under the square root, $b^2 - 4ac$, is called the discriminant, D, and it determines the nature of the roots.

- If $D > 0$, the equation has two distinct real roots.
- If $D = 0$, it has one repeated real root.
- If $D < 0$, the equation has two complex conjugate roots.

When $D < 0$, the square root of a negative number introduces the imaginary unit i, where $i = \sqrt{-1}$.

Example 1: Find Roots of $x^2 + 4x + 8 = 0$ in the domain of complex numbers.

$x^2 + 4x + 8 = 0$ $x = \frac{-b \pm \sqrt{b^2 - 4ac}}{2a}, a = 1, b = 4, c = 8$ $x = \frac{-4 \pm \sqrt{4^2 - 4(1)(8)}}{2(1)} = \frac{-4 \pm \sqrt{-16}}{2}$	$x = \frac{-4 \pm 4i}{2} = -2 \pm 2i$ Two complex conjugate roots are the solution of the given equation.

Example 2: Find Quadratic Equation with Roots as $\frac{1 \pm 3i}{2}$

Important Strategy as Shortcut Compare the roots with $x = \frac{-b \pm \sqrt{b^2 - 4ac}}{2a}$ $x = \frac{1 \pm 3i}{2} = \frac{1 \pm \sqrt{-9}}{2}$ $a = 1, b = -1, b^2 - 4ac = -9$	$(1)^2 - 4(1)c = -9$ $4c = 9 + 1 = 10 \Rightarrow c = \frac{10}{4} = \frac{5}{2}$ The equation is $x^2 - x + \frac{5}{2} = 0$ **Note:** Also solve using alternate methods
Alternate Method 1: expand and simplify: $\left(x - \frac{1+3i}{2}\right)\left(x - \frac{1-3i}{2}\right) = 0$ **Alternate Method 2:** $x^2 - (sum\ of\ zeros)x + product\ of\ zeros = 0$	

STEM Project: Explore and Learn More

Beyond Exactness: Navigating the Infinite on the Number Line

In mathematics, we often deal with numbers that are either easily defined or seem impossible to pin down. Take **Rational Numbers** like -2.5, 3/4, or 1.25 for example. We can place the exact values of these numbers on the number line. Each of these numbers corresponds to a specific point on the number line, and we can visualize their exact location with ease.

But what about Irrational Numbers, like $\boldsymbol{\pi(pi) or \sqrt{3}}$? These numbers don't have exact value ($\pi \approx 3.141592654, \sqrt{3} = 1.732050808$), yet they still occupy a specific place on the number line.

Project 1: Display Exact Value of Numbers in the Given Set of Real Numbers

Draw a number line and describe method(s) to place following set of real numbers exactly on it.

Set of numbers is $\left\{-\sqrt{2}, 0, \frac{1}{2}, \sqrt{5}, 1.\bar{7}, \pi\right\}$

SCAN ME

Anil Kumar Video Display $\sqrt{3}$ on Number Line

Project 2: Determine Value of Pi (π)

- Describe the experimental set-up to find the value of π.
- List limitations and constraints
- Suggest ways of improvement for better result
- Find the percentage error as compared with the value of 3.14159

Challenge Problems

Write equation of a parabola passing through the points $(-4, 21), (4, 5),\ and\ (6, 11)$.

$$y = \frac{1}{2}x^2 - 2x + 5$$

Evaluate

$$\frac{(4 \times 7 + 2)(6 \times 9 + 2)(8 \times 11 + 2) \ldots (100 \times 103 + 2)}{(5 \times 8 + 2)(7 \times 10 + 2)(9 \times 12 + 2) \ldots (99 \times 102 + 2)}$$

Hint: $(n(n + 3) + 2) = (n + 1)(n + 2)$

STEM Project: Lines and Regions — Quadratics in a Circle

Title: Cutting Through the Circle — Discovering Quadratic Patterns

Objective:

Investigate how the maximum number of **regions** a circle can be divided into by drawing **n straight lines** and discover how this relates to **quadratic equations.**

Investigation Task:

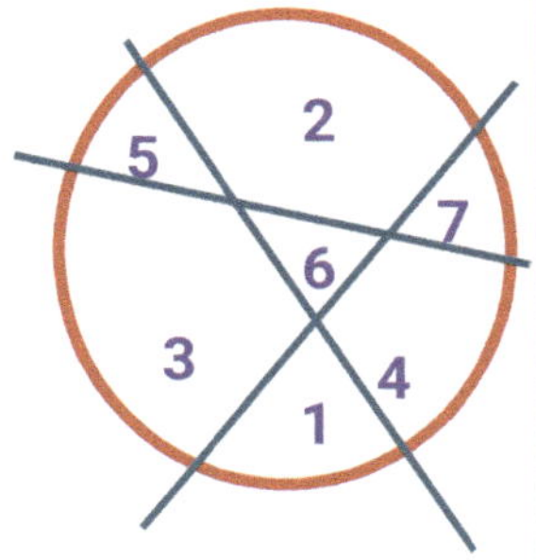

1. Start with **no line**: How many regions in the circle? ➤ ✅ 1
2. Draw **1 line**: How many regions? ➤ ✅ 2
3. Draw **2 lines** (intersecting at one point): ➤ ✅ 4
4. Draw **3 lines** (maximum intersections): ➤ ✅ 7
5. Continue up to 5 or 6 lines — count carefully!

Record your findings in a table:

Lines	0	1	2	3	4	5
Regions						

Verify that the second difference is constant. This indicates quadratic growth.

....

Derive quadratic formula to relate the lines with regions

....

🧭 **Suggested Extension:**

➡️ Investigate similar quadratic growth in:

- Handshakes among n people
- Triangular Numbers
- Diagonals in polygons

💡 ***Reflect in your Learning Journal:***

....

AMC 10 Algebraic Skills Practice Test

Q1	Sam estimates that he spends $\frac{1}{4}$ of his day at school, $\frac{1}{3}$ of it sleeping, and $\frac{1}{6}$ of it on homework. What fraction of the day is left for other activities?
Q2.	The sum of two numbers is 9. Sum of their reciprocals is $\frac{1}{2}$. Find the numbers.
Q3	Two numbers a and $b$are in the ratio of 2:5. If 16 is added to both the numbers the ratio becomes 1:2. Find the sum of numbers a and b.
Q4.	Every atom has negatively charged particles called electrons and positively charges particles called protons. Each electron has mass of $9.11 \times 10^{-28} g$ and each proton has mass of $1.67 \times 10^{-24} g$. How many times as heavy is a proton than an electron, when rounded to hundreds?
Q5	Use the patterns to find the last or unit digit of 2^{2025}
Q6.	All the prime numbers less than 100 are multiplied together to get the product of N. What will be the last digit of N.
Q7.	A retailer increases the price of an item by 10% and the sell it at 10% discount. What percent of profit or loss will the retailer make by doing so?
Q8.	Find the largest number that can exactly divide $(n-2)(n^2-n)$
Q9.	Factor the number 2024 using the concept of difference of squares. Find the largest prime factor of 2024. Hint: 2024 = 2025 − 1
Q10.	Any six-digit number of the form "abcabc" is always divisible by what number? Hint: Write in Expanded Form and then Factor.

Answers AMC 10 Algebraic Skills

Q1	$\frac{1}{4}$	Q2	3, 6	Q3	112	Q4	1800	Q5	2
Q6	0	Q7	1% loss	Q8	6	Q9	23	Q10	1001

STEM Explorers Challenge Questions

<table>
<tr><td>Q1</td><td>Find the condition for the values of a and q in the equation $y = a(x-p)^3 + q$ to have
<table><tr><td>a.</td><td>One real root</td><td>b.</td><td>Two real roots</td><td>c.</td><td>No real root</td></tr></table></td></tr>
<tr><td>Q2</td><td>Ben bought some pizzas for $900 to raise funds for the charity. He kept 10 pizzas for the volunteers and sold the rest for $1040. There were 8 slices in each pizza. The profit was 50 cents per slice. How many pizzas were in the original order. $x = 75$</td></tr>
<tr><td>Q3</td><td>Write equation to represent the sequence: 0, 8, 20, 36, 56, 80, …</td></tr>
<tr><td>Q4.</td><td>A car averaged 12 L/100 km in city driving and 8 L/100 km on the highway. In a week of mixed driving, the car used 48 L of fuel and travelled 500 km. Determine the distance travelled in highway driving.
$0.12x + 0.08y = 48; x + y = 500; y = 300\ km$</td></tr>
<tr><td>Q5</td><td>Find the volume of a cube with side lengths of $2\sqrt{3} - \sqrt{2}$ $36\sqrt{3} - 86\sqrt{2}$</td></tr>
<tr><td>Q6.</td><td>Prove that $n^2 + n + 1$ will not always produce prime number.</td></tr>
<tr><td>Q7</td><td>Explain the differences between usage of the terms quadratic function and quadratic equation.</td></tr>
</table>

IB Math Challenge

Find the number of pairs of positive integers (x, y) where x and y are prime numbers and $x^2 - 2y^2 = 1$.

Show that the line $y = 2x - 5$ is tangent to the circle $x^2 + y^2 = 5$. Find the point of tangency. $(2, -1)$

Find quadratic equation whose roots are $1 \pm i\sqrt{2}$ $x^2 - 2x + 3 = 0$

SCAN ME

Quadratic Equation for Complex Roots: 3 Methods

Challenge Test: Thinking and Applications

Q1	4 men and 6 boys can finish a piece of work in 5 days while 3 men and 4 boys can finish in 7 days. Find the time taken by a man alone to complete the work.
Q2	A factory produces gadgets (g) and sensors (s). Each gadget takes 2 hours, each sensor 3 hours. Total time is 60 hours for 25 total items. Form a system and find g and s.
Q3	For what value of k will the following set of linear equations have infinitely many solutions? $kx + ky = 12, (k - 3)x + 3y = k$
Q4	the sum of numerator and denominator of a fraction is 4 more than twice the numerator. If 3 is added to each, their ratio becomes 2:3. Find the fraction.
Q5	a motorboat whose speed is 24 km/h in still water takes an hour more to go upstream than to return downstream to the same spot. Find the speed of the stream.
Q6	The total cost of a certain length of a copper wire is \$200. If it was 5 cm longer each metre of copper wire costs \$2 less, the cost of the piece would have remained unchanged. How long is the length of the piece.
Q7	a farmer wants to build a regular hexagonal animal pen next to his barn. One of the sides will border the barn, so that only five sides of the fence needs to be built. The pen will hold a total area of 3000 m^2 of space. What length of fence will the farmer need?
Q8	Anil bought some pens for \$20. If each pen was 20 cents less, he could have bought five more for the same amount. How many pens did he buy?
Q9	Determine the value of p such that line $y = 3x + p$ intersects the parabola $y = 2x^2 - 5x + 3$ exactly once.
Q10	Factor $x^4 + 4$ in the domain of real numbers.
Q11	How many diagonals are there in a polygon of 12 sides?
Q12	Determine the values of k for which $y = 3x^2 - 2kx + 12$ has two real roots.

Answers

Q1	35 days	Q2	15, 10	Q3	6	Q4	5/9	Q5	30 km/h
Q6	20 m	Q7	170 m	Q8	20 pens	Q9	-5	Q10	
Q11	54	Q12	± 6						

Chapter 5: Geometry Fundamentals

Angles Formed by Two Intersecting Lines

An Angle

- Two rays with a common point form an angle

Vertically Opposite Angles

- Two intersecting lines form four angles at the point of intersection.
- Directly opposite angles are called vertically opposite angles
- These non-adjacent opposite angles are always equal
- $\angle 1 = \angle 3$ and $\angle 2 = \angle 4$

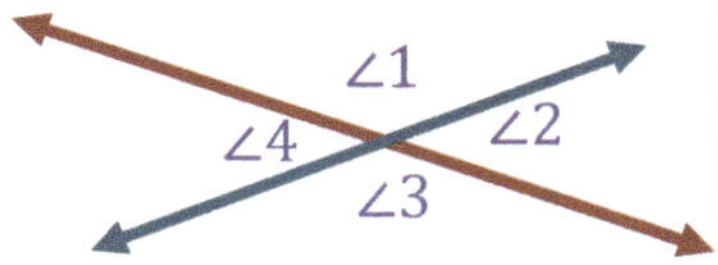

Adjacent Angles

- Angle that shares a common vertex and common arm are adjacent angles.
- $\angle 1\ and\ \angle 2$ are adjacent angles

Straight Angles

- Angle that measures $180°$ is called straight angle.
- $\angle 1\ and\ \angle 2$ form a straight angle.
- These angles should be adjacent angles
- Identify other set of angles that form straight angle

Supplementary Angles

- Two angles that add to $180°$ are called supplementary angles.
- These angles may not be adjacent angles

Complementary Angles

- Two angles that add to $90°$ are called complementary angles.
- These angles may not be adjacent angles

Linear Pairs

- Two adjacent angles whose non-common arm forms a straight line.
- Linear pairs are always supplementary.

Angles Around a Point

- Sum of angles around a point is always 360°.

Exercise 1: Find Measure of Each Angle

Q1	The ratio of the measure of two complementary angle is 4:5. Find each angle.
Q2.	The ratio of the measure of two supplementary angle is 4:5. Find each angle.
Q3	The ratio of the measure of the angles in a triangle is 2:3:5. Find each angle
Q4.	The measure of an acute angle (angle less than $90°$) in a right triangle is $37°$. Find the measure of the other acute angle.
Q5.	The vertically opposite angles are $2x - 32$ and $x + 38$. Find the measure of these vertical angles.
Q6	The figure shows intersecting lines. If $m\angle 1 = 145°$ Find the following 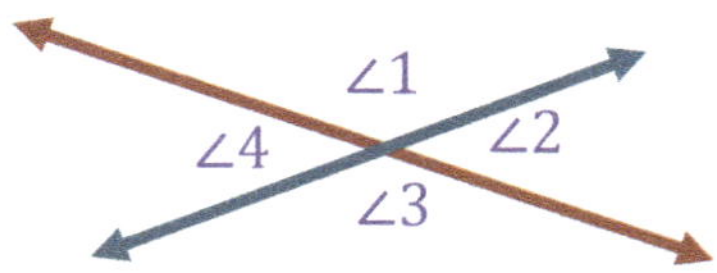 a. $m\angle 2 + m\angle 4$	b. $m\angle 3 - m\angle 4$	c. $m\angle 2 + m\angle 4 - m\angle 3$
Q7.	An angle is 20° more than its complement. Find the angle. Hint: $\boldsymbol{x - (x - 90) = 20}$
Q8	The supplement of an angle is 30° less than twice the angle. Find the angle. Hint: $\boldsymbol{180 - x = 2x - 30}$
Q9	In a triangle, two angles are equal, and the third is 54°. Find each angle.
Q10	Three angles form a straight line. Two angles are equal and the third is 30°. Find all three angles.
Q11	In an isosceles triangle one angle is 32°, find the other angles. Explain why we have two sets of answers in this case.

Parallel and Transverse Lines

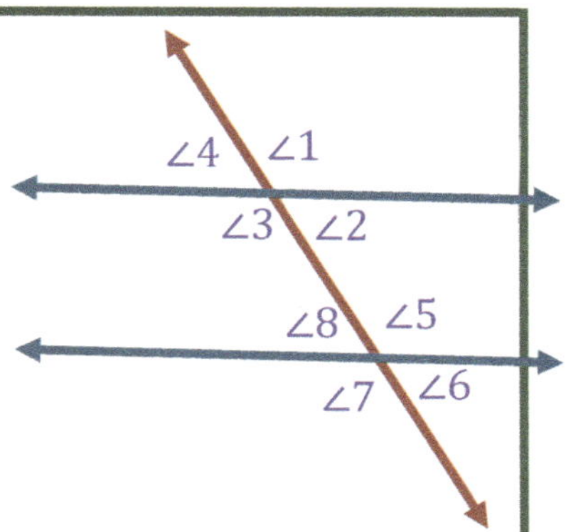

Transverse line intersects two or more line.
These lines may or may not be parallel.

Parallel lines do not intersect.
When a transverse line intersects two parallel lines, eight angles are formed as shown in the figure.
Here are special properties of these angles:

Corresponding Angles

Angles with same relative position at each intersection of transverse line with parallel line are called **corresponding angles**. These are always equal. Set of corresponding equal angles for the given figure are:
$\angle 1 = \angle 5; \angle 2 = \angle 6; \angle 4 = \angle 8; \angle 3 = \angle 7$

Alternate Interior Angles

Angles formed on the opposite side of the transverse line and on the inside of parallel lines are **alternate interior angles**. These are always equal. Set of alternate interior angles here are:

$\angle 2 = \angle 8; \angle 3 = \angle 5$

Alternate exterior Angles

Angles formed on the opposite side of the transverse line and on the outside of parallel lines are **alternate exterior angles**. These are always equal. Set of alternate exterior angles here are:

$\angle 1 = \angle 7; \angle 4 = \angle 6$

Co-interior Angles

Co-interior angles are pair of angles on the same side of the transverse line between the parallel lines. Their sum is always $180°$.

$\angle 2 + \angle 5 = 180°; \angle 3 + \angle 8 = 180°$

Converse of alternate interior angles theorem states that if two lines are cut by a transversal line and **the alternate interior angles are congruent, then the two lines are parallel**.

Exercise 2: Intersection of Lines

Find all the missing angles in the given figure with transverse line cutting across two parallel lines.

Q1.

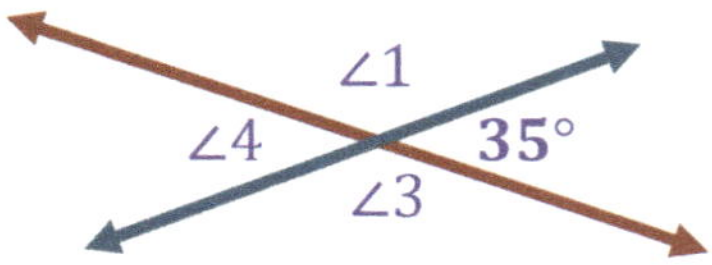

Q2.

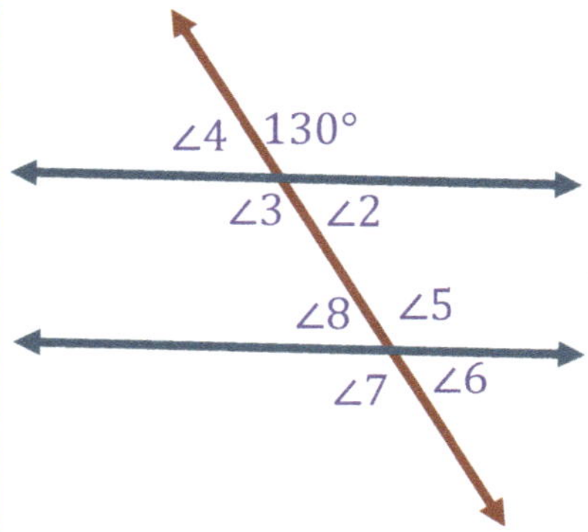

Q3.

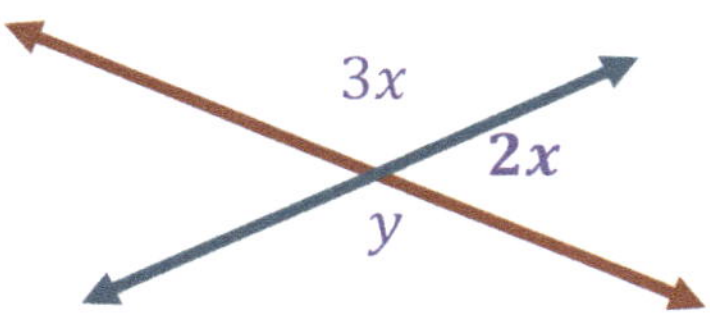

Q4.

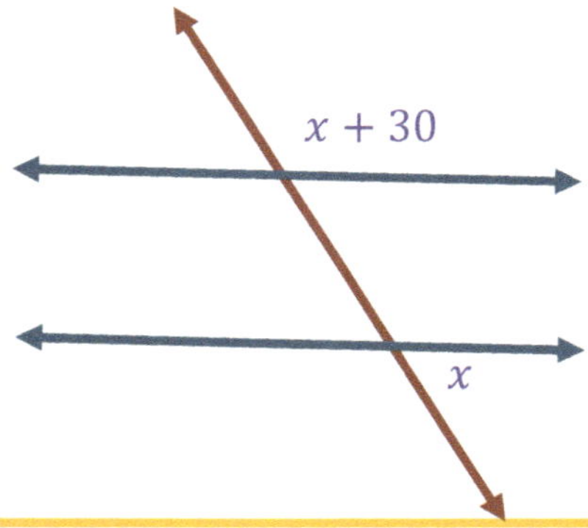

Q5

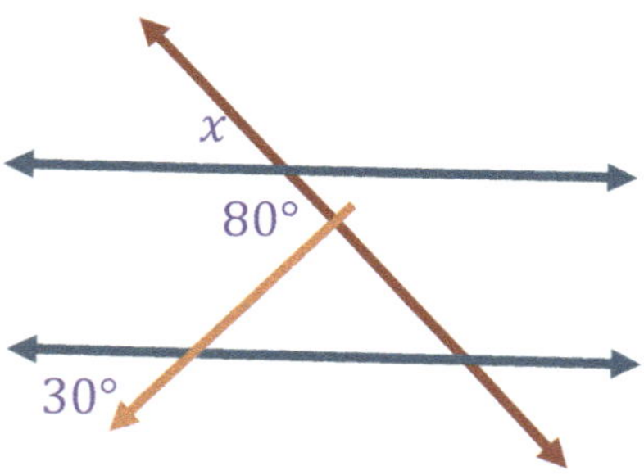

Q6.

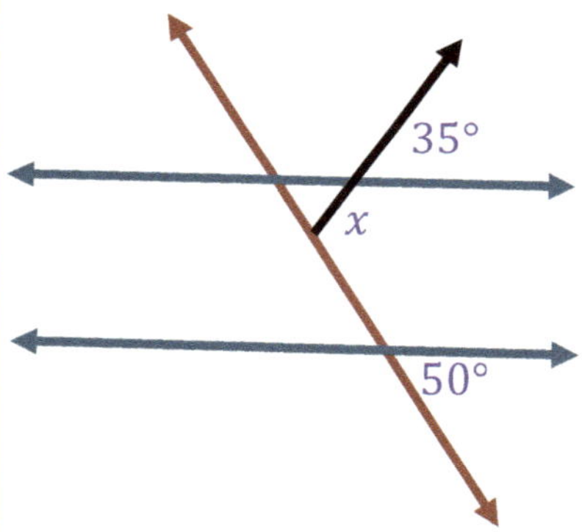

Answers

Q1	$\angle 1 = \angle 3 = 145°$	Q2	$\angle 2 = \angle 8 = 50°$	Q3	$y = 108°$
Q4	$x = 75°$	Q5	$x = 50°$	Q6	$x = 85°$

Explore Regular Polygons

1	Prove that sum of exterior angles of a closed polygon is $360°$
2	Prove that the sum of internal angles of a closed n-sided polygon is $180°(n-2)$ and sum of exterior angles is $360°$.
3	Describe different ways to finding the interior angle of a regular polygon.
4.	Find the interior angles of a regular polygon with 12 sides.
5.	Calculate the number of sides of a regular polygon whose interior angle is $108°$
6.	Calculate the number of sides of a regular polygon whose exterior angle is $12°$
7	Find the number of diagonals in a regular decagon (10-sided polygon). Hint: Formula $\frac{n(n-3)}{2}$
8.	The interior angle of a regular polygon is five times its exterior angle E. Find the number of sides of the polygon. Hint: $E + 5E = 180 \Rightarrow E = 30$ Number of sides $n = \frac{360}{E}$
9	Each exterior angle of a regular polygon is smaller than each interior angle by 120°. Find the number of sides of the polygon. Hint: $I - E = 120; I + E = 180$ SCAN ME Watch Video **Regular Polygons Internal/External Angles**

Olympiad Challenge

A square is inscribed in a regular pentagon as shown in the figure.

Find the sum of angles α and β

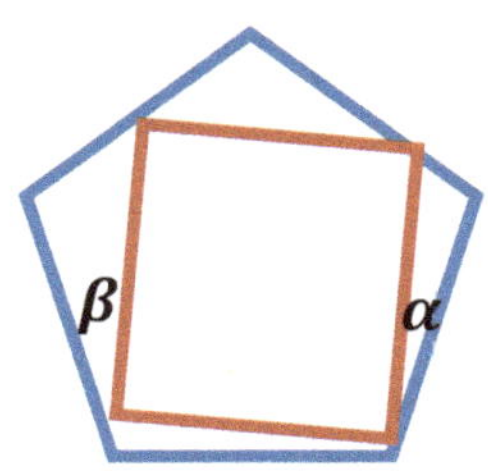

a.	$30°$	b.	$36°$	c.	$45°$	d.	$48°$

Chapter 5 Test: Geometry Essentials

Q1. Find the value of unknowns in the given figures.

a.

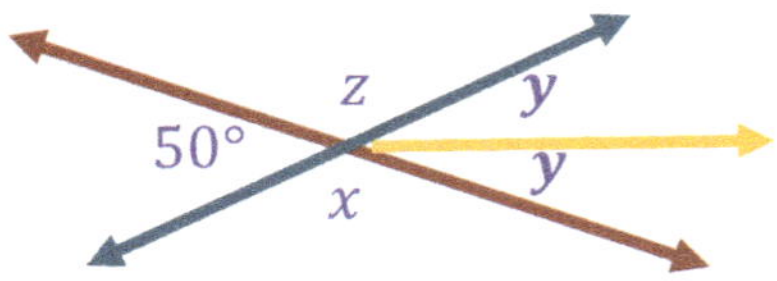

b.

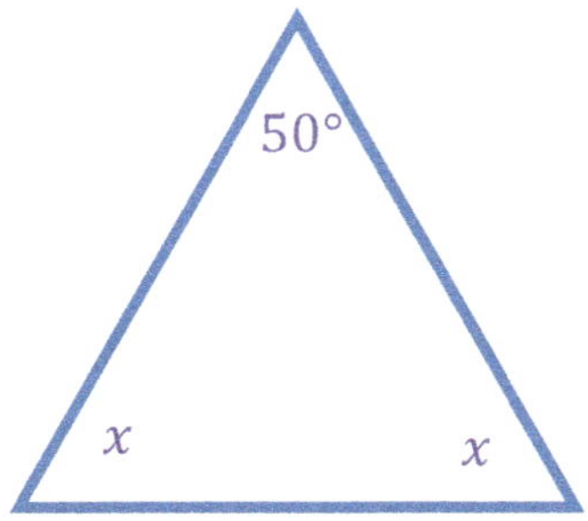

c.

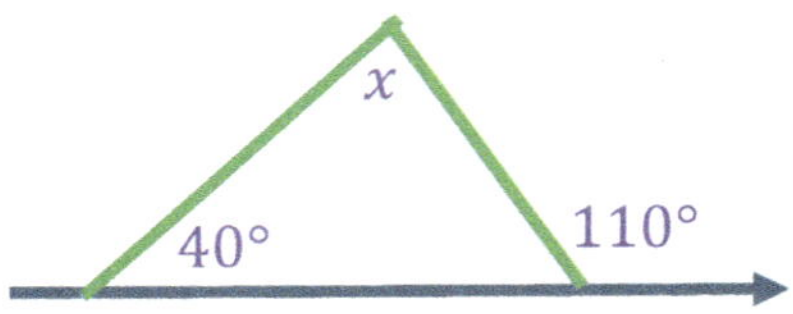

d.

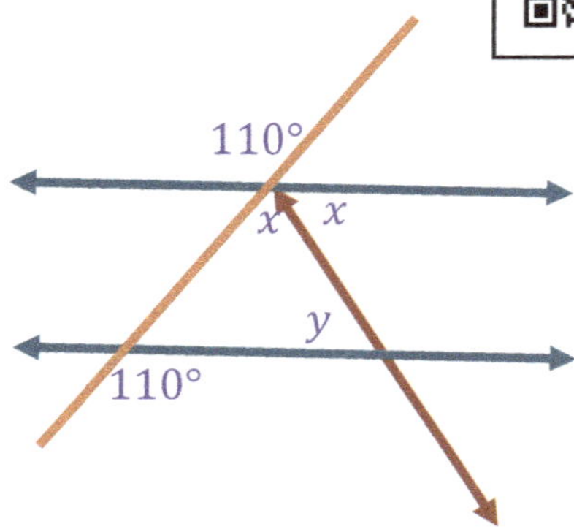

Q2	Angles of a triangle are in the ratio of 5:6:7. Find the angles.
Q3	A right triangle has an angle of $\mathbf{35}°$. Write all the three angles.
Q4	Isosceles triangle has two equal sides. Find all possible angles of an isosceles triangle in which one of the angles is $\mathbf{50}°$.
Q5	An equilateral triangle has all equal sides. Explain why its angles should be $\mathbf{60}°$.
Q6.	Prove that the sum of interior angles of a triangle is always $180°$.

Summary: Angles Lines and Triangles

Angle Properties

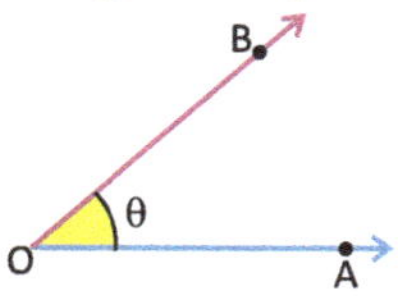

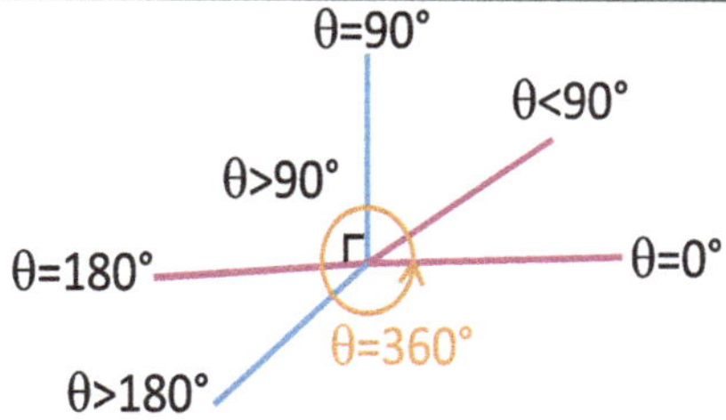

Angle, θ, with vertex O, is formed by the rays OA and OB.

The two rays are referred to as initial arm (OA) and terminal arm (OB).

How to Name Angles

Angle θ or $\angle AOB$ or $A\vec{O}B$

How to Write Rays

Ray $\overrightarrow{OA}$ and Ray $\overrightarrow{OB}$

How Classify Angles

$0° < \theta < 90°$	acute angle
$\theta = 90°$	right angle
$\theta > 90°$	obtuse angle
$\theta = 0°\ or\ \theta = 180°$	straight angle
$\theta > 180°$	reflex angle

Supplementary angles sum is 180°	Complementary angles sum is 90°	Opposite angle are equal
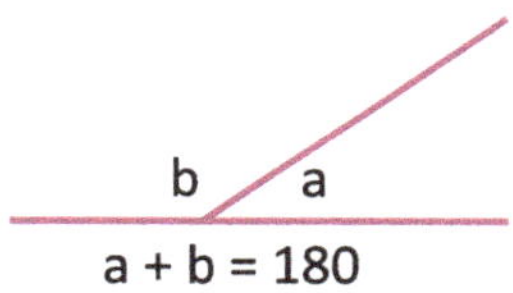	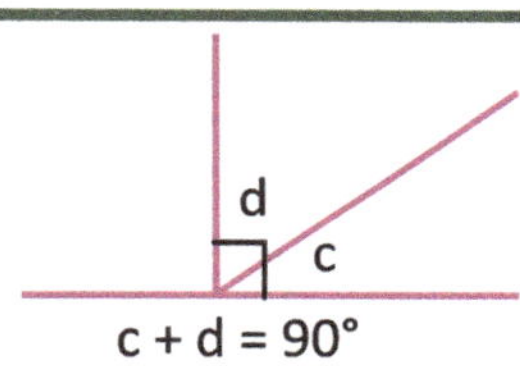	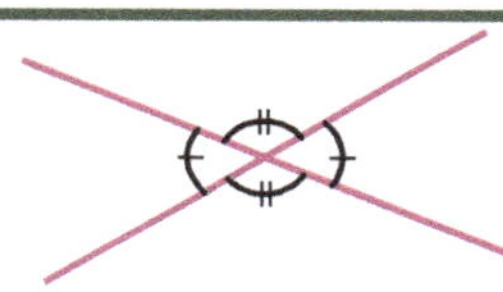

Sum of interior angles of a triangle is $180°$.

Here is a short proof

Consider the two parallel sides and two transverse lines as shown in the figure.

$a, b, and\ c$ are interior angles of the given triangle between two parallel lines.

$\angle 1 = b\ and\ \angle 2 = c$ since these are interior alternate angles

$\angle 1 + c + \angle 2 = 180°$ as it forms a straight line

That implies $a + c + b = 180°$

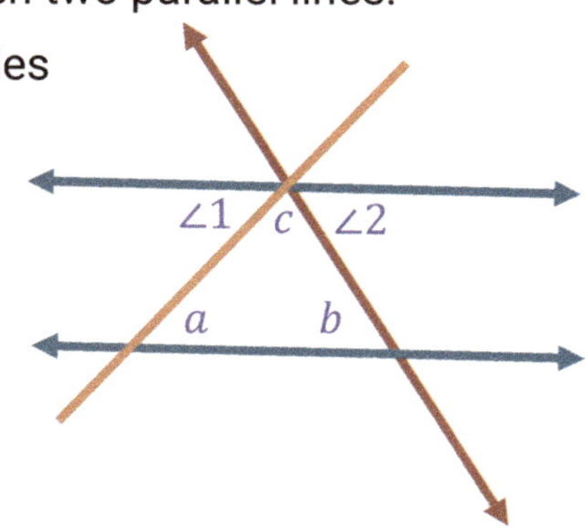

Summary: Transverse and Parallel Lines

Alternate angles are equal	Corresponding angles are equal	Co-interior angles are supplementary
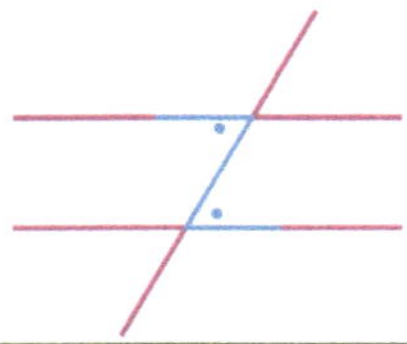	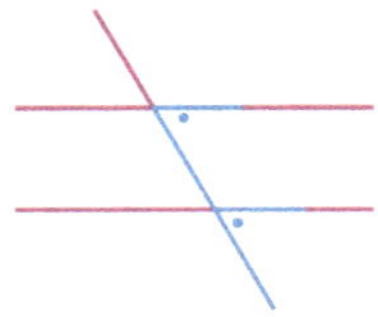	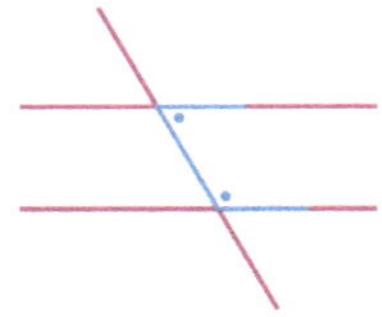

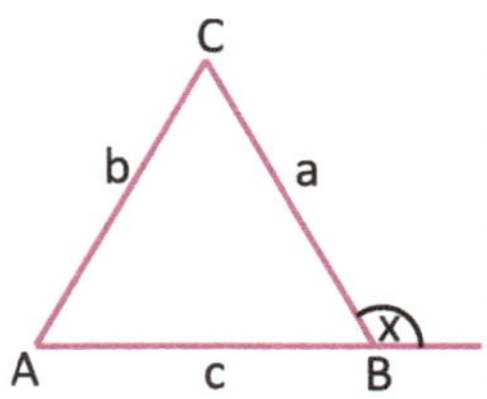

Sum of two side $\geq$ third side

Sum of **interior angles** is $180°$ or $\angle A + \angle B + \angle C = 180°$

External angle = sum of remote interior angles or $\angle x = \angle A + \angle C$

Sum of **exterior angles** is 360°

Isosceles triangles	:	Two equal sides corresponding equal angle
Equilateral triangle	:	All equal sides All angles $60°$
Right Triangle	:	One right angle, $\angle C = 90°$ $a^2 + b^2 = c^2$ $\angle B = 90° - \angle A$

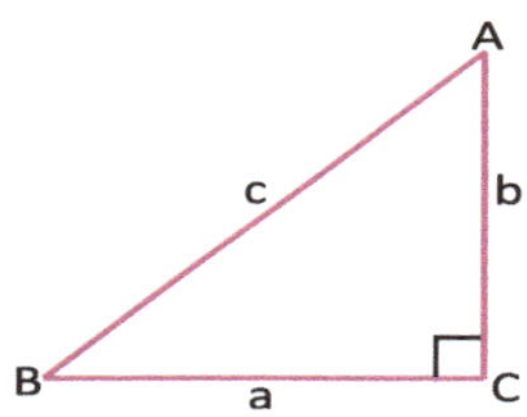

$\angle A$ and $\angle B$ are **complementary pairs** or $\angle A + \angle B = 90°$

Olympiad Challenge

A regular pentagon is inscribed in a regular hexagon as shown in the figure.

Find the sum of angles α and β

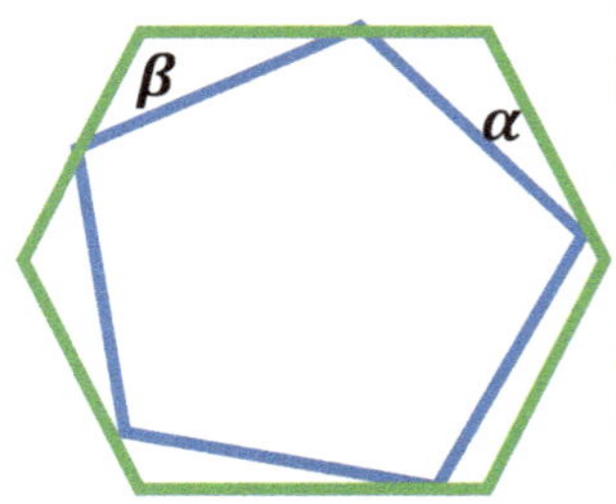

a.	$52°$	b.	$36°$	c.	$45°$	d.	$48°$

SAT Practice: Angles, Parallel Lines and Triangles

Hints are provided to solve the questions 1 – 10. Complete solution for question 11.

Review Angle Properties. Write solutions in the space provided

1. Identify all supplementary pairs. How many supplementary pairs do you get?

   List all supplementary pairs to confirm your answer.

   A. 4
   B. 6
   C. 7
   D. 8
   E. 9

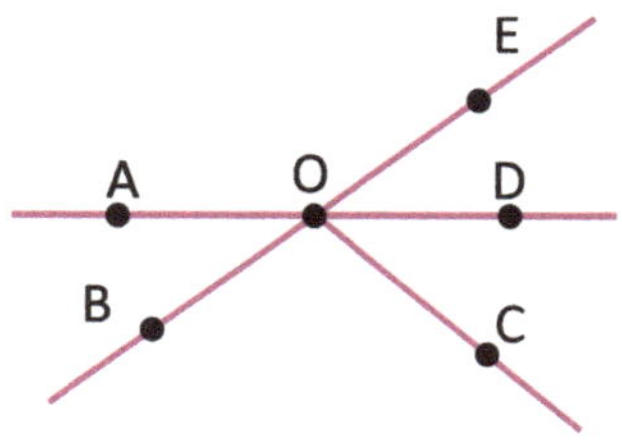

2. Find the measure of angle **a** in degrees. Provide complete solution with reason for each statement.

   A. 20°
   B. 40°
   C. 60°
   D. 80°
   E. 100°

   [Justify that $a = 20°$]

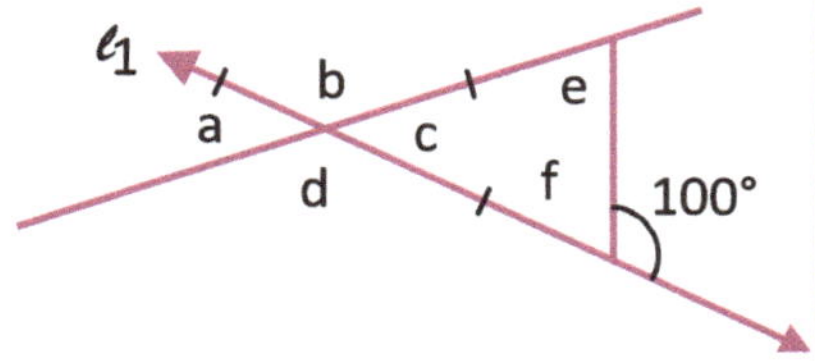

3. Find the value of the product, xy.

   [complementary angle and vertically opposite angles]

   A. 150°
   B. 200°
   C. 250°
   D. 300°
   E. 350°

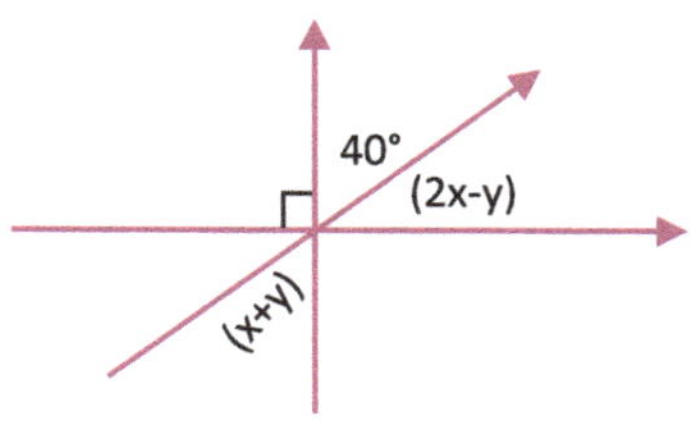

4. Bearing angles are provided in the figure below. Find the value of angles: $x + y$.

   [co-interior angles sum is 180°]

   A. 148°
   B. 178°
   C. 182°
   D. 238°
   E. 180°

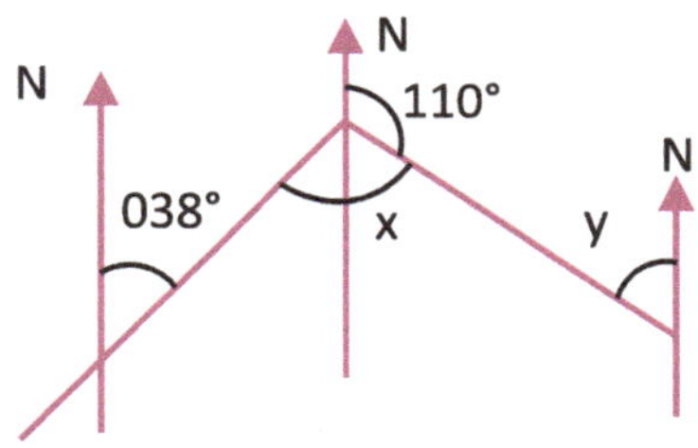

5. Find value of $x + y$.

   [equal corresponding angles and sum of interior angles in a triangle is 180°]

   A. 105°
   B. 120°
   C. 125°
   D. 130°
   E. 135°

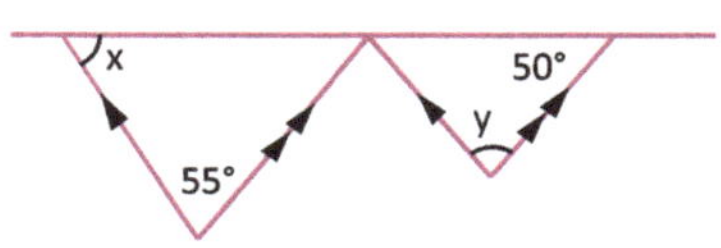

6. Which of the following set of side lengths will not form a triangle?

   [sum of two sides is greater than the third]

   A. 4, 3, 5
   B. 12. 15. 26
   C. 4, 5, 10
   D. 3, 4, 5
   E. 10, 11, 12

Watch Video **Angle Properties Solutions** **SCAN ME**

7. Find the value of $x - y$.

[exterior angle concept]

A. 5°
B. 10°
C. 15°
D. 20°
E. 25°

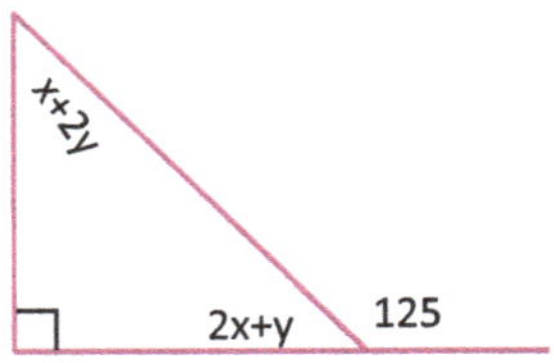

8. Find the value of x from the measure of angles provided below.

[exterior angle concept]

A. 50°
B. 54°
C. 56°
D. 58°
E. 60°

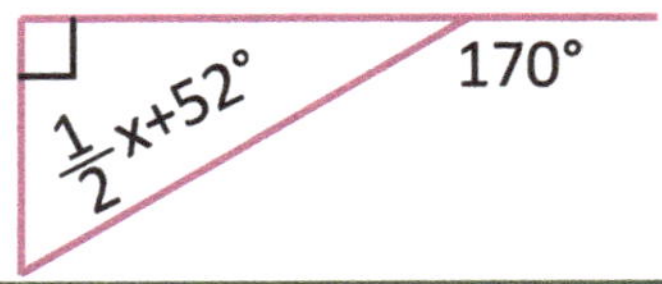

9. Angles of a triangle are in the ratio of 1:2:3. Find the measure of largest angle.
A. 60°
B. 70°
C. 90°
D. 100°
E. 120°

10. Explain co-interior angles.
Create a problem that can be solved using co-interior angles.

Answers to the Questions 1 – 10

Q1.	D.	Q2	A.	Q3.	D	Q4.	B.	Q5.	D.
Q6.	C.	Q7.	D.	Q8.	C.	Q9.	C.	Q10.	D

Provide reason and justification for your solution

11a. Given $PQ = PR$, $\angle PRS = 36°$ and $PQ\ parallel\ to$ ST.

Find value of x and y.

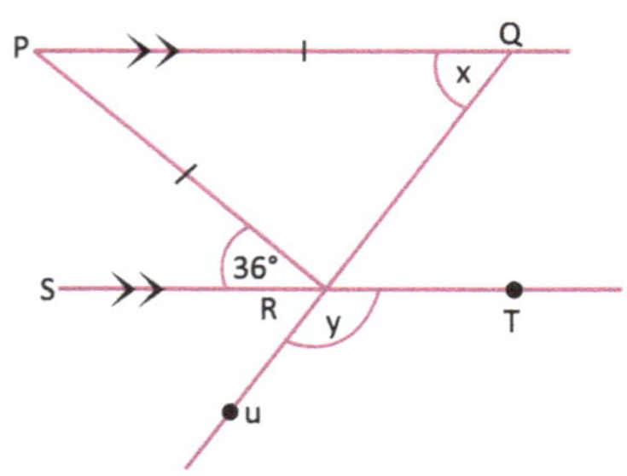

11b. Find the angle x and y in the figure. Given AB || CD, $\angle IGH = 60°$, $\angle IHB = 110°$

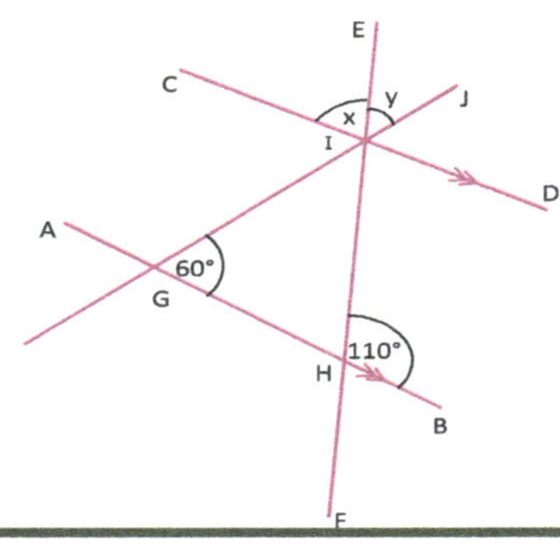

11c. Find angle x and y in the figure.

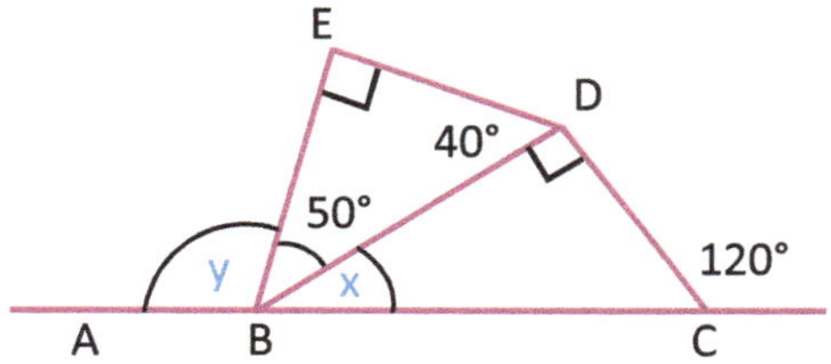

11d. Given $PQ \parallel RS$ with transverse line UV. AB bisects $\angle UBQ$ and AC bisects $\angle SCV$. Find $\angle BAC$.

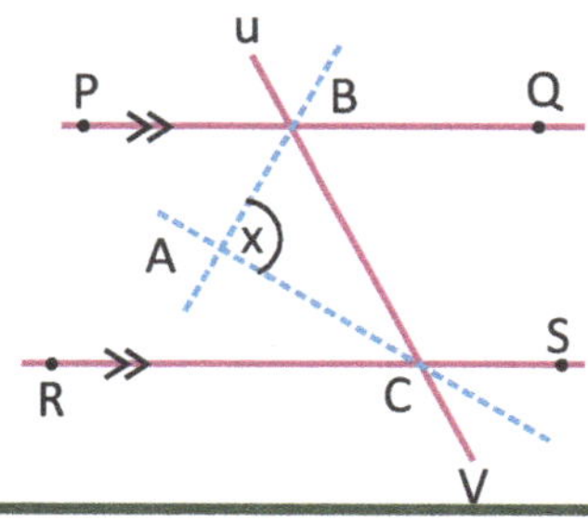

11e. Find the angle x in the figure below.

Given $\angle DAB = \angle DAC = 26°$

$\angle ACD = \angle BCD$

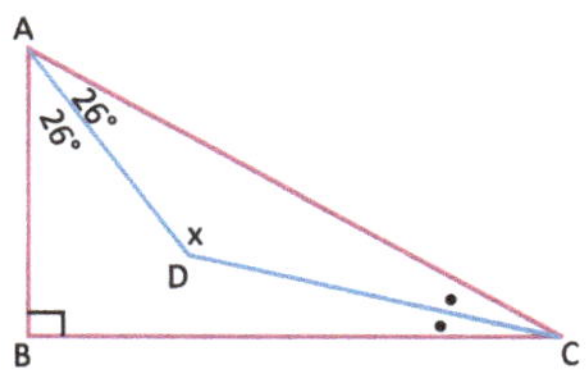

11f. In the figure above, find the change in angle x if $\angle DAC$ is increased from 26 to 31°.

a) $0°$
b) $1.5°$
c) $2.0°$
d) $2.5°$
e) $3.0°$

Solutions to the Question 11

A11a.

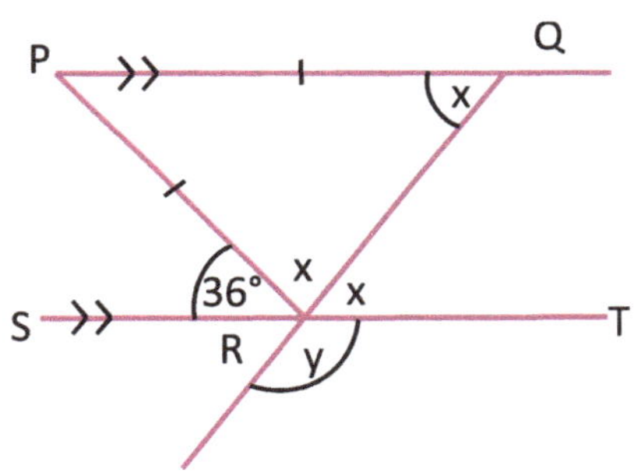

ΔPQR is isosceles $PQ = PR$

$\angle PRQ = \angle PQR = x$; also

$\angle QRT = \angle PQR = x$ alternate angles

$36° + x + x = 180°$ supplementary angles

$2x = 180° - 36° = 144°$

$x = \frac{144°}{2} = 72°; y = 36° + 72° = 108°$

A11b.

$\angle HIG = 110° - 60°$

$y = 50° = \angle HIG$

(opposite angle)

$\angle JID = 60°$

(corresponding angles)

$x = 180° - 50° - 60° = 70°$

(Linear or straight angle)

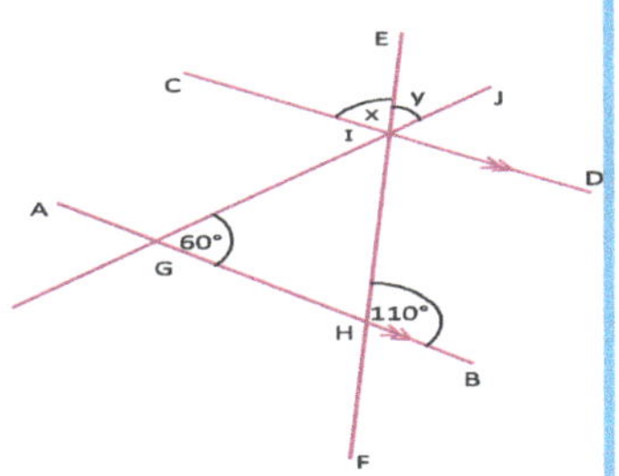

Exterior Angle Theorem

Exterior angle = sum of remote interior angles.

$e = a + b$

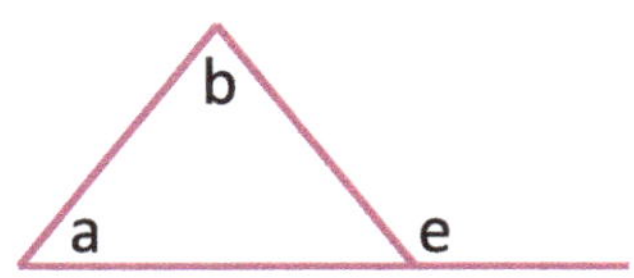

A11c. Exterior Angle Theorem $x = 120° - 90° = 30°$

$\angle DBE = 180° - 90° - 40° = 50°$

$y = 180° - 50° - 30° = 100°$

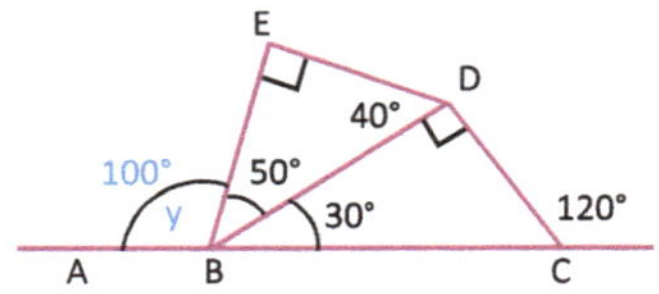

A11d. $\left.\begin{aligned} m &= \angle PBA = \angle ABC \\ n &= \angle BCA = \angle RCA \end{aligned}\right\}$ opposite to equal angles.

$2m + 2n = \angle PBC + RCB = 180°$

$m + n = \angle ABC + \angle ABC = \frac{1}{2}(\angle PBC + \angle RCB)$

$= \frac{1}{2}(180°) = 90°$

$\angle BAC = 180° - (m + n) = 180° - 90° = 90°$

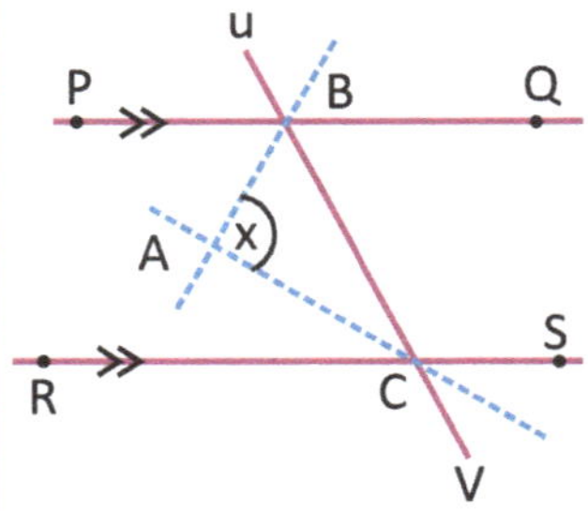

co-interior angles are supplementary

A11e. $\angle ACB = 90° - \angle BAC = 90° - 52° = 38°$

$\angle ACD = \frac{1}{2}\angle ACB = \frac{1}{2}(38°) = 19°$

$\angle ADC = x = 180° - 26° - 19° = 135°$

A1f. Let $\angle A = a$ then $\angle C = 90° - a$

$\angle x = 180° - \frac{1}{2}(a + 90° - a) = 180° - \frac{1}{2}(90)$
$= 135°$

So, always $\angle ADC = x$ will be 135°

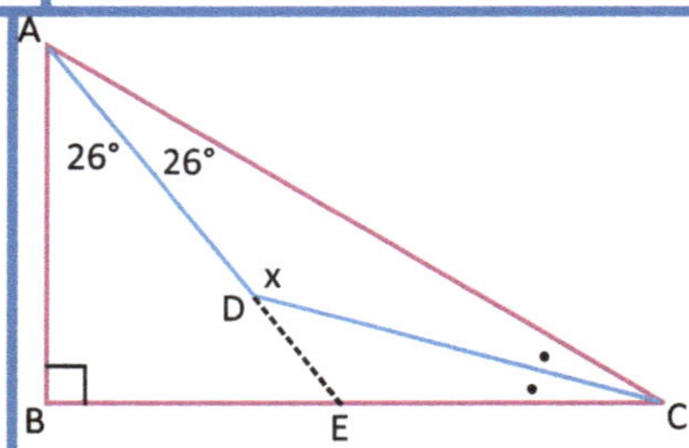

NOTE: **In a right triangle, bisector of acute angles intersect to form 135°**

Reflections and Notes

Chapter 6: Similar Triangles and Circle Geometry

Similar figures are identical shape with different sizes.

Similar triangles have equal angles with proportionate corresponding sides.

Properties of Similar Triangles ($\backsim \Delta s$)

$\Delta ABC \backsim \Delta DBE$

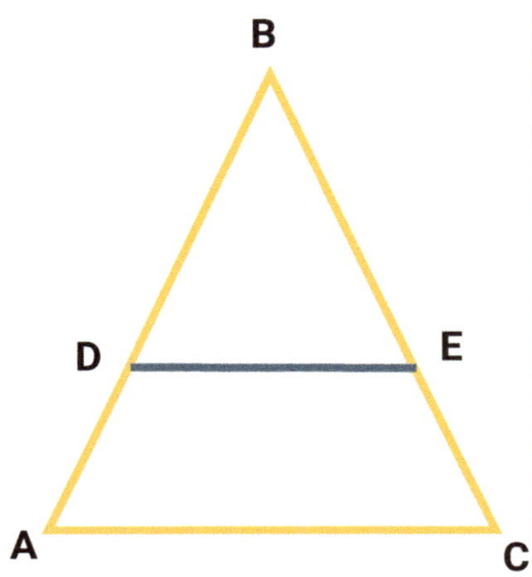

Congruent corresponding angle

$\angle A = \angle D;\ \angle B = \angle B;\ \angle C = \angle E$

Proportional corresponding sides.

Ratio of corresponding sides is a constant, k

$$\frac{AB}{DB} = \frac{BC}{BE} = \frac{AC}{DE} = k$$

Proportional height

Ratio of corresponding sides is a constant, k

$$\frac{height\ of\ \Delta ABC}{height\ of\ \Delta DBE} = k$$

Area of triangles are k^2

$$\frac{\Delta ABC}{\Delta DBE} = k^2$$

Perimeter of triangles are $3k$

$$\frac{Perimeter\ \Delta ABC}{Perimeter\ \Delta DBE} = 3k$$

Similar Triangles Postulate

AA Similarity Postulate

If two angles of one triangle are congruent to two angles of another triangle, then the triangles are similar.

Example 1: Rectangle side extended

Width of a rectangular field ABCD is extended to E as shown in the figure.

Given $AP = 36\ ft, PB = 20\ ft, BE = 15\ ft$

Find the width AD of the rectangle.

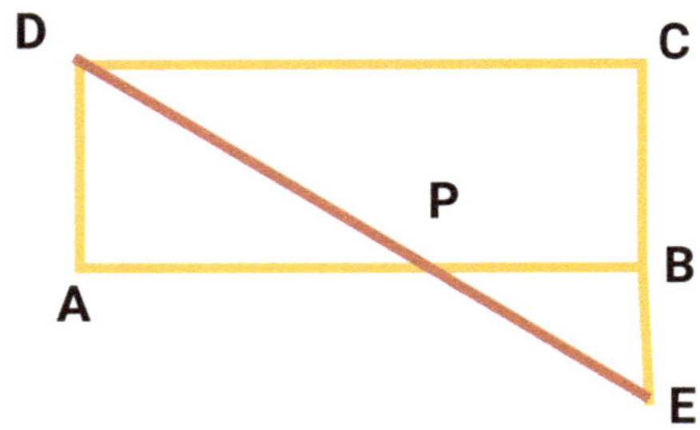

Solution

Consider ΔAPD and ΔBPE

$\angle A = \angle D$ right angles of rectangle

$\angle APD = \angle BPE$ opposite angles

$\Delta APD \backsim \Delta BPE$ for $\backsim AA\ Postulate$

$$\frac{AD}{BE} = \frac{AP}{PB}$$ Corresponding sides are proportional

$$\Rightarrow AD = \frac{AP}{PB} \times BE$$

$$AD = \frac{36}{20} \times 15 = 27$$

Width of the rectangle, AD is 27 ft.

Example 2: Shadow and Height Application

On a sunny day Anil's shadow measures 2.1 m. The school's flagpole has a shadow of 13.5 m. If Anil's height is 175 cm, find the height, h, of flagpole.

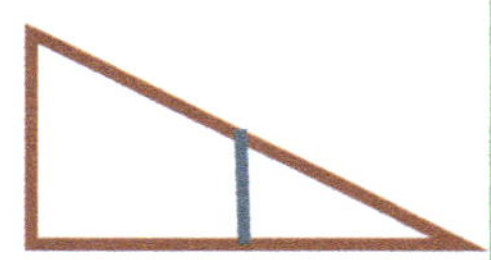

Short Solution:

Students need to label the diagram and provide reasoning

$$\frac{h}{1.75} = \frac{13.5}{2.1} \Rightarrow h = \frac{13.5}{2.1} \times 1.75 = 11.25\ m$$

Practice Problem

A loading ramp is built so that it rises 20 cm vertically for each metre horizontally. What is the vertical height of the ramp with horizontal distance of 4 m?

$80\ cm$

Similar Triangles Extended Application

Example 2:

Two poles, PQ and RS, of height a and b are c units apart. The guy wire from top of one to the base of other intersect at height h. Show that $\boldsymbol{h} = \frac{\boldsymbol{ab}}{\boldsymbol{a+b}}$

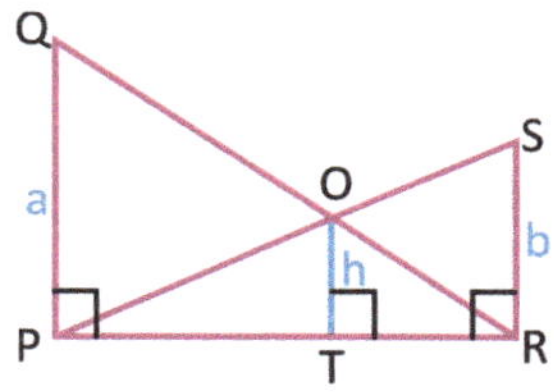

Given: $PQ = a, RS = b,\ TO = h, PR = c$

$\Delta PRS \sim \Delta PTO$ $\frac{RS}{TO} = \frac{PR}{PT}$ $PT = PR \times \frac{TO}{RS}$ .... (1)	$\angle PRS = \angle PTO = 90°$ $\angle SPR = \angle OPT$ common $AA \sim symmetry$
$\Delta\, RPQ \sim \Delta\, RTO$ $\frac{PR}{TO} = \frac{PR}{TR}$ $TR = PR \times \frac{TO}{PQ}$ .... (2) $PR = PT + TR$ $= PR \times \frac{TO}{RS} + PR \times \frac{TO}{PQ}$	$\angle RPQ = \angle RTO = 90°$ $\angle QRP = \angle ORT = common$ $AA \sim symmetry$
$PR = PR \times TO \left(\frac{1}{RS} + \frac{1}{PQ}\right)$ $\frac{1}{TO} = \frac{1}{RS} + \frac{1}{PQ}$ $\frac{1}{TO} = \frac{PQ + RS}{PQ \times RS}$	$TO = \frac{PQ \times RS}{PQ + RS}$ $\boldsymbol{h} = \frac{\boldsymbol{ab}}{\boldsymbol{a+b}}$ $\frac{1}{h} = \frac{1}{a} + \frac{1}{b}$ also written as half of harmonic mean

Challenge Altitude Problem

Determine the length of the altitude AD in the right triangle ABC where the shorter sides AB = 6 cm and AC = 8 cm.

4.8 cm

Example 3: Trapezoid Area Applications

ABCD is a trapezoid

Find ratio of the area of

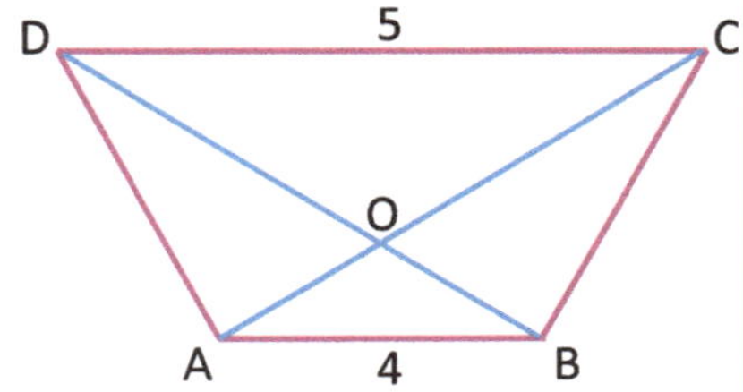

a.	ΔCOD and ΔAOB
b.	ΔCOD and ΔAOD

Solution

a. ΔCOD and ΔAOB: $\Delta COD \sim \Delta AOB$

Ratio of the side is $\frac{5}{4}$

Ratio of the area is $\left(\frac{5}{4}\right)^2$

$$\frac{Area\ of\ \Delta COD}{Area\ of\ \Delta AOB} = \frac{25}{16}$$

b. ΔCOD and ΔAOD

Draw a perpendicular from D to the extended line CA at P. Height of both the triangles is now DP (Same).

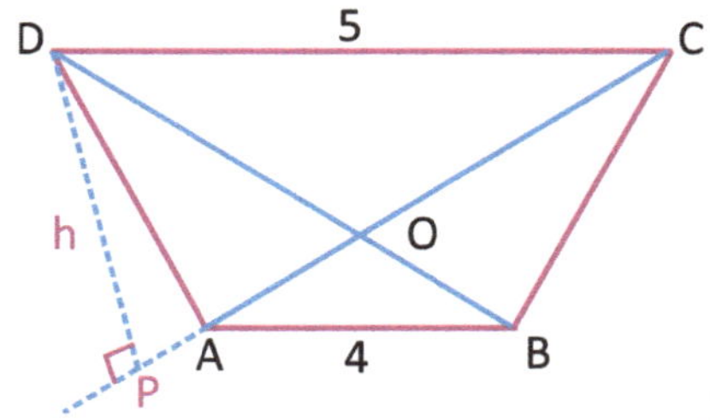

Ratio $\Delta\ COD \sim \Delta\ BOA$

$$\frac{CD}{BA} = \frac{CO}{OA}$$

$$\frac{5}{4} = \frac{CO}{OA}$$

$$\frac{Area\ of\ \Delta COD}{Area\ of\ \Delta AOD} = \frac{\frac{1}{2}\ CO \times h}{\frac{1}{2}\ OA \times H} = \frac{CO}{OA} = \frac{5}{4}$$

Note: If the triangles have the same height, then the ratio of the area is proportional to the length of their base.

When triangle have same base linear ratio of their area is proportional to the heights.

▲ Unit of Inquiry – Extend Your Thinking

Similar triangles show that shape can remain the same even when size changes — a balance between proportion and geometry.

Think about: How does the equality of ratios preserve angles and create similarity?

Try this: Measure the shadow of a tall object (like a flagpole or tree) and compare it with your own shadow to find its height using similar triangles.

Reflect: How do these geometric patterns help architects, engineers, and scientists make measurements when direct access isn't possible?

Chapter 6 Test: Similar Triangles

Similar Triangles

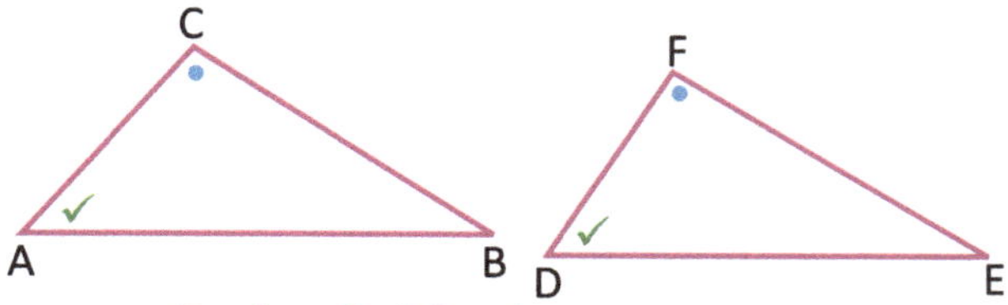

- Ratio of sides is constant
- Corresponding angles are congruent

ΔABC is similar to ΔDEF

$(\Delta ABC \sim \Delta DEF)$ $\sim AA$ Postulate

$\angle A = \angle D, \angle C = \angle F$

$$\frac{AB}{DE} = \frac{BC}{EF} = \frac{AC}{DF} = k$$

$$\frac{Area\ of\ \Delta ABC}{Area\ of\ \Delta DEF} = k^2$$

Provide reasoning and write complete solution. Figure are not to the scale.

1. Find the value of x.
Are the given triangles similar?

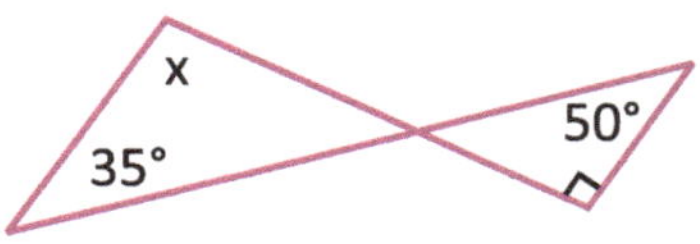

2. Given $\Delta\ MNP \sim \Delta TUV$, find the missing side and all angle measures.

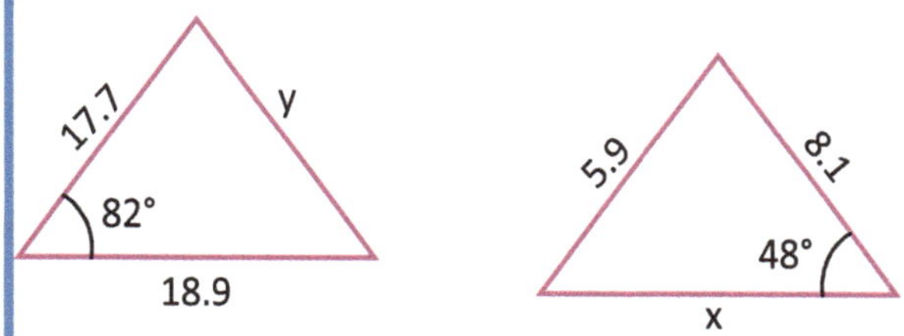

3. Find the unknown sides. Round to nearest tenth.

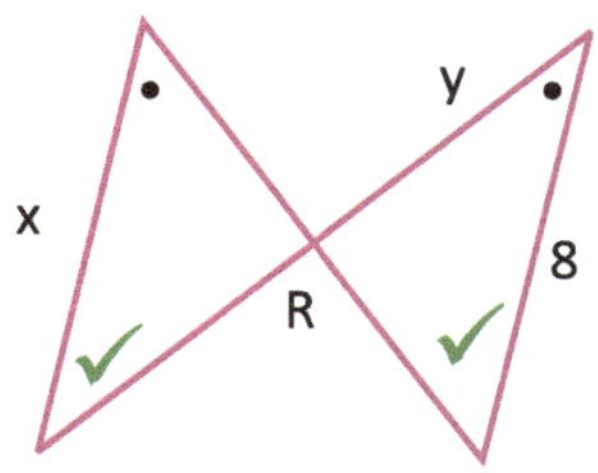

4. Find the unknown sides. Round to nearest tenth.

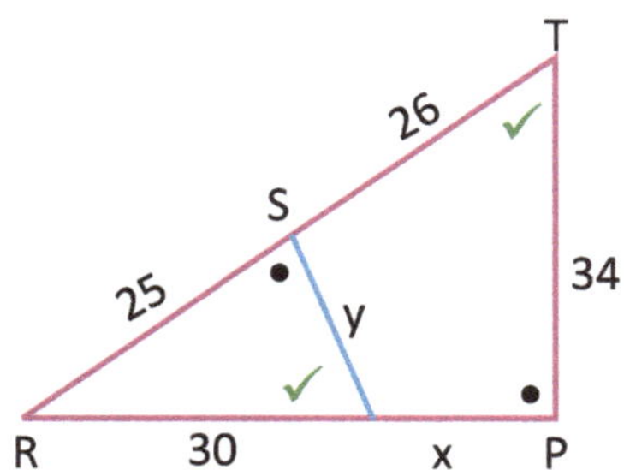

5. Find the unknown sides. Round to nearest tenth.

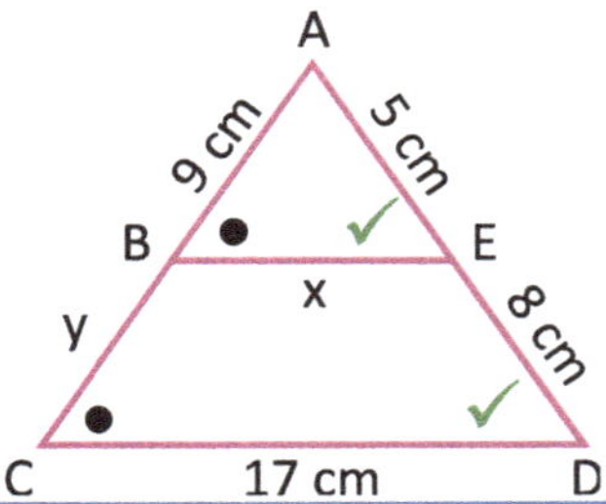

6. A circle with radius 10 cm and a circle with diameter of 40 cm are similar. Area of smaller circle is $100\ \pi cm^2$. Find the area of the larger circle.

7. A triangle has area of $160\ m^2$. A similar triangle is drawn using a scale factor of $\frac{1}{4}$. What is the area of the similar triangle?

8. Surveyors measured the length of a lake, DE, using concepts of similar triangles. Sketch of their measurements is given in the figure. Calculate the length, DE.

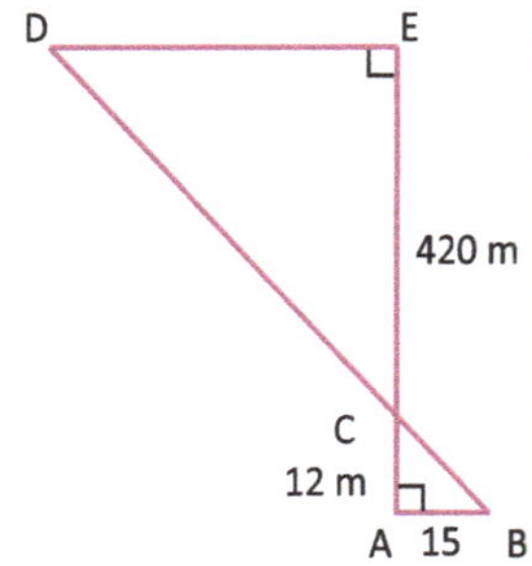

9 To find the height of a tree, Anil placed a minor on the ground at m, and performed measurements as shown. Find the height of the tree if position of Anil's eye level as $1.5\ m$ from level ground.

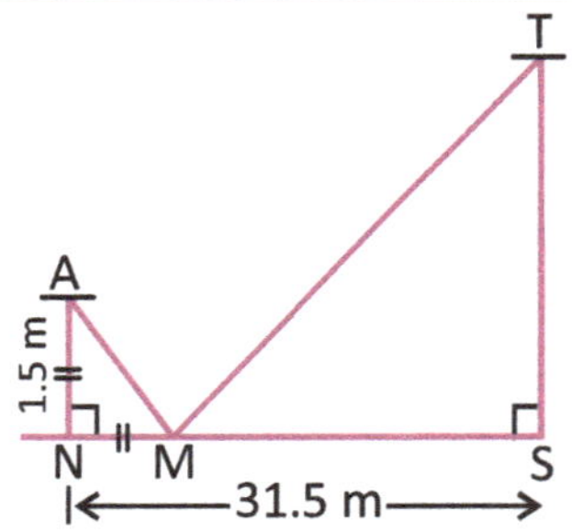

Answers to Chapter 6 Test

1.	$105°$	2.	$x = 63, \quad y = 24.3, \quad a = 50°$	3.	$x = 10.4, \ y = 3$
4.	$x = 12.5 \quad y = 20$	5.	$x = 6.5, \quad y = 14.4$	6.	$400\ \pi$
7.	$10\ m^2$	8.	525 m	9.	30 m

STEM Project: Similar Triangles

10 m ladder making an angle of θ with the ground slides along a vertical wall and comes to rest when the angle with the ground is $90° - \theta$.

Figure is provided for the situation.

Prove that $\Delta PAR \sim \Delta BAQ$ and find value of a and b the given proportionality relation.

$$\frac{h_2}{x} = \frac{\boxed{a}}{\boxed{b}}$$

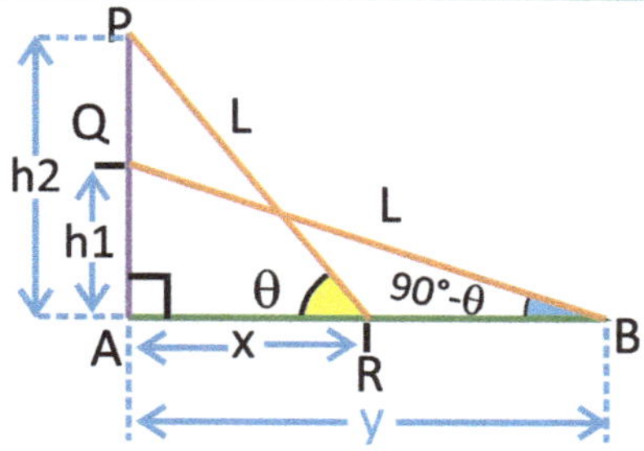

Challenge: Optimization Application

A rectangle is inscribed in an isosceles triangle with height of 8 and base of 10. Find the greatest area of the rectangle.

Hint:

Drop a perpendicular and then use similarity of triangles.

SAT Challenge Question

In the triangle ABC, $\angle ABC = \angle ADEF$

Find the length x and y.

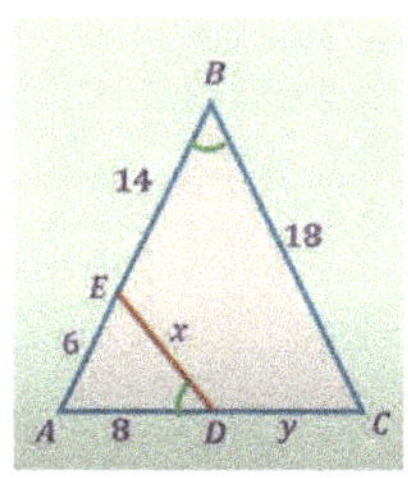

Answer: $x = 7.2, y = 7$

Watch Video

SAT Challenge: Similar Triangle

SCAN ME

Rediscover Circle, Arc and Sector

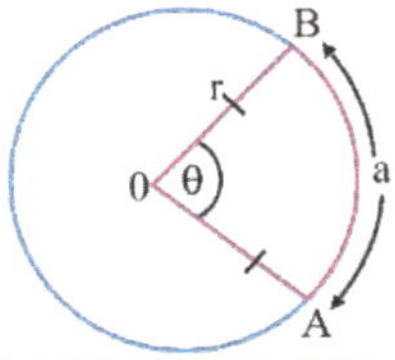

Radius (r)	**Diameter (d)**
Distance from the center to any point on the circle. **Example:** If r=7 cm, every point is 7 cm away from center.	Line passing through the center joining two points on the circle. Formula: d=2r **Example:** If r=7 cm, then d=14 cm.
Circumference (C)	**Area (A)**
The distance around the circle. Formula: C=2πr or πd, π≈3.14 **Example:** For r=7 cm, C≈44 cm.	The region enclosed by the circle. Formula: A=πr² **Example:** For r=7 cm, A≈154 cm².
Arc Length (a)	**Sector Area (S)**
Distance along a part of the circle. Formula: a=(θ/360)×2πr **Example:** For θ=90° and r=7 cm, a≈11 cm.	Region bounded by two radii and the arc. Formula: S=(θ/360)×πr² **Example:** For θ=90° and r=7 cm, S≈38.5 cm².

Example with Short Solution:

Q1	A circle has an area of $16\pi\ unit^2$. Find its circumference. $A = \pi r^2 = 16\pi \Rightarrow r = 4; C = 2\pi r = 2\pi(4) = 8\pi\ units$
Q2	A circle has circumference of $10\pi\ units$. Find its area. $C = 2\pi r = 10\pi \Rightarrow r = 5; A = \pi r^2 = \pi(5)^2 = 25\pi\ unit^2$
Q3	Find sector area if the arc of 10π subtends and angle of $60°$ $a = \frac{\theta}{360°} \times 2\pi r \Rightarrow r = a\frac{360°}{2\pi\theta} = 10\pi\ \frac{360°}{2\pi \times 60°} = 30;$ $S = \frac{\theta}{360°}\ \pi r^2 = \frac{60°}{360°}\ \pi(30)^2 = 150\pi$

Quiz: Circle, Circumference, Area, Arc Length and Sector

Q1	A circle has radius $r = 7$ cm. Find its circumference.
Q2	In a circle of radius 10 cm, find the arc length subtended by a central angle of 60°.
Q3	A circle has diameter 14 cm. Find the area of a sector with angle 90°.
Q4	The arc length of a circle is 22 cm, and the radius is 7 cm. Find the angle subtended at the center.
Q5	Find the radius of a circle if an arc of 72° has length 25.12 cm.
Q6	A circle has radius 21 cm. Find the area of a sector with angle 120°.
Q7	A wheel of radius 35 cm rotates through 90°. How far does a point on the rim travel?
Q8	A circular field has radius 28 m. Find the length of the fence needed to enclose half the field along its arc.
Q9	In a circle of radius 14 cm, a sector has angle 45°. Find both arc length and area of the sector.
Q10	A sector of a circle with angle 150° has area 75π cm². Find the radius of the circle.

Answers: Round to 2 decimal places

Q1	43.98 cm	Q2	10.47 cm	Q3	7 cm	Q4	180°
Q5	20 cm	Q6	461.81 cm²	Q7	54.98 cm	Q8	87.96 m
Q9	Arc length = $(45/360) \times 2\pi \times 14 = 1/8 \times 28\pi = 3.5\pi \approx 10.99$ cm Sector area = $(45/360) \times \pi \times 14^2 = 1/8 \times \pi \times 196 = 24.5\pi \approx 76.97$ cm²						
Q10	$A = (\theta/360) \times \pi r^2 \Rightarrow 75\pi = (150/360) \times \pi r^2 \Rightarrow 75 = (5/12)r^2$ $r^2 = 180 \Rightarrow r = \sqrt{180} \approx 13.42$ cm						

3 Theorems on Chords, Secants, and Tangents

Extended Application of Similar Triangles.

A chord of a circle is a line segment whose endpoints lie on the circle.

If a chord AB is cut at point P (not necessarily at the midpoint), then the two parts, AP and PB, are called the segments of the chord AB.

Theorem 1

When two chords intersect inside a circle, the product of the segments of one chord equals the product of the segment of the other chord.

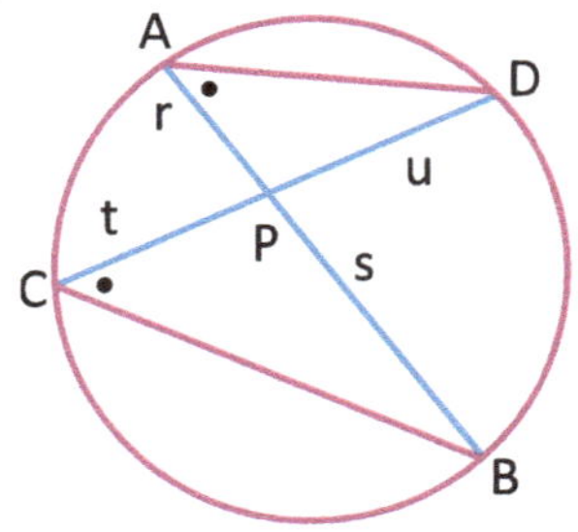

Understand Concept

Segments are from common point P.

Chords AB and CD interest at P.

Prove: $r.s = t.u$

Proof

Join AD and CB $\angle P$ is common $\angle A = \angle C$ subtends same arc BD $\Delta\, APD \sim \Delta\, CBP$ *~AA Postulate*	$\frac{r}{t} = \frac{u}{s} \Rightarrow$ $r.s = t.u$ **r. s = t. u**

✍ **Practice: Refer to the figure above to answer the questions.**

Q1	If $AP = 8, PB = 3, CP = 6$, find DP	4
Q2	If $AP = 9, PB = 2, CP = x, and\ PD = x - 3$, find x	3
Q3	If $AP = 8 - x, PB = x, CP = 3, and\ PD = 4$, find x	6 or 2

Theorem 2

When two secants are drawn from external point, the product of one secant segment and its external segment equals the product of the other set of secant segment and its external segment.

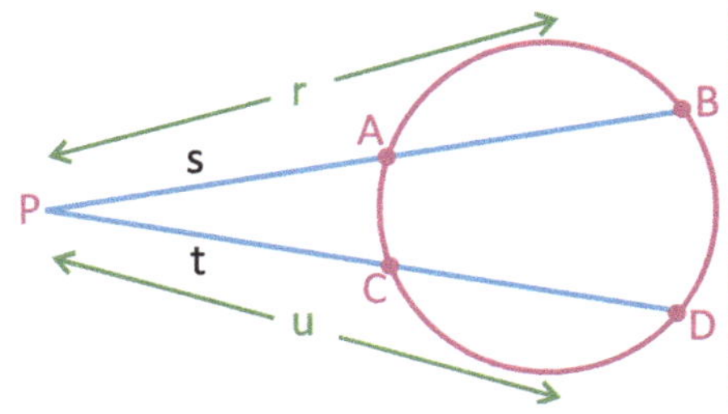

Important: lengths from the common point, P, should be taken.

You can prove that: $r.s = t.u$

r. s = t. u

Proof of these theorems is based on similar triangles. It is an exercise for the students.

Theorem 3

If you have a secant and a tangent from the external point P as shown, then also the product property holds.

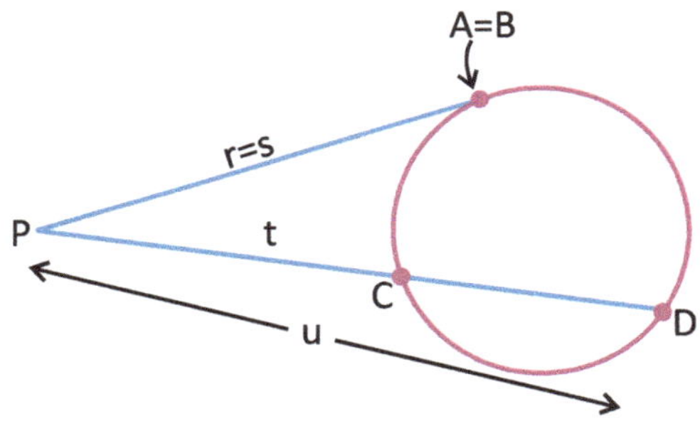

r. s = t. u

Practice: Refer to the figure above to answer the questions.

Q3	**Theorem 2 — Two Secants from P** If $AP = 6, PB = 20, CP = 8$, find DP 15
Q4	**Theorem 3 — Secant & Tangent from P** If $PB = 8, CP = 10$ find tangent length PA 12

Important Topic for Olympiad

Chord Segments Relation in the Circle

1. Two chords AB and CD intersect at P such that $AP = 4\ cm$, $PB = 6\ cm$ and $DP = 8\ cm$. Find the value of x
   a) 1.0
   b) 2.0
   c) 3.0
   d) 3.5

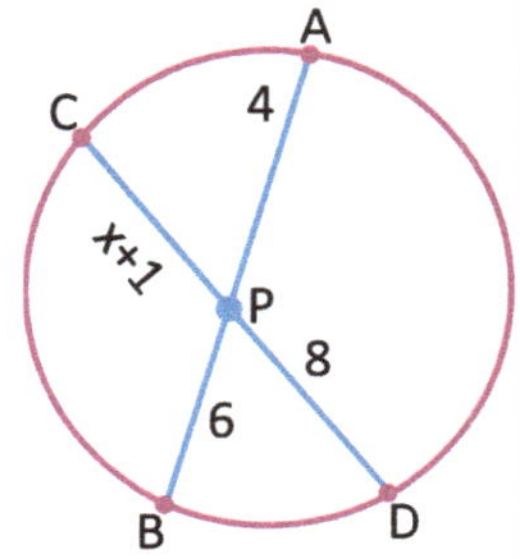

2. **Segments of Chords Theorem Application**
   Tangent and the secant intersect in the exterior of a circle
   at A as shown in the figure below. Find the value of x.
   a) 2.0
   b) 2.5
   c) 3.0
   d) 4.0

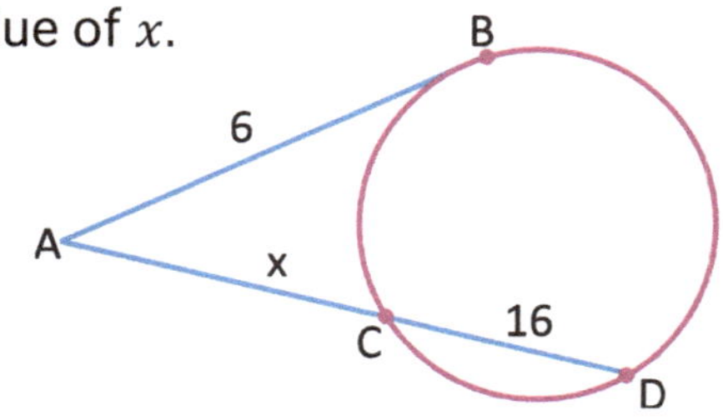

3. **Intersection of chords and Radius Relations**
   Two mutually perpendicular chords AB and CD Intersect at P such that $AP = 2\ cm$, $PB = 6\ cm$ and $DP = 4\ cm$.
   Find the radius of the circle
   a) $\frac{1}{2}\sqrt{56}\ cm$
   b) $\frac{1}{2}\sqrt{65}\ cm$
   c) $\frac{1}{4}\sqrt{56}\ cm$
   d) $\frac{1}{4}\sqrt{65}$ cm

SCAN ME

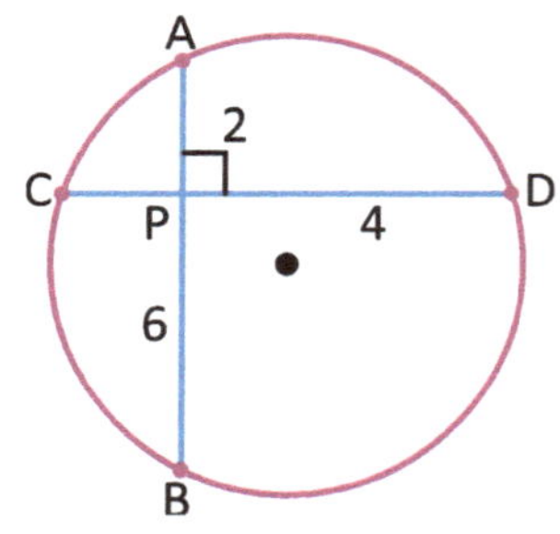

Answers: Chord Segment Relations

Q1	b	Q2	a	Q3	b.

Challenge Problem: Distance to Horizon

Horizon is where the line of sight is tangent to the earth. This can help find the distance to horizon. Show that the distance to horizon is $3600\sqrt{lookout\ height}$.
Diameter of the earth is 12,800,000 m.

Radius and Perpendicular Chords Challenge

Extremely Important Relation for Competitions

Prove:	$4r^2 = a^2 + b^2 + c^2 + d^2$ where r is radius of a circle with two perpendicular chords forming chord segments a, b, c and d as shown in the figure.

Product of lengths of chords is equal, $ab = cd$ Chord Segments Theorem

Construct right bisectors of chords to find centre O.

M and N are midpoints of AB and CD respectively

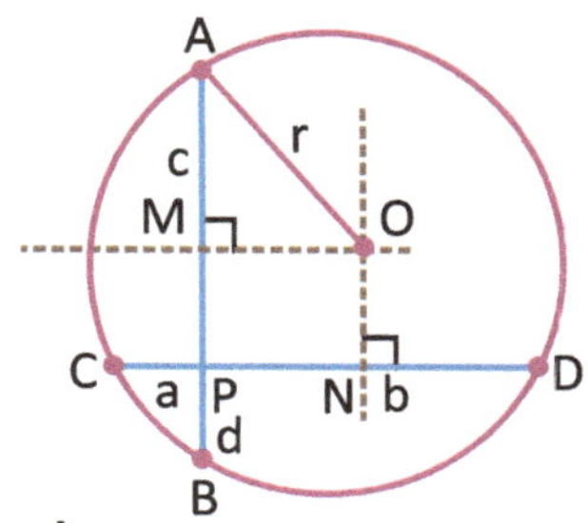

$$AM = \frac{c+d}{2} ; CN = \frac{a+b}{2}$$

$$PN = CN - CP = \frac{a+b}{2} - a = \frac{a+b-2a}{2} = \frac{b-a}{2}$$

$$OM = PN = \frac{b-a}{2}, \ AM = \frac{c+d}{2}$$

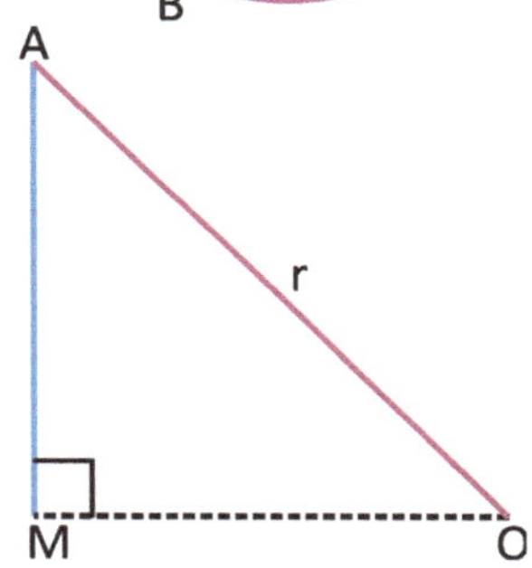

$$r^2 = AM^2 + OM^2$$

$$r^2 = \left(\frac{c+d}{2}\right)^2 + \left(\frac{b-a}{2}\right)^2$$

$$r^2 = \frac{1}{4}(c^2 + d^2 + 2cd + b^2 + a^2 - 2ab)$$

$$r^2 = \frac{1}{4}(c^2 + d^2 + b^2 + a^2 + 2(cd - ab))$$

$$r^2 = \frac{1}{4}(c^2 + d^2 + b^2 + a^2)$$

$ab = cd$

Chord Segments Theorem

$$4r^2 = a^2 + b^2 + c^2 + d^2)$$

✍ Practice Problem

Prove that lengths of tangents drawn from an external point to a circle are equal.

Draw a diagram and provide reason for each step.

Olympiad Challenge: Square and Semicircle

Square $ABCD$ has side-length of $a\ cm$. A semicircle with diameter AB is constructed inside the square with its centre at O. The tangent to the semicircle from C intersects the side AD at E. Find the length of CE.

Hint: Let P be the point of tangency. The length of tangents from a point is equal.

Answer: $CE = \frac{5a}{4}$

SCAN ME

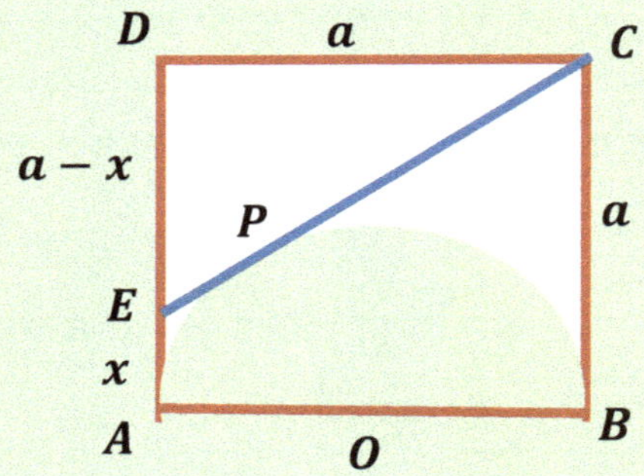

SAT Challenge: Area Concept

Circle is inscribed in the right triangle as shown. The circle touching the hypotenuse at P divides it in the lengths x and y. Show that the area of the triangle is xy.

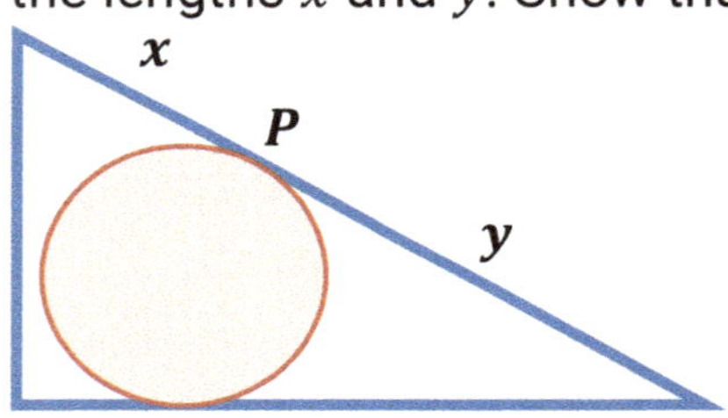

Watch Video **SAT Challenge: Area Concept** SCAN ME

SAT Challenge

I have a piece of wire in the shape of a square with the each side length of $x\ cm$. If I bend the wire to form a circle, by how many times the area increases?

a.	$\frac{1}{\pi}$	b.	$\frac{2}{\pi}$	c.	$\frac{4}{\pi}$	d.	$\frac{8}{\pi}$

Chapter 7: Right Triangles

Right Triangles Basics

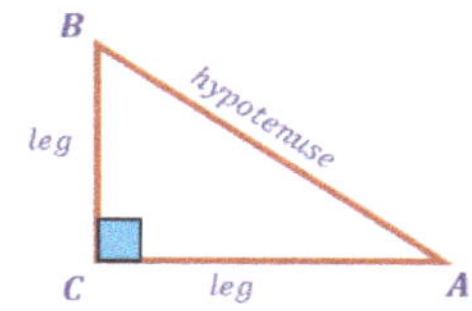

- *Right triangle has two acute angles and a right angle*
- *Hypotenuse, is the longest side opposite to right angle*
- *Sides opposite to the acute angles are called legs*
- *Acute angles form complementary pairs (sum is* $90°$*)*
- *The three sides of right triangle are related with Pythagorean Theorem*
- *Midpoint of hypotenuse, circumcenter, is equidistant from all vertices*

Let us explore and master strategies to understand effective use of right triangles and extend it as a foundation for Trigonometry and advanced Algebra.

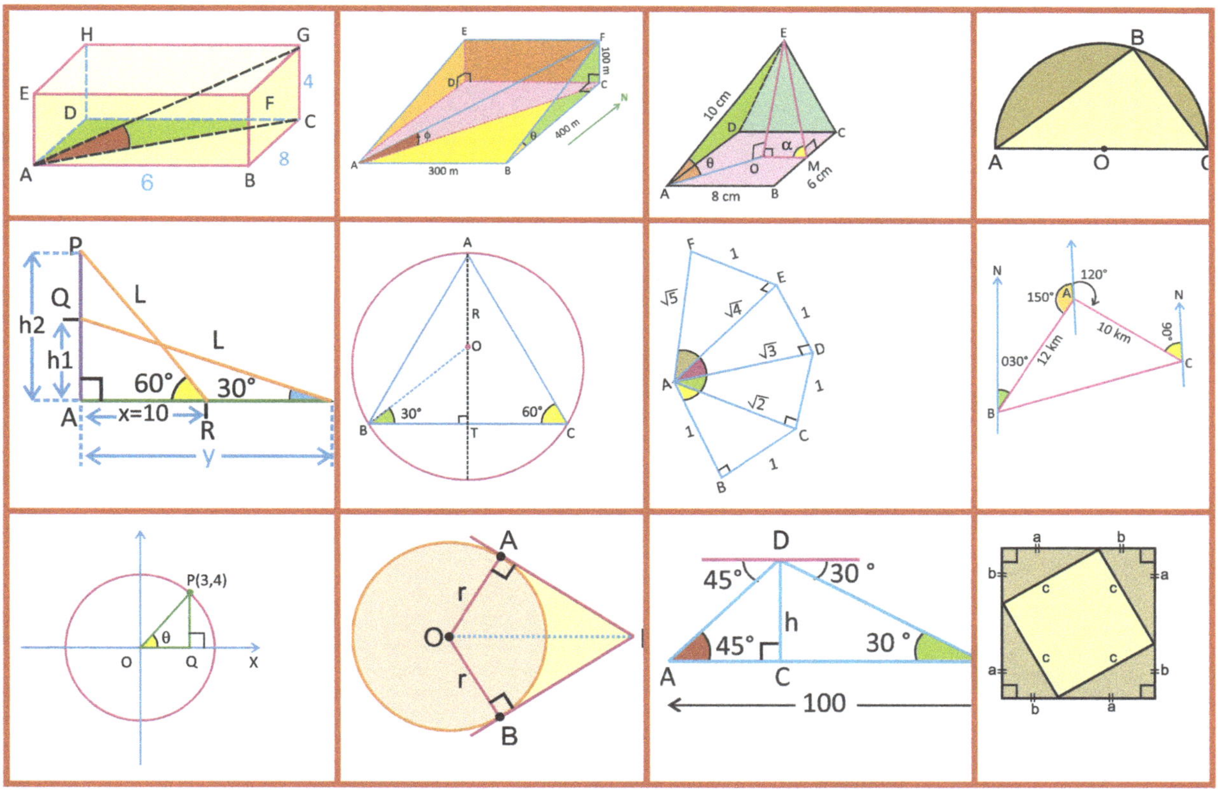

Right Triangle Applications SCAN ME

The Pythagorean Theorem Applications

Pythagorean theorem relates the three sides of a right triangle. Sum of the squares of the smaller sides, called legs, is equal to the square of hypotenuse.

The terms – legs and hypotenuse will also be used to represent their length.

In a right triangle ABC $a^2 + b^2 = c^2$ or $leg^2 + leg^2 = hypotenuse^2$

Sum of two sides is greater than the third, $a + b > c$

Here, a and b represents the legs and c is the hypotenuse of right triangle ABC

Find Length of Third Side

Example 1: Find the missing side x in each triangle

a.

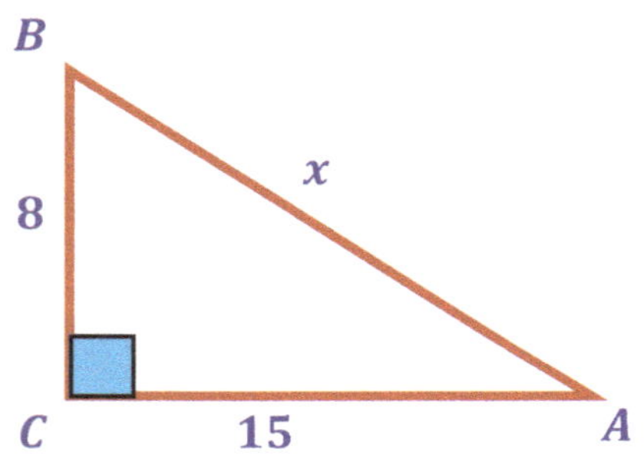

$8^2 + 15^2 = x^2$

$64 + 225 = x^2$

$289 = x^2$

$\sqrt{289} = x$

$x = 17$

Sum of squares for hypotenuse

b.

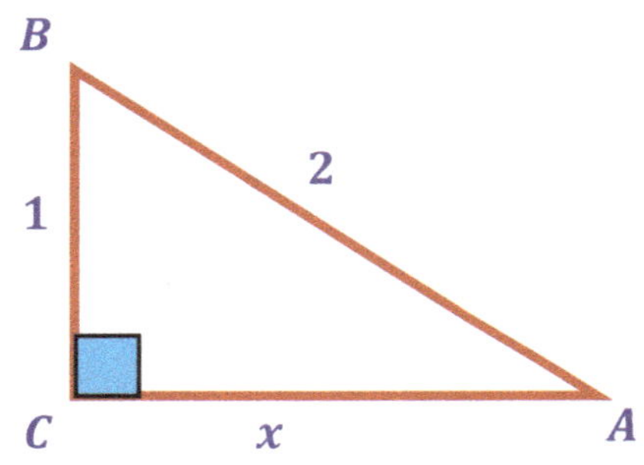

$1^2 + x^2 = 2^2$

$1 + x^2 = 4$

$x^2 = 4 - 1$

$x^2 = 3$

$x = \sqrt{3}$

Difference of squares for legs

Converse of Pythagorean theorem states that if the square of longest side is equal to the sum of the squares of shorter sides in a triangle, then the triangle is a right-angle triangle.

Q1	Find the diagonal length of a square with an area of 144 cm^2 $12\sqrt{2}cm$

Practice Questions based on Example 1

Pythagorean Theorem Application: Find Missing Side Length

Find the missing side x in each triangle

Q 1

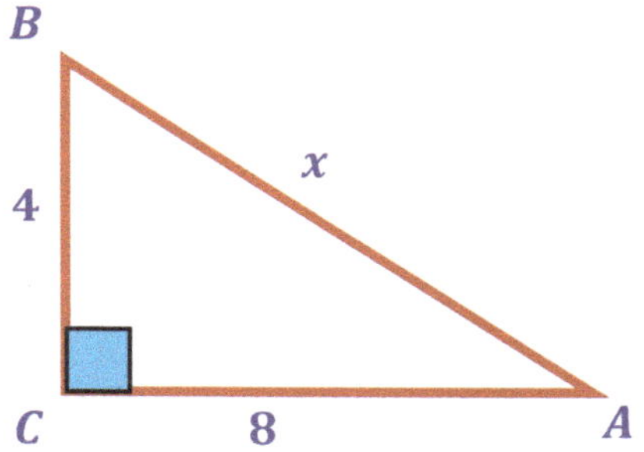

$x = 4\sqrt{5}$

Q 2

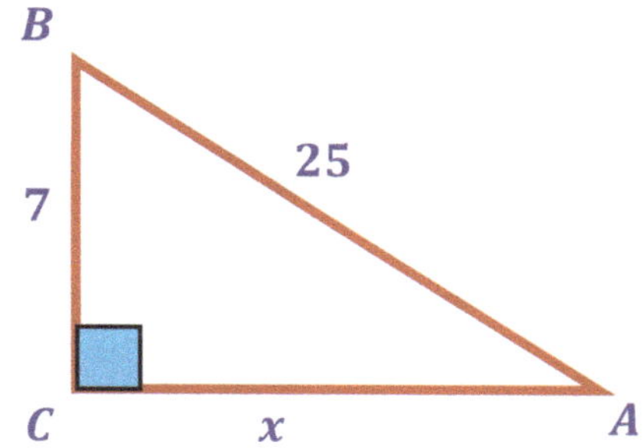

$x = 24$

Q 3

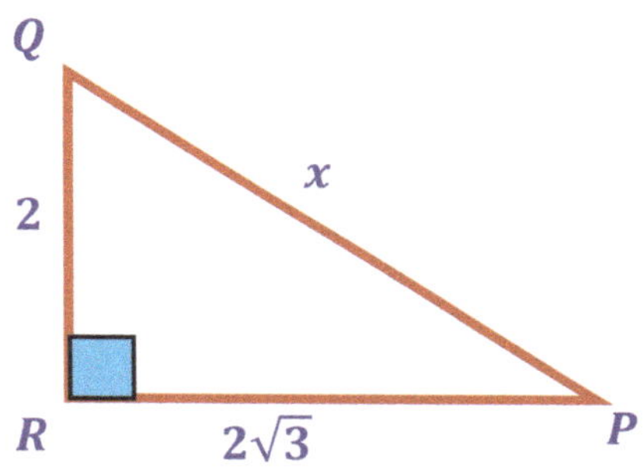

$x = 4$

Q 4

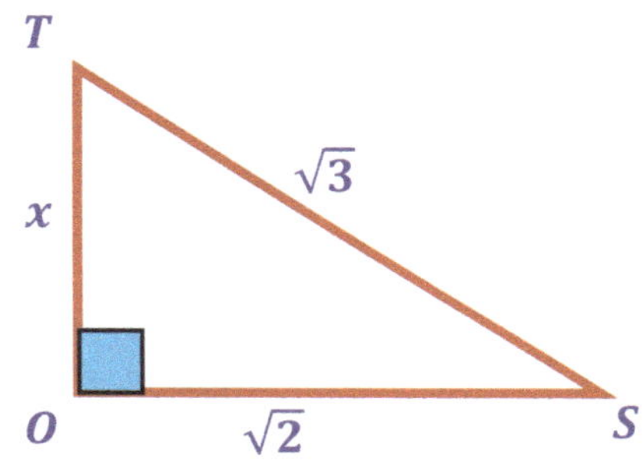

$x = 1$

Additional Information: Irrational Numbers

$\sqrt{n}$ is an irrational number if n is not a perfect square. These are non-repeating non-terminating decimal numbers and so we cannot find their exact value or even write as a rational number. Rational numbers can be written in the form $\frac{p}{q}$, where p and q do not have any common factor and $q \neq 0$.

Write exact answer if not specified. $= 4\sqrt{5}$ is the exact value

Some right triangles have integer values for all the side lengths.

These are called Pythagorean Triples. Triangle in Q2 is an example (7, 24, 25)

Activity 1: Represent Exact Value of Irrational Number

$\sqrt{5}$ is an irrational number. $\sqrt{5} \approx 2.236067977$. Since we know approximate value only, how can we draw a segment that measures exactly $\sqrt{5}\ cm$.

Show $\sqrt{5}$ on a number line.

Verifying Right Triangle

Example 2: Verify if the given triangles are right triangles or not.

Strategy: Check if $a^2 + b^2 = c^2$ or not.

a.

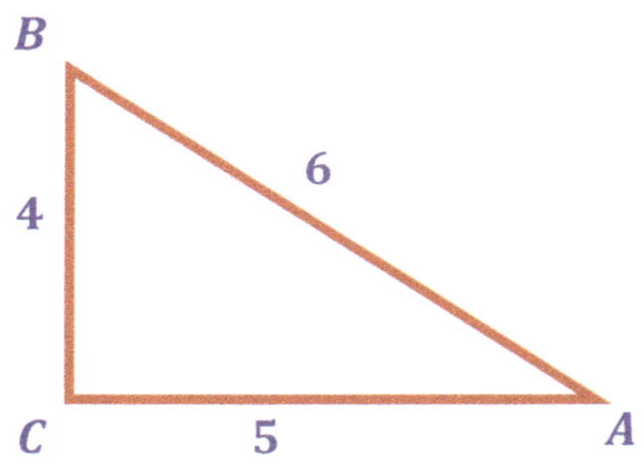

$4^2 + 5^2 \cdots 6^2$

$16 + 25 \cdots 36$

$16 + 25 \cdots 36$

$41 \neq 36$

Check sum of squares of the leg with hypotenuse square

Not a right triangle

b.

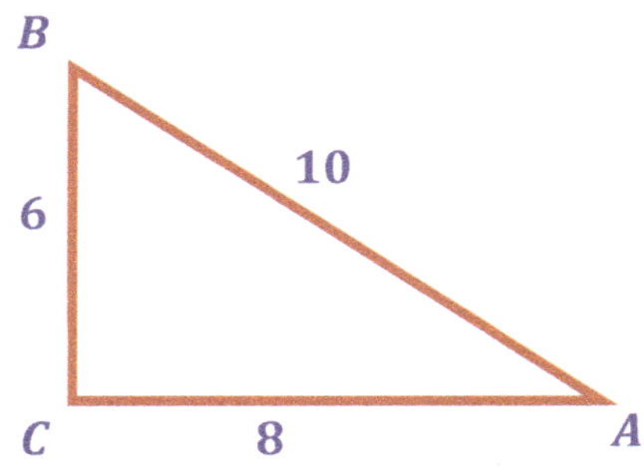

$6^2 + 8^2 \cdots 10^2$

$36 + 64 \cdots 100$

$100 \cdots 100$

$100 = 100$

Check sum of squares of the leg with hypotenuse

It is right triangle

NOTE:

Pythagorean theorem is used to verify if the triangles are right triangle or not.

$leg^2 + leg^2 = hypotenuse^2$: Right Triangle

$leg^2 + leg^2 \neq hypotenuse^2$: Not a Right Triangle

Practice Questions based on Example 2

Pythagorean Theorem Application: Verification of Right Triangle

Verify if the given triangles are right triangles or not.

Strategy: Check if $a^2 + b^2 = c^2$ or not.

Q 5

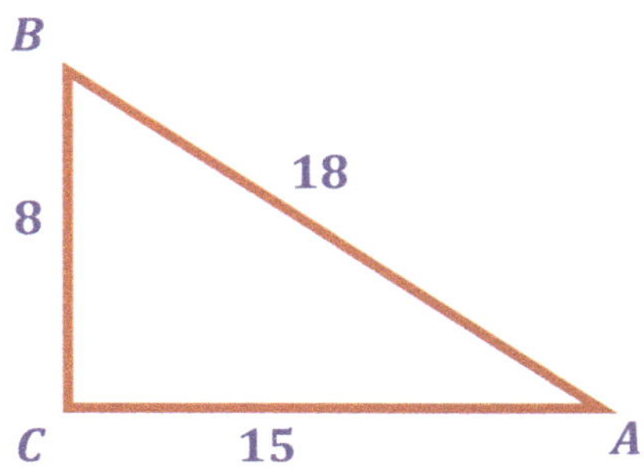

Q 6

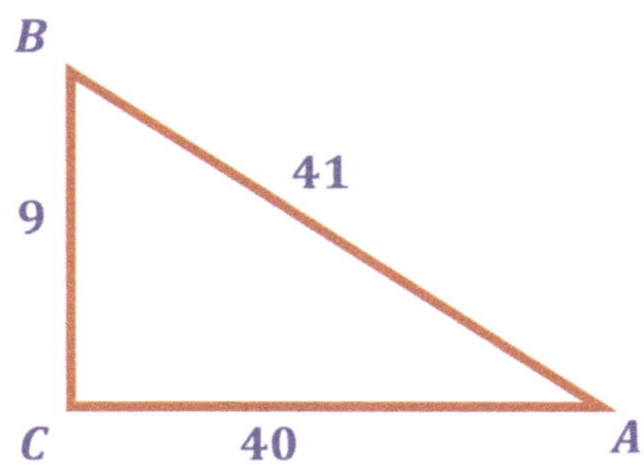

Q 7

In the ΔPQR,

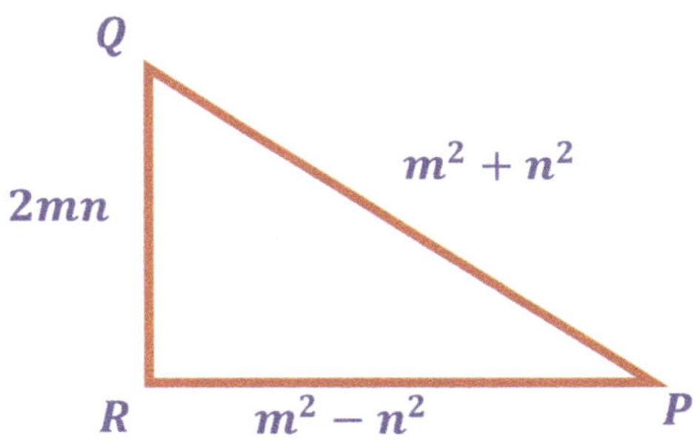

$m > n > 0$

Activity:

Prove that this triange is an right triangle.

Calculate 3 set of possible values for the given triangle.

Note:

Pythagorean theorem is used to verify if the triangles are right triangle or not.

Methods to Create Pythagorean Triple

Euclid's Formula for Pythagorean Triple

Solution of Q7: Verify if the given triangles are right triangles or not.

Strategy: Check if $a^2 + b^2 = c^2$ or not.

Q7

Let us verify if the given triangle, PQR, is a right angle triangle or not.

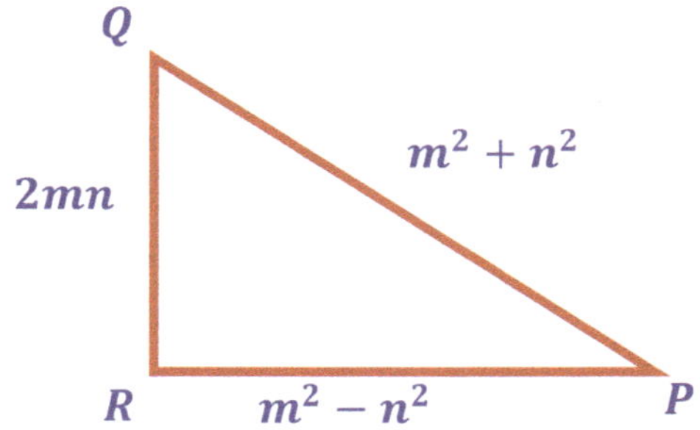

For right triangle

$PR^2 + QR^2 = PQ^2$

$PR = m^2 - n^2, QR = 2mn$

$PQ = (m^2 + n^2)$

Let us simplify the expression

$PR^2 + QR^2$

$= (m^2 - n^2)^2 + (2mn)^2$

$= m^4 - 2m^2n^2 + n^4 + 4m^2n^2$

$= m^4 + n^4 + 2m^2n^2$

$= (m^2 + n^2)^2 = PQ^2$

Since

$\boldsymbol{PR^2 + QR^2 = PQ^2}$ ***the triangle is a right triangle***

Try to generate Pythagorean Triples for few values

NOTE: This method of finding Pythagorean Triples is by Euclid's Formula

Primitive Pythagorean triple is one in which all the side lengths are coprime.

$\boldsymbol{(3, 4, 5)}$ ***and*** $\boldsymbol{(5, 12, 13)}$ are examples of primitive triples.

There are 16 primitive triples with hypotenuse less than 100.

Infinitely many other triples can be produced by multiplying primitive triple with any number. $(6, 8, 10)$ *and* $(10, 24, 26)$ are not primitive triples.

Pythagorean Triples: Odd Length as the Base

Example 3: Find the Pythagorean Triples

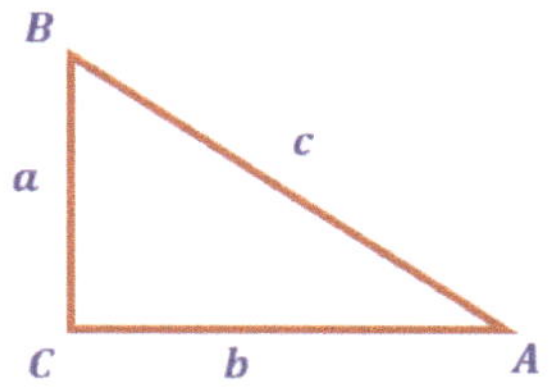

Create a Table of values to show that $a^2 + b^2 = c^2$ for $n = \{1, 2, 3, 4, 5, 6\}$ where

$$a = 2n + 1, b = \frac{a^2 - 1}{2}, and\ c = \frac{a^2 + 1}{2}$$

n	a	b	c
1	$2(1) + 1 = 3$	$\frac{3^2 - 1}{2} = 4$	$\frac{3^2 + 1}{2} = 5$
2	$2(2) + 1 = 5$	$\frac{5^2 - 1}{2} = 12$	$\frac{5^2 + 1}{2} = 13$
3	$2(3) + 1 = 7$	$\frac{7^2 - 1}{2} = 24$	$\frac{7^2 + 1}{2} = 25$
4	$2(4) + 1 = 9$	$\frac{9^2 - 1}{2} = 40$	$\frac{9^2 + 1}{2} = 41$
5	$2(5) + 1 = 11$	$\frac{11^2 - 1}{2} = 60$	$\frac{11^2 + 1}{2} = 61$
6	$2(6) + 1 = 13$	$\frac{13^2 - 1}{2} = 84$	$\frac{13^2 + 1}{2} = 85$

NOTE: This method of finding Pythagorean Triples is credited to Pythagoras

Primitive Pythagorean triples calculated above have hypotenuse less than 100.

$$(3, 4, 5), (5, 12, 13), (7, 24, 25), (9, 40, 41), (11, 60, 61), (13, 84, 85)$$

Challenge Question

If area of the green square is 9 m^2. Find the area of the red square.

Question 8: Find the Pythagorean Triples for Even Base Length

Create a Table of values to show that $a^2 + b^2 = c^2$ for

$n = \{1, 2, 3, 4, 5, 6, 7\}$ where

$a = 2n, b = \left(\frac{a}{2}\right)^2 - 1, and\ c = \left(\frac{a}{2}\right)^2 + 1$

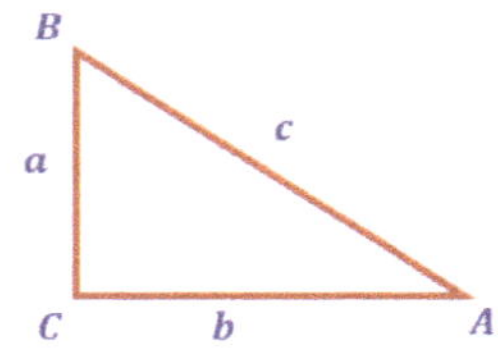

n	$a = 2n$	$b\left(\frac{a}{2}\right)^2 - 1$	$c\left(\frac{a}{2}\right)^2 + 1$
1			
2			
⋮			
7			

NOTE: This method of finding Pythagorean Triples is credited to Plato

*Some of the triples calculated above are **not prime as they do have common factors.** Extend the series to get all the 16 prime Pythagorean triples with hypotenuse less than 100.*

Question 9: Find the Pythagorean Triples for Odd Base Length

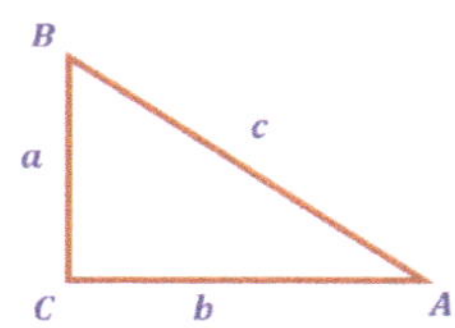

Create a Table of values to show that $a^2 + b^2 = c^2$ for

$n = \{1, 2, 3, 4, 5, 6, 7\}$ where

$a = 2n + 1, b = 2n^2 + 2n, and\ c = 2n^2 + 2n + 1$

$\boldsymbol{n}$	$\boldsymbol{a} = 2n + 1$	$\boldsymbol{b} = 2n^2 + 2n$	$\boldsymbol{c} = 2n^2 + 2n + 1$
1			
2			
⋮			
7			

NOTE: This method of finding Pythagorean Triples is credited to Plato

STEM Project to Create Pythagorean Triple

Pythagorean triples connect arithmetic patterns with geometric truth — but there's more than one way to create them.

Think about: What happens when you take the reciprocals of two consecutive even or two consecutive odd numbers and use them to form sides of a right triangle?

Try this: Choose pairs like (2, 4) or (3, 5). Add their reciprocals and explore how the result relates to a Pythagorean triple.

Reflect: Why do patterns built from reciprocals still lead to right-triangle relationships? What does this reveal about harmony between numbers and geometry?

Example:

$$\frac{1}{3}+\frac{1}{5}=\frac{5+3}{15}=\frac{8}{15}$$

$$8^2+15^2=289=17^2$$

Pythagorean Triple is: $\mathbf{8, 15, 17}$

Show algebraically that this method always produces triples.

General Equation is of the form:

$$\frac{1}{n}+\frac{1}{n+2}=\frac{A}{B}$$

Challenge Problem: Right Triangle

Determine the unknown side of the right triangle whose hypotenuse is 10 cm and area is 25 square units. $5\sqrt{2}$

Challenge Question: Pythagorean Application

If area of the red square is 9 m^2. Find perimeter of the green square.

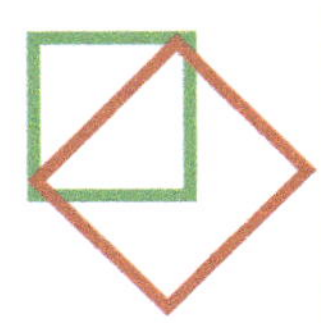

The Midpoint of the Hypotenuse Theorem

In any right triangle, the **midpoint of the hypotenuse** is equidistant from all three vertices. Equivalently, the midpoint of the hypotenuse is the **circumcenter** of the triangle, and the circumradius equals **half the hypotenuse**.

Prove the theorem.

Pythagorean Triples Applications

Knowledge of **Pythagorean triples** (three positive integers that satisfy right triangle relation) help solve problems quickly. This is an extremely useful strategy when time is a major constraint.

Let us first list all the **16 primitive Pythagorean triples with hypotenuse less than 100**. Infinitely many other triples are created by multiplying with common factor.

Set of primitive triples starting with odd number as length of the smallest side: $(3,4,5), (5,12,13), (7,24,25), (9,40,41), (11,60,61),$ $(13,84,85), (33,56,65), (39,80,89), (65,72,97)$
Set of triples starting with even number as length of the smallest side: $(8,15,17), (12,35,37), (16,63,65), (20,21,29),$ $(28,45,53), (36,77,85), (48,55,73)$
Multiples of primitive triples are also triples. $k(3,4,5), k\ is\ Natural\ number$ Example 10c uses this concept to find the solution.

Example 10: Pythagorean Triples for Quick Application Solution

a.	Nick walks **24 m** east and then **7 m** North. How far is he from the starting point?
b.	A **13 feet** ladder is placed against a vertical wall. Find the height to which it will reach if the foot of the ladder is **5 feet** from the base of the wall.
c.	Find the hypotenuse of a right triangle whose legs are 9 cm and 12 cm.
d.	A rectangular sports field is **36 m** by **15 m**. How long is the shortest diagonal path across it?
e.	A pole is **20 ft** tall. A stake is placed **21 ft** from the base. What length of wire is needed from the stake to the top?

Solve and then check the solution on the next page

Solution 10: Pythagorean Triples for Quick Solution

Students are encouraged to sketch figure and then solve each question using Pythagorean theorem. Shortcut solution is provided as TIP.

a.	Nick walks 24 m east and then 7 m North. How far is he from the starting point? Reference triple: $(7, 24, 25)$ Find hypotenuse as 24 and 7 are the legs Nick is 25 m from the starting point
b.	A 13 feet ladder is placed against a vertical wall. Find the height to which it will reach if the foot of the ladder is 5 feet from the base of the wall. Reference triple: $(5, 12, 13)$ Height is one of the legs with given hypotenuse of 13 and a side of 5 Height reached by the ladder is 12 feet
c.	Find the hypotenuse of a right triangle whose legs are 9 cm and 12 cm. We do not have any triple with (9, 12, ?) to find the hypotenuse. Notice that 9 and 12 have a common factor of 3 and so it can be related with the triple $3(3, 4, 5) = (9, 12, 15)$ Now, we have the solution. Hypotenuse should be 15 cm for the legs as 9 cm and 12 cm.
d.	Hint: **triple:** $(5, 12, 13) \equiv 3(5, 12, 13)$
e.	Hint: **triple:** $(20, 21, 29)$

Mastering above technique can help to beat the competetion

Pythagorean Triples SCAN ME

Pythagorean Triple Advantage

Question 10: Pythagorean Triples Shortcut Solution

Write the Pythagorean triple that can help find the solution and then answer.

a.	A ship sails 5 km due North, then 12 km due East. How far is it from its starting point. Provide exact value in kilometers.
b.	Find diagonal length of a rectangular box with length 4 cm, width 3 cm and height 12 cm.
c.	Find the hypotenuse of a right triangle whose legs are 9 cm and 12 cm.
d.	Find the hypotenuse of a right triangle whose legs are 3003 and 4004.
e.	Two cars start from an intersection at the same time. One travels East at a constant speed of 20 miles per hour and the other travels South at an average speed of 21 miles per hour. How far apart are they after 2 hours?
f.	Diagonal of a rectangle is 52 cm. Find the width if the length is 48 cm.
g.	A rescue boat heads **79 km N**, **90 km E**, **63 km S**, **27 km W**. How far from the start?
h.	Two cars leave the same intersection at the same time. One travels **East at 30 miles per hour**, and the other travels **North at 40 miles per hour**. **How far apart are they after 2 hours?**
i.	The base of a right triangle is **7 cm less** than its height. If the hypotenuse is **17 cm**, find the base and height.
j.	Find **all** Pythagorean triples with hypotenuse of $\sqrt{65}$.

Answers

a.	13 km	b.	13 cm	c.	15 cm
d.	5005	e.	58 miles	f.	20 cm
g.	65 km	h.	100 miles	i.	8 cm, 15 cm

Real Life Applications

Example 12a: Rectangular Box

Find the length of the longest diagonal in a rectangular box with length of 6 cm, width of 8 cm and height of 4 cm.

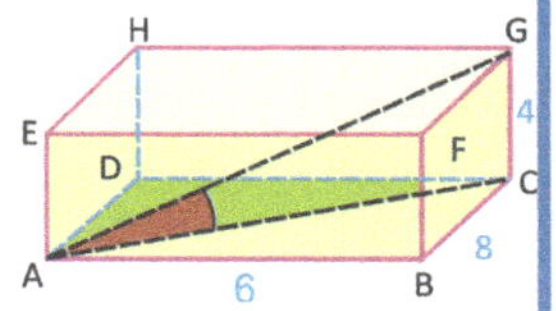

Solution 12a: Rectangular Box

First find the diagonal AC in rectangle $ABCD$

$AC^2 = AB^2 + BC^2 = 6^2 + 8^2 = 100$

$\boldsymbol{AC^2 = 100}$

No need to find $\boldsymbol{AC}$

Now, solve the right triangle ACG for the diagonal AG

$AG^2 = AC^2 + CG^2 = 100 + 4^2 = 100 + 16 = 116$

$AG = \sqrt{116} = 2\sqrt{29}$

The longest diagonal is $2\sqrt{29}\ cm$ long.

Example 12b: Algebraic Relation

The diagonal of a rectangle is two more than the length. If width of the rectangle is 10 cm, find the length.

Solution 12b: Algebraic Relation

Width of rectangle is 10 cm

Let the length be x

Then diagonal will be $x + 2$

Apply Pythagorean Theorem

$\boldsymbol{(x + 2)^2 = x^2 + 10^2}$

$\boldsymbol{x^2 + 2x + 4 = x^2 + 100}$

$\boldsymbol{2x = 100} - 4 = 96$

$x = \frac{96}{2} = 48$

Length of the rectangle is $48\ cm$.

Q11. Given a cube with sides of 10 cm as shown in the figure. Find the length of the diagonal AG.

Hint: Calculate diagonal of the base, AC

$10\sqrt{3}cm$

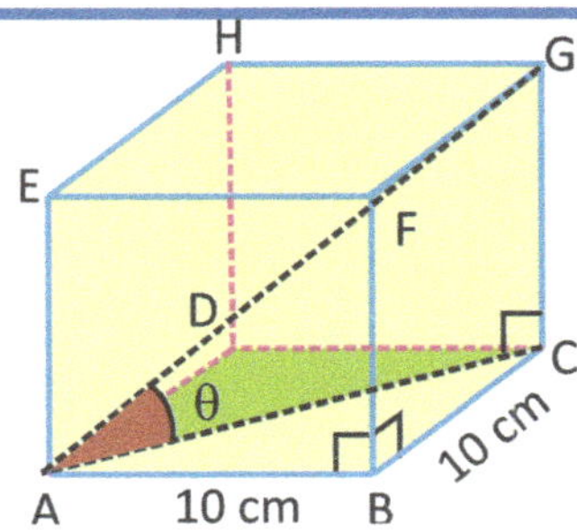

Example 12c: Coordinate Point

Given points $P(1, 2)$ and $P(5, 7)$ on the coordinate plane. Find the exact distance from point $P\ to\ Q$.

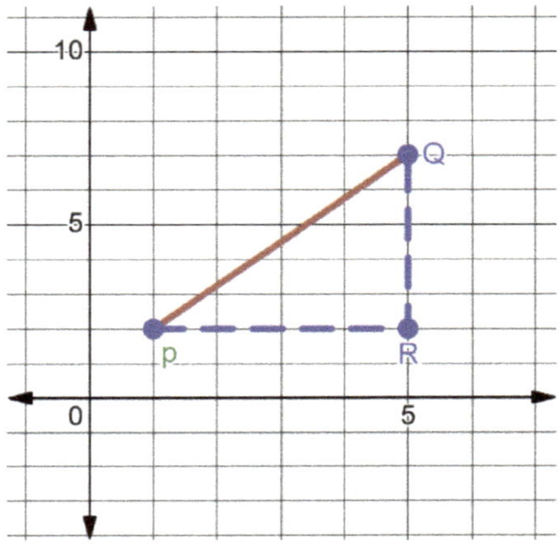

Solution 12c:

Construct a right triangle PQR as show,

$PR = 5 - 1 = 4, QR = 7 - 2 = 5$

$PQ^2 = PR^2 + QR^2 = 4^2 + 5^2 = 41$

$PQ = \sqrt{41}$

exact distance from point $P\ to\ Q$ is $\sqrt{41}\ units$

Example 12d: Navigation

A ship sails 14 km due South, then 10 km due East, and finally 10 km south. How far is it from its starting point. Provide exact value in kilometers.

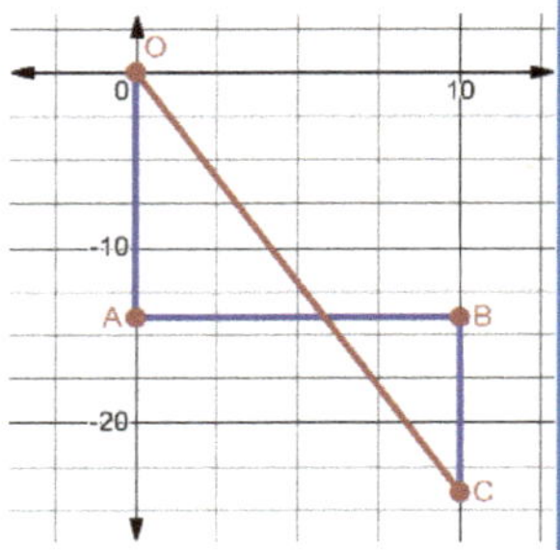

Solution 12d

Sketch figure to illustrate the situation. Coordinate plane may be used as shown

A ship sails 14 km due South OA, then 10 km due East AB, and finally 10 km south BC. We need to find the distance OC.

It travels 19 km East and total of 14 km South to reach C from O.

$OC^2 = 10^2 + 24^2 = 676$

$OC = \sqrt{676} = 26$ km

Note: Distance Formula is application of Pythagorean Theorem

Distance d between coordinate points $A(x_1, y_1)$ and $B(x_2, y_2)$ is :

$$d = \sqrt{(x_2 - x_1)^2 + (y_2 - y_1)^2}$$

Q12	Find tangent length AP from the point $P(0, -10)$ to the circle $x^2 + y^2 = 25$ $5\sqrt{3}$

Q13	A loading ramp is 3.0 m long. One end rest on a loading dock 0.8 m above the ground, and the other end is on the truck 1.2 m above the ground. Find the horizontal distance between the back of the truck and the loading dock in centimetres. Sketch diagram and define variable. Short solution is given below $d^2 = 3^2 - (1.2 - 0.8)^2 = 8.84$ $d = \sqrt{8.84} = 2.97\ m$ **Distance is $\mathbf{2.97\ m}$**

Alternate Proof of Pythagorean Theorem

Student's Activity: Explore and Learn

Enhance your algebraic skills and provide alternate proof of Pythagorean theorem

Take four copies of a triangle. Let the sides be a, b, c.

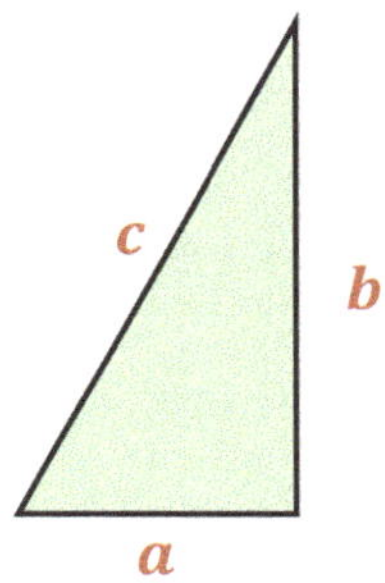

Rearrange the four triangles as shown in the next figure. You will get a square within a square.

The inside square will have side length of $(b - a)$.

Now adopt similar procedure as we did in our original proof and then prove that $a^2 + b^2 = c^2$

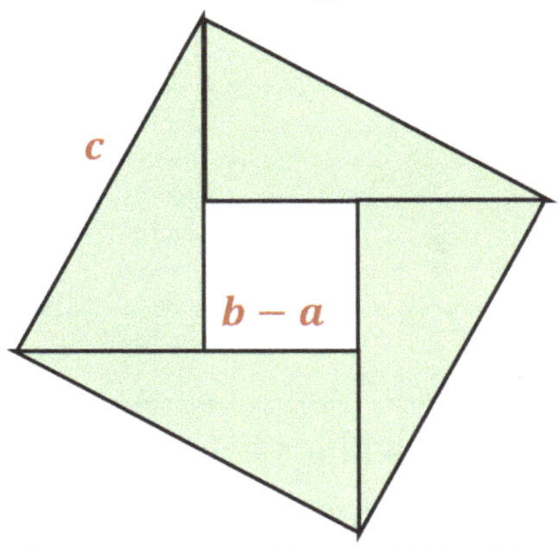

Hint:

Area of outside square is c^2

This is equal to the area of 4 triangles and inside square.

Think and Prove

Prove that the sum of squares of the length of any parallelogram equals the sum of the squares of the diagonals.

Verify this for the parallelogram with vertices A(0, 0), B(4, 0), C(6, 8) and D(2, 8).

Proving of Pythagorean Theorem: 2 Methods

Proof:

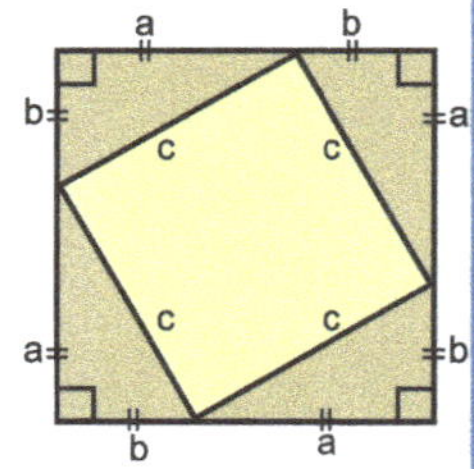

Consider a square with sides length of c inscribed in a square of $(a+b)$ sides as shown in the figure.

Observe that we have 4 right triangles of sides a, b and c.

Here a and b are the legs and c is the hypotenuse

To prove Pythagorean Theorem, we need to show that $a^2 + b^2 = c^2$

Proof is provided by considering the area of the square.

Area of the outside square is $(a+b)^2$

This area of the square can also be written as combination of the area of 4 triangles and inscribed square.

$4\left(\frac{1}{2}ab\right) + c^2$

Provide reasoning and details

Since both are equal, we can equate them

$(a+b)^2 = 4\left(\frac{1}{2}ab\right) + c^2$

$a^2 + 2ab + b^2 = 2ab + c^2$

$a^2 + b^2 = c^2$

Hence Proved

$\boldsymbol{(a+b)^2 = a^2 + 2ab + b^2}$

Pythagorean Theorem:

$\boldsymbol{leg^2 + leg^2 = hypotenuse^2}$

Q14	The diagonal of a rectangle is two more than the length. If width of the rectangle is 10 cm, find the length. Define variable: let the length be $\boldsymbol{l}$ $(l+2)^2 = l^2 + 10^2$ $l^2 + 2l + 4 = l^2 + 100$ $2l = 100 - 4 = 96 \Rightarrow l = \mathbf{48}$ 48 cm

STEM Project 1: Spiral Right Triangles

Projects have been designed to inculcate creative thinking and encourage students to connect Science, Technology, Engineering, and Mathematics (STEM)

Students are encouraged to find alternate ways of completing the project.

Project Description

Earlier the concept of irrational numbers was introduced and students were encouraged to represent $\sqrt{5}$ on a number line. Some of you must have discovered method to do so.

An isosceles right triangle with unit side length will have hypotenuse of $\sqrt{2}$. Adding another right triangle with a side length of one unit will provide hypotenuse of $\sqrt{3}$, as shown in the figure. This series of spiral right triangles, can be extended to get accurate length for $\sqrt{n}$. With every other triangle the value of n increases by one. As you extend, the spiral will overlap. Now, **we are interested in finding the maximum value of $\sqrt{n}$ that can be represented without overlapping of the triangles.**

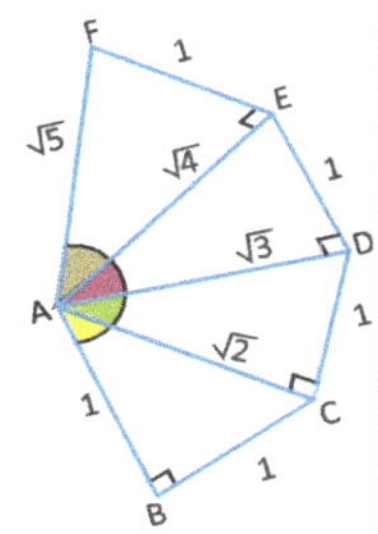

Let us build this project with straws. Engineering design skills will be required to accurately get the result.

This project can be shared with many students in a group.

Describe your method and provide the solution. You may also mention the reasons for error or better ways of conducting this project.

We will extend this project after learning trigonometric ratios and adopt technology to find solution.

British Flag Challenge

Given a rectangle $ABCD$ with point O inside it. Prove that

$$OB^2 + OD^2 = OC^2 + OA^2$$

SCAN ME

Chapter 7 Test: Pythagorean Theorem

Q1	Find the length of diagonal for 8 cm by 6 cm rectangle.
Q2	A ladder is leaning against a wall. The foot of the ladder is 6 m away from the wall and the ladder is 10 m long. How high up the wall does it reach?
Q3	A rectangular field has a diagonal of 50 m and one side 30 m. Find the other side.
Q4	A boat sails 8 km east and then 15 km north. How far is it from the starting point?
Q5	The base of a right triangle is 5 cm less than its height. If the hypotenuse is 13 cm, find the base and height in centimeters.
Q6	A rectangular TV screen has a width of 48 cm and a diagonal of 80 cm. Find its height.
Q7	A kite string is 25 m long and is tied to the ground. If the kite is 15 m above the ground, how far is the kite horizontally from the point on the ground where the string is tied?
Q8	The sides of a right triangle differ by 7 cm. If the hypotenuse is 13 cm, find the sides.
Q9	A ramp is 5 m long and reaches a height of 1.5 m. Find the horizontal distance from the base of the ramp to the wall in metres.
Q10	A park is in the shape of a right triangle. The shortest side is 120 m, and the hypotenuse is 169 m. Find the length of the other side.
Q11	Determine altitude of a triangle whose shorter sides are 6 cm an 8 cm.

Answers

Q1	10 cm	Q2	8 m	Q3	40 m	Q4	17 km	Q5	7 cm
Q6	64 cm	Q7	20 m	Q8	5, 12	Q9	4.77	Q10	119 m
Q11	4.8								

Extended Word Problems for Competitions

Two Similar Right Triangles

Example 11: Two Right Triangles

In the figure two right triangles are shown.

$CE = 8\ cm, BC = 6\ cm, AB = 8\ cm$

$\angle ECD = \angle BAC$. Find the length of BD

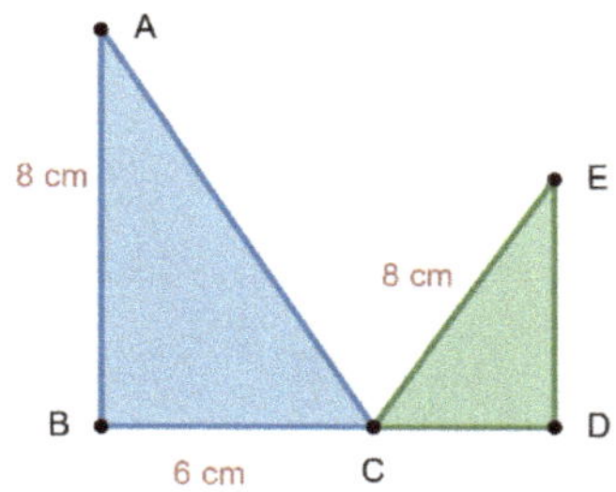

Solution 11: Two Right Triangles

Strategy: $BD = BC + CD$

Relate the two triangles and find CD

ΔABC is right triangle $AC^2 = AB^2 + BC^2$ $AC^2 = 8^2 + 6^2 = 100$ $AC = \sqrt{100} = 25$ $\angle ECD = \angle BAC$ and $\angle ABC = \angle EDC = 90°$	$\Delta ABC \sim \Delta CDE$ $\frac{CD}{BC} = \frac{CE}{AC} \Rightarrow \frac{CD}{6} = \frac{8}{10}$ $\frac{CD}{6} = \frac{8}{10} \times 6 = 4.8$ $BD = BC + CD = 6 + 4.8 = 10.8$ BD is $10.8\ cm$

Practice Question 11: Two Right Triangles

In the figure two right triangles are shown.

$AC = 20\ cm, BC = 12\ cm, DE = 10\ cm$

$\angle D = \angle A$. Find the length of AF

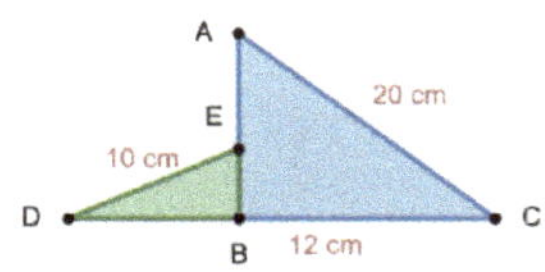

Watch Video **Two Right Triangles** **SCAN ME**

Chapter 8: Advanced Applications and Special Triangles

Working efficiently to provide solution in minimum time is key to beat the competition. In this section we will understand strategies to solve special triangles which help to provide solution in minimum time for such cases.

We will consider two special triangles:

- $45° - 45° - 90°$ ***Isosceles Right Triangle also written as*** $45 - 45 - 90$
- $30° - 60° - 90°$ ***Split Equilateral Triangle also written as*** $30 - 60 - 90$

Special Triangle: $45° - 45° - 90°$

Split Square Cut Diagonally

Consider a square $ABCD$ cut in half diagonally as shown in the figure.

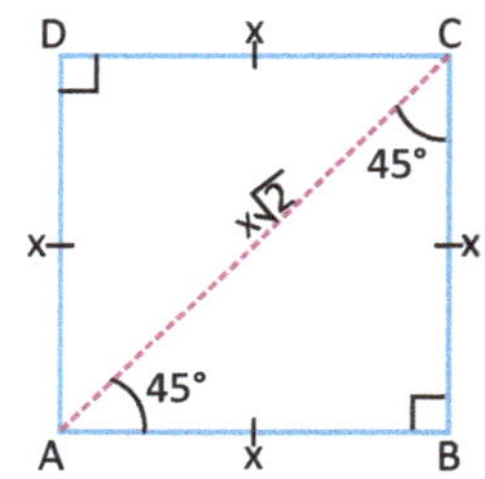

If each side is of length x then the diagonal will be

$AC = \sqrt{x^2 + x^2} = x\sqrt{2}$ *Pythagorean Theorem*

This gives us our first special triangle ΔABC with

$45° - 45° - 90°$.

The side lengths of triangle ABC are in the ratio of $x: x: x\sqrt{2}$

In simplified terms we will refer it as

special triangle $\mathbf{45 - 45 - 90}$

with ratio of sides as $\mathbf{1: 1: \sqrt{2}}$.

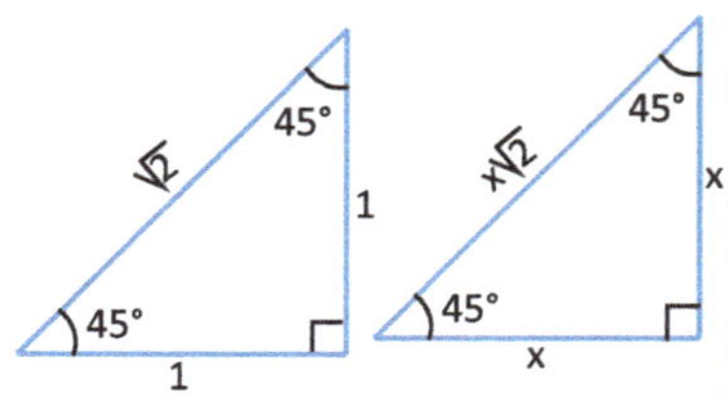

Gererally the ratio can be written as

$\boldsymbol{n: n: \sqrt{2}n}$ where $\boldsymbol{n}$ is any real number. Hypotenuse is $\sqrt{2}$ times the legs.

All the triangles with the following side lengths will be isosceles right triangles

$2: 2: 2\sqrt{2}$	$7: 7: 7\sqrt{2}$	$10: 10: 10\sqrt{2}$

Example 1: Usage of special triangle $45° - 45° - 90°$

Find perimeter of the right isosceles triangle with hypotenuse as $10\sqrt{2}\ cm$.

Solution 1: Usage of special triangle $45° - 45° - 90°$

Right isosceles triangle is special triangle $45° - 45° - 90°$

Ratio of the sides is $n: n: \sqrt{2}n$

Since hypotenuse is $10\sqrt{2}$,

the side lengths will be $10: 10: 10\sqrt{2}$

Perimeter, sum of all the three sides

$10\sqrt{2} + 10 + 10 = 20 + 10\sqrt{2}\ cm$

$or\ 10(2 + \sqrt{2})\ cm$

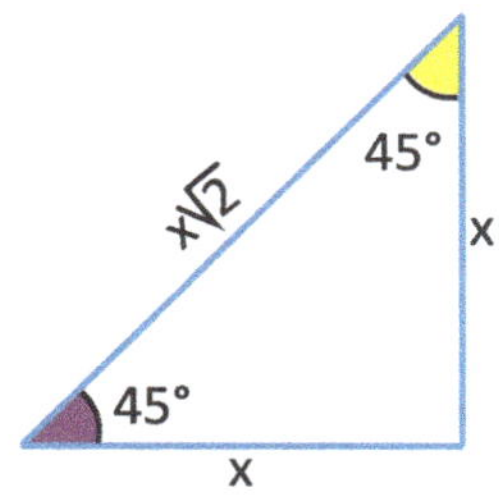

Example 2: Application of special triangle $45° - 45° - 90°$

Peter is flying a kite which is 500 m above the ground. The angle of elevation of the kite is $45°$. Find the length of the string.

Solution 2: Application of special triangle $45° - 45° - 90°$

Since angle of elevation is $45°$ we have the right isosceles triangle

$45° - 45° - 90°$ with the ratio of the sides as $n: n: \sqrt{2}n$

Height of 500 m above the ground is the smaller side of the triangle.

Hypotenuse will be $\sqrt{2}$ times this value.

Length of the string, the hypotenuse in this case, will be $500\sqrt{2}\ m$.

Hypotenuse is $\sqrt{2}$ times the leg in $45 - 45 - 90$ Right Triangle

Practice Problem

Q1	In a 45°−45°−90° triangle, the hypotenuse is **20** cm. Find the perimeter.
Q2	Find diagonal length of an isosceles right triangle whose area is 16 cm^2.
Q3	Midpoints of a square are joined to form an inscribed square with area of 64 cm^2. Find perimeter and area of the outer square.

Special Triangle $30° - 60° - 90°$

Split Equilateral Triangle for Special Triangle $30° - 60° - 90°$

Consider an equilateral tringle ABC with length of each side as $2x$. Drop a perpendicular, CD, from the vertex C to the opposite side as shown in the figure. This will divide ΔABC in two congruent right triangles. The base gets divided in two, $AD = DB = x$. The altitude, CD, will be $x\sqrt{3}$

Pythagorean Theorem: $CD = \sqrt{(AC)^2 - (AD)^2} = \sqrt{(2x)^2 - (x)^2} = x\sqrt{3}$

This gives us our second special triangle: $30° - 60° - 90°$ whose sides are in the ratio of $x: \sqrt{3}x: 2x$.

We will refer it as **equilateral split triangle, $30 - 60 - 90$ with ratio of the sides as $1: \sqrt{3}: 2$.**

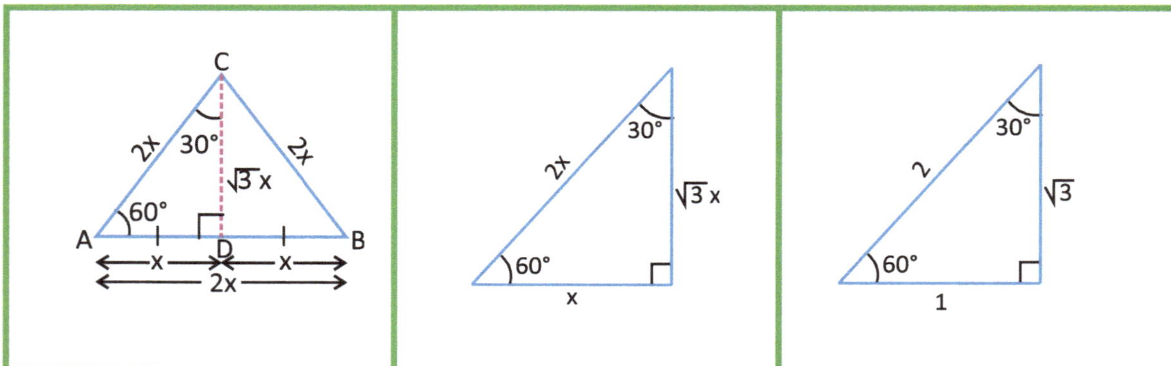

Alternate ways for $30 - 60 - 90$ special triangle

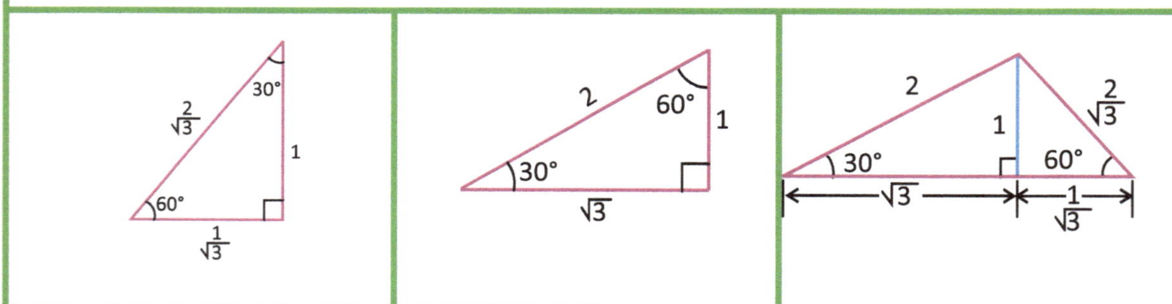

Hypotenuse is 2 times smallest side leg in $30 - 60 - 90$ Triangle

✍ Practice Problem

Find the perimeter of right triangle with hypotenuse of 12 and acute angle of $30°$.

Example 2: Usage of special triangle $30° - 60° - 90°$

Ratio of angles of the triangle ABC is 1:2:3. Find perimeter of the ΔABC if the hypotenuse is $4\ cm$.

Solution 2: Usage of special triangle $30° - 60° - 90°$

Sum of angle of a triangle is 180°.

Since the ratio of angles is 1:2:3 the angles will be $x, 2x, 3x$

$x + 2x + 3x = 180$

$6x = 180$

$x = \frac{180}{6} = 30$

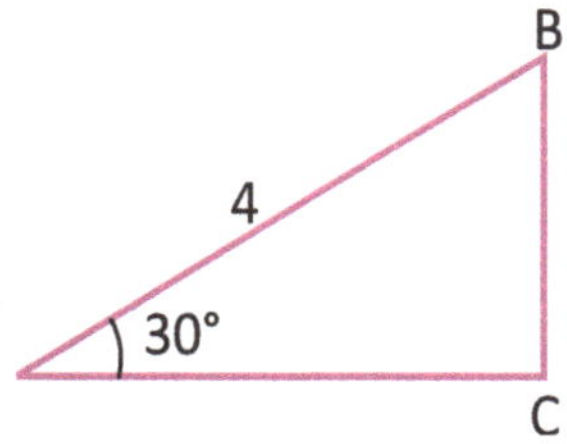

The three angles are $30:60:90$

ratio of the sides will therefore be $x:\sqrt{3}x:2x$

since the hypotenuse is 4 the exact length of the sides will be $2:2\sqrt{3}:4$

The perimeter is $2 + 2\sqrt{3} + 4 = 6 + 2\sqrt{3}cm$

Special Triangle Challenge

Given a ΔABC with height BD of $12\ cm$.

Find the exact area of the triangle.

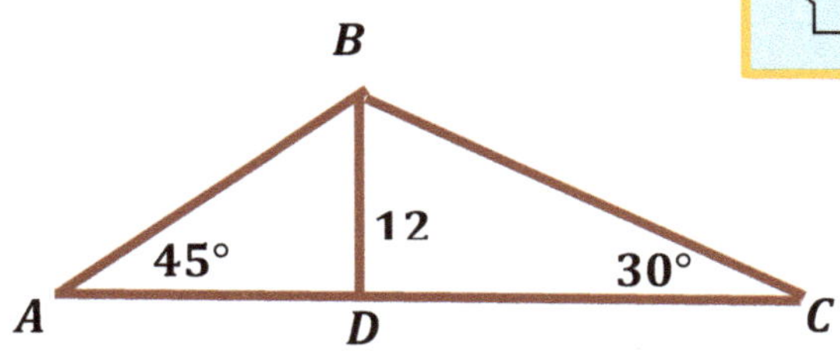

Special Triangle Challenge

SCAN ME

Combination of Special Triangles

Combination of special triangles can be solved easily.

Application : Special Triangles Proportional Sides

Example 4: Solve Related Special Triangles

Given right ΔADC in which $AD = DC$ and right ΔBDC in which $BD = 10$, $\angle DBC = 30°$

Find the length of AC.

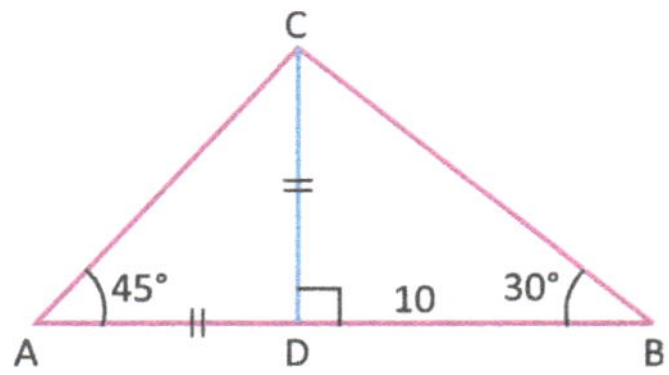

Solution 4: Solve Related Special

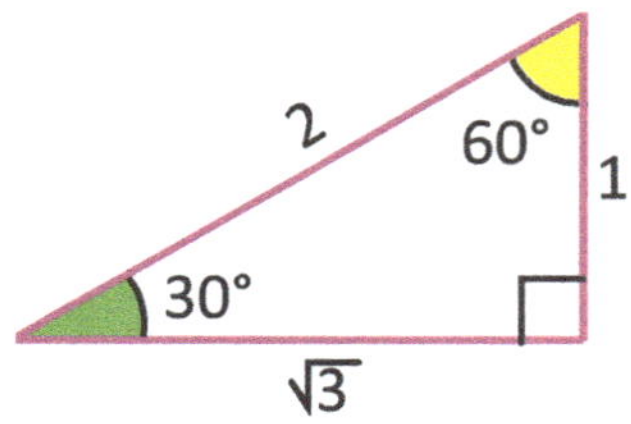

First consider the ΔBCD, $30° - 60° - 90°$

Sides are in the ratio of $1: \sqrt{3}: 2$

$DB = 10\ cm$, smallest side is CD

$$CD = \frac{DB}{\sqrt{3}} = \frac{10}{\sqrt{3}}$$

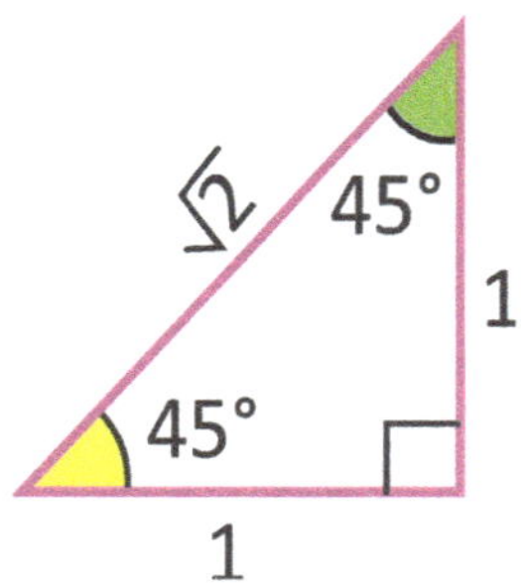

Now consider the ΔADC, $45° - 45° - 90°$

Sides are in the ratio of $1: 1: \sqrt{2}$

$CD = \frac{10}{\sqrt{3}}\ cm$, hypotenuse is $CD \times \sqrt{2}$

$$AC = CD \times \sqrt{2} = \frac{10}{\sqrt{3}} \times \sqrt{2}$$

$$AC = \frac{10}{\sqrt{3}} \times \sqrt{2} \times \frac{\sqrt{3}}{\sqrt{3}} = \frac{10\sqrt{6}}{3}\ cm$$ Rationalize Denominator

NOTE:

No Calculations required since the sides ratios is known for special triangles

- *For* $45° - 45° - 90°$ *triangle is* $1: 1: \sqrt{2}$
- *For* $30° - 60° - 90°$ *triangle is* $1: \sqrt{3}: 2$

Use proportional length to scale the related triangles.

Inscribed Geometrical Shape

Thinking and Communication

Example 5: Application Special triangle

Find the area of regular hexagon inscribed in circle of diameter 10 cm.

Solution 5: Application Special triangle

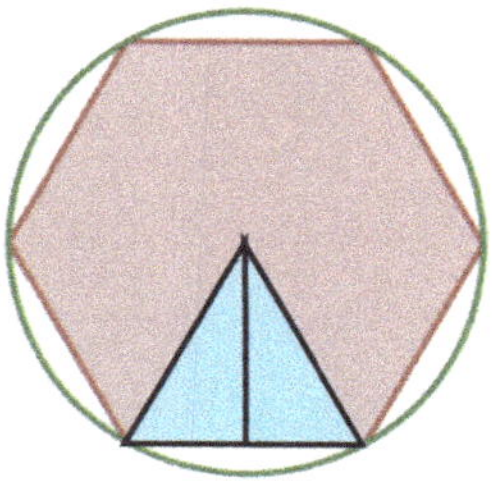

Internal angle of hexagon will be $120°$.

Hexagon forms 6 equilateral triangles.

Diameter is $10\ cm$. Radius will be $5\ cm$.

Each side of the equilateral traingle will be $5\ cm$.

Height will be calculated with $30° - 60° - 90°$ triangle

Ratio of the sides being $1 : \sqrt{3} : 2$

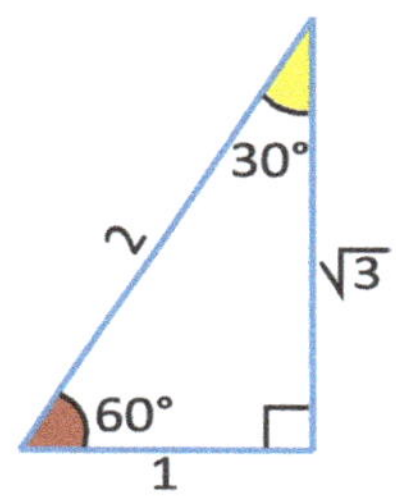

Height will be $\sqrt{3}$ times half of hypotenuse

$$h = 5 \times \frac{\sqrt{3}}{2}$$

Area of the hexagon equals area of 6 equilateral triangles with each side of 5.

$$Area\ of\ Hexagon = 6 \times \frac{1}{2} \times 5 \times 5 \times \frac{\sqrt{3}}{2} = \frac{75\sqrt{3}}{2} cm^2$$

Height h is also called apothem. $\boldsymbol{Area\ of\ Hexagon = \frac{1}{2} \times apothem \times perimeter}$

Engaging Minds and Creative Thinking

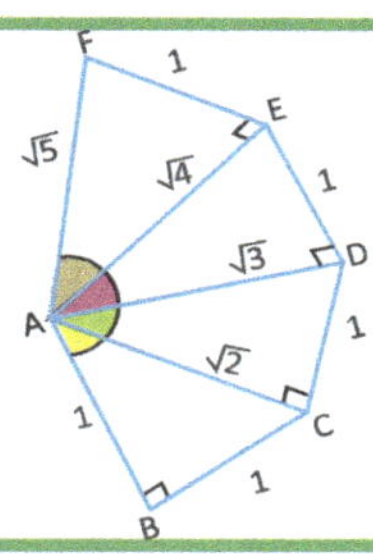

A series of right triangles is added to the isosceles triangle ABC with a common vertex A. Here we have shown a method of represent irrational number with exact length. Verify algebraically.

AMC Challenge: Octagon from a Rectangle

From Rectangle to Octagon

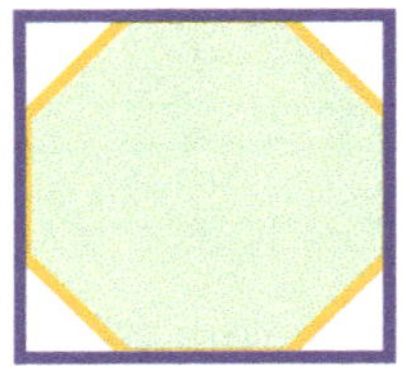

A rectangle measuring **8 cm by 6 cm** has each of its four corners cut off in the same way to form a **regular octagon** (all sides equal).

If each new side of the octagon is of length a,
Find the perimeter of the octagon.

Hint: Consider the side length a of octagon at the corner of the rectangle.

$$\left(4-\frac{a}{2}\right)^2+\left(3-\frac{a}{2}\right)^2=a^2$$

Answer: $24\sqrt{11}-56$

Watch Video **Rectangle to Octagon** SCAN

AMC Challenge: Octagon from a Square

From Square to Octagon

A **square** of side s has each corner cut off identically to form a **regular octagon** (all sides equal).
If each new octagon side has length a, **find the perimeter of the octagon**.

Hint: Let the cut-off at each corner be an isosceles right triangle with leg x.
Then the new side is the hypotenuse $a=x\sqrt{2}$, and the remaining middle segment on each edge is $s-2x$, which must also equal a:

$$\begin{cases} a = x\sqrt{2} \\ a = s-2x \end{cases}$$

$$x\sqrt{2}=s-2x$$

$$x\sqrt{2}+2x=s$$

$$x(\sqrt{2}+2)=s$$

$$x=\frac{s}{\sqrt{2}+2}$$

$$a=x\sqrt{2}=\frac{s\sqrt{2}}{\sqrt{2}+2}$$

$$a=\frac{s\sqrt{2}}{\sqrt{2}+2}\times\frac{\sqrt{2}-2}{\sqrt{2}-2}$$

$$a=\frac{s\sqrt{2}(\sqrt{2}-2)}{2-4}$$

$$a=s(\sqrt{2}-1)$$

$$Perimeter=8a$$

$$P=8s(\sqrt{2}-1)$$

Chapter 8 Test: Special Triangles

Q1	In a 45°–45°–90° triangle, the hypotenuse is $\mathbf{10}\sqrt{2}$ cm. Find the perimeter.
Q2	In a 30°–60°–90° triangle, the shortest side is 8 m. Find the hypotenuse.
Q3	An equilateral triangle has side length 10 cm. Find the height.
Q4	In a 30°–60°–90° triangle, the hypotenuse is 12 cm. Find the area of the triangle
Q5	A square has a diagonal length of 8 cm. Find the area of the square.
Q6	Find the exact distance of point (3, 3) from origin.
Q7	Ratio of the angles of a triangle is 1:2:3. Find perimeter of the triangle if the hypotenuse is 4.
Q8	Peter is flying a kite which is 500 m above the ground. Find the length of the string in kilometers if the angle of elevation is 60°.
Q9	Mark exact value of $\sqrt{6}$ on a number line
Q10	A **square** of side 10 has each corner cut off identically to form a **regular octagon** (all sides equal). Find perimeter of the regular octagon. Hint: example on the last page.
Q11	Given isosceles triangle ABC where $AB = AC = 10, and\ BC = 12$. Find circumradius of the triangle ABC. R = 25/4

Answers

Q1	$30\sqrt{2}$	Q2	16 m	Q3	$10\sqrt{2}$	Q4	$36\sqrt{3}$	Q5	32cm^2
Q6	$3\sqrt{2}$	Q7	$6 + 2\sqrt{3}$	Q8	$\frac{\sqrt{3}}{3}$	Q10	$80(\sqrt{2} - 1)$		

Chapter 9: Trigonometric Ratios

Slope Tangent to Tangent Ratio

Connecting Slope with Trigonometry

Algebra, Geometry, and Trigonometry are deeply interconnected. When we appreciate these connections, we unlock new ways of thinking — and ignite curiosity.

Extend the Concept of Slope

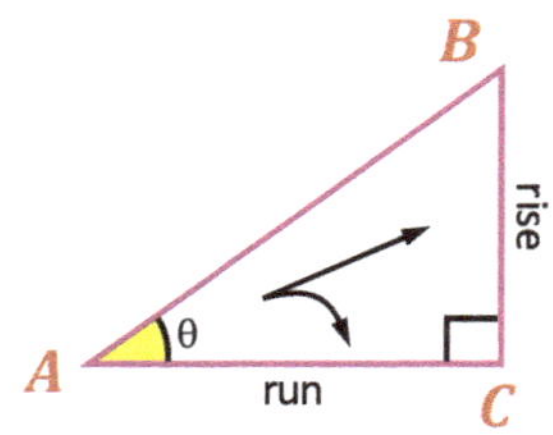

We've already learned that the **slope** (or **gradient**) of a line is defined as the ratio of **rise over run** – that is, the vertical change divided by the horizontal change between two points on a coordinate plane.

Now, imagine a right triangle formed by this rise and run.
You'll notice that no matter how large or small the triangle is, if the **angle** θ at the base remains the same, the **slope remains constant**. That's because slope is determined by the **angle of inclination** of the line – not the size of the triangle.

This insight is powerful:
A constant angle means a constant ratio of side lengths.
And that's where **trigonometry begins.**

From Slope to Tangent – A Natural Transition

The **tangent of an angle** $\boldsymbol{\theta}$, written as $\mathbf{tan}\,\boldsymbol{\theta}$, is defined as

$$\tan\theta = \frac{opposite}{adjacent} = \frac{rise}{run}$$

Yes – **the slope is simply the tangent of the angle**.
This shows that the angle controls the ratio, and not the triangle's size. That's why the angle becomes the input (or argument) in trigonometric functions.

Greek alphabets are widely used in mathematics. Lower case alphabets, $(\boldsymbol{alpha})$, $\boldsymbol{\beta}(\boldsymbol{beta}), \boldsymbol{\gamma}\ (\boldsymbol{gamma}), \boldsymbol{\theta}(\boldsymbol{thita}),\ \boldsymbol{\phi}(\boldsymbol{phi})$, *are commonly used to represent angles.*

Primary Trigonometric Ratios

Relating Acute Angle with Sides of Triangle

SOH CAH TOA a common mnemonic to remember the trigonometric ratios

Definition of Trigonometric Ratios

We will begin with giving special names to the three sides of the right triangle in reference to the acute angle.

- **O**pposite side is across the acute angle
- **A**djacent side is between the acute and right angle
- **H**ypotenuse is across the right angle

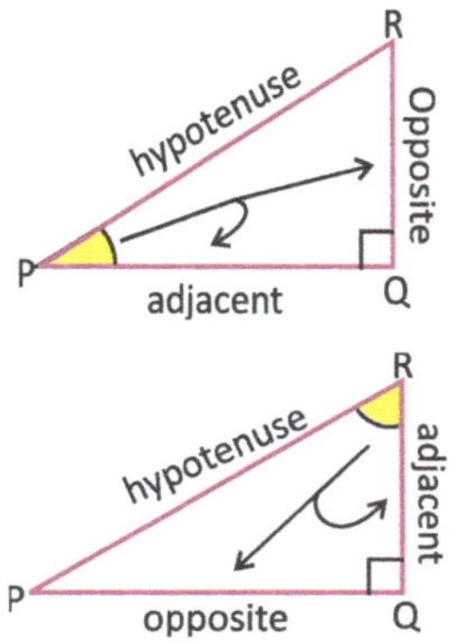

Figure shows these sides from each acute angle. Definition of the trigonometric ratios along with easy to remember mnemonic (SOH CAH TOA) and our innovative symbol is shown in the table below.

$\sin\theta = \dfrac{opposite}{hypotenuse}$	$\cos\theta = \dfrac{adjacent}{hypotenuse}$	$\tan\theta = \dfrac{opposite}{adjacent}$
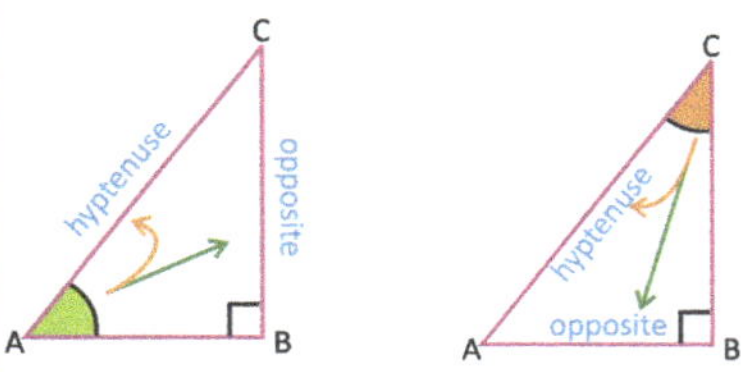	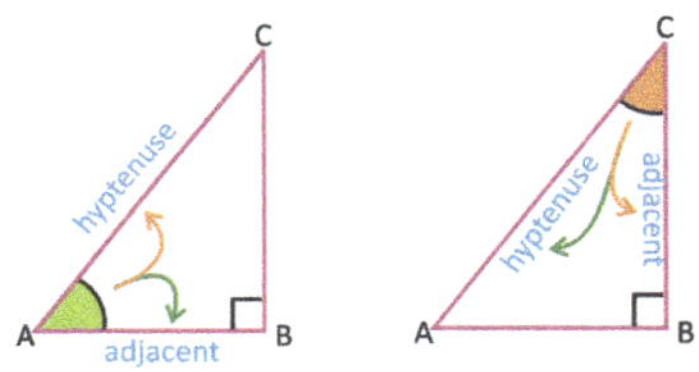	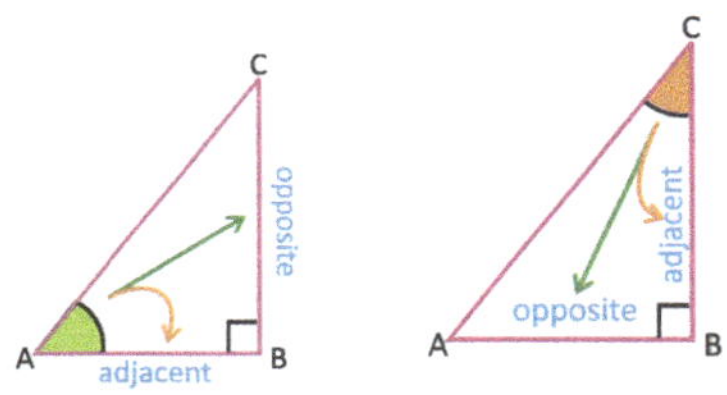
SOH $\quad$ (S^O_H) C^A_H T^O_A	CAH $\quad$ S^O_H (C^A_H) T^O_A	TOA $\quad$ S^O_H C^A_H (T^O_A)

These ratios stay the same for a given angle, no matter how large or small the triangle — a key property that allows trigonometry to model real-world situations, from the slope of a roof to the height of a mountain.

Trigonometry comes from Greek terms **trigonom** which means triangle and **metrikos** which means measure. In true sense, trigonometry begins with measurement of both sides and angles in any triangle and extends to helps us even explore the celestial bodies in our universe.

Watch Video **SOH CAH TOA** **SCAN ME**

Investigate Trigonometric Ratios

Connecting Pythagorean Theorem – Similar Triangles – Trigonometry

Investigate: Are the trigonometric ratios same for equal angles in similar triangles?

Investigate Similar Right Triangles

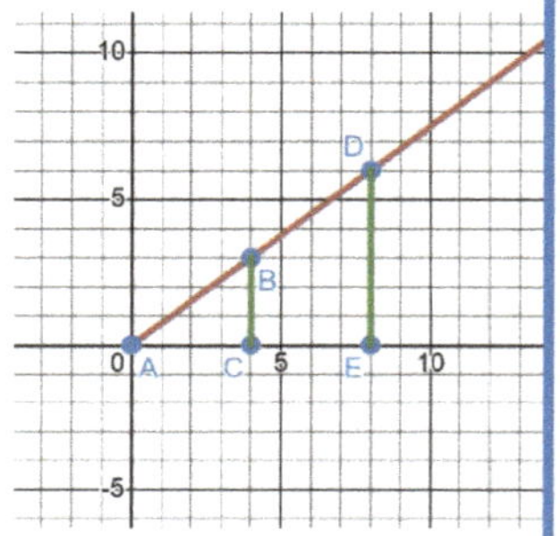

The figure shows two similar right triangles, ΔABC and ΔADE, on a coordinate plane. Segment BC and DE are perpendicular to the line segment AE.

Complete the table below to evaluate each trigonometric ratios and then check your work from next page.

What do you observe? Explain the result.

	ΔABC	ΔADE	
Length of the side opposite to $\angle A$			
Length of the side side adjacent to $\angle A$			
Length of the hypotenuse (Pythagorean Theorem)			
$\sin A$ S^O_H C^A_H T^O_A			
$\cos A$ S^O_H C^A_H T^O_A			
$\tan A$ S^O_H C^A_H T^O_A			
Slope of the line AD			
Reflect on how each ratio changes in the similar triangles with different sizes.			

Investigate Similar Right Triangles: Our Observations

$\Delta ABC \sim \Delta ADE$ (right triangles with common angle A)

Coordinate points are $A(0,0), B(4,3), D(8,6)$

$AC = 4, BC = 3, AE = 8, DE = 6$

$\angle C = \angle E = 90°$

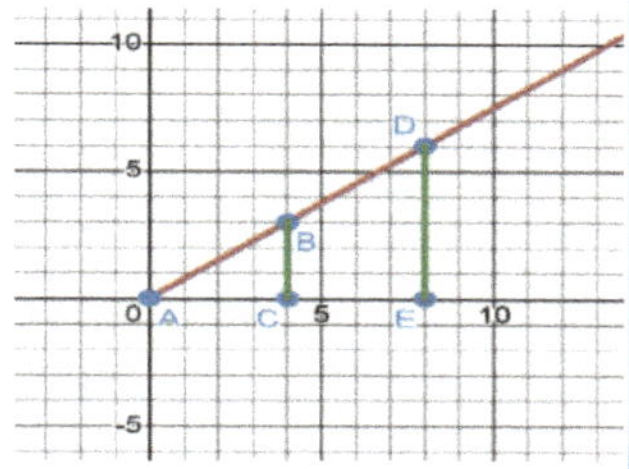

	ΔABC	ΔADE	
Length of the side opposite to $\angle A$	$BC = 3$	$DE = 6$	
Length of the side side adjacent to $\angle A$	$AC = 4$	$AE = 8$	
Length of the hypotenuse (Pythagorean Theorem)	$AB = \sqrt{3^2 + 4^2} = 5$	$AD = \sqrt{6^2 + 8^2} = 10$	
$\sin A$ S^O_H C^A_H T^O_A	$\frac{3}{5}$	$\frac{6}{10} = \frac{3}{5}$	
$\cos A$ S^O_H C^A_H T^O_A	$\frac{4}{5}$	$\frac{8}{5} = \frac{4}{5}$	
$\tan A$ S^O_H C^A_H T^O_A	$\frac{3}{4}$	$\frac{6}{8} = \frac{3}{4}$	
Slope of the line AD	$\frac{3}{4}$	$\frac{3}{4}$	

- Trigonometric ratios are dependent on angle only and not the size of the triangle.
- Slope of the line is equal to the tangent ratio.

Additional Exercise: Investigate and find how the trigonometric ratios change if the angle increases. Take coordinate points as $A(0,0), B(3,4), D(6,8)$ in the above case.

Finding Trigonometric Ratios in Triangle

Connecting Pythagorean Theorem – Similar Triangles – Trigonometry

All trigonometric ratios can be calculated with two sides of right triangle

Example 1: Find Values of Primary Trigonometric Ratios

Given: $\Delta PQR,\ PQ = \sqrt{10}\ cm,\ QR = 1\ cm,\ \angle PQR = 90°.$

Evaluate the length of side PQ and then evaluate the trigonometric ratios $\sin\theta$, $\cos\theta$, *and* $\tan\theta$

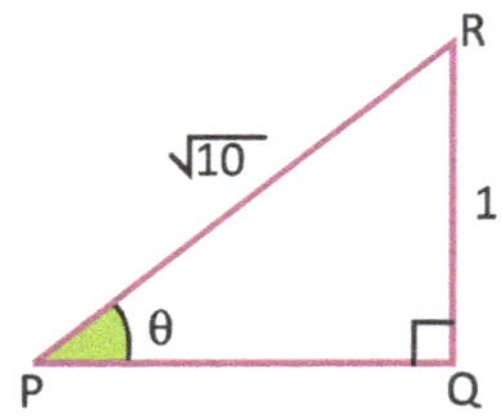

Solution 1: Find Values of Primary Trigonometric Ratios

First find the length of the missing side PQ using Pythagorean Theorem

$$PR^2 = PQ^2 + QR^2$$

$$PQ^2 = PR^2 - QR^2$$

$$PQ^2 = \left(\sqrt{10}\right)^2 - (1)^2 = 10 - 1 = 9$$

$$PQ = \sqrt{9} = 3$$

Now write each ratio $S_H^O\ C_H^A\ T_A^O$

$\sin\theta = \frac{1}{\sqrt{10}} = \frac{\sqrt{10}}{10}$	$\cos\theta = \frac{3}{\sqrt{10}} = \frac{3\sqrt{10}}{10}$	$\tan\theta = \frac{1}{3}$

Rationalization of denominator is important as most answers should not have radical in the denominator.

✍ Practice

Draw a right triangle for the given trigonometric ratio. Then write exact value of other primary trigonometric ratios.

a.	$\sin\theta = \frac{2}{5}$	b.	$\tan\theta = 2$
c.	$\cos\theta = \frac{3}{5}$	d.	$\tan\theta = 1.5$

✍ Exercise 1: Exact Value of Primary Trigonometric Ratios

Write exact value of the primary trigonometric ratios $(\sin\theta, \cos\theta, \tan\theta)$

Q1.1	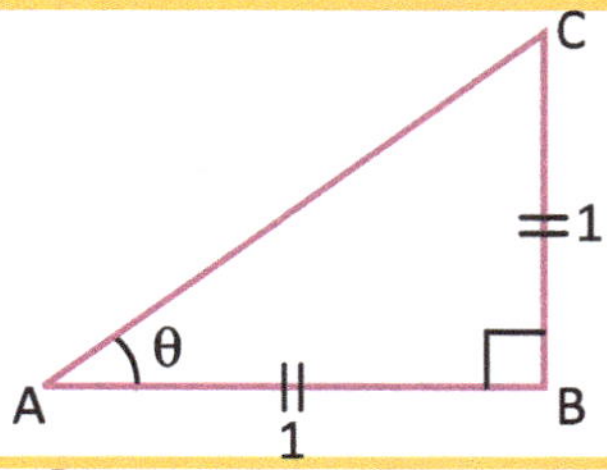	Q1.2	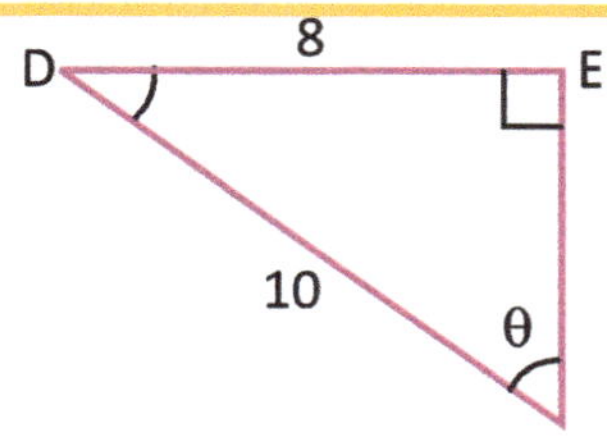
Q1.3	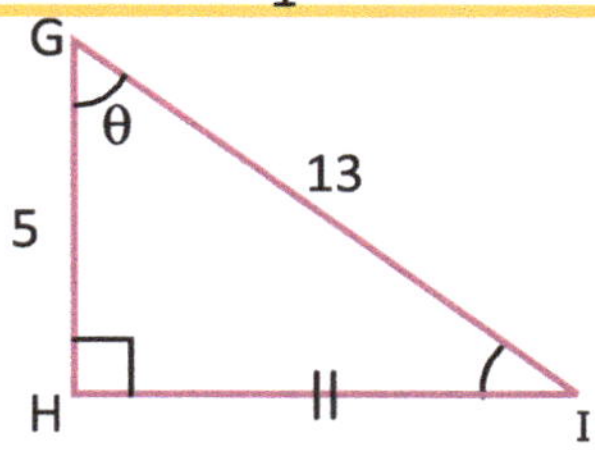	Q1.4	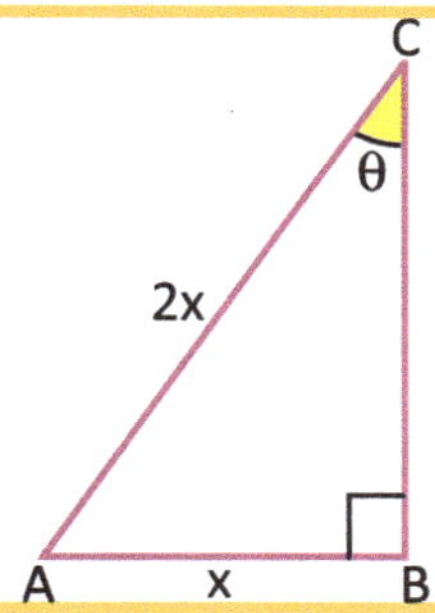
Q1.5	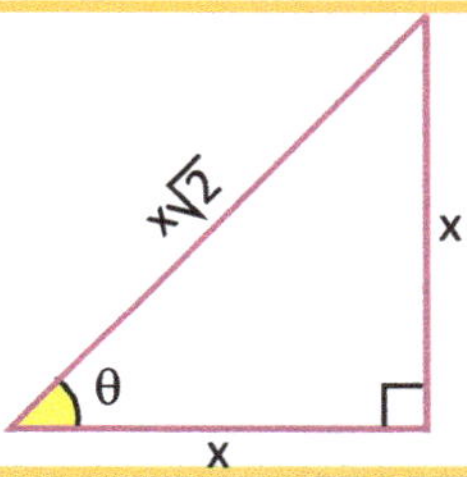	Q1.6	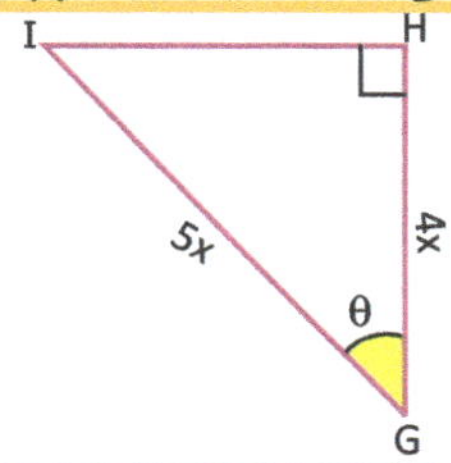

Solution of only Q1.1 to Q1.3. Rest Students Should Solve

$AC = \sqrt{1^2 + 1^2} = \sqrt{2}$	$EF = \sqrt{10^2 - 8^2} = 6$	$HI = \sqrt{13^2 - 5^2} = 12$
$\sin\theta = \frac{O}{H} = \frac{1}{\sqrt{2}} = \frac{\sqrt{2}}{2}$ $\cos\theta = \frac{A}{H} = \frac{1}{\sqrt{2}} = \frac{\sqrt{2}}{2}$ $\tan\theta = \frac{O}{A} = \frac{1}{1} = 1$	$\sin\theta = \frac{8}{10} = \frac{4}{5}$ $\cos\theta = \frac{6}{10} = \frac{3}{5}$ $\tan\theta = \frac{8}{6} = \frac{4}{3}$	$\sin\theta = \frac{12}{13}$ $\cos\theta = \frac{5}{13}$ $\tan\theta = \frac{12}{5}$

Note: Rationalize the denominator is required to provide answers in proper form.

Calculator Section: Ratios and Angles

Trigonometric Ratios for Given Angle with Scientific Calculator

Scientific calculators can be used to evaluate primary trigonometric ratio for any angle. Angles can be in degrees or radians. We will consider all the angles in degrees for now.

Example 1: Trigonometric Ratio to 3 Significant Value if not exact

Evaluate each of the following to 3 significant figures.

Note: Set calculator to degree mode for all these calculations

a.	$\sin 25°$	b.	$\cos 56°$	c.	$\tan 55°$

Solution 1a:

Evaluate $\sin 25°$

Press the keys as follows

sin	25	=

You get $0.4226 = 0.423\ (to\ 3\ sf)$

Solution 1b:

Evaluate $\cos 56°$

Press the keys as follows

cos	304	=

You get $0.5591 = 0.560\ (to\ 3\ sf)$

Solution 1c:

Evaluate $\tan 55°$

Press the keys as follows

tan	55	=

You get $1.44281 = 1.44\ (to\ 3\ sf)$

Significant figures are the number of digits in each value. All digits after decimal including zeros are counted. Leading zero before decimal is not counted. Examples of $3\ sf$ are:$1.23, 0.141, 0.120, 1.00, 0.500$

Extension:

a. Label right triangle ABC if $\cos A = \frac{\sqrt{3}}{2}$
b. Identify all angles and side lengths.
c. Write exact value of $\sin A$

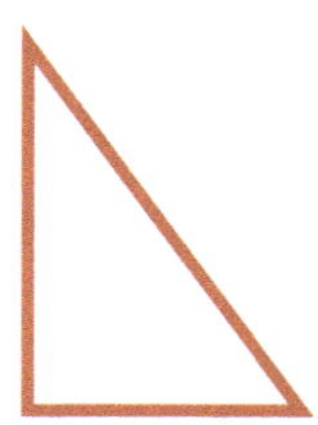

Exercise 2: Evaluate Trigonometric Ratios

Q1.1

Use calculator to evaluate Trigonometric Ratios.

Round to 3 Significant Value if not exact.

explain why sine ratio increases and cosine decreases for the increasing angles in first quadrant.

a.	$\sin 0°$	b.	$\sin 30°$	c.	$\sin 45°$	d.	$\sin 60°$
e.	$\sin 90°$	f.	$\cos 0°$	g.	$\cos 30°$	h.	$\cos 45°$
i.	$\cos 60°$	j.	$\cos 90°$	k.	$\tan 0°$	l.	$\tan 90°$
m.	$\sin 240°$	n.	$\tan 225°$	o.	$\tan 405°$	p.	$\tan(-600°)$

Q1.2a

Evaluate $\sin 240°$

Press the keys as follows

sin	240	=

You get $-0.8660 = -0.866\ (to\ 3\ sf)$

Note: Value is negative

The principal angle lies in quadrant III and sine value is negative.

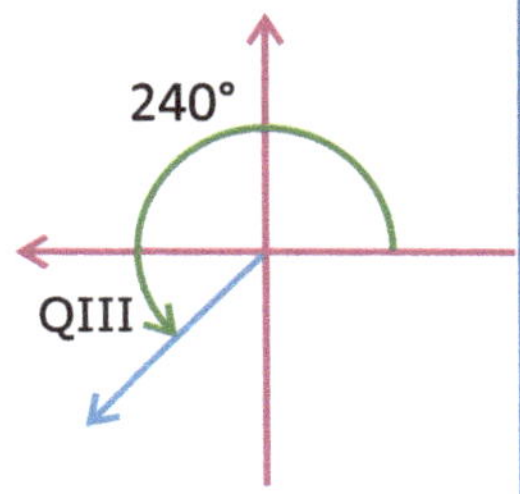

Q1.2j

Evaluate $\tan -600°$

Press the keys as follows

tan	-120	=

You get $1.732 = 1.73\ (to\ 3\ sf)$

Note: Clockwise for negative angle

The principal angle lies in quadrant III and tangent value is positive.

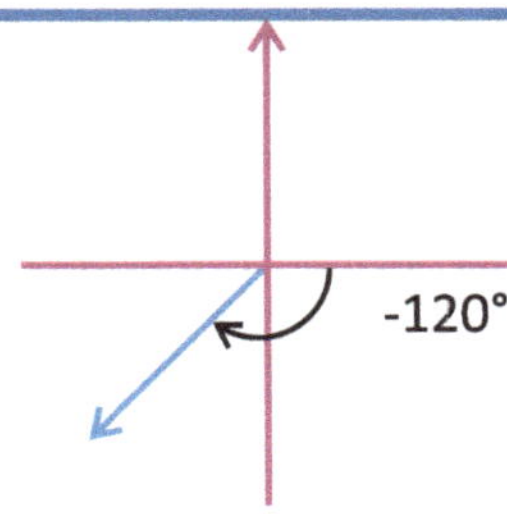

Watch Video **How to Use Calculator To Find Angle** SCAN ME

Find Unknown Angle from Primary Trigonometric Ratio

Inverse of Trigonometric Ratio

Inverse trigonometric function is used to find the angle

Example 2: Evaluate Angle for the Trigonometric Ratio to one decimal place

a.	Find $\angle A$ if $\tan A = \frac{1}{2}$ $0° < A < 90°$
b.	Find acute angle θ if $\sin\theta = 0.3245$
c.	Find acute angle α if $\cos\alpha = \frac{\sqrt{3}}{2}$

Solution 2a: $\tan A = \frac{1}{2} \Rightarrow \angle A = \tan^{-1}\left(\frac{1}{2}\right)$

Press the shift key for inverse as in sequence below

SHIFT	tan	(	1	÷	2	)	=

You will get the measurement of angle A: $\angle A = 25.56 = 25.6°$

Solution 2b: $\sin\ \theta = 0.3245 \Rightarrow A = \sin^{-1}(0.3245)$

Press the shift key for inverse as in sequence below

SHIFT	sin	(	0.3245	)	=

You will get the measurement of angle A: $\theta = 18.93 = 18.9°$

Solution 2c: $\cos\alpha = \frac{\sqrt{3}}{2} \Rightarrow \alpha = \cos^{-1}\left(\frac{\sqrt{3}}{2}\right) = 30$

$$\alpha = 30°$$

✍ **Practice:**

Find acute angle θ if $\sin\theta = 0.8480$

Practice: Find Angle for the Trigonometric Ratio to one decimal place

a.	Find $\angle A$ if $\tan A = \sqrt{3}$ $0° < A < 90°$
b.	Find acute angle θ if $\sin\theta = 0.3245$
c.	Find acute angle α if $\cos\alpha = \frac{\sqrt{3}}{2}$
d.	Find acute angle α $\cos\alpha = -\frac{\sqrt{3}}{2}$
e.	Find acute angle A if $\tan A = 2$
f.	Find obtuse angle θ if $\sin\theta = 0.5$
	Use trigonometric ratios to find the angles in right triangles given below:
	g. 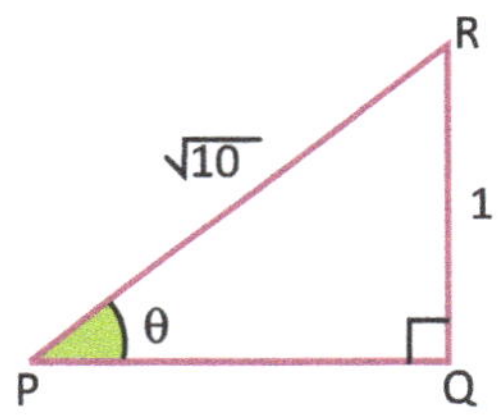 h. 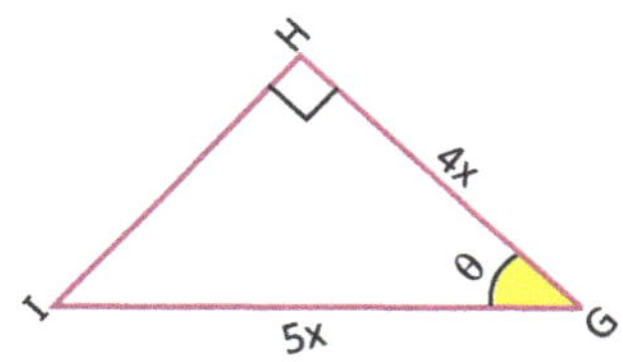

Investigate and explain why the following statements are always true for any angle:

a.	$\sin x = \sin(360° + x)$	b.	$\sin x = \sin(180° - x)$
c.	$\tan x = \tan(180° + x)$	d.	$\cos x = \sin(90° - x)$
e.	$\cos x = \cos(360° - x)$	f.	$\sin(-x) = -\sin x$
g.	$\cos(-x) = \cos x$	h.	$\cos x = -\sin(90° + x)$
Practice Question: Apply learnings to find value of the following:			
Q1	$\sin 40° = 0.6428$, find $\sin 140°$	Q2	$\cos 70° = 0.3420$, find $\sin 160°$

Consolidate Learnings

Connecting Pythagorean Theorem – Similar Triangles – Trigonometry

How do we find the angles of a right triangle from the given sides?

Let us explore and investigate from what is known to us.

Investivation: Pythagorean Theorem & Similar Triangles

Given a right triangle AFG with BC and DE parallel to FG. Find side lengths x, y, z and $\angle BAC$ in the figure.

Probe Investigation: Pythagorean Theorem & Similar Triangles

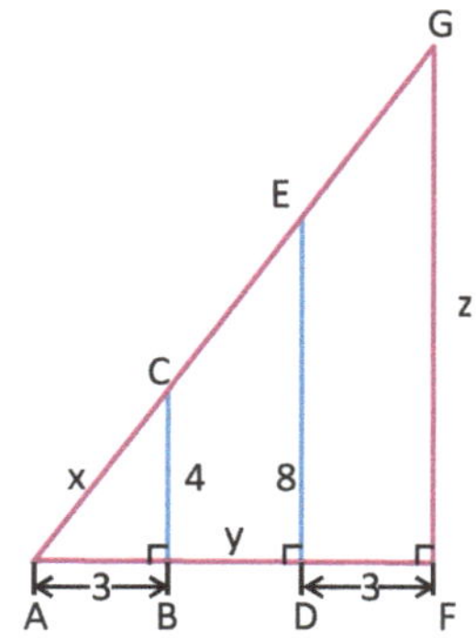

$\Delta\, ABC$ is right triangle. Pythagorean Theorem can be applied.

$x^2 = 3^2 + 4^2$

$x^2 = 9 + 16 = 25$

$x = \sqrt{25} = 5$

ΔADE is similar to ΔABC

$\angle A$ is common; $\angle\, ABC = \angle ADE = 90°$

$\Delta ADE \sim \Delta ABC \sim AA$ similarity

Ratio of sides is constant for similar triangles

$\frac{DE}{BC} = \frac{DA}{AB}$

$\frac{8}{4} = \frac{3+y}{3}$

$3(8) = 4(3+y)$

$24 = 12 + 4y$

$24 - 12 = 4y$

$\frac{12}{4} = y$ or $y = 3$

Also $\Delta\, AFG \sim \Delta\, ABC$

$\frac{FG}{BC} = \frac{FA}{AB}$

$\frac{z}{4} = \frac{3+3+3}{2}$

$z = \frac{9}{3} \times 4 = 12$

So, side lengths are: $\boldsymbol{x = 5}$, $\boldsymbol{y = 3}$ $\boldsymbol{and\ z = 12}$

Note:

We could find the side lengths with our knowledge of Pythagorean theorem and similar triangles but not the angle.

Explore and Learn

Find $\angle BAC$ *and* $\angle BCA$ in the triangle ΔABC

Angle Measurement from the given sides of a right triangle

Now we are ready to close our investigation with all the concepts learned.

Investigation: Pythagorean Theorem & Similar Triangles

Given a right triangle AFG with BC and DE parallel to FG. Find side lengths x, y, z and $\angle BAC$ **in the figure.**

Investigation Probing: Pythagorean Theorem & Similar Triangles

Earlier we calculated length of the sides as

$x = 5\,,\ \ y = 3\ \ and\ z = 12$

$\Delta\ ABC$ is a right triangle with

opposite side $BC = 4\ and\ \ adjacent\ side\ AB = 3$

Tangent ratio can be used to find the angle

$\tan A = \dfrac{opposite}{adjacent} = \dfrac{4}{3} \Rightarrow \angle A = \tan^{-1}\left(\dfrac{3}{4}\right) = 53.1°$

$\boldsymbol{\angle BAC = 53.1° \approx 53°}$

S^O_H C^A_H T^O_A

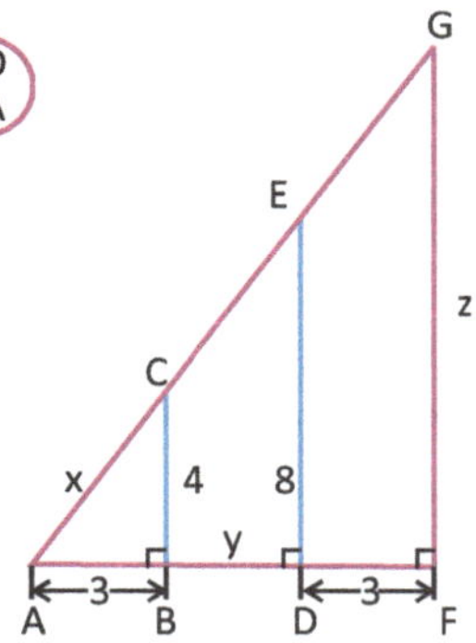

✍ Exercise 4: Evaluate Angles

Q.	Evaluate Angle for the Trigonometric Ratio to one decimal place
a.	Find $\angle A$ if $\tan A = \frac{3}{2}$, $0° < A < 90°$
b.	Find acute angle θ if $\sin\theta = 0.50$
c.	Find acute angle α if $\cos\alpha = \frac{1}{2}$
d.	Find $\angle A$ if $\tan B = 1$, $0° < A < 90°$
e.	Find acute angle θ if $\sin\theta = \frac{3}{5}$
f.	Find acute angle θ if $\cos\theta = \frac{\sqrt{2}}{2}$
g.	Find acute angle θ if $\cos\theta = \sin\theta$
h.	Find range of acute angle when if $\cos\theta < \sin\theta$

Secondary Trigonometric Ratios

Triangle has 3 sides that gives 6 combinations of ratio of the sides.

Three primary ratios and their reciprocals are called secondary ratios.

Reciprocal Relationship of Trigonometric Ratio

Secondary ratios are reciprocal of the primary ratios:

- cosecant is reciprocal of sine.
- secant is reciprocal of cosine.
- cosecant is reciprocal of sine.

$\csc\theta = \dfrac{1}{\sin\theta}$	$\sec\theta = \dfrac{1}{\cos\theta}$	$\cot\theta = \dfrac{1}{\tan\theta}$

Example 3: Identify six trigonometric ratios for angle P

Given ΔPQR , $PQ = 8\ cm,\ QR = 6\ cm\ \angle PQR = 90°$.

Write the six trigonometric ratios in lowest terms for angle P.

Solution 3: Identify six trigonometric ratios for angle P

Step 1: First find the length of the missing side PR.

Pythagorean Theorem

$PR^2 = PQ^2 + QR^2 = 8^2 + 6^2 = 64 + 36 = 100$

$\boldsymbol{PR = \sqrt{100} = 10}$

Now write each ratio $S_H^O\ C_H^A\ T_A^O$

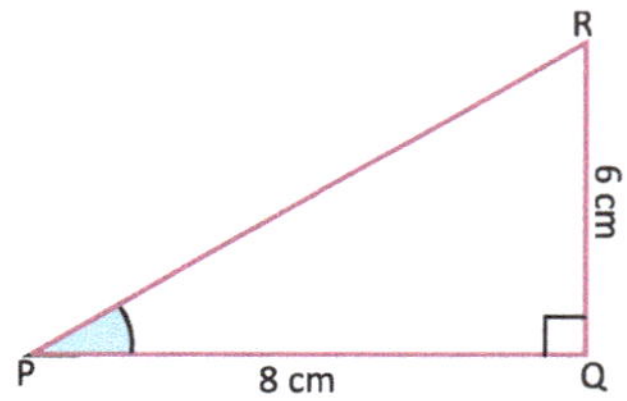

Step 2: Write each trigonometric ratio

a.	$\sin P = \dfrac{6}{10} = \dfrac{3}{5}$	b.	$\cos P = \dfrac{8}{10} = \dfrac{4}{5}$
c.	$\tan P = \dfrac{6}{8} = \dfrac{3}{4}$	d.	$\csc P = \dfrac{10}{6} = \dfrac{5}{3}$
e.	$\sec P = \dfrac{10}{8} = \dfrac{5}{4}$	f.	$\cot P = \dfrac{8}{6} = \dfrac{4}{3}$

Find Unknown Angle: Secondary Trigonometric Ratio

Reciprocal trigonometric identities connection

Secondary trigonometric functions are normally not available on calculators. Find equivalent primary trigonmetic ratio and then get the solution.

$\csc\theta = \dfrac{1}{\sin\theta}$	$\sec\theta = \dfrac{1}{\cos\theta}$	$\cot\theta = \dfrac{1}{\tan\theta}$

Example 4: Evaluate Angle for the Trigonometric Ratio to one decimal place

a.	Find acute angle θ if $\cot\theta = 4$
b.	Find acute angle θ if $\sec\theta = 3.2$

Solution 4: Find equivalent primary trigonometric ratio and then the angle.

a.	$\cot\theta = 4$ $\tan\theta = \dfrac{1}{\cot\theta} = \dfrac{1}{4}$ $\theta = \tan^{-1}\left(\dfrac{1}{4}\right) = 14.0°$	b.	$\sec\theta = 3.2$ $\cos\theta = \dfrac{1}{\sec\theta} = \dfrac{1}{3.2} = 71.8°$ $\theta = \cos^{-1}\left(\dfrac{1}{3.2}\right) = 14.0°$

✍ Practice Based on Example 4

Q1 Given right triangle where $\angle PQR = 90°$ *and* $PQ = x, QR = 2x$

a. Sketch ΔPQR.
b. Find all the angles of the triangle.
c. Write the six trigonometric ratios.

$\sin P =$	$\cos P =$	$\tan P =$
$\csc P =$	$\sec P =$	$\cot P =$
$\csc R =$	$\sec R =$	$\cot R =$

Q2 Find each of the following to one decimal place.

a.	angle $\boldsymbol{\theta}$.
b.	$\sin\theta$
f.	$\sec\theta$
g.	$\cot\theta$
j.	$\tan K$

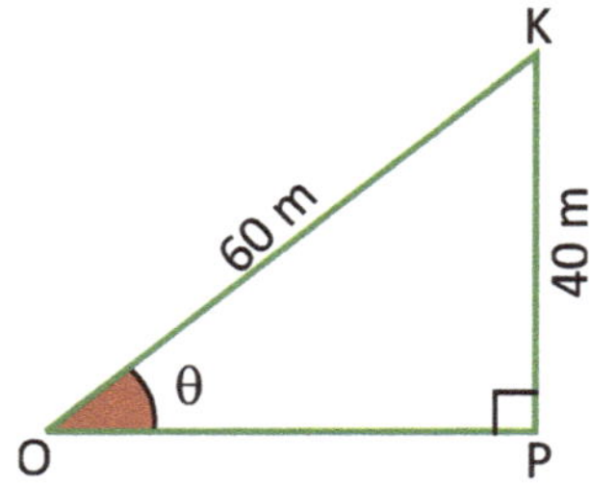

Q3 **Communication and Thinking**

Consider a point $P(3,4)$ on a coordinate plane making an angle θ with positive $x - axis$ as shown in the figure.

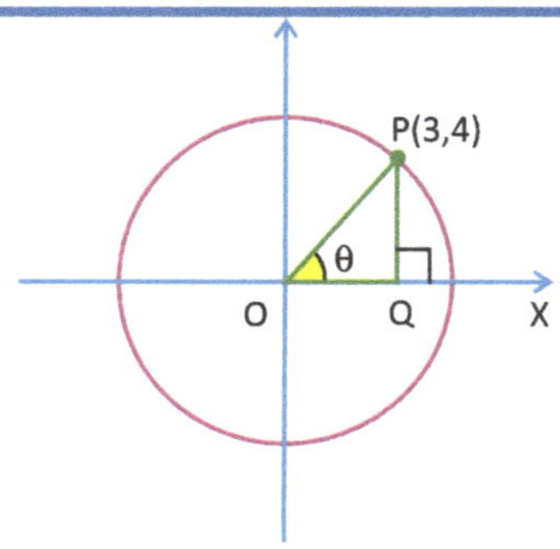

- Find angle θ
- Find the length of each side of the ΔPOQ
- Write the value of the six trigonometric ratios given in the table below.
- Provide reason for some values to be equal.

$\sin\theta =$	$\cos P =$	$\tan\theta =$
$\csc\theta =$	$\sec P =$	$\cot P =$

Q4. Explain why $|\sin\theta| \leq 1$

Q5 Explain why $|\sec\theta| \geq 1$

Q6 What is the value of $\sec 0°$

Q7 Arrange $\sin 30°$, $\cos 30°$ and $\tan 30°$ in increasing order.

Q8 If coordinate point $(a, 0.4)$ lies on a unit circle find the value of a.

Q9 If coordinate point $\left(\frac{\sqrt{6}}{3}, b\right)$ lies on a unit circle find the value of b. $\pm\frac{\sqrt{3}}{3}$

Q10 Draw a right triangle where $\sin\theta > \cos\theta$, then find the value of $\tan\theta$

Using Given Ratios to Find Others

Trigonometric Ratios Are All Related

Trigonometric ratio relates quotient of two sides of a right triangle. Third side length can be calculated using Pythgorean theorem.

Example 5: Relate Trigonometric Ratios

Given right triangle where $\angle PQR = 90°$ *and* $\mathbf{cos}\, \boldsymbol{P} = \frac{\sqrt{3}}{2}$

Sketch ΔPQR

Write all the six trigonometric ratios.

Solution 5: Relate Trigonometric Ratios

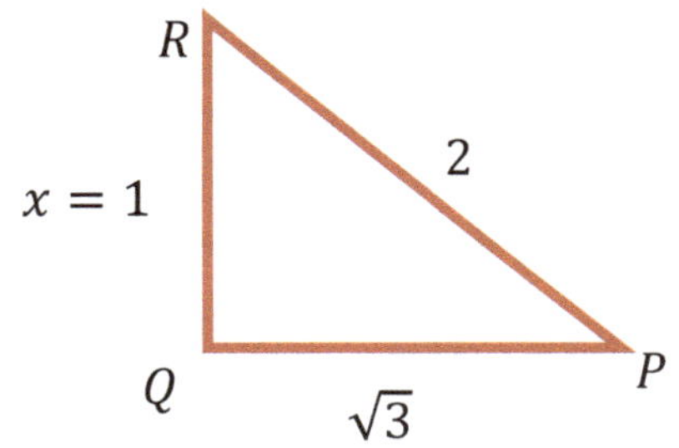

Given ΔPQR where $\angle PQR = 90°$ *and* $\cos P = \frac{\sqrt{3}}{2}$

$$\cos P = \frac{\sqrt{3}}{2} = \frac{adjacent}{hypotenuse} = \frac{PQ}{PR}$$

This implies: $PQ = \sqrt{3}, PR = 2$

Pythagorean Theorem to find the third side

$$PR^2 = QR^2 + PQ^2$$

$$QR = \sqrt{PR^2 - PQ^2} = \sqrt{(2)^2 - \left(\sqrt{3}\right)^2} = 1$$

ΔPQR

$PQ = \sqrt{3}, PR = 2$

$QR = 1$

$\sin P = \frac{1}{2}$	$\cos P = \frac{\sqrt{3}}{2}$	$\tan P = \frac{1}{\sqrt{3}}$
$\csc P = 2$	$\sec P = \frac{2}{\sqrt{3}}$	$\cot P = \sqrt{3}$

Practice:

Given $\sin \theta = \frac{1}{4}$

- Draw a right triangle with acute angle of θ.
- Find the value of other five trigonometric ratios for θ being an acute angle.

Example 6: Find sine and cosine ratio from the given tangent ratio

Given $\tan x = \frac{2}{5}, 0° < x < 90°$. Find $\sin x$ *and* $\cos x$

Solution 6:

Strategy:

- Sketch the right triangle with given ratio.
- Find the missing side using Pythagorean theorem
- Write required ratios from the triangle.

Sketch the right triangle with given ratio

$$\tan x = \frac{2}{5} = \frac{\text{opposite}}{\text{adjacent}} = \frac{BC}{AC}$$

In the triangle $BC = 2, AC = 5$

Hypotenuse $AB = \sqrt{2^2 + 5^2} = \sqrt{29}$

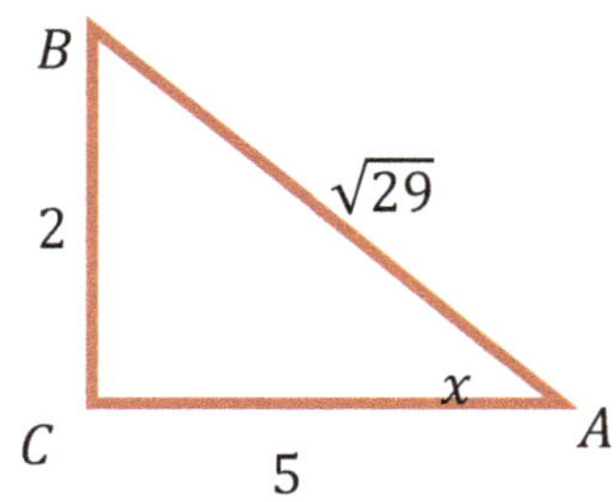

SOH CAH TOA

$$\sin x = \frac{\text{opposite}}{\text{hypotenuse}} = \frac{BC}{AB} = \frac{2}{\sqrt{29}}$$

Rationalize denominator

$$\sin x = \frac{2\sqrt{29}}{29}$$

SOH CAH TOA

$$\cos x = \frac{\text{adjacente}}{\text{hypotenuse}} = \frac{AC}{AB} = \frac{5}{\sqrt{29}}$$

Rationalize denominator

$$\cos x = \frac{5\sqrt{29}}{29}$$

It is a good practice to rationalize denominator. $\frac{a}{\sqrt{b}} = \frac{a}{\sqrt{b}} \times \frac{\sqrt{b}}{\sqrt{b}} = \frac{a\sqrt{b}}{b}$

Practice: Based on Example 5

Q5.1 Given right triangle PQR where: $\angle PQR = 90°$ *and* $\mathbf{\sin P} = \frac{\sqrt{5}}{3}$

- Sketch ΔPQR and then Write the six trigonometric ratios.

$\sin P =$	$\cos P =$	$\tan P =$
$\csc P =$	$\sec P =$	$\cot P =$

Q5.2	Given right triangle PQR where: $\angle PQR = 90°$ *and* $\mathbf{cot}\, \boldsymbol{R} = \mathbf{2}$ • Sketch ΔPQR • Write the six trigonometric ratios.

$\sin P =$	$\cos P =$	$\tan P =$
$\csc P =$	$\sec P =$	$\cot P =$

✍ Practice 6: Based on Example 6

Q6.1	Given $\sin x = \frac{3}{5}, 0° < x < 90°$. Find $\tan x$ *and* $\cos x$
Q6.2	Given $\tan x = \sqrt{3}, 0° < x < 90°$. Find $\sin x$ *and* $\cos x$
Q6.3	Given $\cos x = \frac{5}{12}, 0° < x < 90°$. Find $\sin x$ *and* $\tan x$
Q6.4	Given $\sin x = \frac{\sqrt{2}}{2}, 0° < x < 90°$. Find $\sin x \quad \cos x + \tan x$
Q6.5	Given $\cos x = \frac{7}{24}, 0° < x < 90°$. Find $\sin x + \cos x + \tan x$
Q6.6	Use special triangles to find the exact value of the following

a.	$\sec 60°$	b.	$\cot 30°$	c.	$\csc 45°$

Student Activity 1

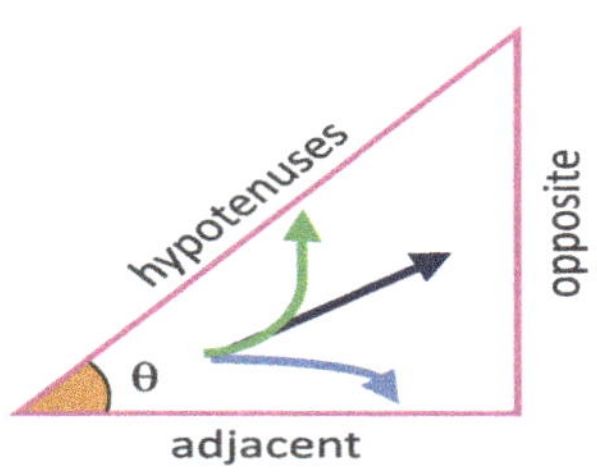

Quotient Identity

Prove the quotient identity

$$\tan\theta = \frac{\sin\theta}{\cos\theta}$$

in the right triangle.

Finding Ratio from Secondary Ratio

Example 7: Find Trigonometric Ratio from Given Secondary Ratio

Find other trigonometric ratios if $\mathbf{sec}\,\boldsymbol{A} = \frac{\mathbf{13}}{\mathbf{12}}$.

Solution 7: Find Trigonometric Ratio from Given Secondary Ratio

Given: $\sec A = \frac{13}{12}$

$$\sec A = \frac{hypotenuse}{adjacent} = \frac{AB}{AC}$$

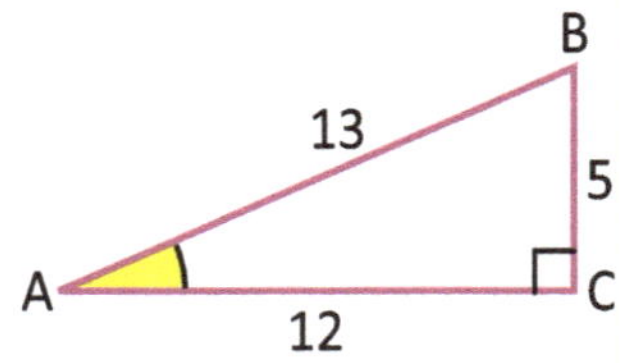

Steps:

- Sketch ΔABC with hypotenuse of 13 and adjacent side of 12
- Calculate length of opposite side, BC, using Pythagorean Theorem
- Write all the trigonometric ratios

$BC^2 = AB^2 - AC^2 = 13^2 - 12^2 = 169 - 144 = 25$ $BC = \sqrt{25} = 5$ Now write each ratio $S_H^O\ C_H^A\ T_A^O$	$\sin A = \frac{5}{13}$ $\cos A = \frac{12}{13}$ $\tan A = \frac{5}{12}$	$\csc A = \frac{13}{5}$ $\sec A = \frac{13}{12}$ $\cot A = \frac{12}{5}$

✍ Practice based on Example 7

Q7.1 Given right triangle PQR where:

$$\angle PQR = 90° \text{ and } \mathbf{csc}\,\boldsymbol{P} = \frac{\mathbf{7}}{\mathbf{2}}$$

- Sketch ΔPQR
- Find the angle $\boldsymbol{P}$
- Write the six trigonometric ratios.

$\sin P =$	$\cos P =$	$\tan P =$
$\csc P =$	$\sec P =$	$\cot P =$

Complementary Angle Theorem (Cofunction Identity)

Sum of Acute Angles in Right Triangle 90°

Acute angles in a right triangle are complementary.

Equivalent Trigonometric Ratios

Consider right triangles being viewed from the complementary angles.

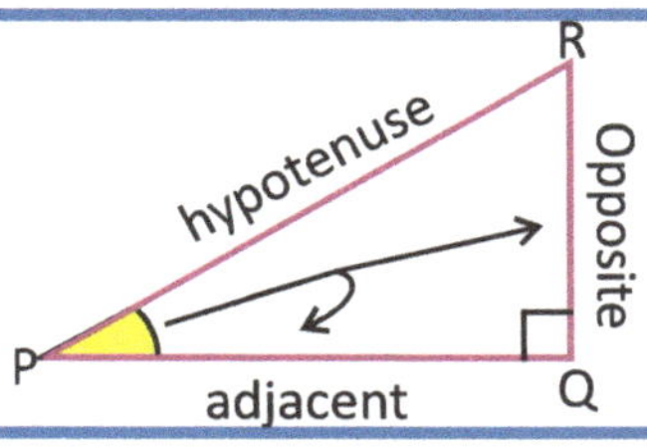	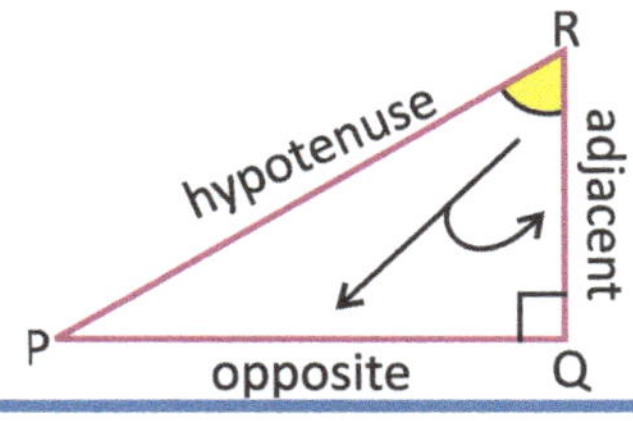	Opposite and adjacent sides flip when we look from the other acute angle.
$\tan P = \dfrac{opposite}{adjacent} = \dfrac{QR}{PQ}$	$\cot R = \dfrac{adjacent}{opposite} = \dfrac{QR}{PQ}$	$P = 90° - Q$ $Q = 90° - P$ $\tan P = \cot R$ $\sin P = \cos R$
$\sin P = \dfrac{opposite}{hypotenuse} = \dfrac{QR}{PR}$	$\cos R = \dfrac{adjacent}{hypotenuse} = \dfrac{QR}{PR}$	

Complementary Angle Theorem

$\sin\theta = \cos(90° - \theta)$	$\cos\theta = \sin(90° - \theta)$
$\tan\theta = \cot(90° - \theta)$	$\cot\theta = \tan(90° - \theta)$
$\csc\theta = \sec(90° - \theta)$	$\sec\theta = \csc(90° - \theta)$
$\sin 30° = \cos 60°$	$\cot 23° = \tan 67°$
$\tan 28° = \cot 62°$	$\csc 49° = \sec 41°$
$\csc 12° = \sec 78°$	$\cos 42° = \sin 48°$

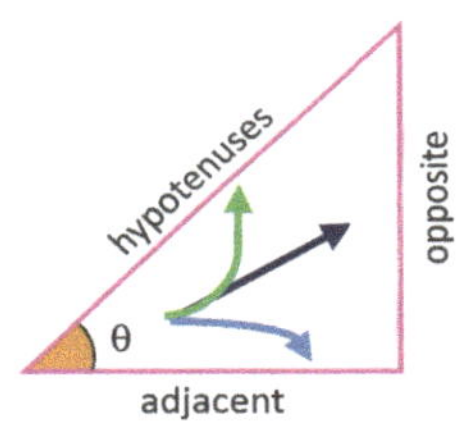

Practice Question: Find the value of K that makes

Q1	$sin\,42° = cos\,K$	Q2	$\tan 23° = \cot K$
Q3	$\csc 29° = \sec K$	Q4	$\sec 58° = \csc 2K$

Example 8: Cofunction Identity

Consider the ΔOPK in the figure to answer the following questions.

a.	verify that $sin\,\theta = cos\,K$.
b.	Explain if $cos\,\theta = sin\,K$
c.	$\sin 30° = 0.5$ Find the value of $\cos 60°$
d.	Find exact value of $\csc 70° - \sec 20°$
e.	Find exact value of $\frac{\cos 80° - \sin 10°}{\tan 80°}$, given $\tan 80° = 5.67$

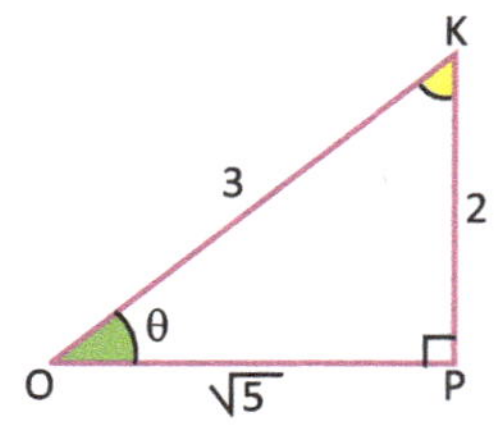

Solution 8: Cofunction Identity

a.	$\sin\theta = \frac{2}{3}$, $\cos K = \frac{2}{3}$ $\Rightarrow$ $\sin\theta = \cos K = \frac{2}{3}$
b.	ΔOPK is a right triangle. $\angle K = 90° - \theta$ $\cos\theta = \sin(90° - \theta)$ Cofunction Identity $\cos\theta = \sin K$
c.	Given: $\sin 30° = 0.5$ $\cos 60° = \cos(90° - 30°) = \sin 30° = 0.5$
d.	$\csc 70° - \sec 20°$ acute angle in right triangle $= \csc 70° - \sec(90 - 70)°$ apply Cofunction Identity $= \csc 70° - \csc 70° = 0$
e.	$\frac{\cos 80° - \sin 10°}{\tan 80°} = \frac{0}{\tan 80°} = 0$

SAT Challenge Problem:

Determine exact value of b if $\sec\left(2b - \frac{45}{2}\right) = \csc\left(2b + \frac{45}{2}\right)$

SCAN ME

✍ Practice Based on Example 8

Q8.1

a.	$\tan 40° = 0.8390$ Find the value of $\cot 40°$ Find the value of $\cot 50°$	b.	Find exact value of $\frac{\sin 52°}{\cos 38°} + (\sec 64° - \csc 26°)$

Q8.2

In a right triangle ABC, with $\angle C = 90°$,

$\angle A$ and $\angle B$ form a complementary pair.

If $\cos A = 0.75$, then find the exact value of $\tan(90° - A)$.

Complementary Pairs: sum of the angles is 90°.

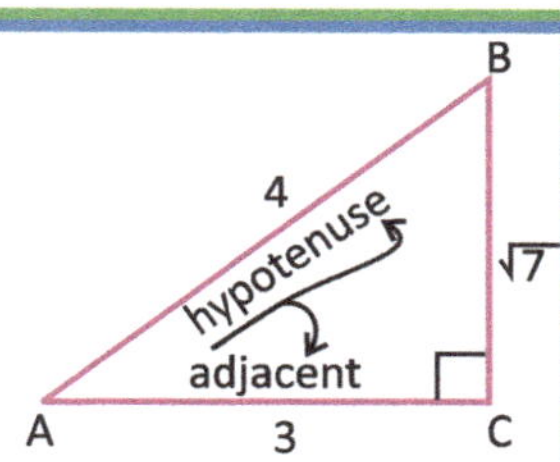

Hint: Refer to the triangle ABC

$\cos A = 0.75 \ = \frac{75}{100} = \frac{3}{4}$

Calculate BC, the missing side length

$BC = \sqrt{4^2 - 3^2} = \sqrt{7}$

$\tan(90° - A) = \tan B = \frac{opposite}{hypotenuse} = = \frac{3}{\sqrt{7}} = \frac{3}{\sqrt{7}} \times \frac{\sqrt{7}}{\sqrt{7}} = \frac{3\sqrt{7}}{7}$

Rationalize to write the correct solution

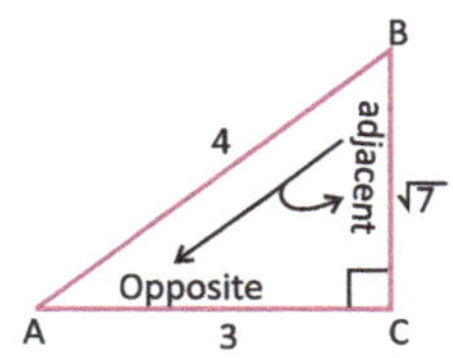

Q8.3

In a right triangle ABC, with $\angle C = 90°$, $\angle A$ and $\angle B$ form a complementary pair.

If $\sin A = 0.5$, then find the exact value of $\cot(90° - A)$.

Q8.4

In a right triangle ABC, with $\angle C = 90°$, $\angle A$ and $\angle B$ form a complementary pair.

If $\cos A = 0.5$, then find the exact value of $\sec(90° - A)$.

Q8.5

Given $\csc T = \frac{25}{24}$ in the right triangle RST where $\angle S = 90°$

a.	Draw a right triangle RST with given information
b.	Find exact value of all the sides of the triangle
c.	Find all the angles of the triangle to 3 sf.

Chapter 9 Test: Trigonometric Ratios

Q1	Find the acute angle for which sine and cosine value is equal.
Q2	If $\sin x = \frac{1}{2}$ find the value of x
Q3	Write exact value of all the six trigonometric ratios for $x = 30°$ using special triangle.
Q4	If $\sin x = \frac{3}{4}$ find exact value of $\sec x$
Q5	Find exact value of: $\frac{\sin 27°}{\cos 63°} + (\sec 44° - \csc 46°)$
Q6	$\cos 38° = \sin x$, Find the value of x
Q7	Given right triangle PQR where: $\angle PQR = 90°$ and $\tan P = \frac{1}{2}$ • Sketch ΔPQR • Find the hypotenuse • Write the six trigonometric ratios. SCAN ME
Q8.	If $\cot\theta = 3$ find the acute angle θ to two decimal places.
Q9.	Prove that $\sin(90° - x)\cot(90° - x) = \sin(90° - x)$
Q10	Explain why $\sin(360n + x) = \sin(x)$ where n is an integer.

Things to Remember

- $\sin\theta$, $\cos\theta$ and $\tan\theta$ are called **primary trigonometric ratios**
- $\csc\theta$, $\sec\theta$ and $\cot\theta$ are called **secondary trigonometric ratio**
- secondary trigonometric ratios are reciprocal of the primary trigonometric ratios: $\sin\theta \times \csc\theta = 1$, $\cos\theta \times \sec\theta = 1$, $\tan\theta \times \cot\theta = 1$
- trigonometric ratios have no units
- all trigonometric ratios are written in lower case
- most of the time we work with primary trigonometric ratios
- Greek symbols, $\alpha, \beta, \gamma, \theta, \delta$ are commonly used for angles
- Mnemonic to remember primary ratios: $SOH\ CAH\ TOA$ or $S_H^O\ C_H^A\ T_A^O$
- Inverse of primary trigonometric ratio is used to find the angle

Key Concepts

Trigonometric Relations

$$\sin\theta = \frac{opposite}{hypotenuse} = \frac{y}{r}$$

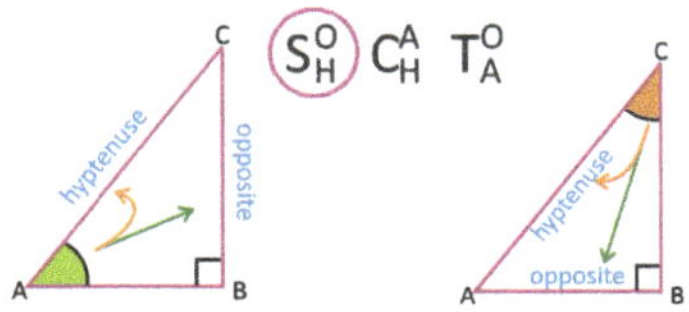

$$\cos\theta = \frac{adjacent}{hypotenuse} = \frac{x}{r}$$

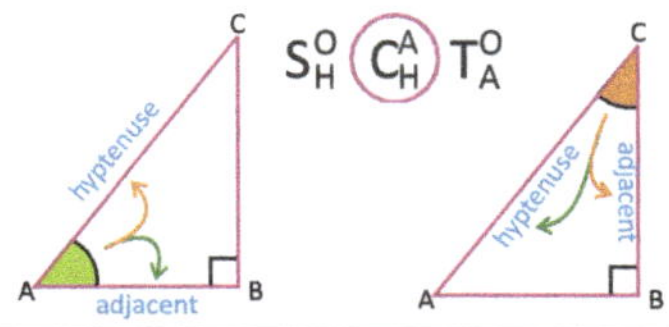

$$\tan\theta = \frac{opposite}{adjacent} = \frac{y}{x}$$

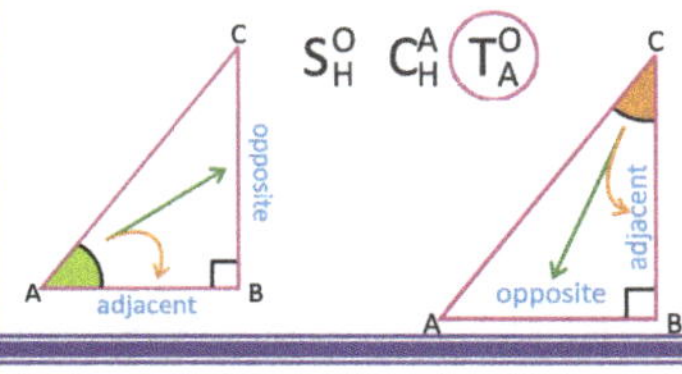

$$\csc\theta = \frac{1}{\sin\theta} = \frac{x}{y}$$

$$\sec\theta = \frac{1}{\cos\theta} = \frac{r}{x}$$

$$\cot\theta = \frac{1}{\tan\theta} = \frac{x}{y}$$

Co-function Identities

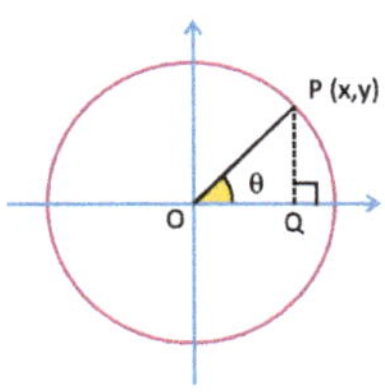

$$\sin(90^\circ - \theta) = \cos\theta$$

$$\cos(90^\circ - \theta) = \sin\theta$$

$$\tan(90^\circ - \theta) = \cot\theta$$

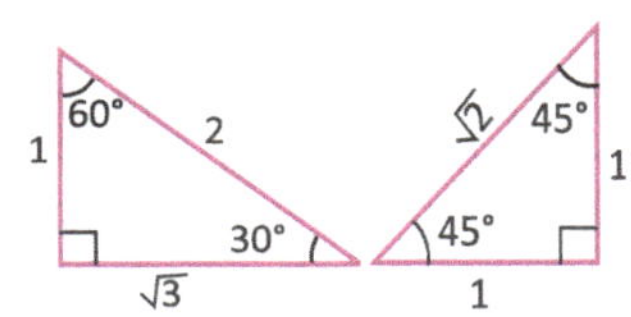

Key Learnings

- *Rationalize denominator, if required*
- $tan\ \theta = \frac{sin\ \theta}{cos\ \theta}$ *Proof:* $\frac{sin\ \theta}{cos\ \theta} = \frac{opp/hyp}{adj/hyp} = \frac{opp}{adj} = tan\ \theta$
- *Since adjacent or opposite sides are not longer than the hypotenuse, sine and cosine ratios are not greater than one*
- *Absolute value of reciprocal of sine and cosine ratio is greater than or equal to one*
- *Tangent and cotangent have range of all real numbers*
- *Trigonometric ratios of similar triangles are same.*

Chapter 10: Coordinates and Trigonometry

What is the Unit Circle?

The unit circle is a circle centered at the origin $O(0,0)$ on the coordinate plane with a radius of 1 unit.

Equation of the unit circle: $x^2 + y^2 = 1$

Point on the Circle

Let $P(x, y)$ be a point on the unit circle.
Draw a perpendicular from $P$to the x-axis, meeting it at point Q.
Now, triangle △POQ is a right triangle, where:

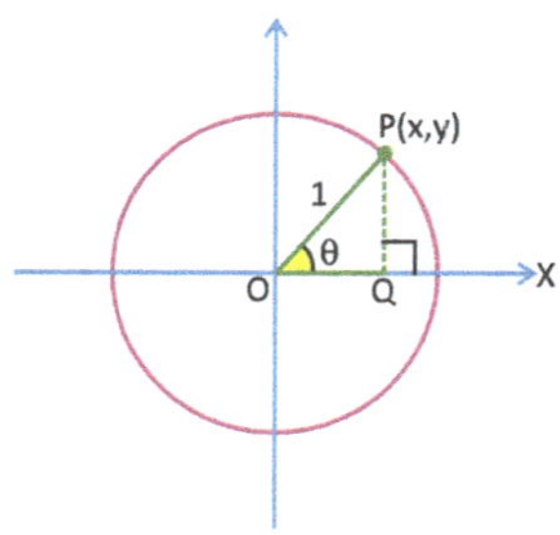

- $OP = 1$ (radius of the unit circle)
- $OQ = x$ (adjacent side)
- $PQ = y$ (opposite side)
- Angle θ is measured between the positive x-axis and the segment OP

SCAN ME

Trigonometric Definitions:

Using the triangle △POQ we define:

$\cos\theta = \dfrac{adjacent}{hypotenuse} = \dfrac{x}{1} = x$	$\sin\theta = \dfrac{opposite}{hypotenuse} = \dfrac{y}{1} = y$

Thus, any point on the unit circle can be written as: $P(x, y) = P(\cos\theta, \sin\theta)$

Intercepts and Extremes

- Top Point: $(0,1) \rightarrow \sin\theta = 1$ (maximum value of sine)
- Bottom Point: $(0,-1) \rightarrow \sin\theta = -1$ (minimum value of sine)
- Right Point: $(1,0) \rightarrow \cos\theta = 1$
- Left Point: $(-1,0) \rightarrow \cos\theta = -1$

These values show the maximum and minimum values of sine and cosine.

Why Do Signs Change?

As the angle θ rotates around the circle, the coordinates (x, y)— and therefore $\cos\theta$ and $\sin\theta$— change sign depending on the quadrant. Coordinates in:

quadrant I are $(+, +)$: All are positive	quadrant II are $(-, +)$: sine is positive
quadrant III are $(-, -)$: tangent is	quadrant IV are $(+, -)$: cosine is positive

CAST stands for:

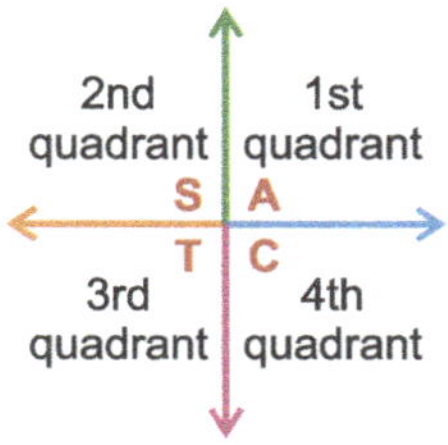

C (Cosine) – In Quadrant IV, only cosine is positive.

A (All) – In Quadrant I, all trigonometric ratios are positive.

S (Sine) – In Quadrant II, only sine is positive.

T (Tangent) – In Quadrant III, only tangent is positive.

Mnemonic Tip: Counter-clockwise from quadrant (IV): **C – A – S – T** Rule

Example 1: Trigonometric Ratios for Coordinate Points

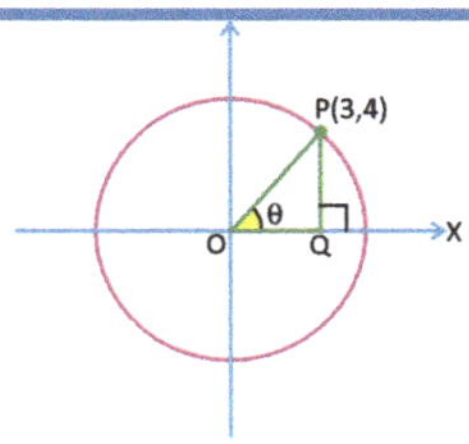

Point P $(3, 4)$ lies on a circle with centre at origin, O, as shown in the figure. Radius OP makes an angle θ with horizontal $x - axis$.

- Find angle θ correct to one decimal place.
- Write exact value of all the six trigonometric ratios for the angle θ

Solution 1: Trigonometric Ratios for Coordinate Points

Let Q be projection of P on the $x - axis$. ΔPOQ is right triangle with acute angle θ.

$$\tan\theta = \frac{4}{3} \Rightarrow \theta = \tan^{-1}\left(\frac{4}{3}\right) = 53.13° \approx 53.1°$$

	Primary Ratios	**Secondary Ratios**
$OP^2 = OQ^2 + QP^2$ $OP^2 = 3^2 + 4^2 = 25$ $OP = \sqrt{25} = 5$ In ΔPOQ: PQ = 4, OQ = 3, OP = 5	$\sin\theta = \frac{4}{5}$ $\cos\theta = \frac{4}{5}$ $\tan\theta = \frac{4}{3}$	$\csc\theta = \frac{5}{4}$ $\sec\theta = \frac{5}{3}$ $\cot\theta = \frac{3}{4}$

Practice Questions based on Example 1

For Point P on a circle with centre at origin, O radius OP makes an angle θ with horizontal $x - axis$. For each of the given points find:

- Find length of the radius OP
- Find angle θ correct to one decimal place.
- Write exact value of all the six trigonometric ratios for the angle θ.

a.	$P(12, -5)$	b.	$P(-8, 6)$	c.	$P(-3, -4)$

Unit Circle – Test Questions

Q1	What is the **equation of the Unit Circle** centered at the origin?
Q2	A point $P(a, b)$ lies on the Unit Circle and makes an angle θ with the positive x-axis. If $a = \frac{12}{13}$ and $b = -\frac{5}{13}$, find: i. $\sin\theta$ ii. $\cos\theta$ iii. $\tan\theta$
Q3	If $\theta = 90°$, what are the coordinates of the point on the unit circle corresponding to this angle?
Q4	State the **maximum and minimum values** of sine and cosine functions on the unit circle. Also, identify the **coordinates** at which they occur.
Q5	What are the signs of **sine, cosine, and tangent** in **Quadrant II** of the unit circle? Use the **CAST Rule** to justify your answer.
Q6	Find coordinate point on the unit circle if $\theta = 60°$
Q7	Explain how the **unit circle** helps in understanding the periodicity and sign changes of **trigonometric functions** as the angle increases from $0°$ to $360°$. Use a labeled sketch of the circle to support your answer.
Q8	A point $P(0.9, b)$ lies on the Unit Circle. Find the value of b.
Q9	What should be the sign of values for cosine and sine if angle is in Quadrant III. Explain your answer.

Unit Circle Challenge

Prove the identity $\cos^2\theta + \sin^2\theta = 1$

Hint: Use a point on the unit circle.

Summary:

- Unit circle links trigonometry and geometry beautifully.
- Coordinates on the unit circle are $(\cos\theta, \sin\theta)$
- The CAST rule helps determine which trig ratios are positive in each quadrant.
- Signs change as the point P moves around the circle — a powerful idea that builds understanding for identities, graphs, and more.

Range of Trigonometric Ratios in Quadrant I

Ratio of sides restricts the range of values the trigonometric ratios can have

Example 1: Communication & Thinking

Explain why $\sin\theta$ is never greater than one and $|\sec\theta|$ is never less than one.

$|\sec\theta| \equiv$ Absolute value of $\sec\theta$

Solution 1: Communication & Thinking

$$\sin\theta = \frac{opposite\ side}{hypotenuse}$$

In a right triangle hypotenuse is never less than the other legs. As a special case, for 90° opposite side will be equal to the hypotenuse. So, the maximum value of $\sin\theta$ is one.

$$\sec\theta = \frac{hypotenuse}{adjacent\ side}$$

In a right triangle hypotenuse is greater than the length of any leg (cathetus) of the triangle. For zero degree the adjacent side length and the hypotenuse is same. Therefore, the absolute value of ratio of hypotenuse to adjacent will be at least one.

Since hypotenuse ≥ adjacent; $|\boldsymbol{\sec\theta}| \geq \mathbf{1}$

Example 2: Variation in Trigonometric Ratio with Increasing Angle

a.	How will $\sin\theta$ change as θ increases from $0°$ to $90°$?
b.	How will $\cos\theta$ change as θ increases from $0°$ to $90°$?
c.	How will $\tan\theta$ change as θ increases from $0°$ to $90°$?

Solution 2: Variation in Trigonometric Ratio with Increasing Angle

a.	How will $\sin\theta$ value change as θ increases from $0°$ to $90°$? $\sin\theta = \frac{opposite}{hypotenuse}$ 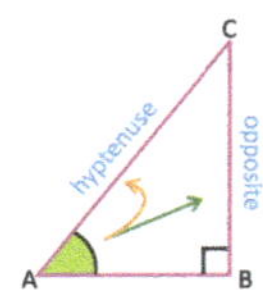 As θ increases from $0°$ to $90°$, opposite side increases from zero to hypotenuse. Therefore, value of $\sin\theta$ increases from zero to one.
b.	How will $\cos\theta$ value change as θ increases from $0°$ to $90°$? $\cos\theta = \frac{adjacent}{hypotenuse}$ 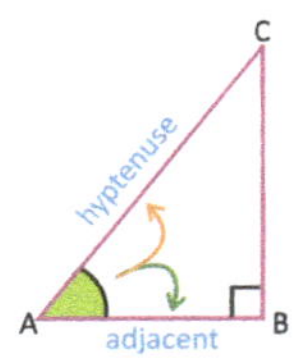 As θ increases from $0°$ to $90°$, adjacent side decreases from hypotenuse to zero. Therefore, value of $\cos\theta$ decreases from one to zero.
c.	How will $\tan\theta$ value change as θ increases from $0°$ to $90°$? $\tan\theta = \frac{opposite}{adjacent}$ 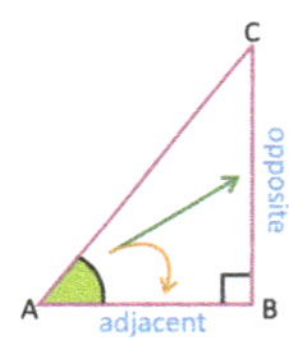 As θ increases from $0°$ to $90°$, opposite side increases and the adjacent side decreases. Therefore, value of $\tan\theta$ increases from zero to infinitely large value.

✍ Practice Questions based on Example 1 and 2

Q1a	How will $\csc\theta$ value change as θ increases from $0°$ to $90°$?
Q1b	How will $\cot\theta$ value change as θ increases from $0°$ to $90°$?
Q1c.	When is $\cos\theta = \sin\theta$ as θ increases from $0°$ to $90°$?

STEM Project: Explore Trigonometric Ratios for Angles Beyond 90°

Explore Beyond 90°

In right triangles all trigonometric ratios are positive. When we extend angles greater than 90°, in our coordinate plane, we will see that they do have negative values. Consider coordinate points in the four quadrants $P(x, y)$ and explore trigonometric ratios.

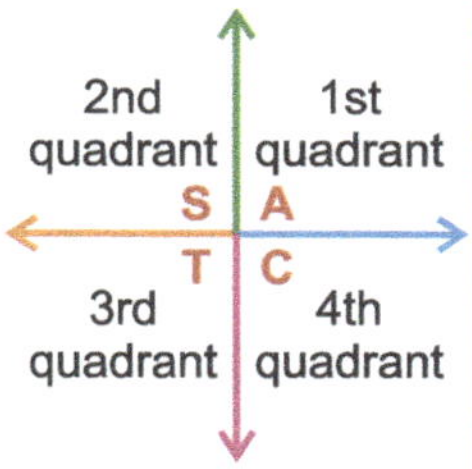

Show that

- All ratios are positive in quadrant I
- Only sine and cosecant are positive in quadrant II
- Only tangent and cotangent are positive in quadrant III
- Only cosine and secant are positive in quadrant IV

This is called **CAST** Rule

Student Activity: Analyse Points on Unit Circle

Part 1:

Prove that in a unit circle for any coordinate point $P(x, y)$, the value of coordinate points are:

$x = \cos\theta$ and $y = \sin\theta$

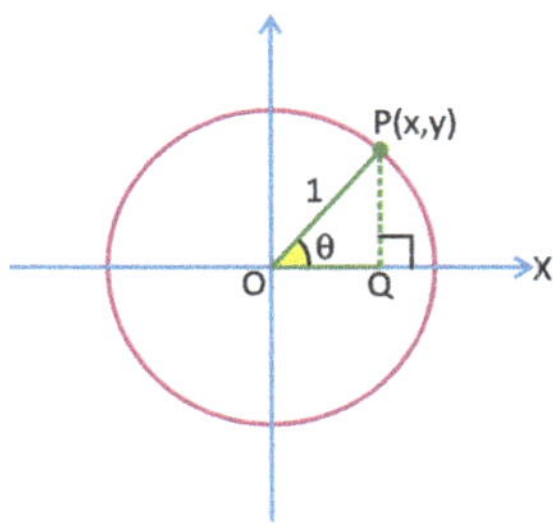

Part 2:

Explain from Part 1 that the values of $\cos\theta$ and $\sin\theta$ will be positive only in two quadrants. Identify the quadrants where:

Part 3:

Write the value of $\cos\theta$ and $\sin\theta$ for the point $A(-1, 0)$ on the unit circle.

Part 4: Quotient Identity

Prove that $\tan\theta = \frac{\sin\theta}{\cos\theta}$

Evaluate Trigonometric Expressions Without Calculator

Exact Trigonometric Ratios from **special triangles** (30°–60°–90° and 45°–45°–90°) allow us to solve trigonometric problems **without using a calculator**.
These triangles provide **precise values** for sine, cosine, and tangent of key angles using simple ratios.

45° – 45° – 90° Isosceles Right Triangle

Hypotenuse is $\sqrt{2}$ times the length of legs.

Example 1: Right Isosceles Triangle

Find exact trigonometric ratios for the acute angle in the right isosceles triangle.

Solution 1: Right Isosceles Triangle

Let the acute angle in the right isosceles triangle be A each

$A + A + 90° = 180°$ Then sum of all the angles is **180°**

$2A = 180° - 90° = 90°$

$A = \frac{90°}{2} = 45°$

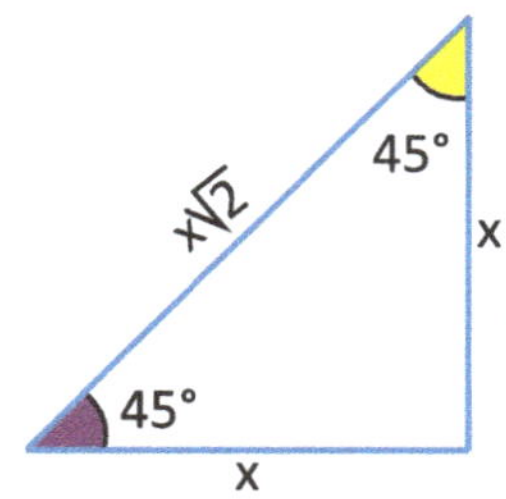

The angles in a right isosceles triangle are $45° - 45° - 90°$

If we consider each leg of length x

Then hypotenuse will be $\sqrt{x^2 + x^2} = \sqrt{2x^2} = x\sqrt{2}$ as shown in the figure.

$\sin 45° = \frac{x}{x\sqrt{2}} = \frac{1}{\sqrt{2}}$	$\cos 45° = \frac{x}{x\sqrt{2}} = \frac{1}{\sqrt{2}}$	$\tan 45° = \frac{x}{x} = 1$
$\csc 45° = \frac{x\sqrt{2}}{x} = \sqrt{2}$	$\sec 45° = \frac{x\sqrt{2}}{x} = \sqrt{2}$	$\cot 45° = \frac{x}{x} = 1$

$\sin 45° = \cos 45° = \frac{1}{\sqrt{2}} = \frac{1}{\sqrt{2}} \cdot \frac{\sqrt{2}}{\sqrt{2}} = \frac{\sqrt{2}}{2}$ Note: It is better to rationalize denominator

Practice Question: Find the value of k in the given conditions:

Q1	$k \sin 45° = 3$	Q2	$k \sec 45° = 4$	Q3	$2k \cot 45° = 5$

The triangle with $30° - 60° - 90°$ has angles in the ratio of 1:2:3

Example 2: Special $30° - 60° - 90°$ Triangle

Find exact trigonometric ratios for the acute angles in the right triangle shown in the figure.

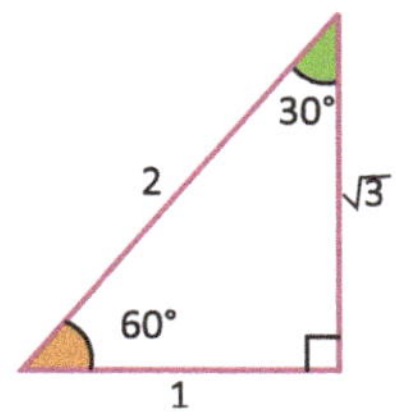

Solution 2: Special $30° - 60° - 90°$ Triangle

$30° - 60° - 90°$ is a special triangle

Angles are in the ratio of $1: 2: 3$ and

Sides are in the ratio of $1: \sqrt{3}: 2$

$\sin 30° = \frac{1}{2}$	$\cos 30° = \frac{\sqrt{3}}{2}$	$\tan 30° = \frac{1}{\sqrt{3}}$
$\csc 30° = 2$	$\sec 30° = \frac{2}{\sqrt{3}}$	$\cot 30° = \sqrt{3}$
$\sin 60° = \frac{\sqrt{3}}{2}$	$\cos 60° = \frac{1}{2}$	$\tan 60° = \sqrt{3}$
$\csc 60° = \frac{2}{\sqrt{3}}$	$\sec 60° = 2$	$\cot 60° = \frac{1}{\sqrt{3}}$
$\sin 90° = 1$	$\cos 90° = 0$	$\tan 90° = Not\ Defined$
$\sin 0° = 0$	$\cos 0° = 1$	$\tan 0° = 0$
Rationalize and then re-write the value of all the trigonometric ratios.		

✍ Practice: Evaluate and find the exact value of the following expressions.	
$2 \cos 30° \sin 30° \tan 30°$	$\tan 45° - \sin^2 60° - \cos^2 60°$
$\sin 30° \cos 60° - \sin 60° \cos 30°$	$\cos 60° \cos 30° - \sin 60° \sin 30°$

Thinking and Communication

Example 3: Solve Related Special Triangles

Given right ΔABC in which $AD = DC$.

Find the length of AC, if $DB = 10\ cm$.

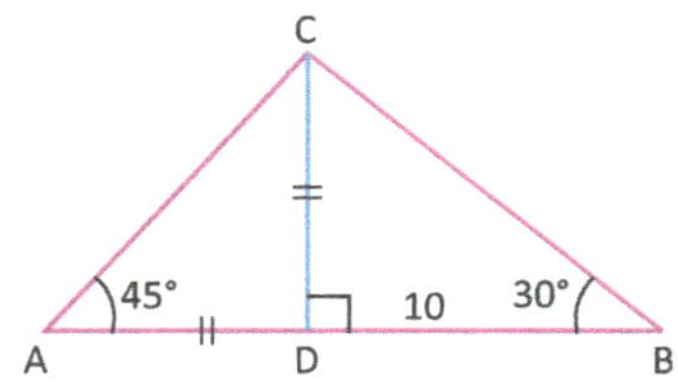

Solution 3: Solve Related Special

First consider the ΔBCD, $30° - 60° - 90°$

Sides are in the ratio of $1 : \sqrt{3} : 2$

$DB = 10\ cm$, smallest side is CD

$$CD = \frac{DB}{\sqrt{3}} = \frac{10}{\sqrt{3}}$$

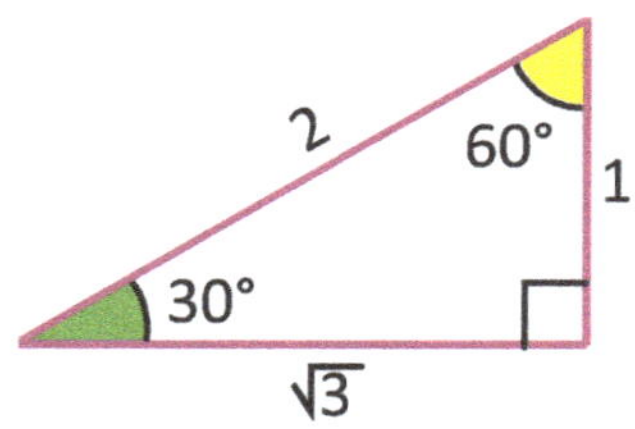

Now consider the ΔADC, $45° - 45° - 90°$

Sides are in the ratio of $1 : 1 : \sqrt{2}$

$CD = \frac{10}{\sqrt{3}}\ cm$, hypotenuse is $CD \times \sqrt{2}$

$$AC = CD \times \sqrt{2} = \frac{10}{\sqrt{3}} \times \sqrt{2} = \frac{10}{\sqrt{3}} \times \sqrt{2} \times \frac{\sqrt{3}}{\sqrt{3}}$$

Rationalize

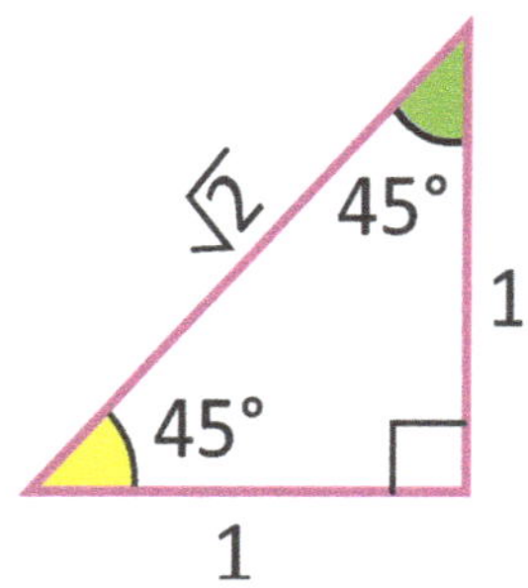

$$AC = \frac{10\sqrt{6}}{3}\ cm$$

✍ Practice Questions based on Example 1 and 2

Q1 Check if the following equations are true.

a.	$\sin 30° + \cos 60° = 1$
b.	$\cos^2 30° + \sin^2 30° = \tan 45°$
c.	$\sin 45° \cos 45° = \cos 60°$
d.	$\tan 60° \cos 60° = \sin 60°$
e.	$(\sin 60° + \cos 60°)^2 = 1$
f.	$\cos 60° \cos 30° - \sin 60° \sin 30° = 1$

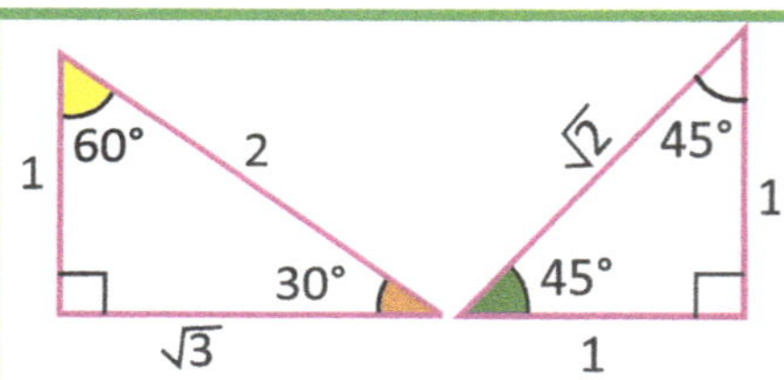

Q2	An 8 m ladder is leaning up against the side of a building wall at an angle of $60°$.	
	a.	How high will it reach on the wall?
	b.	How far is its base from the bottom of the building?

Answers

Q1a	F	Q1b	T	Q1c	T	Q1d	T	Q1e	F
Q1f	T	Q2a	$4\sqrt{3}$ m	Q2b	4m				

Evaluate Trigonometric Expressions

Example 3: Find Exact Value (Solution in the next page)

a.	$\cos 45° \sin 30° \tan 60°$
b.	$\tan 30° - \cos 60°$
c.	$\tan 45° - \sin 60°$
d.	$\sec 30° \sin 45° \cot 60°$
e.	$\tan^2 30° + \cos^2 30° + \sec^2 30°$
f.	$\dfrac{\csc 45° + \sec 30°}{\tan 60° - \tan 30°}$
g.	$\dfrac{\tan 45° - \sin 45°}{\tan 30° \sin 30°}$
h.	$\tan 0° + \cos 0° - \sin 0°$

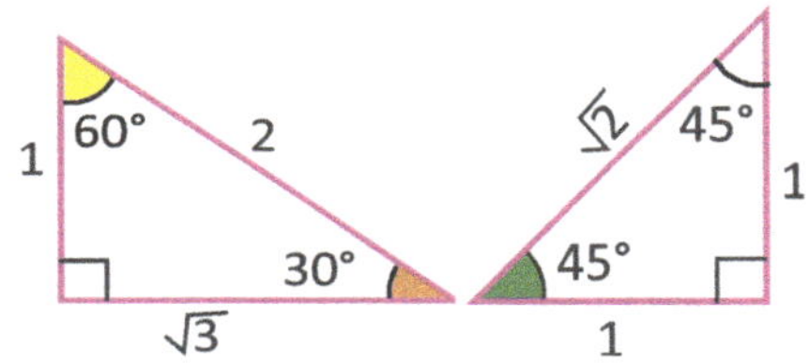

Challenge Question

Find area of a right triangle with hypotenuse of 10 cm and an acute angle of 30°

SCAN ME

Watch Video **Exact Value**

Challenge Question

Ratio of angles of a triangle ABC is 1:2:3. Find the perimeter of the triangle if the longest side is 4 cm.

Solution 3: Find Exact Value

a.	$\cos 45° \sin 30° \tan 60° = \frac{\sqrt{2}}{2} \times \frac{1}{2} \times \sqrt{3} = \frac{\sqrt{6}}{4}$
b.	$\tan 30° - \cos 60° = \frac{\sqrt{3}}{3} - \frac{1}{2} = \frac{2\sqrt{3}-3}{6}$
c.	$\tan 45° - \sin 60° = 1 - \frac{\sqrt{3}}{2} = \frac{2-\sqrt{3}}{2}$
d.	$\sec 30° \sin 45° \cot 60° = 2 \times \frac{\sqrt{2}}{2} \times \frac{3}{\sqrt{3}} = \frac{\sqrt{6}}{3}$
e.	$\tan^2 30° + \cos^2 30° + \sec^2 30°$ $= \left(\frac{1}{\sqrt{3}}\right)^2 + \left(\frac{1}{\sqrt{2}}\right)^2 + (2)^2 = \frac{1}{2} + \frac{1}{3} + 4 = \frac{2+3+24}{6} = \frac{29}{6}$
f.	$\frac{\csc 45° + \sec 30°}{\tan 60° - \tan 30°} = \frac{\sqrt{2}+2}{\sqrt{3}-\frac{1}{\sqrt{3}}} = \frac{\sqrt{3}(\sqrt{2}+2)}{3-1} = \frac{\sqrt{3}\sqrt{2}(1+\sqrt{2})}{2} = \frac{\sqrt{6}(1+\sqrt{2})}{2}$
g.	$\frac{\tan 45° - \sin 45°}{\tan 30° \sin 30°} = \frac{1-\frac{\sqrt{2}}{2}}{\frac{1}{\sqrt{3}}\frac{1}{2}} = \frac{\frac{2-\sqrt{2}}{2}}{\frac{1}{2\sqrt{3}}} = \frac{2-\sqrt{2}}{2} \times 2\sqrt{3} = 2\sqrt{3} - \sqrt{6}$
h.	$\tan 0° + \cos 0° - \sin 0° = 0 + 1 - 0 = 1$

Coordinate points to be considered on Unit Circle for Exact Value Calculations

$(\cos\theta, \sin\theta)$

0°	**30°**	**45°**	**60°**	**90°**
$(1, 0)$	$\left(\frac{\sqrt{3}}{2}, \frac{1}{2}\right)$	$\left(\frac{\sqrt{2}}{2}, \frac{\sqrt{2}}{2}\right)$	$\left(\frac{1}{2}, \frac{\sqrt{3}}{2}\right)$	$(0, 1)$
180°	**210°**	**270°**	**300°**	**360°**
$(-1, 0)$	$\left(-\frac{\sqrt{3}}{2}, -\frac{1}{2}\right)$	$(0, -1)$	$\left(\frac{1}{2}, -\frac{\sqrt{3}}{2}\right)$	$(1, 0)$

Challenge Problem

Draw a rectangle ABCD with side lengths of $AB = \sqrt{3}$ and $BC = \sqrt{3} + 1$

Prove $\tan 15° = 2 - \sqrt{3}$ using the rectangle ABCD.

Define Right Triangle from Given Trigonometric Ratio

A right triangle can be defined from a trigonometric ratio

Inverse trigonometric function is used to find the angle. Other acute angle is the complementary angle and the third side is calculated with Pythagorean theorem.

Example 4: Draw Labeled Right Triangle

Given $\sin A = \frac{12}{13}$ in the right triangle ABC where $\angle B = 90°$

a.	Draw a right triangle to represent $\sin A = \frac{12}{13}$
b.	Find all the sides of the triangle (Pythagorean Theorem)
c.	Find angle accurate to 3 significant figures
d.	Find angle C (complementary to angle A)

Solution 4:

Step 1: Draw Right Triangle from the given trigonometric ratio

$\sin A = \frac{12}{13}$ It ratio of the opposite to hypotenuse in triangle ABC.

$\sin A = \frac{12}{13} = \frac{opposite}{hypotenuse} = \frac{BC}{AC}$

S$^{O}_{H}$ C$^{A}_{H}$ T$^{O}_{A}$

$BC = 12\ and\ AC = 13$

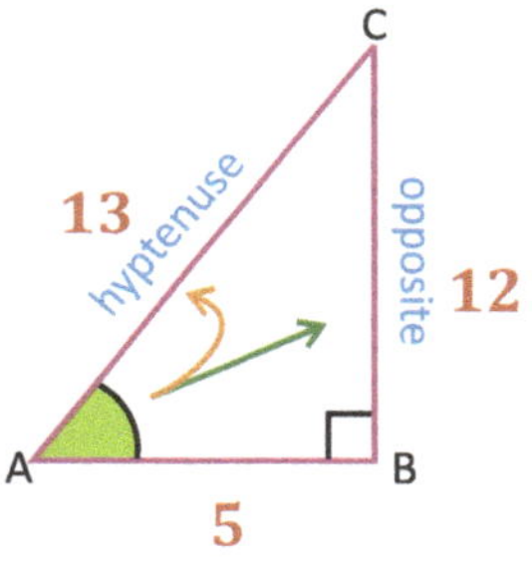

Step 2: Find the third side (adjacent)

using Pythagorean Theorem

Apply Pythagorean theorem to find the adjacent side AB

$AB = \sqrt{AC^2 - BC^2} = \sqrt{13^2 - 12^2} = \sqrt{169 - 144} = \sqrt{25} = 5$

Triangle ABC with all the three side lengths is shown.

Communication and Thinking

Why $\sin A = \frac{5}{4}$ is not possible?

Step 3: Find angle A with inverse function

$$\sin A = \frac{12}{13}$$

$$\angle A = \sin^{-1}\left(\frac{12}{13}\right)$$

Press the shift key for inverse as in sequence below

SHIFT	sin	(	12	÷	13	)	=

You will get the measurement of angle A:

$\angle A = 67.38 = 67.4°$

Step 4: Find complementary angle C

It is a right triangle and so $\angle B = 90°$

$\angle C = 90 - 67.4° = 22.6°$

Step 5: Draw labeled right triangle ABC

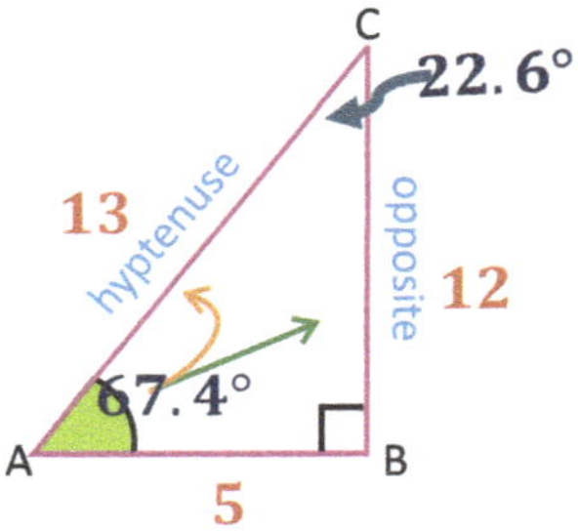

$AB = 5, BC = 12, AC = 13$

$\angle A = 67.4°, \angle B = 90°, \angle C = 22.6°$

Practice Questions based on Example 4

Q4.1	Given $\cos A = \frac{1}{2}$ in the right triangle ABC where $\angle B = 90°$	
	a.	Draw a right triangle ABC with given information
	b.	Find all the sides of the triangle
	c.	Then find angle A to nearest whole number
Q4.2	Given $\tan P = \frac{\sqrt{3}}{3}$ in the right triangle PQR where $\angle Q = 90°$	
	a.	Draw a right triangle PQR with given information
	b.	Find exact value of all the sides of the triangle
	c.	Find all the angles of the triangle to 3 sf.

Find Angles from Given Sides to 3 significant numbers.

Example 5

In the right ΔABC we are given side, $BC = 4$ and $AB = 3$.

a.	Which trigonometric ratio should be used to find $\angle A$?
b.	Determine m$\angle A$ to 3 sf.

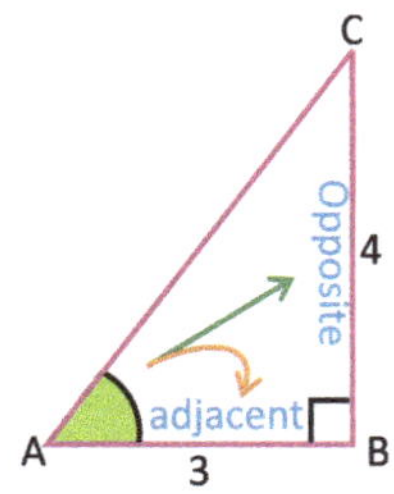

Solution 5

In the ΔABC we are given opposite side, $BC = 4$, and adjacent side $AB = 3$.

Tangent ratio relates these two sides and therefore we can use it to find the measure of angle A.

$\tan A = \frac{4}{3}$ $\quad S^O_H \quad C^A_H \quad T^O_A$

Inverse trigonometric ratio is used to find measure of angle.

m$\angle A = \tan^{-1}\left(\frac{4}{3}\right)$ $\quad$ $m\angle A = \arctan\left(\frac{4}{3}\right) = 53.13010235°$

m$\angle A = 53.13010235°$ $\quad$ is an alternate way for inverse.

m$\angle A = 53.1° \ (3\ sf)$

Example 5 Extended

In the right ΔABC we are given side, $BC = 4$ and $AB = 3$.

c.	Find $\angle C$ using inverse trigonometric function.
d.	Verify that $\angle A$ and $\angle C$ are complimentary.

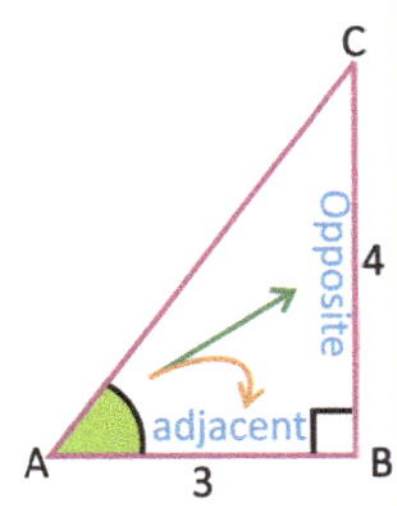

Solution 5c

Now see the sides from angle C as shown in the figure

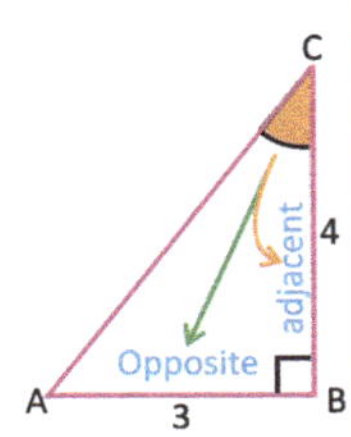

$\tan \angle ACB = \frac{3}{4}$

$\angle ACB = \tan^{-1}\left(\frac{3}{4}\right) = 36.8698°$

$\angle ACB = 36.9°$

Solution 5d.

Complementary angles sum is $90°$

$m\angle A = 53.1°, and\ \angle C = 36.9°$

$m\angle A + m\angle C = 53.1° + 36.9° = 90°$

Hence angle C and angle A form are complementary angles.

Once a trigonometric ratio is applied to find an acute angle, use the concept of complementary angle to find the other acute angle.

Practice Questions based on Example 5: Find the angle θ to 3 s.f.

Q5.1	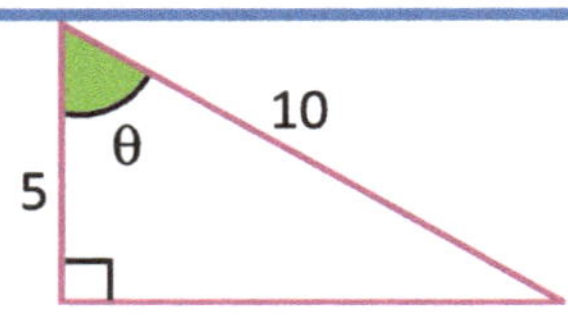	Q5.2	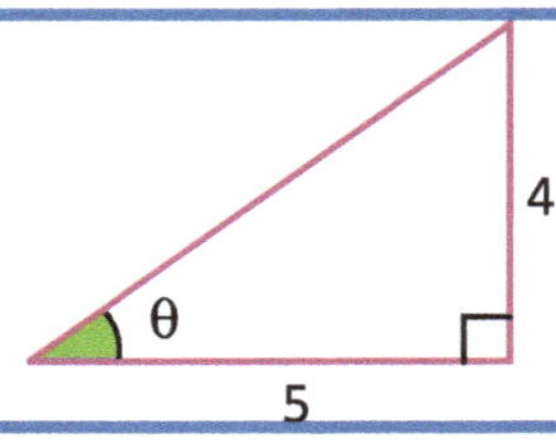
Q5.3	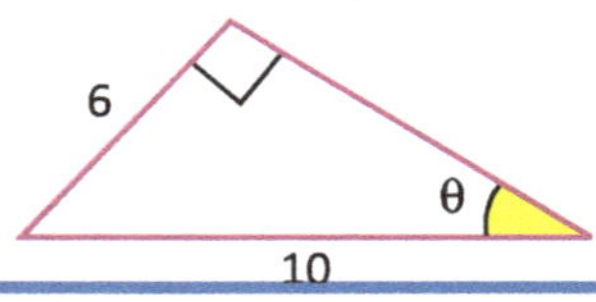	Q5.4	In ΔKLM, $KL = 3, KM = 4\ cm$, $\angle L = 90°$. Find $\angle LKM$
Q5.5	In ΔKLM, $KL = 2, LM = 1\ cm$, $\angle L = 90°$. Find $\angle LKM$	Q5.6	In ΔKLM, $KM = 13, LK = 9\ cm$, $\angle L = 90°$. Find $\angle LMK$

Answers

Q5.1	$\theta = \cos^{-1}\left(\frac{1}{2}\right) = 60.0°$	Q5.2	$\theta = \tan^{-1}\left(\frac{4}{5}\right) = 38.7°$
Q5.3	$\theta = \sin^{-1}\left(\frac{6}{10}\right) = 36.9°$	Q5.4	$\angle LKM = 41.4°$
Q5.5	$\angle LKM = 26.6°$	Q5.6	$\angle LMK = 43.8°$

Watch Video **Extended Trigonometric Inverse** **SCAN ME**

Find Side Lengths from Angle

Example 6: Step by Step Solution

Find the length of the side marked with letters $to\ 3\ significant\ figures.$

All length measurements are in centimeters.

a.

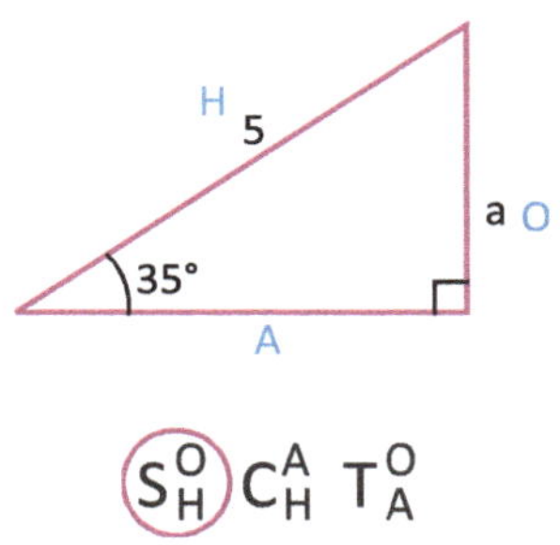

Step 1: Identify correct trigonometric ratio

Step 2: Our interest is in O and H

apply $\sin 35°$ to solve

Step 3: Rearrange and solve using calculator

$$\sin 35° = \frac{O}{H} = \frac{a}{5}$$

$$a = 5 \sin 35° = 2.867$$

Step 4: Round to $3\ s.f$ and include units

$$a = 2.87\ cm\ (to\ 3\ s.f.)$$

b.

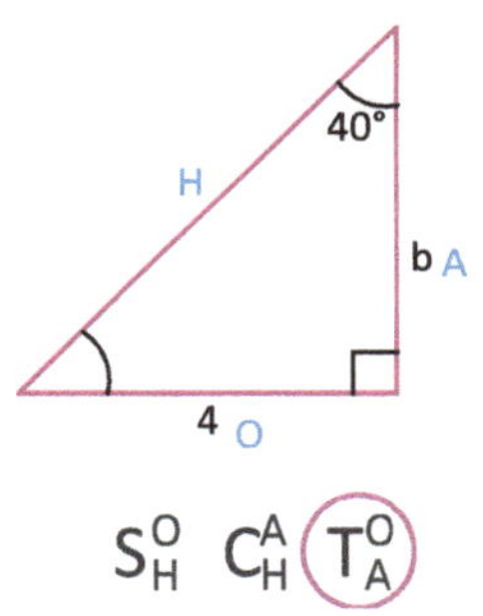

Step 1: Identify correct trigonometric ratio

Step 2: Our interest is in O and H

apply $\tan 40°$ to solve

$$\tan 40° = \frac{O}{A} = \frac{4}{b}$$

Step 3: Rearrange and solve using calculator

$$b = \frac{4}{\tan 40°} = 4.767$$

Step 4: Round to 3 s.f.

$$b = 4.77\ cm \quad (s.f.)$$

c.

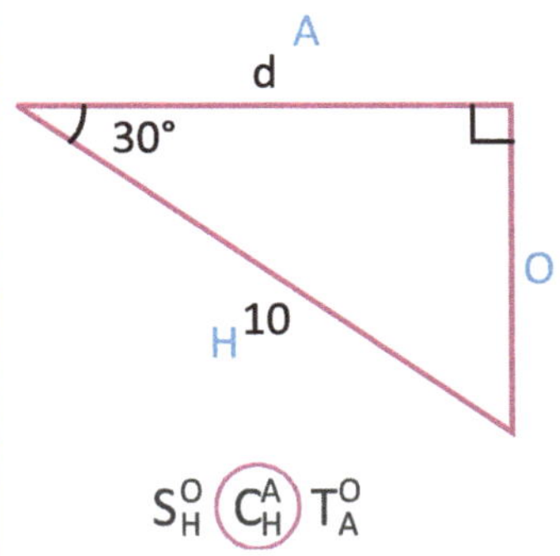

Step 1: Identify correct trigonometric ratio

Step 2: Our interest is in A and H

apply $\cos 30°$ to solve

$$\cos 30° = \frac{A}{H} = \frac{d}{10}$$

Step 3: Rearrange and solve using calculator

$$d = 10 \cos 30° = 8.660$$

Step 4: Round to 3 s.f.

$$d = 8.66\ cm$$

Test: Trigonometric Ratios and Angles

Determine the marked side length to **three significant figures**.			
Q6.1	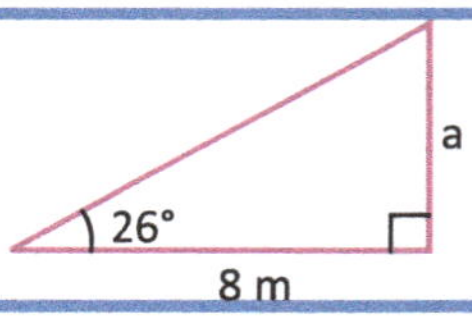	Q6.2	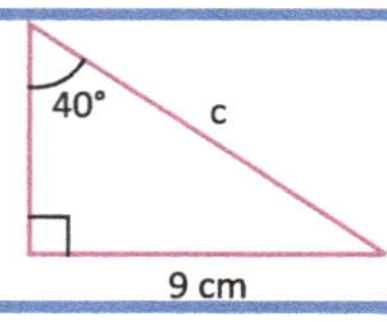
Q6.3	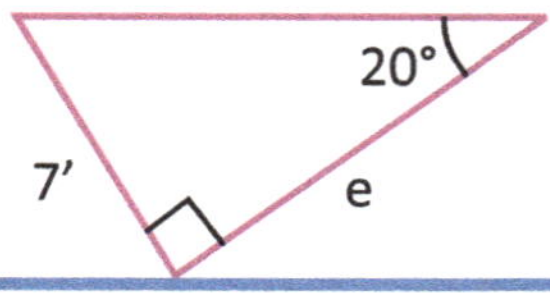	Q6.4	In $\Delta ABC, \angle C = 55°, \angle B = 90°,$ $BC = 5.$ Draw triangle and then find AB
Q6.5	In $\Delta PQR, \angle P = 72°,\ \angle Q = 90°,$ $QR = 10\ cm.$ Draw triangle and then find PR	Q6.6	In $\Delta KLM,\ \angle K = 40°, KM =$ $4\ cm,\ \angle L = 90°$ Draw triangle and then find KL
Q6.7	In $\Delta ABC,\ \angle C = 35°,\ \angle B = 90°,$ $AB = 10. Fnd\ AC.$	Q6.8	In $\Delta DEF,\ \angle D = 70°,$ $\angle E = 90°, EF = 6.$ Find $DE.$
Q6.9	In $\Delta DEF, \angle D = \angle F,\ \angle E = 90°,$ $EF = 6.$ Draw triangle and then find $DF.$	Q6.10	Find the approximate length of AC in the given right triangle A. 13 B. 11 C. 15 D. 12 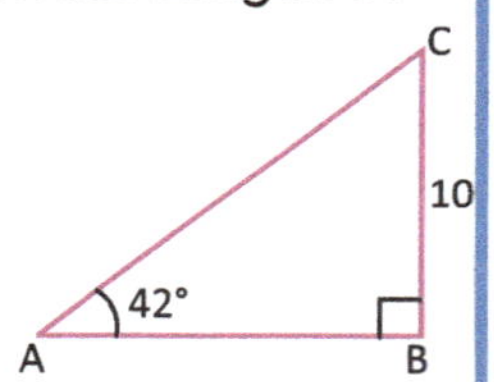
Q6.11	In $\Delta PQR,\ \angle P = 38°, \angle Q = 90°,$ $PQ = 6\ cm$, find QR A. 4.7 cm B. 9.7 cm C. 7.7 cm D. 8.5 cm 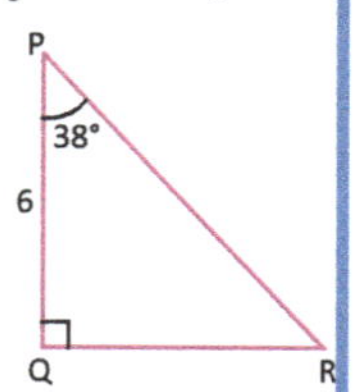	6.12	In ΔABC if $\cos A = 0.7, \angle B = 90°$ then length of CB is A. 7 B. $\sqrt{51}$ C. $5\sqrt{2}$ D. $4\sqrt{3}$

Answers

Q6.1.	3.90 m	Q6.2.	14.0 cm	Q6.3.	19.2	Q6.4.	7.14
Q6.5.	32.4	Q6.6.	3.06	Q6.7.	17.4	Q6.8.	2.18
Q6.9.	$6\sqrt{2}$	Q6.10.	C	Q6.11.	A	Q6.12.	B

Coordinate Geometry Meets Trigonometry

When a straight line rises or falls on the coordinate plane, it makes an angle with the horizontal — the x-axis. But **how can we find that angle**? This is where **trigonometry and slope** come together.

In coordinate geometry, the **slope of a line** tells us how steep it is — the "rise over run." Trigonometry helps us take this a step further by converting slope into an angle.
The key idea:

If a line has slope *m*, then the angle it makes with the x-axis, θ is related as:
$\mathbf{\tan\theta = m, (slope) \Rightarrow \theta = \tan^{-1}(m)}$

This angle is measured from the x-axis up to the line (in the counter-clockwise direction) and is always **acute** for positive slopes and **obtuse** for negative slopes.

Let's apply this to real equations and find the angle each line makes with the x-axis. Round your answers to the nearest whole number.

Example 1: Angle Between Line and X – Axis

Find the acute angle formed by the line $y = x - 1$ and the x-axis.

Solution 1: Angle Between Line and X – Axis

$\mathbf{\tan\theta = m, (slope) \Rightarrow \theta = \tan^{-1}(m)}$

$y = x - 1$

$Slope\ = 1$

$\tan\theta = 1$

$\theta = \tan^{-1}(1)$

$\theta = 45°$

Angle that the line $y = x - 1$ makes with the $x - axis$ is $45°$

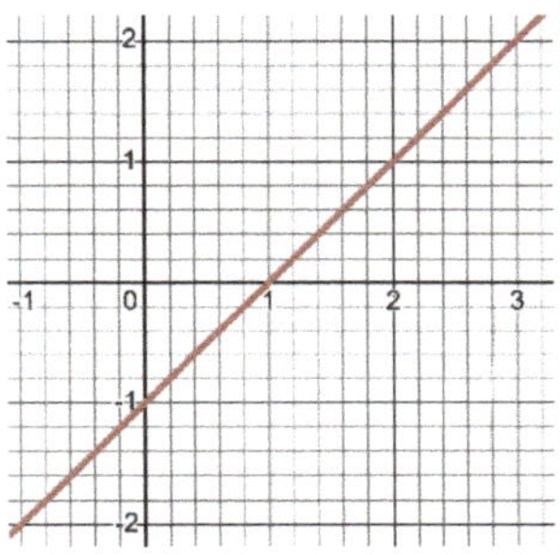

Example 2: Angle Between Line and Y – Axis

Find the acute angle formed by the line $12x - 13y + 6 = 0$ and the y-axis.

Watch Video **Angle Between Lines Extended** SCAN ME

Solution 2: Angle Between Line and Y – Axis

Angle formed by y-axis will be complementary angle as shown in the figure.

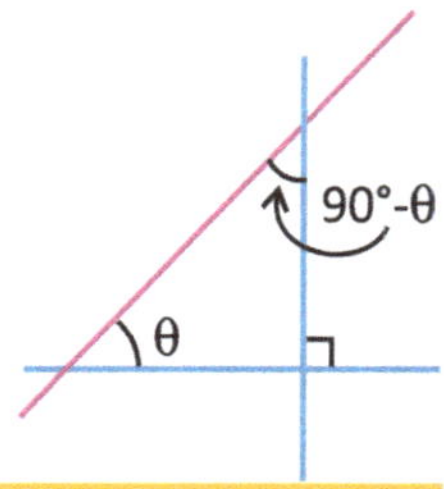

Step 1: Rearrange the equation to find the slope of the line

$12x - 13y + 6 = 0$

$13y = 12x + 6$

$y = \frac{12}{13}x + \frac{16}{13}$

$\boldsymbol{Slope,\ m = \frac{12}{13}}$

Step 2: Find angle line makes with the $x - axis$

$\tan\theta = \frac{12}{13}$

$\theta = \tan^{-1}\left(\frac{12}{13}\right) = 42.7$

Step 3: Find the complementary angle to get the solution

Acute angle formed by y-axis is

$90° - \theta = 90° - 42.7° = \mathbf{47.3}°$

✍ Practice Questions

Q1	Find acute angle which the line $3x - 3y - 6 = 0$ makes with the $y - axis$
Q2	Find acute angle which the line $2x + 3y - 6 = 0$ makes with the $y - axis$
Q3	Find acute angle which the line $2x + y = 0$ makes with the $x - axis$
Q4	Find **obtuse** angle which the line $3y - 6 = 0$ makes with the $y - axis$
Q5	Find acute angle which the line $2x - y - 6 = 0$ makes with the $x - axis$
Q6	Find acute angle which the line $y = x$ makes with the $x - axis$
Q7	Find obtuse angle which the line $2x + y = 0$ makes with the $x - axis$

Example 3: Angle Between Intersection of Lines

Find the acute angle at which these set of lines intersect: $y = \frac{1}{2}x + 1 \ \ and \ \ y = 3x - 1$

Solution 3

Figure is shown to illustrate the situation. θ_1 and θ_2 are the angles that the lines make with the $x - axis$. $\boldsymbol{\phi}$ is the angle between the lines which we need to find.

Let us first find each angle lines make with $x - axis$.

Step 1: Find the angle lines make with the $\boldsymbol{x - axis}$

$y = \frac{1}{2}x + 1$

$\tan\theta_1 = \frac{1}{2}$

$\theta_1 = \tan^{-1}\left(\frac{1}{2}\right)$

$\theta_1 = 26.6°$

$y = 3x - 1$

$\tan\theta_2 = 3$

$\theta_2 = \tan^{-1}(3)$

$\theta_2 = 71.6°$

Step 2: Apply exterior angle theorem.

Exterior angle in a triangle is equal to the sum of remote interior angles.

θ_2 is the exterior angle for the triangle with angles θ_1 and $\boldsymbol{\phi}$.

Then

$\theta_2 = \theta_1 + \boldsymbol{\phi}$

$\boldsymbol{\phi} = \theta_2 - \theta$

$\boldsymbol{\phi} = 71.6° - 26.6°$

$\boldsymbol{\phi} = 45°$

Lines intersect at 45°

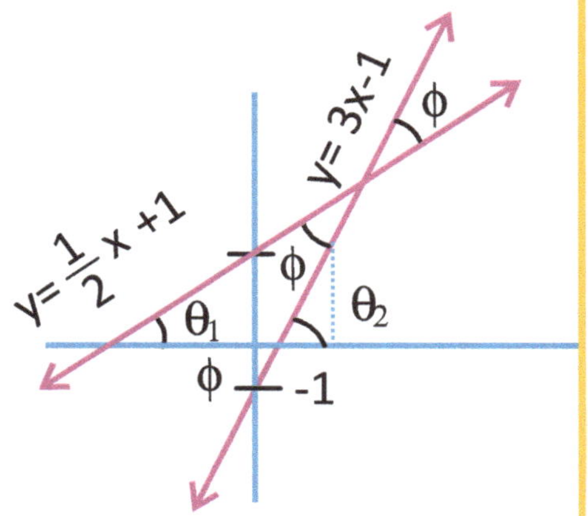

Guided Practice: Angle with the horizontal and vertical axis

Write the equation of the line from the given graph and then find acute angle that the line makes with the $x - axis$ and $y - axis$.

Find the acute angle that the line makes with the horizontal and vertical axis

a. 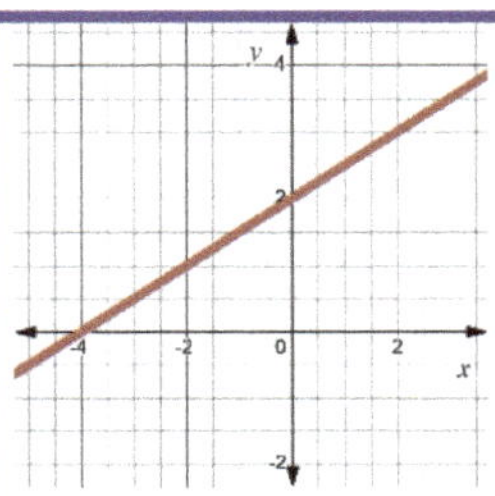

b. 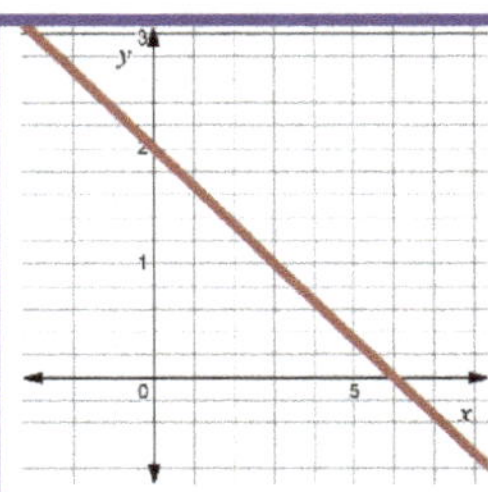

c. 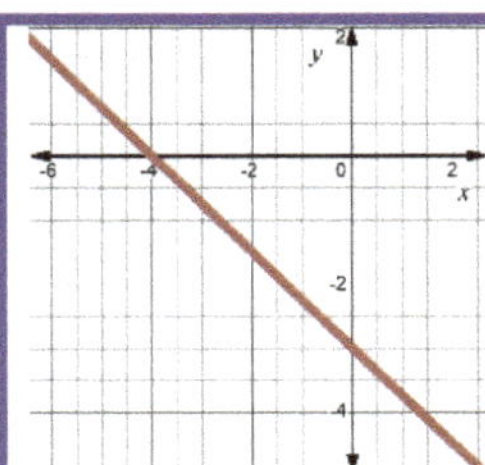

Solve Multiple Triangles

Find Angle and Sides in Connected Triangles

Every Right Triangle with Given Two Parameters Can be Solved

A triangle has three sides and three angles. To solve any triangle, we need information of at lest three of these six parameters. That also implies that at most we can solve for three parameters in a triangle. In a right triangle one of the angle is 90° so this information along with two more is required to solve a right triangle.

Example 1:

Find the value of unknown variable. Round answer to 3 significant figures.

a.

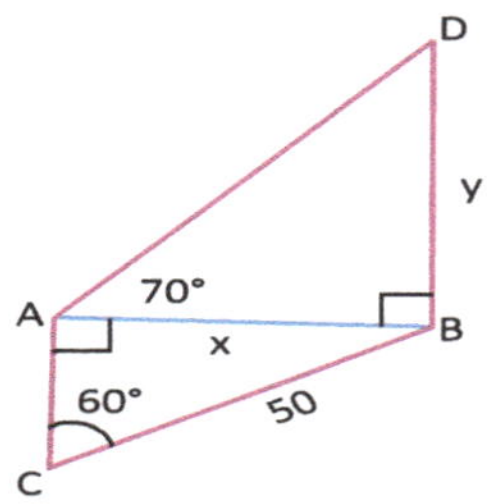

Start with ΔABC

$\sin 60° = \frac{x}{50}$

$x = 50 \sin 60$

$x = 25\sqrt{3}$

$x = 43.3$

Now in ΔABD

$\tan 70° = \frac{4}{x}$

$y = x \tan 70°$

$y = 25\sqrt{3}\ \tan 70°$

$y = 119$

Practice Questions

Find the value of unknown variable. Round answer to 3 significant figures.

a.

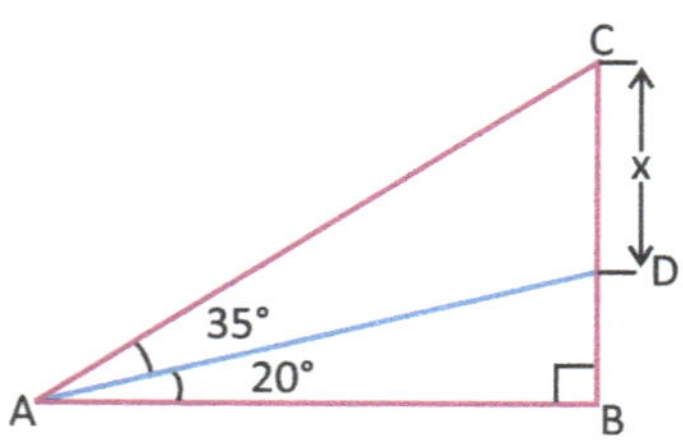

b.

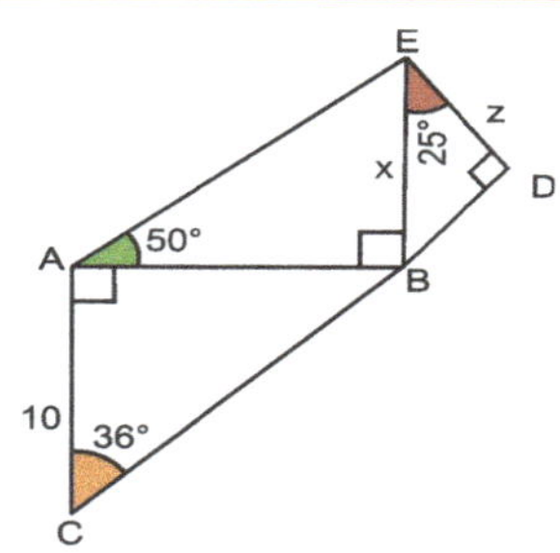

Watch Video **Solve Multiple Triangles** **SCAN ME**

Solution: Practice Questions based on Example 1

Strategy

- Identify the right triangle with given two parameters. Solve this triangle to find all the missing values.
- Move to the next connected triangle whose one of the missing value is calculated.

Find the value of unknown variable. Round answer to 3 significant figures.

Q1.1	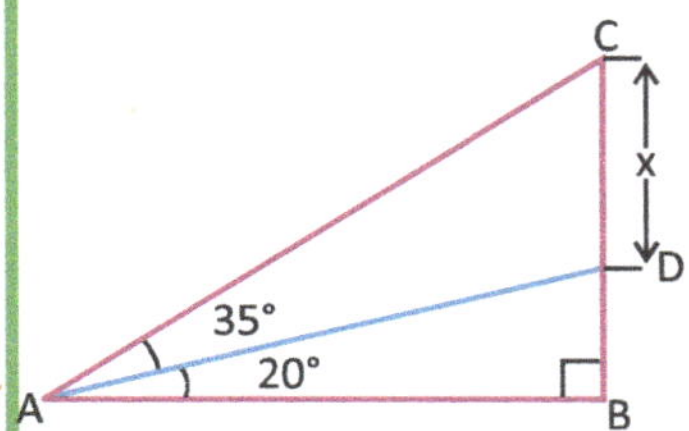 Given: $AB = 27\,m$	Start from ΔABD $\tan 20° = \frac{DB}{27°}$ $DB = 27 \tan 20°$ $DB = 9.83\,m$ Consider ΔCAB $\angle CAB = \angle CAD + \angle DAB$ $= 35° + 20° = 55°$	$\tan 55° = \frac{CB}{27}$ $CB = 27 \tan 55°$ $CB = 38.56$ $CD = CB - DB$ $CD = 38.56 - 9.83$ Length CD is $28.7\,m$ $\boldsymbol{x = 28.7}$ **m**
Q1.2	**Let $\boldsymbol{AB = y}$** 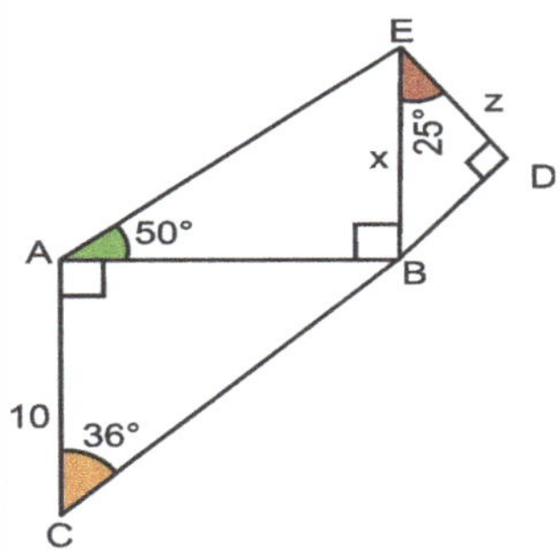	Start from ΔABC $\tan 36° = \frac{y}{10}$ $y = 10 \tan 36°$ $y = 7.2654$ $\tan 50° = \frac{x}{y} = \frac{x}{7.2654}$ $x = 7.265\ \tan 50°$ $\boldsymbol{x = 8.658}$	$\cos 25° = \frac{z}{x}$ $z = x \cos 25°$ $z = 8.658 \times \cos 25°$ $z = 7.8473$ $\boldsymbol{x = 8.66}$ $\boldsymbol{y = 7.27}$ $\boldsymbol{z = 7.85}$

Challenge Problem

The mid-points of the sides of a triangle ABC have coordinates A(3, 1), B(-1, 2) and C(1, -3). Determine the coordinates of the vertices of the triangle.

D(-3, -2), E(5, -4), F(1, 6)

Chapter 10 Unit Test: Solve Multiple Triangles

Q1 Calculate the missing side length with Pythagorean Theorem.
Find exact value of primary trigonometric ratios for each of the given angle.

a.

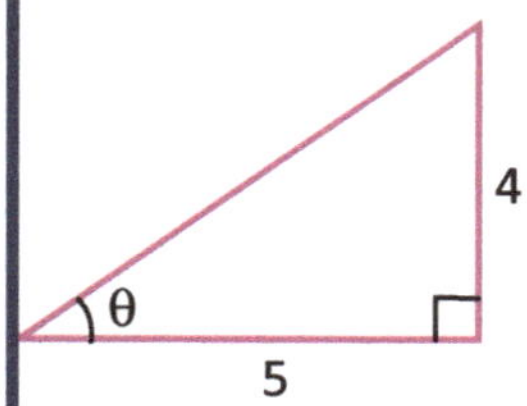

b.

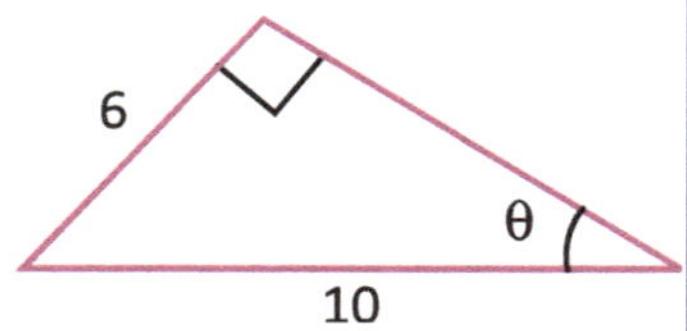

c.

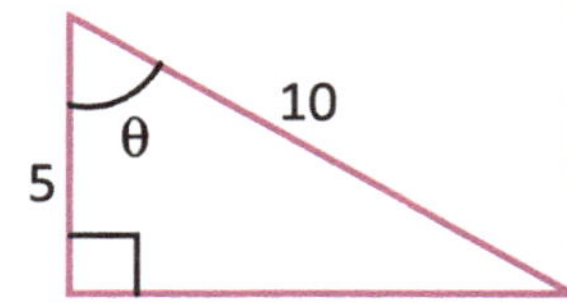

Q2. Write expressions for the three primary trigonometric ratios for the given angle in each of the following figures.

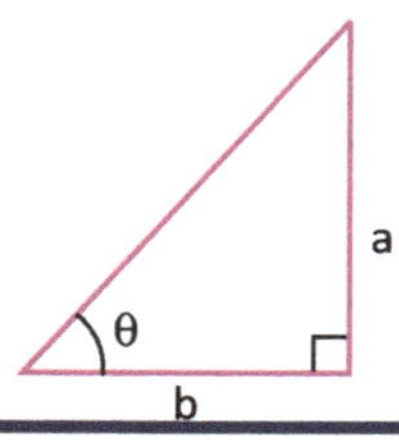

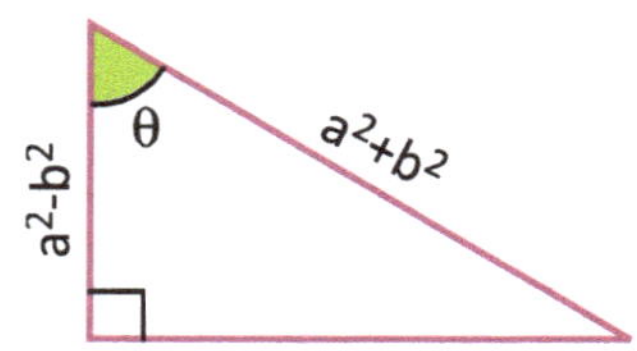

Q3 Calculate the value of each primary trigonometric ratio to two decimal places for the given right triangles.

Q3a

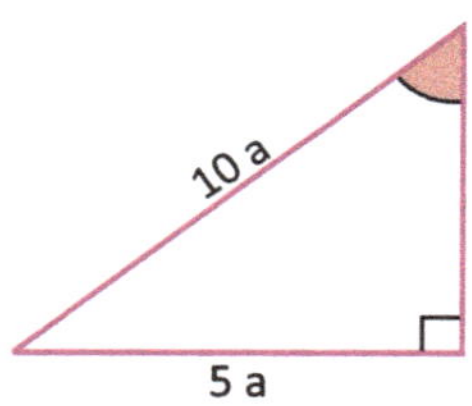

Q3b.

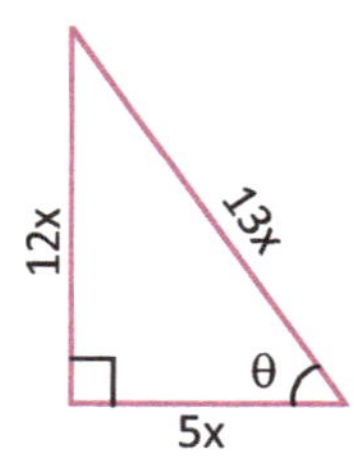

Q4 Sketch right triangles for the given trigonometric Ratios.

a.	$\tan 31° = 0.6$	b.	$\tan 76° = 4.0$
c.	$\sin 12° = 0.2$	d.	$\cos 60° = 0.6$

Q5. Describe how the sides of a right triangle are related for the following conditions.

Q5a. where $\sin A > \cos A$

Q5b. where $\sec\theta = \csc\theta$.

Q6	If $\sec\theta = \frac{5}{3}$ find exact value of other trigonometric ratios.
Q7	A kite is flying 40 km above the ground with string length of 60 m. Calculate the exact value of $\sin\theta$ *and* $\sec\theta$ if θ is angle of elevation.

Unit Test 1: Solutions

Solutions are provided in informal way as hints to solve the problems.

Trigonometric ratios can be remembered as $S_H^O \quad C_H^A \quad T_A^O$

Q1a.	$H = \sqrt{4^2 + 5^2} = \sqrt{41}$ $\sin\theta = \frac{4}{\sqrt{41}}, \cos\theta = \frac{5}{\sqrt{41}},$ $\tan\theta = \frac{4}{5}$	Q1b.	$A = \sqrt{10^2 - 6^2} = 8$ $\sin\theta = \frac{6}{10} = \frac{3}{5}, \cos\theta = \frac{8}{10} = \frac{4}{5},$ $\tan\theta = \frac{6}{8} = \frac{3}{4}$
Q1c.	$O = \sqrt{10^2 - 5^2} = 5\sqrt{3}$ $\sin\theta = \frac{5\sqrt{3}}{10} = \frac{\sqrt{3}}{2}$ $\cos\theta = \frac{5}{10} = \frac{1}{2},$ $\tan\theta = \frac{5\sqrt{3}}{5} = \sqrt{3}$	Q2a.	$H = \sqrt{a^2 + b^2}$ $\sin\theta = \frac{a}{a^2 + b^2}$ $\cos\theta = \frac{b}{a^2 + b^2}$ $\tan\theta = \frac{a}{b}$
Q2b.	$O^2 = (a^2 + b^2) - (a^2 - b^2)$ $O^2 = a^4 + 2a^2b^2 + b^4 - (a^4 - 2a^2b^2 + b^4)$ $O^2 = 4b^2b^2$ $O = \sqrt{4a^2b^2} = 2ab$		$\sin\theta = \frac{2ab}{a^2 + b^2}$ $\cos\theta = \frac{a^2 - b^2}{a^2 + b^2}$ $\tan\theta = \frac{2ab}{a^2 - b^2}$
Q3a.	$\sin\theta = \frac{5a}{10a} = \frac{1}{2}$ $\cos\theta = \frac{5\sqrt{3}a}{10a} = \frac{\sqrt{3}}{2}$ $\tan\theta = \frac{5a}{5\sqrt{3}a} = \frac{1}{\sqrt{3}}$	Q3b.	$\sin\theta = \frac{12}{13}$ $\cos\theta = \frac{15}{13}$ $\tan\theta = \frac{12}{5}$

Q4a.

$\tan 31° = 0.6$

Relate angle to related side in right triangle

$$\tan 31° = \frac{opposite}{adjacent} = 0.6 = \frac{6}{10} = \frac{3}{5}$$

Find complementary angle = $90° - 31° = 59°$

Calculate third side, hypotenuse of the triangle

Hypotenuse = $\sqrt{3^2 + 5^2} = \sqrt{34}$

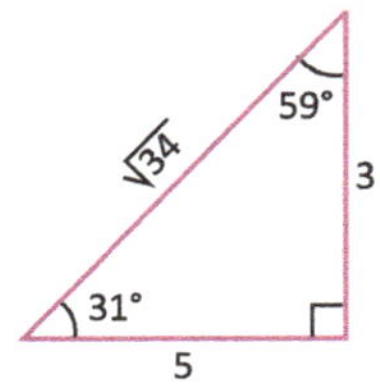

Q4b

$$\tan 76° = 4.0$$

$$\tan 76° = 4.0 \ = \frac{opposite}{adjacent} = \frac{4}{1}$$

$$Hypotenuse \ = \sqrt{4^2 + 1^2} \ = \ \sqrt{17}$$

Complementary angle: $90° - 76° = 14°$

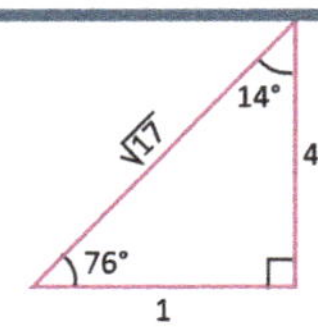

Q4c.

$$\sin 12° = 0.2$$

$$\sin 12° = 0.2 = \frac{opposite}{hypotenuse} = \frac{2}{10} = \frac{1}{5}$$

$$Adjacent \ = \sqrt{5^2 - 1^2} = \sqrt{24} \ = \ 2\sqrt{6}$$

Complementary angle, $96° - 12° = 78°$

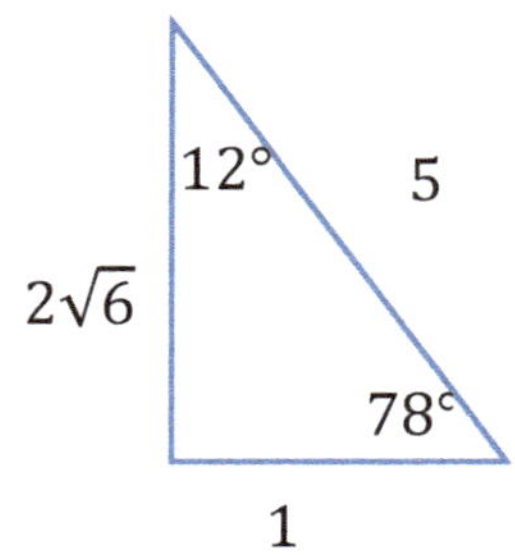

Q4d.

$$\cos 60° = 0.5$$

$$\cos 60° = 0.5 = \frac{adjacent}{hypotenuse} = \frac{5}{10} = \frac{1}{2}$$

$$Opposite = \sqrt{2^2 - 1^2} = \sqrt{3}$$

Complementary angle, $90° - 60° = 30°$

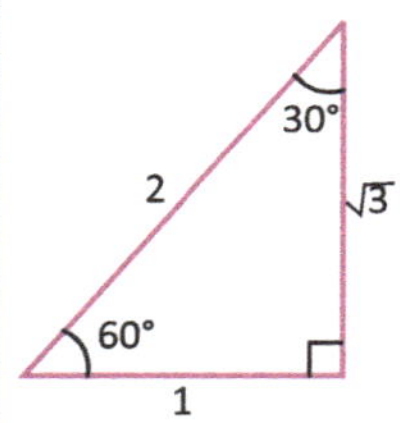

Q5a

triangle where $\sin A > \cos A$

$$\sin A \quad > \quad \cos A$$

$$\frac{opposite}{hypotenuse} > \frac{adjacent}{hypotenuse}$$

$$opposite \ > \ adjacent$$

$$Example: \angle A = 50°$$

Q5b

triangle where

$$\sec\theta = \csc\theta$$

$$\frac{1}{\cos\theta} = \frac{1}{\sin\theta}$$

$$\frac{\sin\theta}{\cos\theta} = 1 \Rightarrow \tan\theta = 1$$

$$\theta = \tan^{-1}(1) = 45°$$

Q6.	$\sec\theta = \frac{5}{3} = \frac{H}{A}$ Draw right triangle and find missing side, $\sin\theta = \frac{4}{5}, \quad \cos\theta = \frac{3}{5}, \quad \tan\theta = \frac{4}{3},$ $\csc\theta = \frac{5}{4}, \cot\theta = \frac{3}{4}$	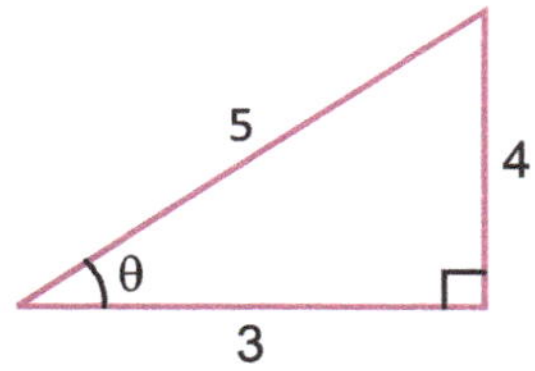
Q7.	Represent situation with diagram: string 60 m is along OK, 40 m height is KP and θ is angle of elevation $\sin\theta = \frac{O}{H} = \frac{40}{60} = \frac{2}{3}$ $OP = \sqrt{60 - 40} = 20\sqrt{5}$ For ratio you may redraw scaled figure. $\sin\theta = \frac{2}{3} \qquad \sec\theta\,\frac{3}{\sqrt{5}} = \frac{3\sqrt{5}}{5}$	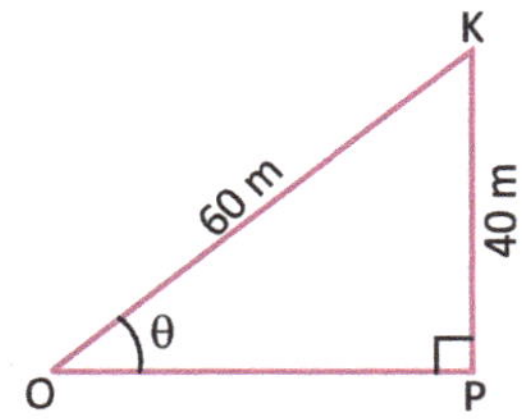

Student's Activity: Exploring Trigonometric Ratios for Cofunction Identity

In the right triangle ABC, angle B is the right angle. For angle ∠A, the opposite side is 12 cm and the adjacent side is 5 cm.

1. Draw triangle ABC
   - Label sides according to the given information.
   - Mark the right angle at C.

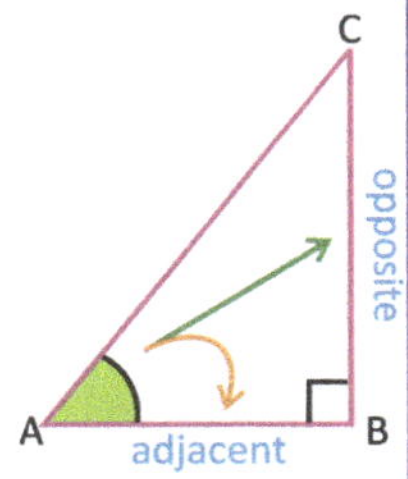

2. Explain how the trigonometric ratios of ∠A are related to those of its complementary angle ∠C:

Challenge Problem

Show that the perimeter of n-side regular polygon inscribed in a circle of radius r is $2rn\sin\frac{180°}{n}$

Trigonometry of Right Triangle: Summary

$$\sin\theta = \frac{opposite}{hypotenuse} = \frac{y}{r}$$

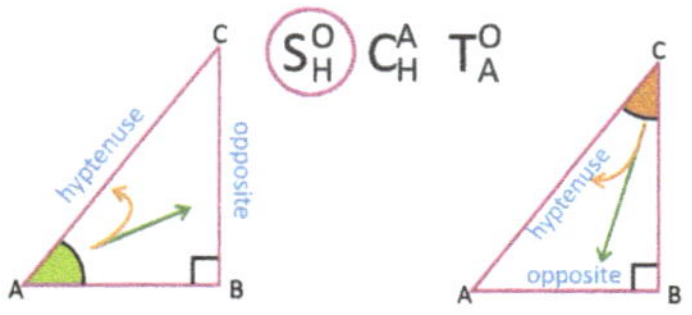

$$\cos\theta = \frac{adjacent}{hypotenuse} = \frac{x}{r}$$

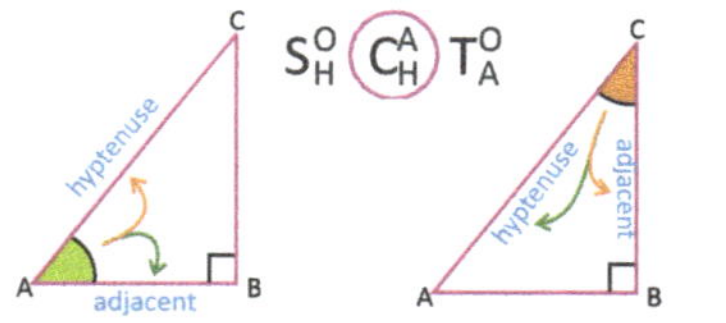

$$\tan\theta = \frac{opposite}{adjacent} = \frac{y}{x}$$

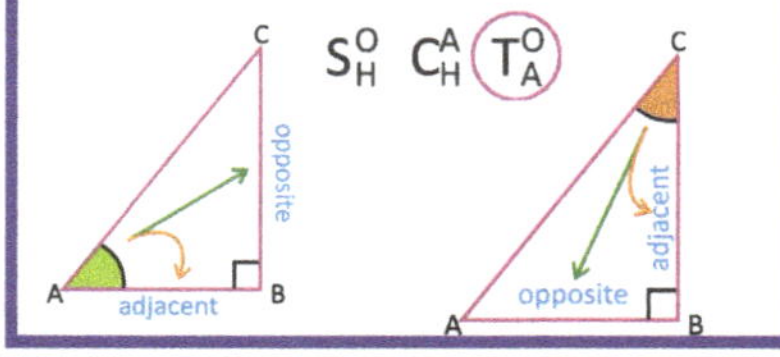

$$\csc\theta = \frac{1}{\sin\theta} = \frac{x}{y}$$

$$\sec\theta = \frac{1}{\cos\theta} = \frac{r}{x}$$

$$\cot\theta = \frac{1}{\tan\theta} = \frac{x}{y}$$

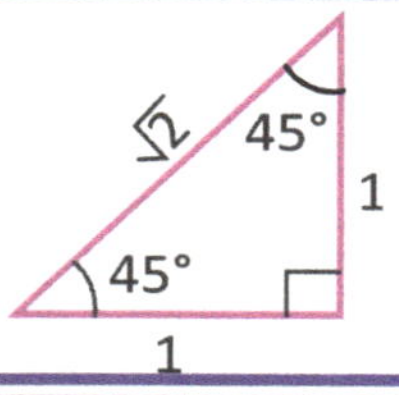

$$\sin(90° - \theta) = \cos\theta$$
$$\cos(90° - \theta) = \sin\theta$$
$$\tan(90° - \theta) = \cot\theta$$

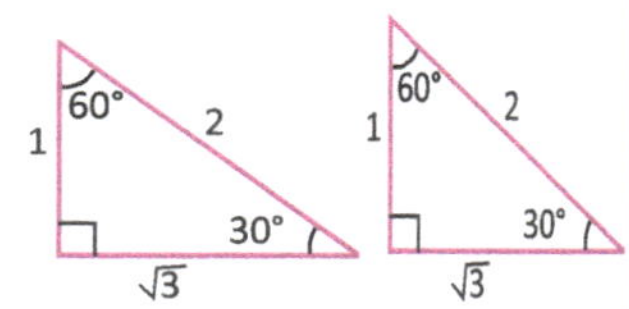

Trig. Ratio	$0°, P(1,0)$	$30°, P(\sqrt{3},1)$	$45°\ P(1,1)$	$60°, P(1,\sqrt{3})$	$90°, P(0,1)$
$\sin\theta$	0	$\frac{1}{2}$	$\frac{1}{\sqrt{2}} = \frac{\sqrt{2}}{2}$	$\frac{\sqrt{3}}{2}$	1
$\cos\theta$	1	$\frac{\sqrt{3}}{2}$	$\frac{1}{\sqrt{2}}$	$\frac{1}{2}$	0
$\tan\theta$	0	$\frac{1}{\sqrt{3}} = \frac{\sqrt{3}}{3}$	1	$\sqrt{3}$	Not defined
$\csc\theta$	Not defined	2	$\sqrt{2}$	$\frac{\sqrt{2}}{\sqrt{3}} = \frac{2\sqrt{3}}{3}$	1
$\sec\theta$	1	$\frac{2}{\sqrt{3}} = \frac{2\sqrt{3}}{3}$	$\sqrt{2}$	2	Not defined
$\cot\theta$	Not defined	$\sqrt{3}$	1	$\frac{1}{\sqrt{3}} = \frac{\sqrt{3}}{3}$	0

Chapter 11: Angle of Elevation and Depression

Angle of Depression

The angle of depression is the angle formed when the line of sight is below the horizontal line, while looking down at an object.

It is also called the **angle of declination**.

Key Points:

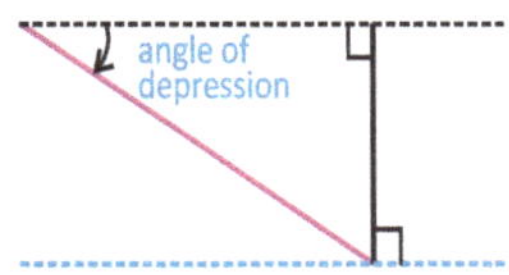

- Always measured from the horizontal line downward.
- In real life, the horizontal line is an imaginary line parallel to the ground, drawn from the observer's eye level.
- The line of sight forms the hypotenuse of a right triangle, with height and distance as the legs.

Example: A person standing on the top of a tower looks down at a car on the road. The angle between the horizontal eye-level line and the line joining the eye to the car is the angle of depression.

Line of Sight

- The **line of sight** is the **straight, unobstructed path** between the observer's eye and the object.
- In trigonometry, it is considered the **hypotenuse** of the right triangle formed in height-and-distance problems.

Angle of Elevation

The angle of elevation is the angle formed when the line of sight is above the horizontal line, while looking up at an object.

It is also called the angle of inclination.

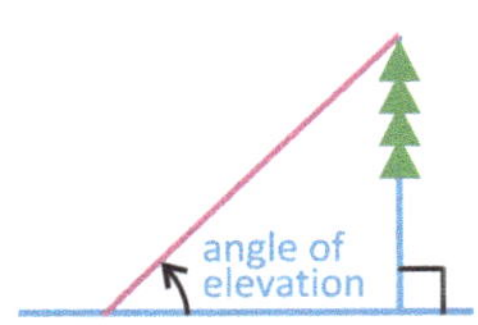

Key Points:

- Always measured from the horizontal line upward.
- Used to calculate heights of towers, trees, mountains, or distances when the observer is at a lower point.

Example: A student looks up at the top of a flagpole. The angle between the horizontal line at the student's eye level and the line joining the eye to the top of the pole is the angle of elevation.

6 Cases for Height and Distance Applications

Following cases help to understand all the situations of using trigonometry to solve height and distance applications.

Case I: Basic Situation. Single point observation	Case II: Two Point Observation along same line
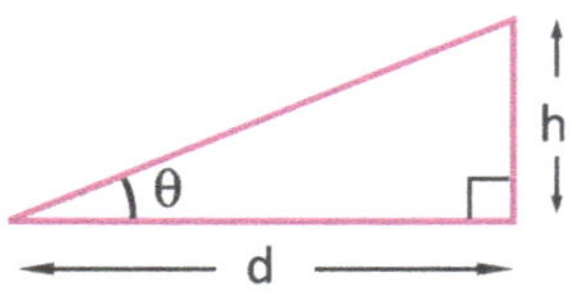	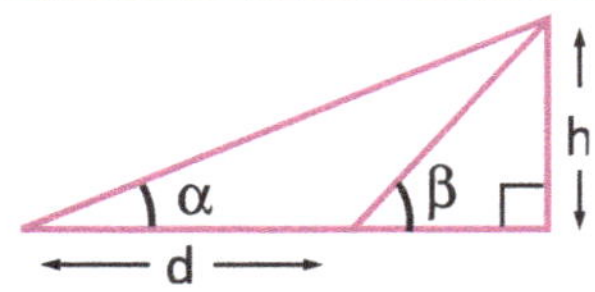
Case III: Both Elevation/Depression	**Case IV: Angle of Elevations. Two observations**
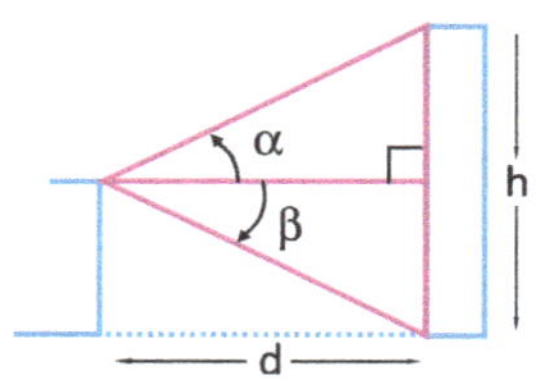	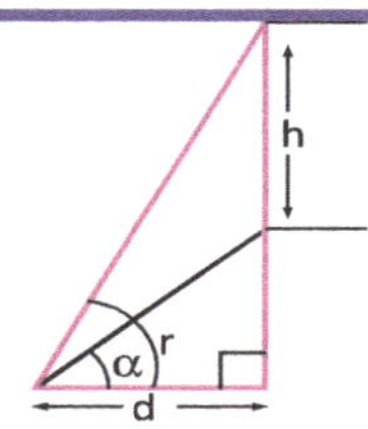
Case V: 3D Accessible Objects	**Case VI: 3D Inaccessible Objects**
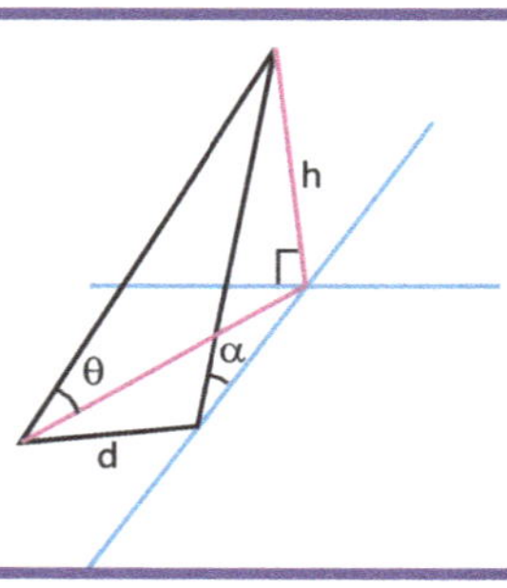	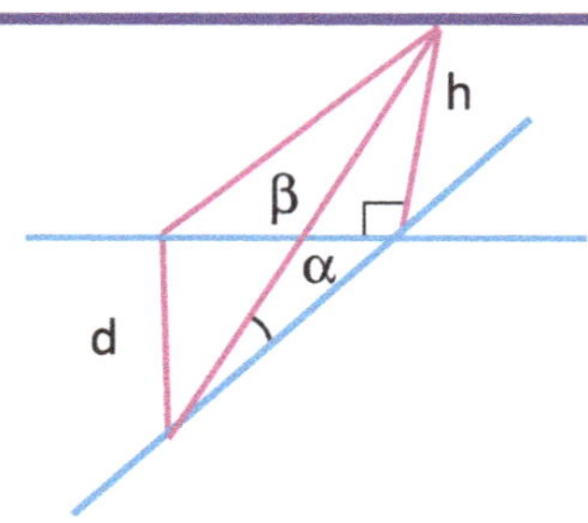

Watch Video **3D Applications with Tangent Ratio** SCAN ME

Case I Example 1: Height from Same Level

The angle of elevation to the top of the tree from 40 m from its foot is 30°. Find the height of the tree.

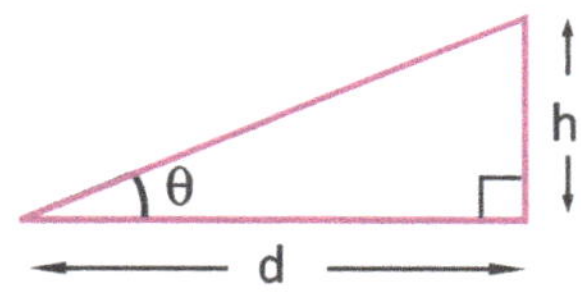

S^O_H C^A_H T^O_A

Solution

Diagram illustrates the situation where $d = 40, \theta = 30°$

Since opposite and adjacent sides are involved we will apply tangent ratio

$\tan\theta = \frac{h}{d} \Rightarrow h = d\tan\theta$

$h = 40\tan 30° = 23.09$

Height of the tree is 23.1 m

CASE I Example 2: Angle of Inclination

For non-self-supporting ladders, **angle of inclination**, the angle that an object makes with the ground, is very important from safety and optional use. At a construction site recommended angle of inclination for 15 m ladder is $68° \pm 4°$ for safety reasons.

a) Find the range of height that can be reached safely with the ladder.
b) Is it safe to climb this ladder if the foot is 5 m from the wall?

Solution Example 2: Angle of Inclination

Ladder length of 15 m is the hypotenuse opposite side to given angle is height. We should apply sine ratio

S^0_H C^A_H T^O_A

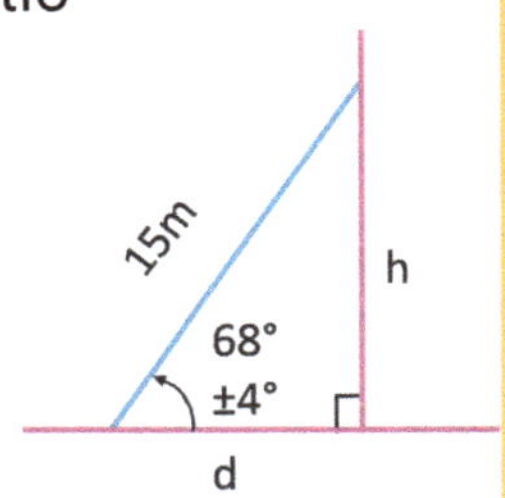

$\sin\theta = \dfrac{h}{15}$

$h = 15\ sin\theta$

For $\theta = 68° - 4° = 64°$

For $= \theta = 68° + 4° = 72$

$h = 15\sin 64° = 13.5\ m\ (3sf)$

$h = 15\sin 72° = 14.3\ m\ (3sf)$

Range of height is 13.5 m to 14.3 m.

Ladder length, $l = 15\ m$

Distance from wall = 5 m

For safety we need to check, θ. θ should be between $64° - 72°$

S^0_H C^A_H T^O_A : Given hypotenuse and adjacent ⇒ apply cosine ratio

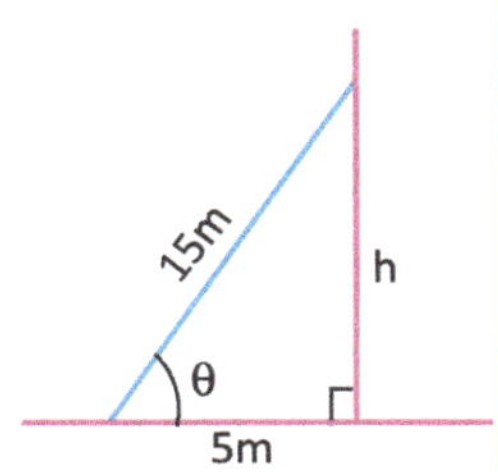

$\cos\theta = \dfrac{5}{15}$

$\theta = \cos^1\left(\dfrac{5}{15}\right) = 70.5°$

Yes, it is safe to climb since the angle of inclination 70.5° is within permitted limits of $64° - 72°$.

Case I: Example 3: Different Level

Height from Eye Level

The angle of elevation of the top of the tree from the eye level of an observer, 1.6 m tall is 50°. Find the height of the tree if the observer is 60 m away from the tree.

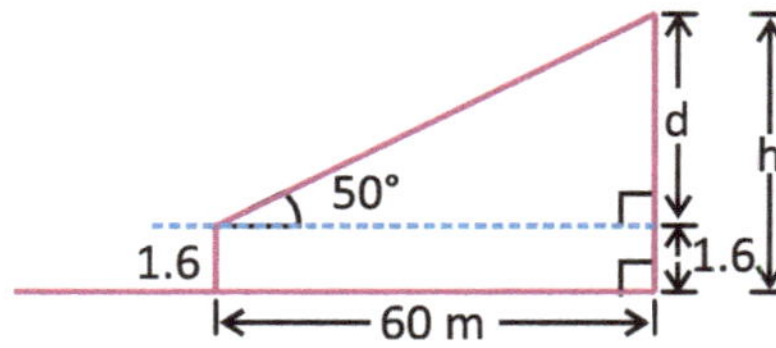

Solution 3

Diagram illustrates the situation where $d = 60, \theta = 50°$

Since opposite and adjacent sides are involved we will apply tangent ratio

$\tan 50° = \frac{d}{60} \Rightarrow$ d$= 60 \tan 50° = 71.5\ m$

We need to add the height of the observer to get the actual height of the tree.

Height of the tree is 71.5 + 1.6 **= 73.1 m**

Case II: Example 4: Two Observation Points Along a Line

Height from Eye Level

The angle of elevation to the top of the tree from the points A and B is 35° and 65° respectively. If these points are 20 m apart along the same line as the base of the tree, then find the height of the tree.

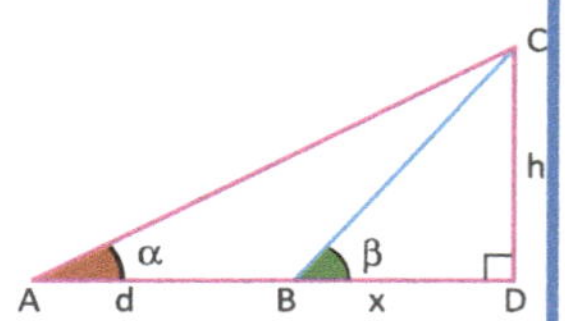

Solution 4: Diagram represents the situation

$\alpha = 35°, \beta = 65°, AB = 20m, Let\ BD = x\ and\ height, CD = h\ °$

$$\tan\beta = \tan 65° = \frac{h}{x}$$

$$x = \frac{h}{\tan 65°} = 0.4663h$$

$$\tan\alpha = \tan 35° = \frac{h}{x+20}$$

$$0.7002 = \frac{h}{0.4663h+20}$$

$$h = 0.7002(0.4663h + 20)$$

$$h = 0.3265h + 14.004$$

$$0.6735h = 14.004$$

$$h = \frac{14.004}{0.6735} = 20.8$$

height of the tree is 20.8 m

Challenge Problem

The angle of elevation of a plane changes from $\mathbf{78}°$ to $\mathbf{42}°$ in 15 seconds when observed from a Radar Station. If the plane is flying at the height of 2 km find the speed of the plane during this time.

STEM Project: Height from 2 Observation Points

Model Real Life Situation

A badger, with poor eyesight, depends on the honey-guide bird to locate beehive and the bird gets treat to sweet honey as the strong claws of badger rips apart the beehive. Normally the bird points at an elevation angle, α, then fly closer to the tree pointing again at an angle β.
Badger has worked out a formula to calculate the distance and height of the beehive from these observations. Badger visualises the situation as shown in the figure and works out the distance using following formula:

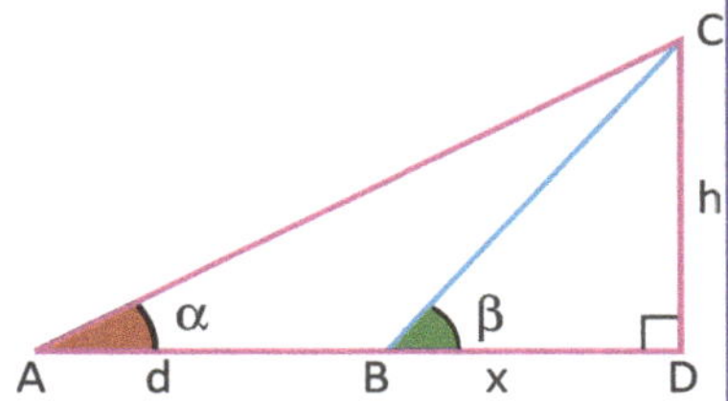

$$x = \frac{d \tan\alpha}{\tan\beta - \tan\alpha}, \qquad h = x \tan\beta$$

Height:

$$\boldsymbol{h = \frac{d \tan\alpha \tan\beta}{\tan\beta - \tan\alpha}}$$

Provide explanation and proof for the above formula for calculating height.

Case III: Example 5: Both Angle of Elevation and Depression

From the top of a building 21.0 m tall, the angle of elevation of the top of taller building is 46°. The angle of depression of the base of the taller 15.39°. What is height of the taller building.

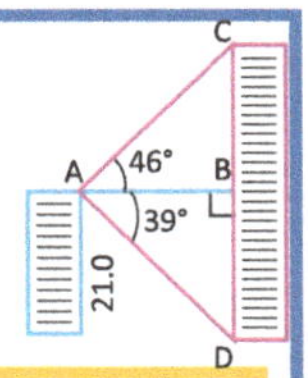

In right triangle ΔABD

$\angle BAD = 39°, BD = 21.0$

Solve for AB

$S^O_H \quad C^A_H \quad T^O_A$

$$\tan 39° = \frac{BD}{AB} = \frac{21}{AB}$$

$$AB = \frac{21}{\tan 39°} = 25.93$$

Now, in right triangle ΔABC

$\angle BAC = 46°, AB = 25.93$

$$\tan 46° = \frac{CB}{AB}$$

$$CB = AB \tan 46°$$

$$CB = 25.93 \tan 46° = 26.85$$

$$Height\ of\ the\ building = CB + BD$$

$$= 26.85 + 21 = 47.85 = \mathbf{47.85\ m}$$

General Solution: Relative Height of Building $CD = AB(\tan 46° + \tan 39°)$

Practice

The angle of elevation and depression of the top and bottom of a tower from the top of a building 80 m high, are $30°$ and $60°$ respectively.

1. Sketch the diagram to illustrate the situation.
2. Find the height of the tower and its horizontal distance from the building.

CASE IV Example 6: Two observations from a point for Vertical Height

Anil wants to find the length of the pole, AB, that is on the roof of a building, CB. The angle of elevation of A and B are 50° and 36°, respectively from D. The distance CD is 50 m. Find the length of the pole.

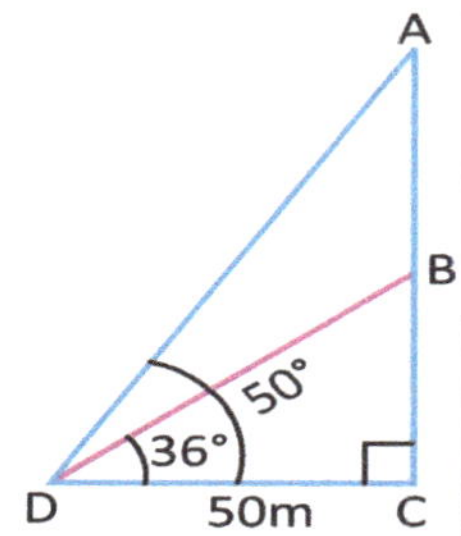

Solution 6:

Right triangle ΔACD represents the situation. Height of the pole AB can be calculated by finding the difference between AC and BC. Since the adjacent side DC is given we can use tangent ratio to calculate the heights.

In right triangle ΔACD	In right triangle ΔBCD	**Pole length, AB**
$\angle ADC = 50°, BC = 50$	$\angle BDC = 36°, BC = 50$	$AB = AC - BC$
$\tan 50° = \dfrac{AC}{CD}$ S$^{O}_{H}$ C$^{A}_{H}$ T$^{O}_{A}$	$\tan 36° = \dfrac{BC}{DC}$	$AB = 59.58 - 36.33$
$AC = DC \tan 50°$	$BC = DC \tan 36°$	$AB = 23.25$
$AC = 50 \tan 50°$	$BC = 50 \tan 36°$	Pole Height: 23.4 m
$AC = 59.58$	$BC = 36.33$	

Explore and Learn

A cuboid is shown. Find plane diagonal AC. Find space diagonal AG. Find angle GAC. Show that the ΔAGH is a right triangle. Find angle AGH.

SCAN ME

Watch Video **Explore Rectangular Prism**

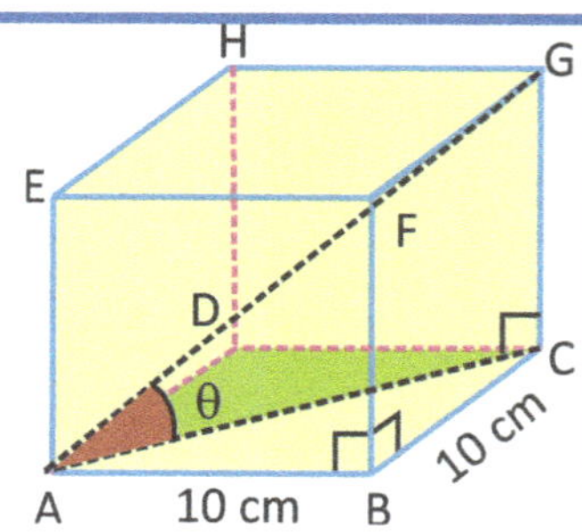

3D Applications

Case V Example 7: 3D Application Accessible Object

Observations are performed from a base line to calculate height of objects

To find the height of CN Tower in downtown Toronto, Canada, Anil took advantage of the perpendicular roads. A crossing where the east-west road intersects the north south was selected as point O. OT was marked as the base line, 50 m long where T is due east of O. Angle of elevation to the top of CN Tower from T is 75°. Find the height of CN Tower if O is 110 m due south of the tower.

Solution 7:

Diagram represents the situation.

In $\Delta NOT, \angle NOT = 90°$

$NO = 110, OT = 50$

Solve for NT

ΔNOT is a right triangle

$NT^2 = NO^2 + TO^2$

$NT^2 = 110^2 + 50^2$

$NT^2 = 146000$

$NT = \sqrt{14600} = 120.83$

Now Solve right triangle ΔCNT

SOH CAH (TOA)

$$\tan 75° = \frac{CN}{TN}$$

$CN = TN \tan 75°$

$CN = 120.83 \tan 75°$

$CN = 450.9 \approx \mathbf{450}$

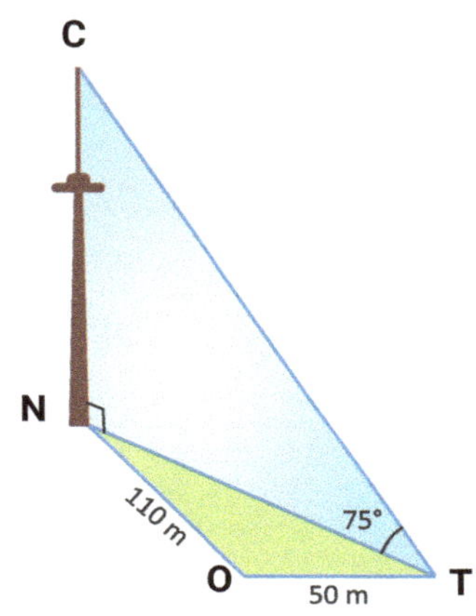

Height of CN Tower is estimated as 450 m (nearest ten of a metre)

Practice

A lighthouse stands on the shore at point **L**. A boat sails in the water and reaches point **B**, which is **120 m east** and **90 m north** of the lighthouse base. The observer on the boat is standing at a deck height of **5 m above water**. From this point, the **angle of elevation** to the top of the lighthouse is 32°.

98.73 m

Case VI Example 8: 3D Application to Find Angle of Elevation and Distance

Suman has let out 80 m string and the kite is at an angle of elevation of $42°$.

Wind is towards north. Her friend Tom stands to the west, 100 m away

a.	Find kite's angle of elevation from Tom.
b.	Find kite's distance from Tom.

Solution 8:

Diagram illustrates the situation

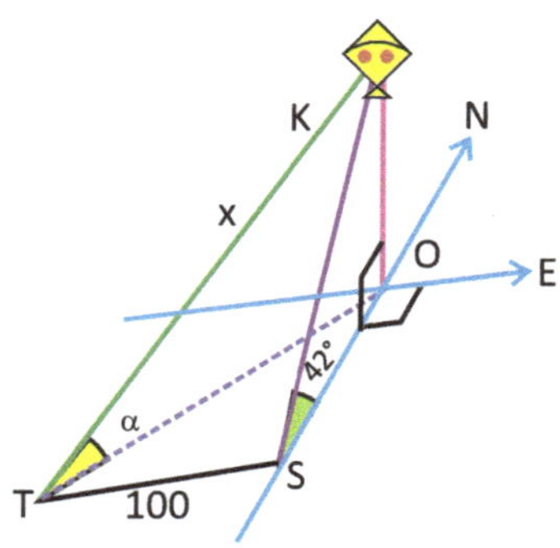

Find kite's angle of elevation from Tom.

Consider right triangle ΔSOK (Vertical Plane)

$SK = 80$ length of the string; $\angle OSK = 42°$

$\sin 42° = \dfrac{OK}{SK} \Rightarrow OK = 80 \sin 42° = 53.53$ (SOH) CAH TOA

$\cos 42° = \dfrac{OS}{SK} \Rightarrow OS = 80 \cos 42° = 59.45$ SOH (CAH) TOA

$TO = \sqrt{100^2 + 59.45^2} = 116.34$ (Horizontal Plane)

$$\tan \angle KTO = \frac{OK}{TO} \Rightarrow \angle KTO = \tan^{-1}\left(\frac{OK}{TO}\right) = \tan^{-1}\left(\frac{53.53}{116.34}\right)$$
$$= \mathbf{24.7}°$$

Angle of elevation: $\angle \boldsymbol{KTO} = \mathbf{24.7}°$

kite's distance from Tom, KT

Consider right triangle ΔTSK:

$KT = \sqrt{100^2 + 80^2} = 20\sqrt{41} = \mathbf{128}\ \boldsymbol{m}$

Kite's distance from Tom is 128 m

Explore and Learn More

Two roads intersect at $90°$. At 12:00 noon, two cars leave the intersection at an average speed of 10 km/h and 80 km/h. At 12:15, a traffic helicopter is 1500 m right above the slower car. Determine the angle of depression and the distance from the helicopter to the faster car. $2.7°, 32\ km$

Case VI Example 9 – 3D Application Inaccessible Object

The angle of elevation to the top of the tower is 32° from a point A due south of it. The angle of elevation to the top of the tower from another point B due west of the tower is 40°. Find the height of the tower if the base line distance AB is 45m

Solution 9: Inaccessible Object, Distance from Foot Cannot be Measured

Sketch figure to illustrate the situation.

(Note how the horizontal plane is represented to provide 3D effect)

Step 1: Relate height, h

Find distance of observation points from foot of the tower in terms of height h.

$$\tan 32° = \frac{h}{OA}$$

$$OA = \frac{h}{\tan 32°} = 1.6h \qquad (1)$$

$$\tan 40° = \frac{h}{OB}$$

$$OB = \frac{h}{\tan 40°} = 1.19h \qquad (2)$$

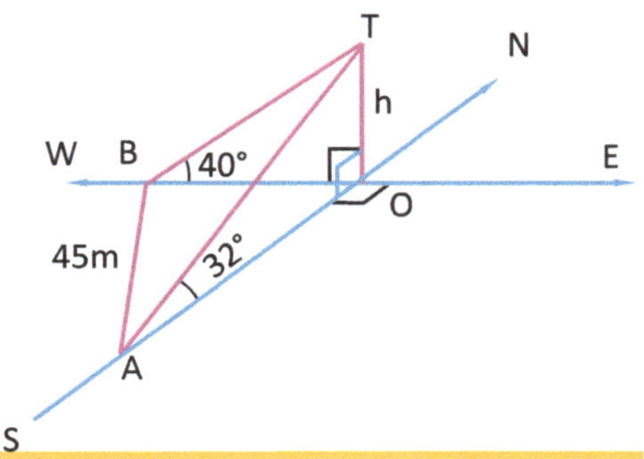

Step 2: Solve for h

Δ AOB is right triangle

(west – south junction)

$$AB^2 = OA^2 + OB^2$$

$$45^2 = (1.6h)^2 + (1.19h)^2$$

$$2025 = 2.56\,\text{h}^2 + 1.42\text{h}^2$$

$$2025 = 3981h^2$$

$$h = \sqrt{\frac{2025}{3.981}}$$

Height of the tower is $\mathbf{22.6\ m}$ $(3\ s.f.)$

Explore and Learn More

A surveyor wants to find the height of the cliff on the other side of the river. He lays the line AB of length 120 m. From A, he selects point C on the base of the cliff and measures angle CAB to be 52°. He then selects a point D on top of the cliff directly above C and measures an angle of elevation of 34°. He walks to the point B and measures angle CBA as 64°. Find the height of the cliff.

81 m

Case VI Example 10: Height of a Bird

Anil and Ben are watching a falcon perched on a nearby branch of the tree. Anil is facing north and has an angle of elevation of 20° to the bird. Ben is facing west. Ben sees the bird at on elevation of 16°. Determine the height of the falcon if the distance between Anil and Ben is 150 m.

Solution 10:

Let us sketch the situation and understand directions.

Anil is facing north; he is on the south side. Ben is facing west; he is on the east side

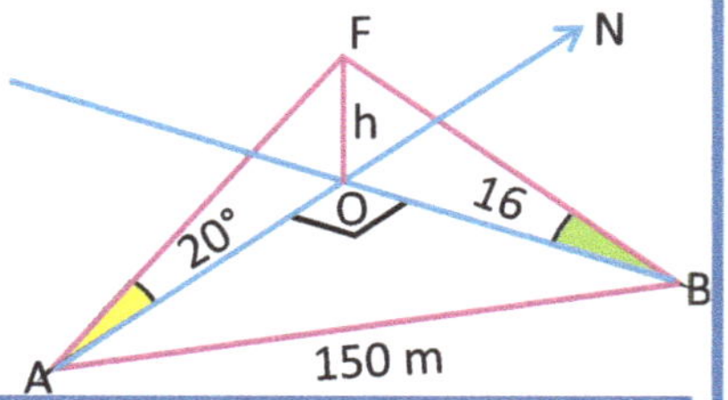

Let the height of the bird be h	$\Delta AOB\ is\ right\ triangle$
$\Delta\ AOF\ is\ right\ triangle$	$AB^2 = AO^2 + OB^2$
$\tan 20° = \frac{h}{AO}$	$150^2 = \left(\frac{h}{\tan 20}\right)^2 + \left(\frac{h}{\tan 16}\right)^2$
$AO = \frac{h}{tan\ 20°}$	$= 7.5486h^2 + 12.162h^2$
$\Delta\ BOF\ is\ right\ triangle$	$150^2 = h^2(19.71)$
$\tan 16° = \frac{h}{BO}$	$h = \sqrt{\frac{150^2}{19.71}} = 33.78 = 33.8\ m$
$BO = \frac{h}{\tan 16°}$	**Height of falcon is $33.8\ m$**

Practice

Anil and Ben are standing on level ground **150 m apart**.

Anil is facing **north** and sees a kite at an **angle of elevation of 20°**.

Ben is facing **west** and sees the same kite at an **angle of elevation of 16°**.

Assume the kite is directly above the point where the north–south line through Anil meets the west–east line through Ben (so they view along perpendicular directions).

Find: The **height** of the kite above the ground.

70 m

STEM Project: Make a Clinometer

Clinometers and Related Applications

A clinometer is a simple instrument used to measure slope. Both angle of depression and elevation can be measured. A narrow piece of wood (or straw) is glued to the base a large protector to act as line of sight. A weighted plumb line hangs from the midpoint of the base line of the protractor.

Example 11: Clinometer

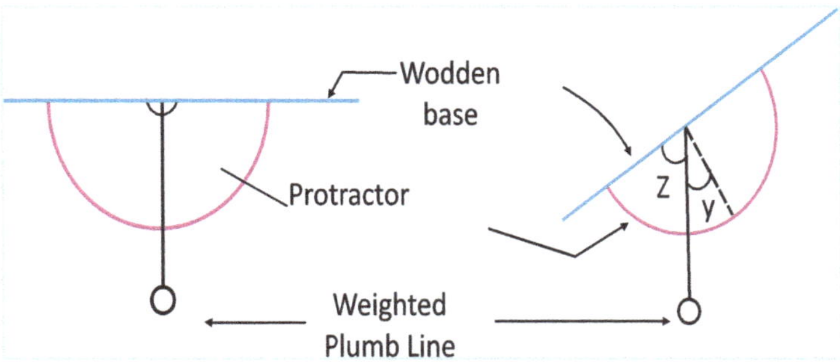

To use the clinometers, tilt the wooden base to align your line of sight to the tip of the object. Someone else should then read the angle, made by the plumb-line on the protractor. Angle of elevation will be $\mathbf{90° - z}$

Application

To measure the height of a tree Anil looked through the straw on the clinometer and Anna read the angle, z, as 40°. Find the height of the tree if Anil is 176 cm tall and is 25 m away from the tree.

Solution 11:

Angle, z is 40, so the angle of elevation is $(90° - 40° = 50°)$

$\tan 50° = \frac{h}{25}$

$h = 25 \tan 50° = 29.79$

Height of the tree above he ground, H

$H = 1.76 + 29.79 = 31.55 \approx 31.6\, m$

Tree Height: 31.6 m

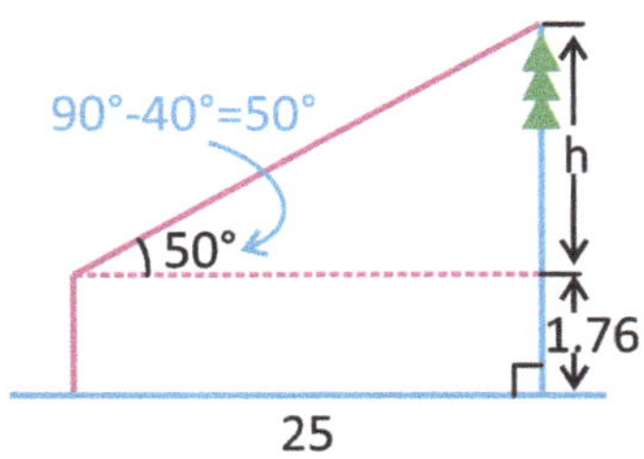

Practice Question

To measure the height of a building, Sam uses a clinometer and observes the top of the building at an angle of elevation of **35°**. If Sam's eyes are **1.7 m above the ground** and he stands **30 m away** from the base of the building, find the **height of the building** to the nearest tenth of a meter.

Real World Applications

The latitude of Toronto, Canada is approximately 44°.

Example 1: Length of Parallel Latitudes, C_p

Find the length of the **44° parallel of latitude**, rounded to the nearest 10 km.
Assume the radius of the Earth is **6380 km**.

Solution 1:

A circle of latitude is an east–west line that connects all locations on Earth sharing the same latitude. These lines are called **parallels** because they run parallel to each other and never meet.

O is the center of the Earth

Q is center of circle for 44° parallel

P is on 44° parallel and

E is on equator

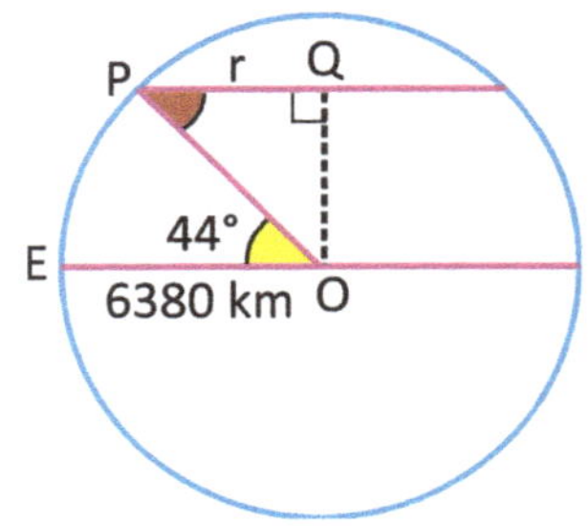

$OP = OE = 6380$ radii of the earth

$\angle OPQ = \angle EOP = 44°$ (alternate angles)

$$\cos 44° = \frac{r}{6380}$$

$r = 6380 \cos 44°$

The length of 44° parallel of latitude is circumference, C_p with radius r

$C_p = 2\pi r = 2\pi(6380 \cos 44°) = 28835.9$

$C_p = 28840\ km\ (round\ to\ 10\ km)$

The length of the **44° parallel of latitude is 28840 km**

Example 2: Pendulum Displacement

A pendulum of length L is d units above the ground when at rest. Find its distance above the ground after a clockwise swing of angle θ degrees.

Solution 2:

Figure illustrates the situation.

Let OP be the position of the pendulum at rest and

OQ be its position after clockwise swing by θ degrees.

Pendulum vertically moves upwards.

Vertical distance $OR = L\cos\theta$

$OP = L$

Change in height = $OP - OR$

$= L - L\cos\theta$

$= L\,(1 - \cos\theta)$

Now, the distance from the ground has increase and this distance is $\boldsymbol{d + L(1 - \cos\theta)}$

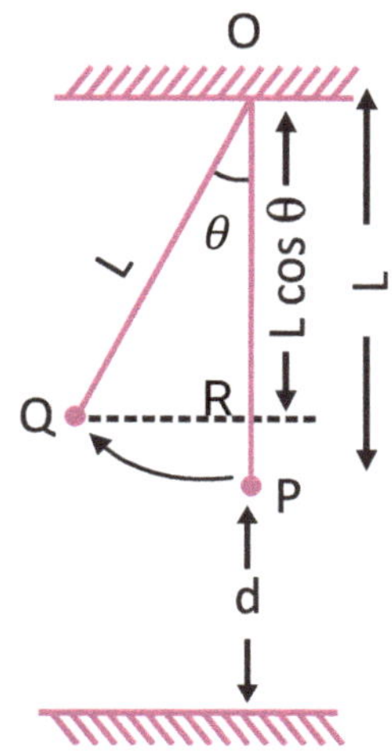

Example 3: Two Trees Supported by Gue Wire (Figure Provided)

Two trees, 15 m apart need support. A gue wire, 11 m long is fastened 7 m from the base of one tree. Another wire, from this point, is fastened to the other tree. Find the length of this wire if it is perpendicular to the first wire.

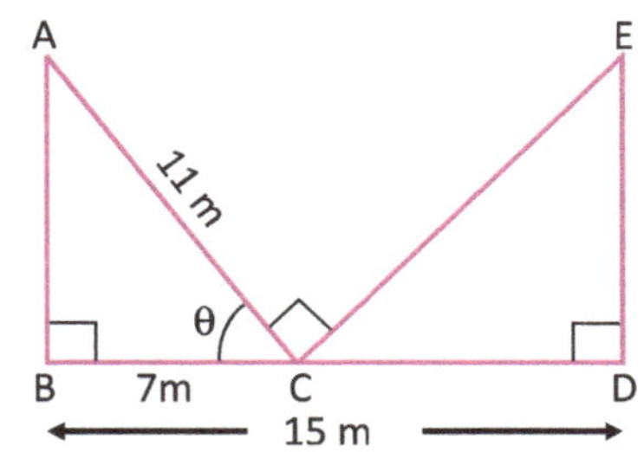

Solution 3:

In $\Delta\,ABC$ adjacent and hypotenuse are given. So, apply cosine ratio.

$S_H^O\ C_H^A\ T_A^O$

$\cos ACB = \frac{BC}{AC} = \frac{7}{11}$

$A\hat{C}B = \cos^{-1}\left(\frac{7}{11}\right) = 50.478$

$\boldsymbol{A\hat{C}B = 50.48°}$

$Solve\ \Delta\,CDE$

Find angle C in ΔDCE

$E\hat{C}D + 90° + A\hat{C}B = 180\ Linear$

$E\hat{C}D = 180 - 90 - 50.48 = 39.52°$

$CD = BD - BD = 5.78\ m$

$\cos E\hat{C}D = \frac{CD}{CE}$

$CE = \frac{CD}{cos E\hat{C}D}$

$CE = \frac{8}{\cos 39.52} = 10.4\ m\ (3\ s.f.)$

Length of the guy wire required is 10.4 m

Theodolite and Related Application

Theodolite is an optimal instrument for measuring angle between visible point in the horizontal and vertical planes. Latest versions have been designed to provide very accurate information about angles and distances which can be electronically fed to the computer.

Example 4: Measurements with Theodolite

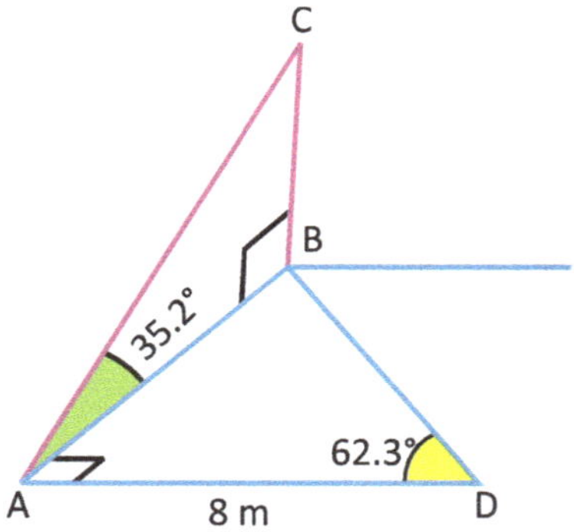

Flowerpots are the rock pillars in Bay of Fundy, Canada. The geologist set up a theodolite at A and measured $\angle BAC$ to be 35.2°. Base line AD, perpendicular to AB, was taken as 8 m, and $\angle ADB$ measured as 62.3°. If the height of the theodolite is 1.2 m, what is the height of the Flowerpot to the nearest tenth of a metre?

Solution 4: Measurements with Theodolite

In right triangle ΔBAD

$\tan \angle BDA = \frac{AB}{AD}$ S^O_H C^A_H T^O_A

$\tan 62.3° = \frac{AB}{8}$

$AB = 8 \tan 62.3 = 15.237$

$AB = 15.2$

In right triangle ΔABC

$\frac{CB}{AB} = \tan \angle BAC$

$CB = AB \tan 35.2°$

$15.237 \tan 35.2 = 10.749$

Since theodolite is set-up 1.2 m high, the height of the Flowerpot is $1.2 + 10.7 = 11.9\ m$

Student's Activity

a) In the above example explain why the $\angle BAD = 90°$.
b) Find the distance of flowerpot, C, from the observation points A and D.
c) Find the angle of elevation of C from D.

$AC = 18.65\ m, CD = 20.28\ m, \angle BDC = 32°$

Watch Video **3D SAT Test Practice** SCAN ME

Equilateral Triangle Inscribed in a Circle

Example 5: Relate Parameters of Inscribed Equilateral Triangle with Radius

Equilateral triangle ΔABC is inscribe in a circle, O, with radius R and altitude AT and base BC as shown in the figure.

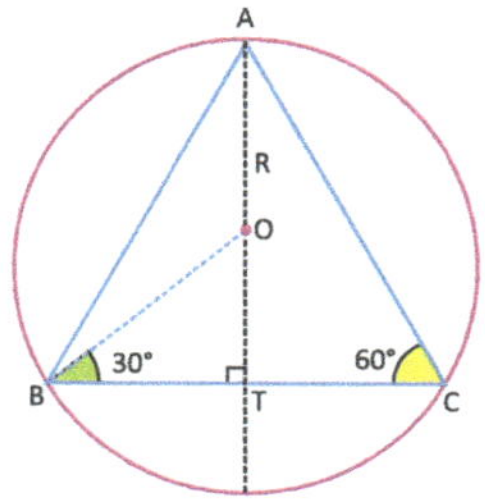

Prove that

a) Height of the triangle, AT = $\frac{3}{2}R$

b) Each side length, BC =$\sqrt{3}R$

c) Area of $\Delta ABC = \frac{3\sqrt{3}}{4} R^2$

Solution 5:

ΔOBT is right triangle $OB = R, BT = \frac{BC}{2}, \angle OBT = 30°$ $\sin 30° = \frac{OT}{OB}$ $OT = OB \sin 30° = \frac{1}{2}R$ $\cos 30° = \frac{BT}{OB}, BT = OB \cos 30° = \frac{\sqrt{3}}{2} R$	**Altitude,** $\boldsymbol{AT = AO + OT = R + \frac{1}{2} R = \frac{3}{2} R}$ **Base,** $\boldsymbol{BC = 2BT = 2 \times \frac{\sqrt{3}}{2} R = \sqrt{3}R}$ Area of $\Delta\ ABC = \frac{1}{2}\ base \times height$ Area of $\Delta\ \boldsymbol{ABC = \frac{1}{2} \times \sqrt{3}R \times \frac{3}{2}\ R = \frac{3\sqrt{3}}{4} R^2}$

Practice Questions

Q1.	What is the area of an equilateral triangle inscribed in the circle of radius 10 cm? A. $50\sqrt{3}\ cm^2$ B. $25\sqrt{3}\ cm^2$ C. $75\sqrt{3}\ cm^2$ D. $75\sqrt{3}\ cm^2$	Q2.	A Circle has an area of $16\pi\ unit^2$. Find the area of an equilateral triangle inscribed in it. A. $12\sqrt{3}\ unit^2$ B. $10\sqrt{3}\ unit^2$ C. $8\sqrt{3}\ unit^2$ D. $18\ unit^2$
C	$\frac{3\sqrt{3}}{4} R^2 = \frac{3\sqrt{3}}{4} 10^2 = 75\sqrt{3}$	A	$16\pi = \pi R^2 \Rightarrow R^2 = 16$ $Area = \frac{3\sqrt{3}}{4} (16) = 12\sqrt{3}$

Chapter 11 Test: Angle of Elevation/Depression

Q1. The angle of elevation of the top of the tree from a distance of 50 m from its foot is 30°. Find the height of the tree.

Q2. The angle of elevation of the top of the tree from the eye level of an observer is 1.2 m tall is 40°. Find the height of the tree if the observer 100 m away from the tree.

Q3. A light house is 100 m above the sea level. The angle of depression of a ship sailing directly towards it changes from 30° to 60°. Find the distance travailed by the ship.

Q4. A ladder inclined at an angle of 60° to the horizontal reaches 3 m short of the top of a tree. Find the length of the ladder if the tree is 10 m high.

Q5. The angle of elevation of the top of the tree from points A and B is 60° and 40° respectively. Find the height of the tree if the distance AB is 6 m. Assume A and B on the same side of the tree.

Q6. The angle of elevation of the top of a building from the top and bottom of a tree is 20° and 60° respectively. Find the height of the building if the tree is 12 m high.

Q7. The angle of elevation of the top of a building from the foot of the tower is 30° and the angle of elevation of the tower is 60° from the foot of the building. If the tower is 40 m high find the height of the building.

Q8. The angle of elevation to the top of the tower is 32° from a point A due south of it. The angle of elevation to the top of the lower from another point b due west of the tower is 40°. Find the height of the tower if the distance AB is 50m.

Q9. Explain why in Q4, tangent ratio is not applied to solve the problem.

Worksheet Answers

Q1.	28.9 m	Q2.	85.1 m	Q3.	115.5 m	Q4.	8.1 m
Q5.	9.8 m	Q6.	15.2 m	Q7.	13.3	Q8.	25 m

Exam Style Wordy Test Problem

Point H on a hot-air balloon is sighted at the same time by two observers. One observer is at the top of a vertical building (Point A) that is 156 m tall. The other observer is at the base of the building (point B). The angle of elevation from point A to H is $40°$, and the angle of elevation from point B to H is 57°. Point X is the ground directly below point H.

a.	Find the size of angle AHB.
b.	Calculate the distance from point B to H.
c.	The hot-air balloon remains at constant height as it moves further away from the building. Describe, in words, the change in the angle of depression from point H to B as the horizontal distance between the balloon and the building increases.
	Angle AHB = 17°. Distance HB is 408.7 m Watch Video **Angle of Elevation** SCAN ME

Quiz: Area of Inscribed Equilateral Triangle

Q1. Figure shows an equilateral triangle inscribed in a circle of radius 10 cm.

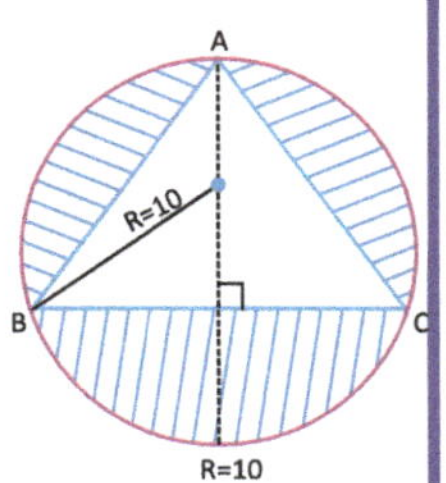

Find the area of the circle excluding the inscribed triangle region.

A. $160\ cm^2$
B. $176\ cm^2$
C. $184\ cm^2$
D. $192\ cm^2$

Q2. An equilateral triangle is inscribed in a circle of radius 10 cm. Find its perimeter.

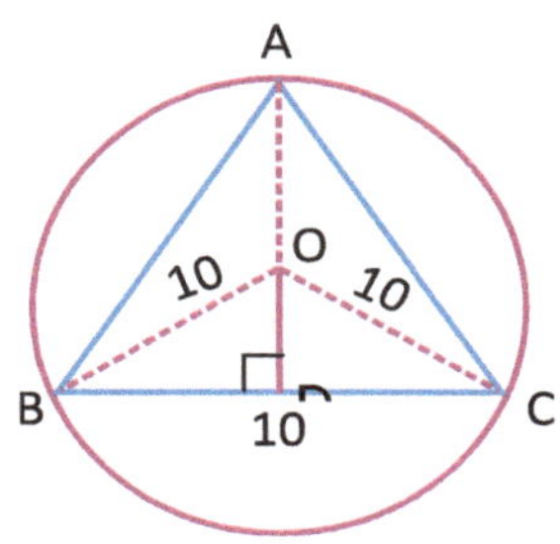

A. $15\sqrt{3}\ cm^2$
B. $20\sqrt{3}\ cm^2$
C. $30\sqrt{3}\ cm^2$
D. $30\ cm^2$

Degrees – Minutes – Seconds Unit Conversion

Degrees, Minutes, Seconds $(D^{\circ}M'S'')$ to Decimal Form

Convert Degree, Minute, Seconds $(D^{\circ}M'S'')$ to Decimal Form

Example 1:	$23^{\circ}\,30' = 23^{\circ} + 30 \times \frac{1^{\circ}}{60} = 23.5^{\circ}$
Example 2:	$42^{\circ}\,15' = 42^{\circ} + 15 \times \frac{1}{60} = 42.25^{\circ}$
Example 3:	$37^{\circ}\,45' = 37^{\circ} + 45 \times \frac{1}{60} = 37.75^{\circ}$
Example 4:	$26.16'30'' = 26 + 16 \times \frac{1}{60} + \frac{30}{3600} = 26.275^{\circ}$
Example 5:	$1^{\circ}\,1'\,1'' = 1 + \frac{1}{60} + \frac{1}{3600} = 1.0169^{\circ}$

$1\ minute = \frac{1}{60}\ degree$

$1\ second = \frac{1}{60}\ minute$

$1\ second = \frac{1}{3600}\ degree$

Practice Questions

Convert Decimal to $(D^{\circ}M'S'')$ Degree, Minute, Seconds Form

Example 1:
Convert 32.375° to $(D^{\circ}M'S'')$ form

$$
\begin{aligned}
32.375^{\circ} &= 32^{\circ} + 0.375\,(60'') \\
&= 32^{\circ} + 22.5' \\
&= 32^{\circ} + 22' + 0.5' \\
&= 32^{\circ} + 22' + 0.5'(60'') \\
&= 32^{\circ} + 21' + 30'' \\
&= 32^{\circ}\!:21'\!:30''
\end{aligned}
$$

Example 2:
Convert $16.234''$ to $(D^{\circ}M'S'')$ form

$$
\begin{aligned}
16.234^{\circ} &= 16^{\circ} + 0.237\,(60') \\
&= 16^{\circ} + 14.04' \\
&= 16^{\circ} + 14' + 0.04' \\
&= 16^{\circ} + 14' + 0.04\,(60'') \\
&= 16^{\circ} + 14' + 2.4'' \\
&= 16^{\circ} + 14' + 2'' \\
&= 16^{\circ}\!:14'\!:2''
\end{aligned}
$$

Practice Questions: Convert Decimal to $(D^{\circ}M'S'')$ Degree, Minute, Seconds Form

a.	64.175°	b.	52.025°	c.	12.327°

Convert Degree, Minute, Seconds $(D°M'S")$ to Decimal Form					
a.	$43° \ 39'$	b.	$33° \ 52' =$	c.	$25° \ 19' \ 24"$

Application Problems	
Q1 a. b.	A boat is **2 km** from a dock at a **bearing of** $26°35'10''$. Find the bearing in decimal form. Find its **east (horizontal)** and **north (vertical)** distances from the dock. East (horizontal) 895 m, North (vertical): 1789 m
Q2	A 10 feet ladder is at an elevation of 43° 39'. Find the maximum height it can reach. 6.9 feet
Q3	A ship sails from point O on a bearing of $32°15'20''$for a distance of **4 km**. Find how far **east** and **north** the ship has travelled. N=3.38 km, E=2.13 km
Q4	From the top of a lighthouse, the angle of depression to a boat is $18°25'40''$. If the lighthouse is **45 m high**, find the **horizontal distance** from the lighthouse to the boat. 136.2 m (approx.)
Q5	A tree casts a shadow of **28 m** when the Sun's elevation is $46°10'15''$. Find the **height of the tree**. 29.0 m

Student Spotlight: Amandeep Saini

"Math was a real struggle for me in high school until I began learning with Mr. Anil Kumar. His patient, step-by-step approach transformed my confidence and helped me consistently score in the 90s. The critical thinking and problem-solving skills I gained continue to guide me today as a Senior Auditor at Crowe Soberman LLP, pursuing my CPA designation."

Chapter 12: Trigonometry in Action

*Everyday navigation, engineering, and exploration rely on one powerful idea — **trigonometry in action**. From measuring the **height of a tree or building** to charting the **course of a ship or airplane**, trigonometry helps us see the invisible: the relationships between **angles, distances, and directions**.*

Direction:

Direction refers to the path along which something moves or faces. It is usually described using cardinal points:

- North (N)
- South (S)
- East (E)
- West (W)
  And combinations like Northeast (NE) or Southwest (SW).

Directions help us describe relative positions — e.g., "The lake is east of the village."

Bearing:

Bearing is a precise way to describe direction using angles, measured clockwise from the North.

There are two main types:

- True Bearings (Three-Figure Bearings)
- Compass Bearings (Quadrantal Bearings):

True Bearings (Three-Figure Bearings):	**Compass Bearings (Quadrantal Bearings):**
• Always measured clockwise from North (0°) • Given as a three-digit number • Example: ○ North → 000° ○ East → 090° ○ Southeast → 135° ○ 045° → NE ○ 270° → due West ○ 180° → due South	• Use cardinal directions like North or South as a starting point • Measure the angle toward East or West • Expressed like: ○ N 30° E (from North 30° East) ○ S 45° W (Southwest) ○ N 135° E (Southeast) ○ S 135° W (Northwest)

Example 1: Bearing from and To

Diagram illustrates bearing of a ship S from port P is $62°$.

a.	Find the bearing of the port P from the ship S. Illustrate your answer and explain the solution.

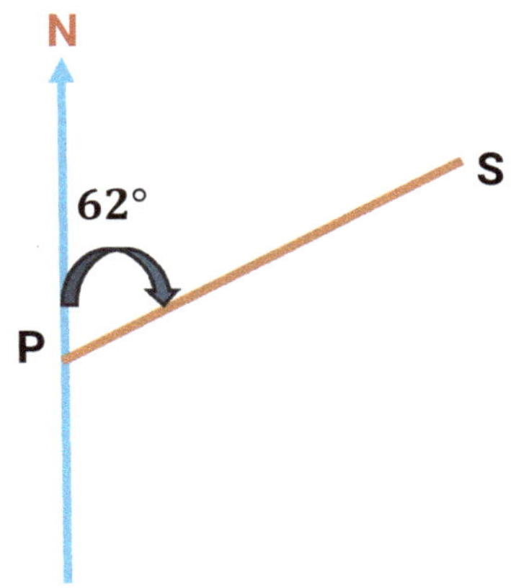

Step by step solution

- Draw Vertical line for North at S
- Return is turning around by $180°$
- Therefore, the bearing of P from B will be $62° + 180° = 242°$

  Diagram illustrates the solution

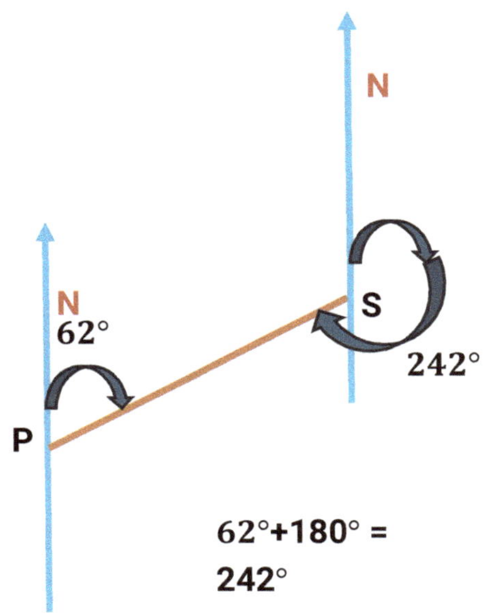

Example 2: Converting Between Bearings

Complete the table below relating true braring and compass bearing.

First four are examples to do the rest.

#	True Bearing	Compass Bearing	#	True Bearing	Compass Bearing
1	$070°$	$N70°E$	2	$290°$	$N70°W$
3	$150°$	$S30°E$	4	$200°$	$S20°W$
5.	$50°$		6		$N50°W$
7	$310°$		8		$S20°E$

Example 3: Horizontal and Vertical Distance from Bearing

A motor boat travel at a bearing of $050°$ fo 2 hours at $30\ km/h$.

How far north and how far east does it travel. Round your answer to 3 significant figures.

Solution

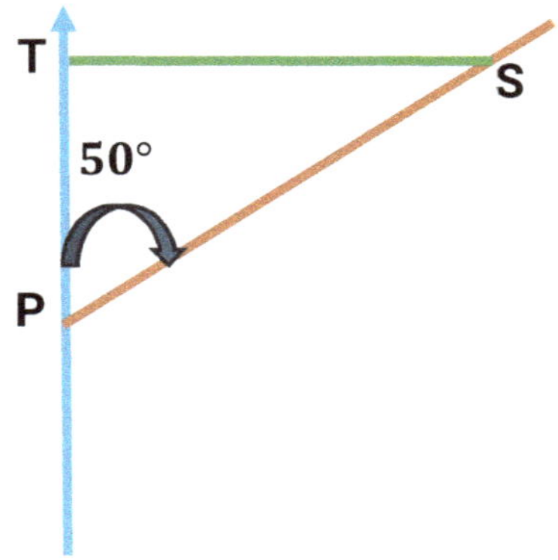

Let boat leaves P at a bearing of $050°$ and reach S in 2 hours.

$distance = speed \times time$

Then $PS = 30 \times 2 = 60\ km$

PT is the distance travelled north and

TS is the distance travelled east.

We know the hypotenuse of the right triangle along with an angle. Trigonometric ratios cosine and sine can now be used to find the shorter sides of this triangle. Use SOH CAH TOA to confirm which ratio for north and which for east.

Vetical (Adjacent to Angle)	Horizontal (Opposite to angle)
$\cos 50° = \frac{PT}{PS}$	$\sin 50° = \frac{TS}{PS}$
$PT = PS \cos 50°$	$TS = PS \sin 50°$
$PT = 60 \cos 50°$	$TS = 60 \sin 50°$
$PT = 38.56$	$TS = 45.96$
$PT \approx 38.6\ km$	$TS \approx 46.0\ km$
The boat has travelled **38.6 km north**.	The boat has travelled **38.6 km east.**

Bearings are used for navigation, mapping, geometry, and STEM applications like robotics, aviation, and marine travel.

Watch Video **Bearing Angle Concept** SCAN ME

Example 4: Horizontal and Vertical Distance

A ship sail 100 km on a bearing of 290° from Port O Port S

a) How far north has it travelled?
b) How far west has it travelled?
c) Find the bearing angle to return to Port O from S.

(Bearing angle is measured clockwise from North)

Solution 4:

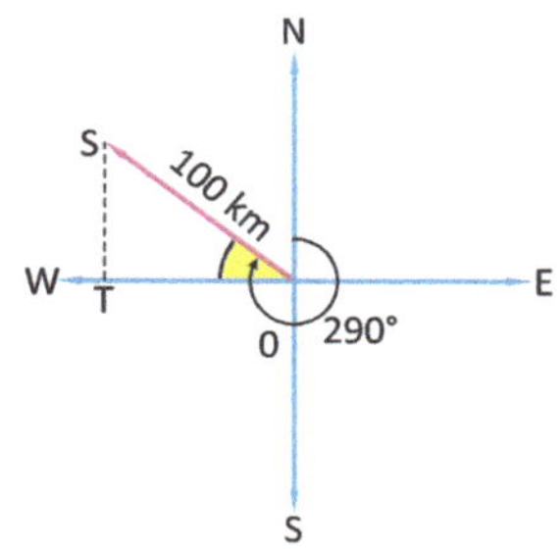

Position of the ship, S is shown in the figure.

Bearing of 290°

$\angle SOT = 290° - 270° = 20°$

$\Delta SOT \text{ is right triangle}$

Distance North, TS	**Distance West, OT**
S_H^O (circled) C_H^A T_A^O	S_H^O C_H^A (circled) T_A^O
$\sin 20° = \frac{TS}{OS}$	$\cos 20° = \frac{OT}{OS}$
$TS = OS \sin 20°$	$OT = 100 \cos 20° = 93.96$
$TS = 100 \sin 20° = 34.20$	$OT = 94.0\ km$
$TS = 34.2\ km$	

Ship sailed 94.0 km west and 34.2 km north

Bearing Angle to Return

Bearing angle to return is $290° - 180° = 110°$

Practice Question

A ship sails **150 km** on a bearing of **230°** from Port A to Port B. How much south did the ship travel. Find the **bearing to return to Port A** from Port B.

96.4 km, 050°

Example 5: Multiple Directions

Question

A boat travels 12 km at a bearing of $030°$ from port B to the island A. It then travels 10 km at a bearing of $120°$ to its final destination in island C. How far is the boat from the starting point?

Figure provided is not to the scale.

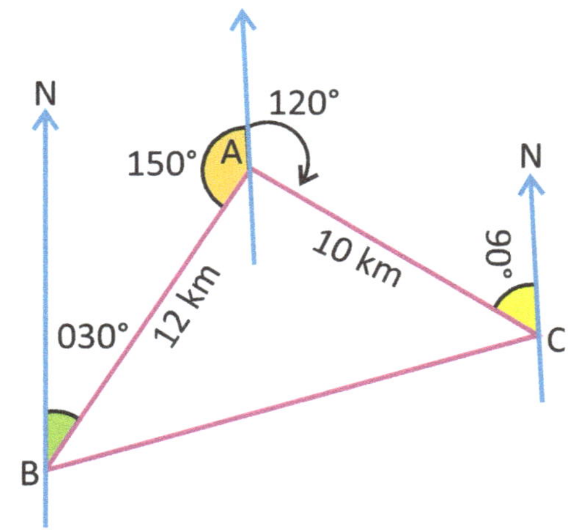

Solution

Figure illustrates the position of the boat as described. Note that the bearing angle is always written in three places ($030°$ *not* $30°$)

The blue vertical lines represent north direction and all of them are parallel.

Find $\angle BAC$

$\angle BAC = 90°$ — Provide reasoning using the concept of parallel lines and supplementary angles.

ΔBAC is a right triangle

$BC^2 = BA^2 + AC^2$ — Pythagorean Theorem

$BC^2 = 12^2 + 10^2 = 244$

$BC = \sqrt{244} = 15.6$

The boat is 15.6 km from the starting point.

Practice

An airplane departs airport **X** and flies **100 km** at a bearing of **045°** to reach checkpoint **Y**.
From **Y**, it changes course and flies **120 km** at a bearing of **150°** to reach its destination **Z**.

Find: The **straight-line distance** from the starting point **X** to destination **Z**.

135 km

🚢 Practice Questions: Directions and Bearings 🌍

Q1. A ship travels 100 km on a bearing of 045°. How far north and how far east has it travelled? (Round to 1 decimal place)

Answer: North = 70.7 km, East = 70.7 km

Q2. An airplane flies at a bearing of 120° for 150 km. How far has it travelled south and east?

Answer: South = 129.9 km, East = 75.0 km

Q3. A ship sails 80 km on a bearing of 210°. How far south and west has it travelled?

Answer: South = 69.3 km, West = 40.0 km

Q4. A plane flies 60 km at a bearing of 300°. Calculate the distance north and west.

Answer: North = 51.9 km, West = 30.0 km

Q5. Convert the following true bearings to compass bearings:
a) 045°
b) 225°
c) 315°

Answer: a) N45°E b) S45°W c) N45°W

Q6. Convert the following compass bearings to true bearings:
a) S60°E
b) N20°W
c) S45°W

Answer: a) 120° b) 340° c) 225°

Q7. A boat sails from a port at a bearing of 135° for 50 km. Find how far it has travelled south and east.

Answer: South = 35.4 km, East = 35.4 km

Q8. A pilot reports a bearing of 315° from airport A to location B. What is the bearing of the airport from location B?

Answer: 135° (reverse bearing)

Q9. A rescue team departs at a bearing of 030° and travels 18 km, then turns to a bearing of 120° for 24 km. How far are they from the starting point?

Answer: ≈ 34.5 km (use cosine rule or vector addition)

Q10. Determine the quadrant of the following bearings:

a) 060°
b) 150°
c) 240°
d) 330°

Answer:
a) NE
b) SE
c) SW
d) NW

Extra Practice: Directions and Bearings

A ship sails from port A on a bearing of 065° for 12 km, then changes direction at B to a bearing of 155° for 15 km to reach at dock C.

- Draw a figure to illustrate the situation.
- Find distance and bearing of C from A.

Sketch diagram to visualize. Solve the problem after learning sine and cosine law.

Q1	The bearing of a lighthouse, L, from a boat at A is 062°. The boat travels 500 m east and the bearing of the lighthouse now from B is 044°. • Draw a labelled diagram to represent the situation. • Find the angle LAB
Q2	An airplane is flying at an airspeed of 500 km/h, on a bearing of 040°. A 150 km/h wind is blowing from a bearing of 120°. Determine the ground velocity of the plane and direction of the flight.
Q3	Two forest fire towers, A and B, are 20.3 km apart. The bearing from A to B is $N70°E$. The ranger in each tower observes a fire and radios the fire's bearing from the tower. The bearing from tower A is $N25°E$. From tower B, the bearing is $N15°W$. How far is the fire from each tower?

The Sine Law and Cosine Law

In earlier chapters we worked with right triangle. Now these concepts will be extended to solve any triangle. Equations relating the sides and angles of any triangle are popularly called **sine law** and **cosine law.** We will derive these equations and see how they can be applied to solve or model many real-life situations. Sine law is also referred as **law of sines** or **sine rule**.

The Law of Sine

The sine law is an equation that relates the length of the sides of a triangle to the sines of the angles opposite to them. In a triangle ABC the sides of a triangle are proportional to the sines of the angles opposite to them. This ratio is equal to the constant d which is same as the diameter of the circumcircle.

Sine Law

$$\frac{a}{\sin A} = \frac{b}{\sin B} = \frac{c}{\sin C} = d$$

This equation is also written as

$$\frac{\sin A}{a} = \frac{\sin B}{b} = \frac{\sin C}{c}$$

Standard convention of ΔABC

a is opposite angle A

b is opposite angle B

c is opposite angle C

$\angle A + \angle B + \angle C = 180°$

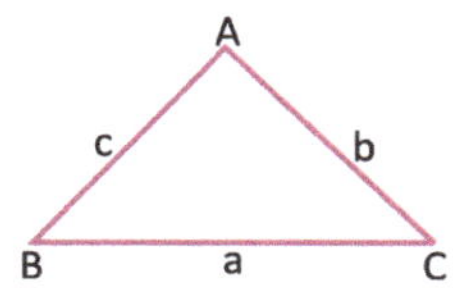

The Cosine Law

The Cosine Law / Cosine Law

The law of cosines relates an angle to the three sides of the triangle

$$a^2 = b^2 + c^2 - 2bc\cos A$$

$$b^2 = a^2 + c^2 - 2ac\cos B$$

$$c^2 = a^2 + c^2 - 2ac\cos C$$

When Can Sine Law be Applied

A triangle has 6 parameters – 3 sides and 3 angles. To solve any triangle, we need information about three of these six parameters. Sine law enable to solve triangle in which one set of angles with opposite side length is given. **Configurations in which sine law can be applied are SAA, SSA and ASA.**

Sine law can solve triangles with given side-angle-angle (SAA), angle-side-angle (ASA) and side-side-angle (SSA) since we have combination of angle and opposite side length	Sine law cannot solve triangles with SSS (three given sides) or SAS (two sides and an include angle since we do not have combination of angle and it opposite side length.
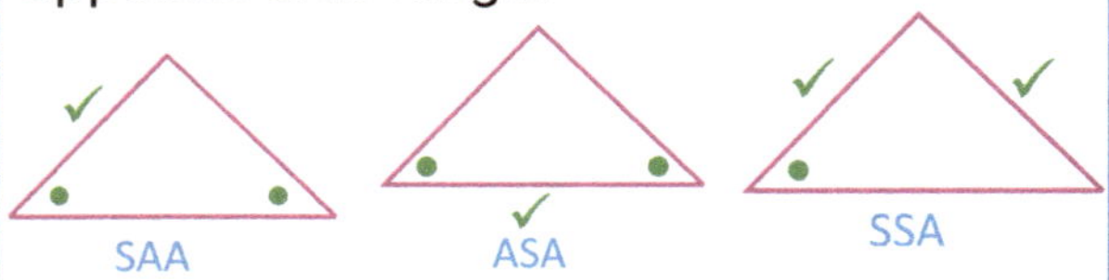	

Quick Proof (Geometric Idea)

Take any triangle ABC. Drop an altitude h from vertex C to side AB.

Using ΔACD: $\sin A = \frac{h}{b} \Rightarrow h = b \sin A$

Using ΔBCD: $\sin B = \frac{h}{a} \Rightarrow h = a \sin B$

$h = a \sin B = b \sin A \Rightarrow \dfrac{a}{\sin A} = \dfrac{b}{\sin B}$

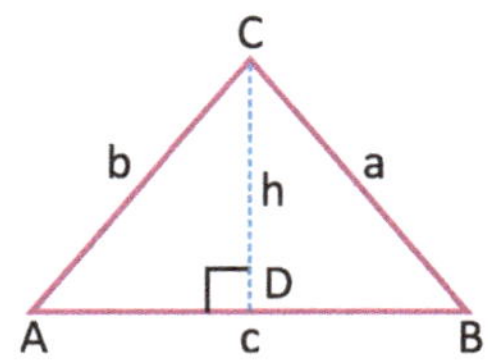

By repeating with altitude from another vertex, we also get $\frac{c}{\sin C}$. Thus

$$\frac{a}{\sin A} = \frac{b}{\sin B} = \frac{c}{\sin C}$$

Relation of Circumcircle Diameter and Sine Law

Consider the triangle ABC inscribed in a circle, O, with radius R as shown in the figure. O is the circumcentre, the intersection point of perpendicular bisectors of the triangle. Join the center O with the vertices B and C. Triangle BOC is an isosceles triangle with OB and OC equal to the radius R.

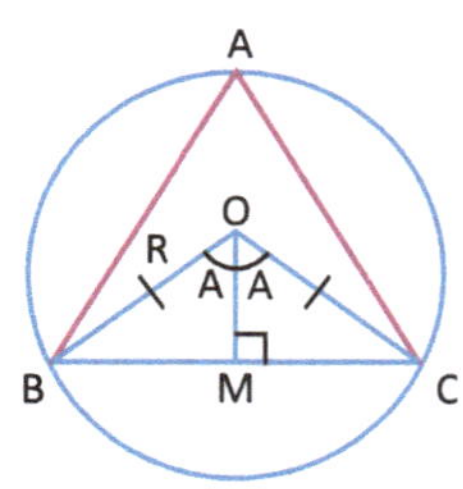

The angle $\angle BOC$ is twice $\angle A$ (circle theorem).

$\angle BOM = \frac{2\angle A}{2} = \angle A$ and $BM = \frac{1}{2}BC = \frac{1}{2}a$ since OM is right bisector

In the triangle BOM: $\sin A = \frac{BM}{R} = \frac{a/2}{R} = \frac{a}{2R} \Rightarrow \frac{\boldsymbol{a}}{\boldsymbol{\sin A}} = \boldsymbol{2R = d}$

$\boldsymbol{\dfrac{a}{\sin A} = 2R = d, where\ d\ is\ the\ diameter\ of\ the\ circumcircle}$

In Sine Law the constant of the ratio of the side length to sine of opposite angle is the diameter of the circumcentre: $\frac{a}{\sin A} = \frac{b}{\sin B} = \frac{c}{\sin C} = d$

Circle Theorem

Angle subtended by an arc at the centre of a circle is twice the angle subtended at any point ion the circumference.

Solve a triangle to find all the missing sides and angles

Example 1: Solve Triangle AAS

Solve the given triangle.

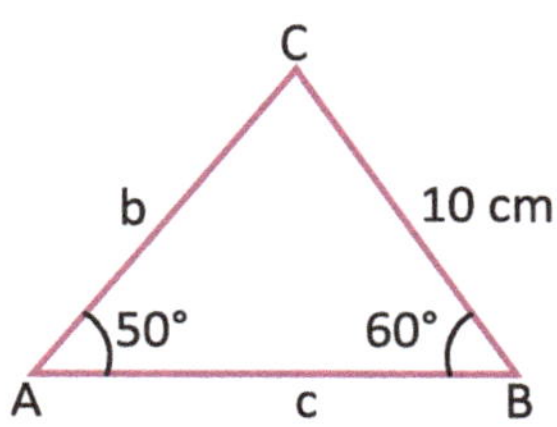

Solution

$\hat{A} = 50°, \hat{B} = 60,$

$a = 10\ cm\ (AAS)$

$$\frac{b}{\sin 60°} = \frac{10}{\sin 50°}$$

$$b = 10\frac{\sin 50°}{\sin 60°} = 8.85$$

Find $\hat{C} = 180° - (50° - 60°)$

$$\hat{C} = 180° - 110° = 70°$$

$$\frac{c}{\sin 70} = \frac{10}{\sin 50°}$$

$$c = 10\frac{\sin 70°}{\sin 50°} = 7.66$$

$b = 8.85\ cm,\ c = 7.66$ and $\hat{C} = 70°$

Example 2: Solve Triangle AAS

Solve the given triangle

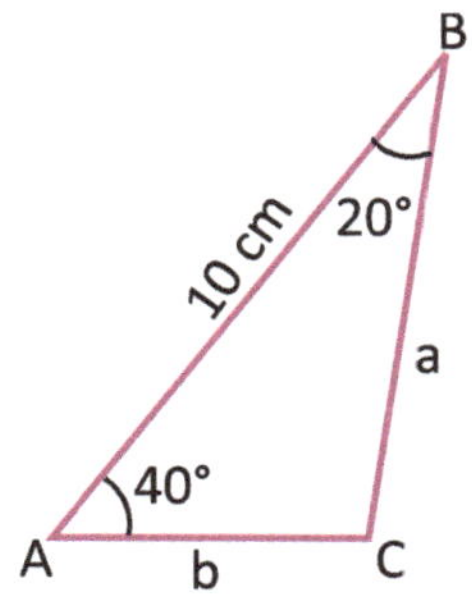

Solution

$\hat{A} = 40°, \hat{B} = 20°,$

$c = 10\ cm\ \ (ASA)$

Find angle C

$$\hat{C} = 180° - (40° + 20°) = 120°$$

Apply sine law

$$\frac{a}{\sin 40°} = \frac{10}{\sin 120°}$$

$$a = \sin 40°\frac{10}{\sin 120°} = 7.42$$

$$\frac{b}{\sin 20°} = \frac{10}{\sin 120°}$$

$$b = \sin 20°\frac{10}{\sin 120°} = 3.95$$

$\hat{C} = 120°, a = 7.42\ cm$ and $b = 3.95\ cm$

Note: $\hat{C}$ is alternate way of writing $\angle C$ or $\angle ACB$

✍ **Practice Problem:** Solve the triangle ABC where $\hat{A} = 50°, \hat{B} = 60,\ \ b = 10\ cm$

Example 3: Two-Dimensional Application

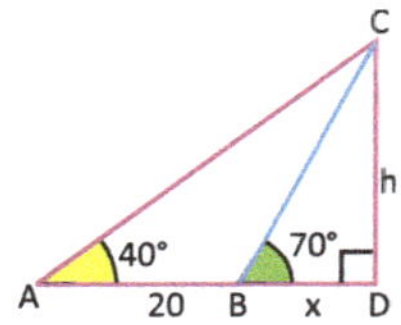

The height of a tree h is observed from two points, A and B, that are 20 m apart as shown in the figure. The angle of elevation at A and B are 40° and 70°, respectively. Calculate the height of the tree nearest to tenth of a metre.

Solution3: Two-Dimensional Application

Find angles in the ΔABC $\angle ABC = 180° - 70° = 110°$ $\angle ACB = 70° - 40° = 30°$ Apply sine law, find BC	$\frac{BC}{\sin 40°} = \frac{20}{\sin 30°}$ $BC = \frac{20}{\sin 30°} \times \sin 40° = 25.7$	$\sin 70° = \frac{h}{BC}$ $h = BC \times \sin 70° = \frac{h}{BC}$ $= 24.16$ Answer: Tree height is 24.2 m

Applications with Sine Law

Example 4: Area of Shaded Region

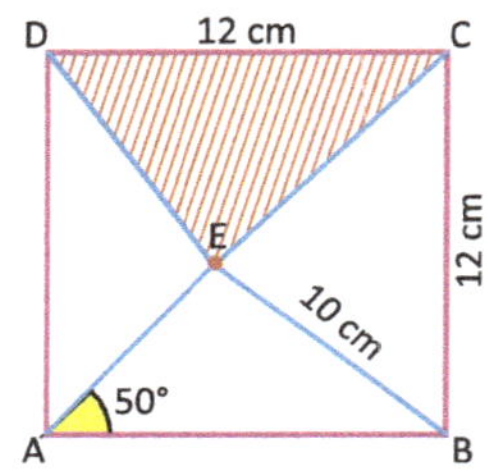

Square ABCD has an area of 144 m^2 with a point E inside, making an obtuse angle AEB. Angle BAE is 50° and the length of the side BE is 10 m. Find the area of the shaded region CDE.

Solution 4: Area of Shaded Region

Square ABCD has an area of 144 m^2, therefore each side is 12 m. We can solve the triangle AEB with sine law to find the angle AEB. SSA configuration

$\frac{\sin \angle AEB}{12} = \frac{\sin 50°}{10}$ $\sin \angle AEB = 12 \times \frac{\sin 50°}{10} = 0.919$ $\angle AEB = \sin^{-1} 0.919 = 66.8°$ The obtuse angle will be $180° - 66.8° = 113.2°$ $\angle ABE = 180° - 50° - 113.2 = 16.8°$	Drop a perpendicular EP on AB $EP = 10 \sin 16.8° = 2.89$ Height of the triangle CDE is $12 - EP = 12 - 2.89 = 9.11$ Area of triangle CDE = $\frac{1}{2} \times 12 \times 9.11 =$ 54.66 m^2 Area of shaded region is 54.7 m^2

Example 5

A line from the point P (25,0) intersects the circle $x^2 + y^2 = 169$ at points A and B. Determine the length of the chord AB. If the line makes an angle of 160° with the positive x-axis.

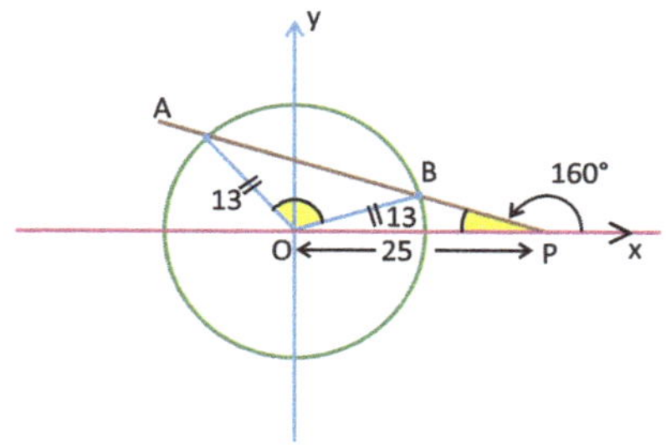

Solution 5:

$x^2 + y^2 = 169 = 13^2, r = \sqrt{13^2} = 13;$

$\angle OPB = 180° - 160° = 20°, OP = 25$

Sine Law can be applied in ΔOAP, SSA configuration.

$$\frac{\sin A}{25} = \frac{\sin 20°}{13}$$

$$\sin A = \frac{25}{13}\sin 20° = 0.6577$$

$$\angle A = \sin^{-1}(0.6577) = 41.1°$$

ΔOAP is an isosceles triangle

$$\angle AOB = 180° - 2(41.1°) = 97.8°$$

To find chord length AB

Apply Sine Law in ΔOAB

$$\frac{AB}{\sin 97.8°} = \frac{13}{\sin 41.1°}$$

$$AB = 13 \times \frac{\sin 97.8}{\sin 41.1} = 19.6$$

Chord length AB is 19.6 units

Challenge Question

Example 9: Rescue at Sea

Coast Guard station A is located 400 m due west of station B. A ship at sea sends SOS call that is received. The call to station A indicates bearing of the ship as N35° E. The call to station B indicates that the bearing or the ship from station B is N56°W. Figure illustrates the situation

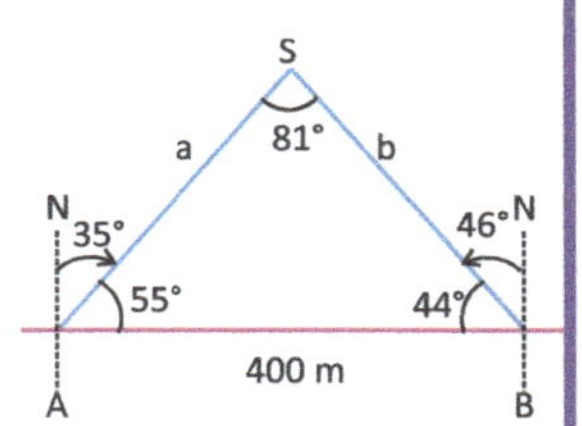

a. How far is the ship from each station?

b. How much time will helicopter, flying at 340 km/h, takes to reach the ship from station A.

Solution provided at the end of the section

Practice Problem: A ship sails on a bearing of $070°$ for $10\ km$, then changes direction to a bearing of $150°$ for $15\ km$. Find the distance and bearing of final position from initial position. 19.4 km

Ambiguous Case for the Sine Law

The **Ambiguous Case** happens when we use the **Sine Law** in a triangle with the information: SSA (two sides and a non-included angle)

1. Why Ambiguity Happens

The sine function has the property:

$\sin\theta = \sin(180° - \theta)$

That means, if you solve for an angle using the sine law, there could be **two possible triangles**:

- One with the angle acute, θ.
- One with the angle obtuse, $180° - \theta$.

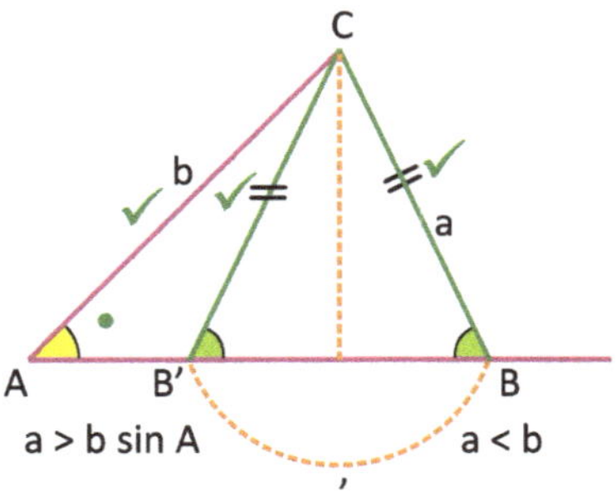

2. Possible Outcomes in SSA Case

When applying the sine law with SSA:

- **No solution**: if the side opposite the given angle is too short to reach the base.
- **One unique solution**:
  - If the side opposite the given angle is exactly the height (right triangle).
  - Or if it's long enough to make only one triangle possible.
- **Two solutions**: one acute and one obtuse angle possible.

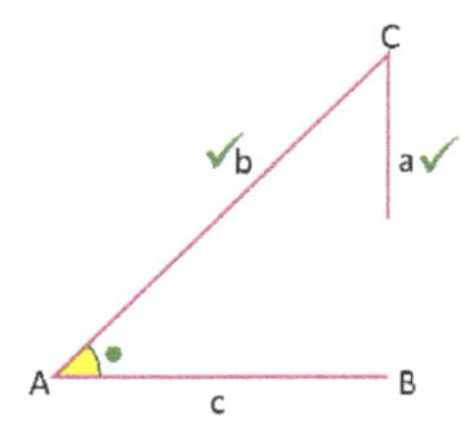

Example 6: SSA – Two Possible Triangles

Solve $\Delta ABC\ where\ \ b = 10\ cm, a = 8\ cm$ and $\angle A = 40°$

Short Solution: Students should provide solution, in details, for both triangles

$$\frac{a}{\sin A} = \frac{b}{\sin B} \Rightarrow \sin B = \frac{b\sin A}{a} = \frac{10\sin 40°}{8} = 0.8035$$

$\angle B = \sin^{-1} 0.8035 = 53.5°$ or $\angle B = 180° - 53.5° = 126.5°$

Hence two possible solutions are:

Case 1: $\angle A = 40°. \angle B = 53.5°\ and\ \angle C = 180° - 40° - 53.5° = 86.5°$

Case 2: $\angle A = 40°. \angle B = 126.5° and\ \angle C = 180° - 40° - 126.5° = 13.5°$

NOTE: Students should find the side lengths to complete the solution.

✍ **Practice Problem:** Solve the $\Delta ABC\ where\ b = 7.0\ cm,\ \ a = 6.2\ cm$ and $\angle A = 50°$

✍ Practice Problem: Connecting Solution to the Problem

Provide reasons and calculation to show that the given figure is the solution for the problem to solve the ΔABC *where* $b = 9.2\ cm,$ $a = 6.5\ cm$ and $\angle A = 41°$

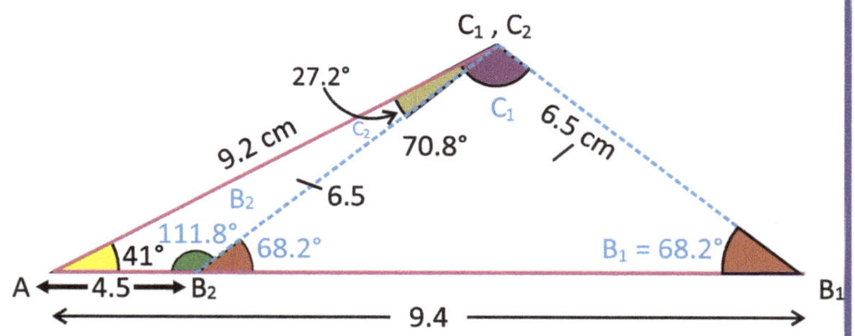

Example 7: SSA – No Solution

Solve ΔABC *where* $b = 10\ cm, a = 3\ cm$ and $\angle A = 40°$.

Short Solution: Students should provide solution, in details, for both triangles

$$\frac{a}{\sin A} = \frac{b}{\sin B} \Rightarrow \sin B = \frac{b \sin A}{a} = \frac{10 \sin 40°}{3} = 2.14$$

$\sin B > 1$ That is impossible

No solution a is smaller than the altitude

Note: If $a < b \sin A$ *there is no solution*

Example 8: SSA – One Solution (Obtuse Angle Triangle)

Find all possible values of $\angle B$ in ΔABC *where* $b = 10\ cm, a = 12\ cm$ and $\angle A = 40°$.

Short Solution: Students should provide solution, in details, for both triangles

$$\frac{a}{\sin A} = \frac{b}{\sin B} \Rightarrow \sin B = \frac{b \sin A}{a} = \frac{10 \sin 40°}{12} = 0.5356$$

$\angle B = \sin^{-1} 0.5356 = 32.4°$

Now the alternate: $\angle B = 180° - 32.4° = 147.6°$ is not possible since

$147.6° + 40° = 187.6° > 180°$

Note: If $a > b$ *only one solution*

Example 9: **Solve ΔABC where $b = 30\ cm, a = 25\ cm$ and $\angle A = 50°$. (SSA)**

Strategy to Solve Ambiguous Case in Details

Given ΔABC where $b = 30\ cm, a = 25\ cm$ and $\angle A = 50°$.

$\angle A$: acute angle, $a < b$. So, first find the height

$b \sin A = 30 \sin 50° = 23.0$

$23 < a < 30 \Rightarrow$ two triangles as shown in the figure

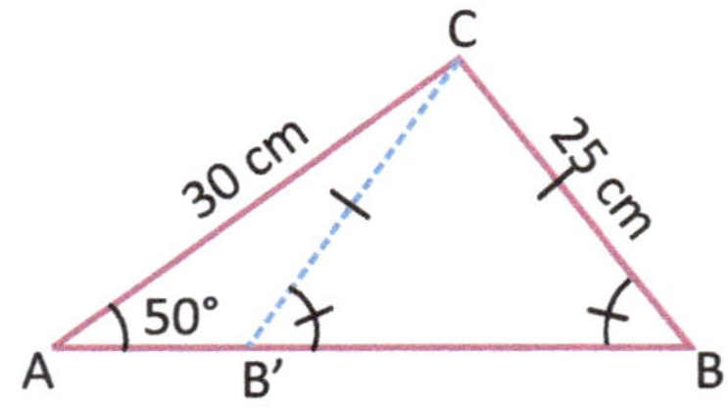

Solve ΔABC

$$\frac{\sin B}{30} = \frac{\sin 50°}{25}$$

$$\sin B = \frac{30}{25} \sin 50° = 0.91925$$

$$\angle B = \sin^{-1} 0.91925 = 67°$$

$$\angle C = 180° - 50 - 67 = 63°$$

Now find the side length AB

$$\frac{AB}{\sin 63°} = \frac{25}{\sin 50°}$$

$$AB = \frac{25 \sin 63°}{\sin 50°}$$

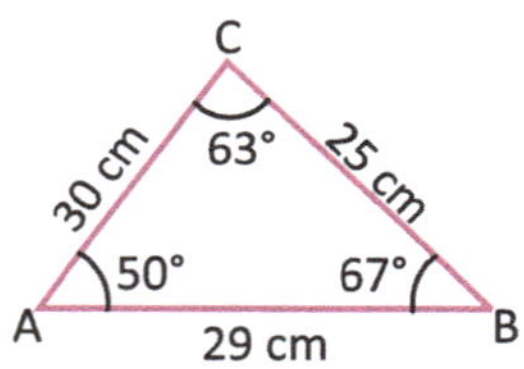

$AB = 29$ cm

Now solve second triangle ACB'

$CB = CB'$ $\therefore \angle C\mathbf{B}B = \angle CBB' = 67°$

$\angle CBA = 180° - \angle CBA = 180 - 67 = 113°$

$\angle ACB' = 180° - 50° - 113° = 17°$

$$\frac{AB'}{\sin 17°} = \frac{25}{\sin 50}$$

$$AB' = \frac{25 \sin 17°}{\sin 50°} = 9.5 \text{ cm}$$

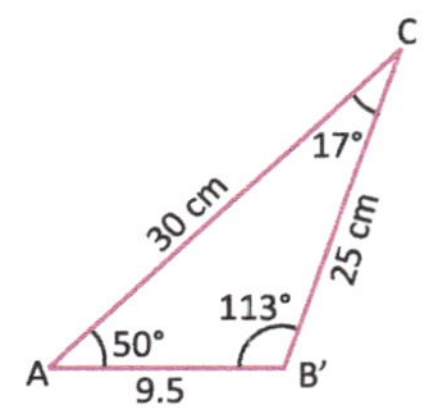

Two Solutions are:

$\Delta ABC: \angle B = 67°, \angle C = 63°, AB = 29$

$\Delta AB'C: \angle B' = 67°, \angle ACB' = 113°, AB = 9.5$

Example 10: **Solve ΔABC where $b = 30\ cm, a = 25\ cm$ and $\angle A = 100°$. (SSA)**

Strategy to Solve

Given ΔABC where $b = 30\ cm, a = 25\ cm$ and $\angle A = 100°$.

$\angle A$: obtuse angle. All other angles in the triangle will be smaller and so the side opposite to the given angle will be longest.

Since we are given $a = 25$, which is smaller than $b = 30$ triangle will not be formed.

Hence there is no solution for the given situation.

Example 11: **Solve ΔABC where $b = 30\ cm, a = 50\ cm$ and $\angle A = 110°$. (SSA)**

Strategy to Solve

Given ΔABC where $b = 30\ cm, a = 50\ cm$ and $\angle A = 110°$.
Diagram represents the situation.
$\angle A$: obtuse angle and $a > b$. This gives one solution

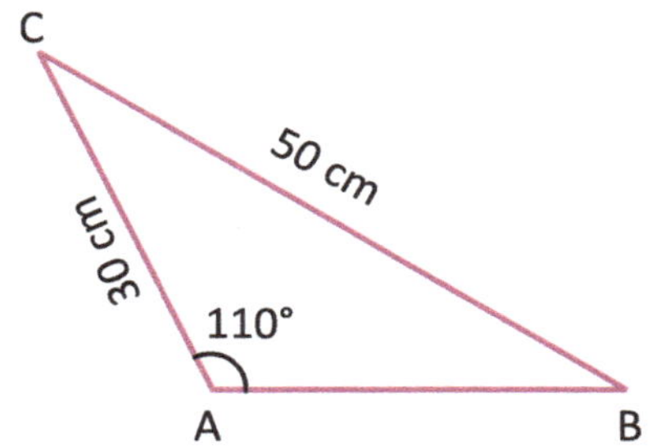

$$\frac{a}{\sin A} = \frac{b}{\sin B} \Rightarrow \sin B = \frac{b \sin A}{a} = \frac{30 \sin 110°}{50} = 0.5638$$

$\angle B = \sin^{-1} 0.5638 = 34.32° \approx 34.3°$

$\angle C = 180° - 110° - 34.3° = 35.7°$

$$\frac{a}{\sin A} = \frac{c}{\sin C} \Rightarrow c = a\frac{\sin C}{\sin A} = 30\frac{\sin 35.7°}{\sin 34.3°} = 31$$

Answer: $\angle B = 34°, \angle C = 36°, AB = 31\ cm$

Practice:

Q1	Given ΔABC where $b = 30\ cm, a = 27\ cm$ and $\angle A = 64.2°$. $AB = 13\ cm,$ $\quad A\hat{C}B = 25.8°, A\hat{B}C = 90°$
Q2	Given ΔABC where $b = 26\ cm, a = 55\ cm$ and $\angle A = 143°$. $\hat{B} = 16.5°, \hat{C} = 20.5°, AB = 32\ cm.$
Q3	Given ΔABC where $b = 40\ cm, a = 35\ cm$ and $\angle A = 110°$. No solution
Q4	Given ΔABC where $b = 7.8\ cm, a = 5.9\ cm$ and $\angle A = 36°$. $Case\ I: AB = 10.3\ cm, A\hat{C}B = 93°, A\hat{B}C = 51°$ $Case\ II: AB = 2.6\ cm, A\hat{C}B = 15°, A\hat{B}C = 129°$

Communication and Thinking

Given ΔABC: $b = 10\ cm$, and $\angle A = 30°$.
How many triangles can be formed with given length of a (BC side opposite to $\angle A$)

a.	$5\ cm$	b.	$6\ cm$	c.	$4\ cm$

Hint: Relate the side with altitude

Key Learnings

Ambiguous Case (SSA Condition with $\angle A$)

When two sides and an **acute angle** are given (**SSA**), we compare the side **a** (opposite ∠A) with the "**height**" $h = b \sin A$

- If $a < b \sin A$ no triangle possible as a is too short
- If $a = b \sin A$ exactly one right triangle is formed as altitude is a
- If $b \sin A < a < b$: two triangles are possible (one acute and other obtuse)
- If $a > b$ Exactly one triangle is formed ($\angle B$ is obtuse)

When two sides and an **obtuse angle** are given (**SSA**), we compare the side **a** (opposite ∠A) with the other given side ***b***

- If $a < b$ no triangle possible as $should$ be the longest side length
- If $a > b$ one triangle is formed

Flow diagram below represents the situation for acute and obtuse angles

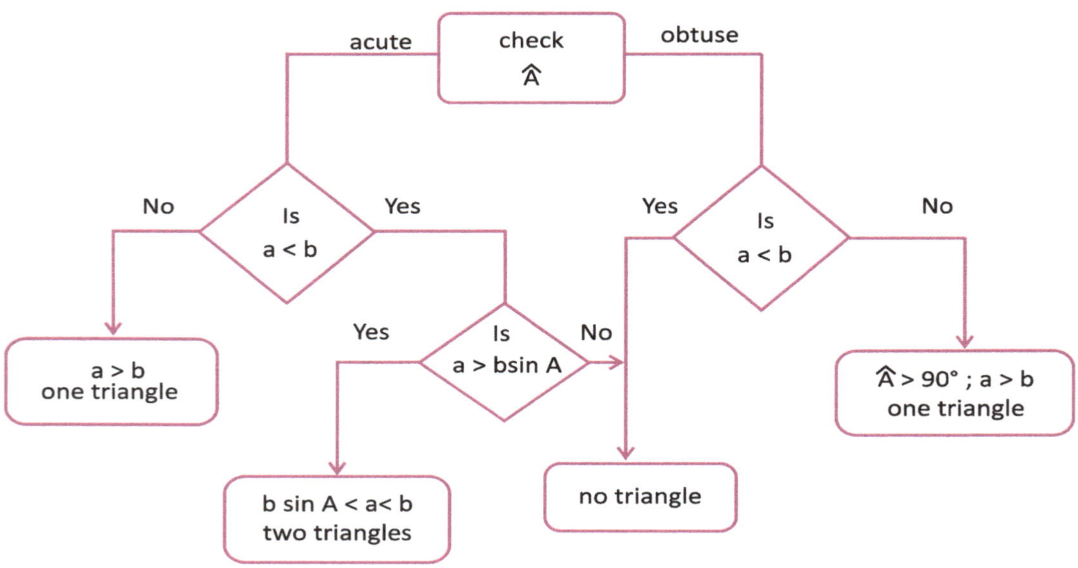

Note: If $a = b \sin A$ exactly one right triangle is formed as altitude is a

Ambigious Case Class SCAN ME

Quiz: Sine Law

Q1.	In a triangle there are six parameters – three sides and three angles. At least how many of these parameters should be known to solve a triangle? A. 1 B. 2 C. 3 D. 4	Q2.	Measure of a side length and two angles (SAA) is known for a triangle ABC. Can you solve this triangle using sine law? A. Sometimes B. Never C. Always
Q3.	Can sine law be applied to solve a right triangle? A. Yes B. No	Q4.	How many triangles can be formed with $a = 10, b = 20, \angle A = 29°$? A. 0 B. 1 C. 2
Q5.	In a triangle ABC, we are given the configuration SSA where length of a side is twice the other, $c = 2a$. Which of the following angles will result in exactly one triangle A. $\angle A = 30°$ B. $\angle B = 30°$ C. $\angle C = 30°$ D. $none\ of\ the\ above$	Q6.	In a triangle ABC, $\hat{A} = 60°, b = 12, a = x$. Find the domain of x so that two tringles can be formed with the given values. A. $a > 3\sqrt{3}$ B. $a > 12$ C. $6\sqrt{3} < a < 12$ D. $3\sqrt{3} < a < 12\sqrt{3}$
Q7.	In a triangle ABC, AD divides BC In the ratio of 3:4 If $\angle B = 30°$ and $\angle C = 45°$, then find $\frac{\sin \angle CAD}{\sin \angle BAD}$ 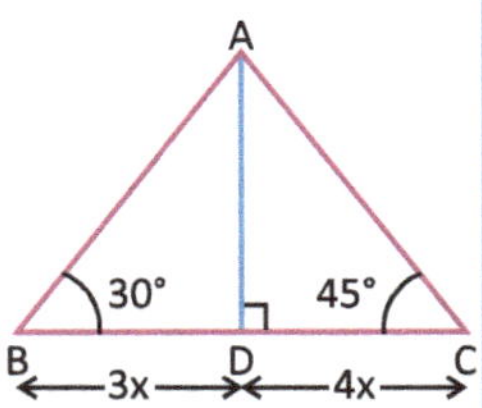 A. $\frac{2\sqrt{2}}{3}$ B. $\frac{4}{3}$ C. $\frac{4\sqrt{2}}{3}$ D. $\frac{3\sqrt{2}}{4}$	Q8.	A loading ramp 5 m long makes an angle of 15° with the horizontal. If it is replaced by another ramp making an angle of 12° with the horizontal, how long 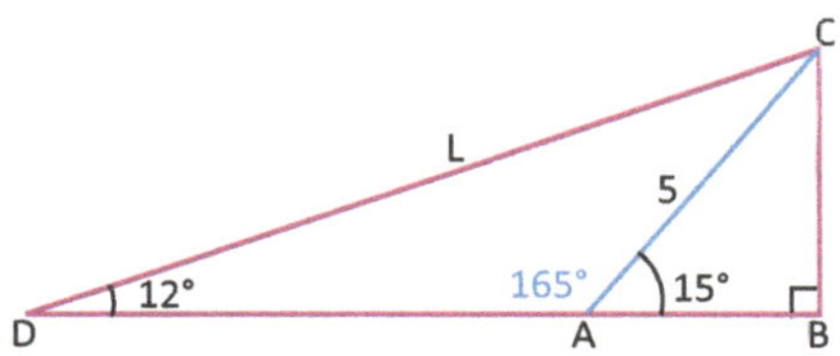 will it be.

Q9. A surveyor in a hot-air-balloon observed that the angle of depression to two points A amd B on the opposite side of a lake are 60° and 40°, respectively.

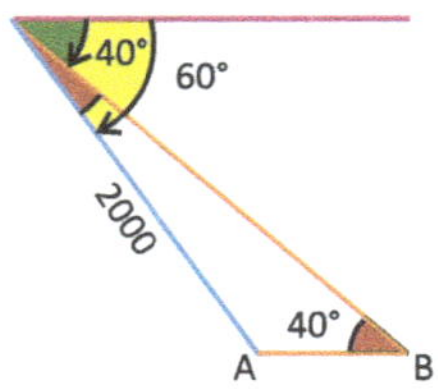

What is the width of the lake, to the nearest meter, if the distance of the surveyour from point A is 2 km.

Q10. The light from a rotating beacon can illuminate effectively up to a distance of 500 m. From a point on the shoreline that is 1 km from the beacon, the sight line to the beacon makes an angle of 15° with the shoreline. Find the length of shoreline that is effectively illuminated by the beacon?

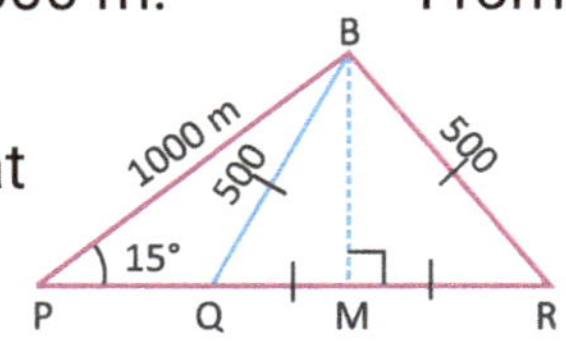

Challenge Question

Example 9: Rescue at Sea

Coast Guard station A is located 400 m due west of station B. A ship at sea sends SOS call that is received. The call to station A indicates bearing of the ship as N35° E. The call to station B indicates that the bearing or the ship from station B is N56°W. Figure illustrates the situation

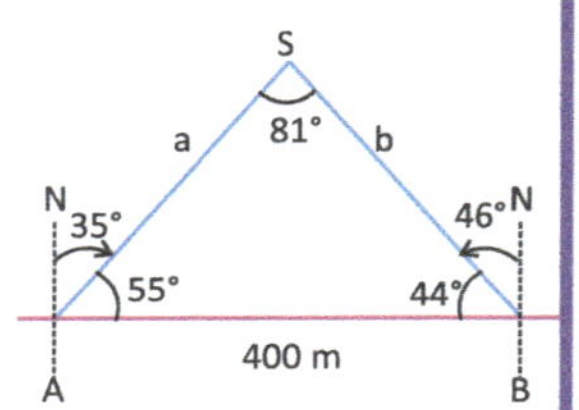

c. How for is the ship from each station?

d. How much time will helicopter, flying at 340 km/h, takes to reach the ship from station A.

Solution 9: Rescue at Sea

Consider the horizontal plane

$\angle ASB = 180° - 55° - 44° = 81°$

$$\frac{a}{\sin 44°} = \frac{400}{\sin 81°} = \frac{b}{\sin 55°}$$

$$a = 400\frac{\sin 44°}{\sin 81°} \quad and \quad b = 400\frac{\sin 55°}{\sin 81°}$$

$$a = 281.32 \qquad b = 331.75$$

Distance of the ship from station A is 281 km and that from station B is 332 km.

Helicopter time from station A.

Speed of helicopter = 340 km/h

Distance = 281 km

$$Time = \frac{distance}{speed} = \frac{281}{340}h = 0.82h = 50\ minutes$$

Helicopter will take 50 minutes to reach ship A.

Answers: Quiz 1

Q1	C	Q2	C	Q3	A	Q4	C	Q5	A
Q6	C	Q7	C	Q8	6.2 m	Q9	1064 m	Q10	8.56 km

Part Solutions

Q6. C

For two triangles $b \sin\theta < a < b$

$12 \sin 60° < a < 12$

$$12\frac{\sqrt{3}}{2} < a < 12 \Rightarrow 6\sqrt{3} < a < 12$$

Q7. C

$$\frac{\sin C\hat{A}D}{DC} = \frac{\sin 45°}{AD} = \frac{1}{\sqrt{2}AD}$$

$$\sin C\hat{A}D = \frac{4x}{\sqrt{2}AD}$$

$$\frac{\sin B\hat{A}D}{PD} = \frac{\sin 30°}{AD} = \frac{1}{2AD}$$

$$\sin BAD = \frac{3x}{2AD}$$

$$\frac{\sin \angle CAD}{\sin \angle BAD} = \frac{\frac{4x}{\sqrt{2}AD}}{\frac{3x}{2AD}} = \frac{4x}{\sqrt{2}AD} \times \frac{2AD}{3x} = \frac{4\sqrt{2}}{3}$$

Q8.

$$\frac{DC}{\sin 165°} = \frac{5}{\sin 12°}$$

$$DC = 5 \times \frac{\sin 165°}{\sin 12°} = 6.22 = 6.2\ m$$

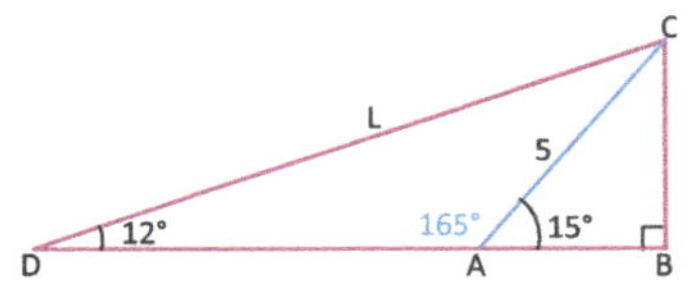

Q9.

$$\frac{AB}{\sin 20°} = \frac{2000}{\sin 40°}$$

$$AB = \frac{2000}{\sin 40°} \times \sin 20° = 1064 \text{ m}$$

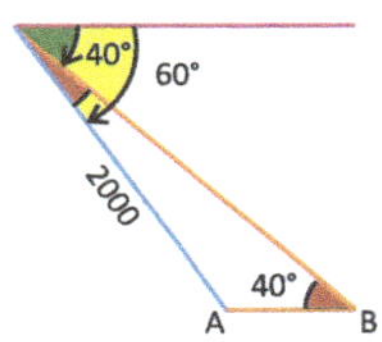

Q10.

Figure below illustrates the situation with beacon at B.

$PB = 1000, \angle P = 15°, BQ = BR = 500$

Drop perpendicular BM. This is right bisector of QR since ΔQBR is an isosceles triangle.

In ΔPBM

$$\sin P = \frac{BM}{PB}$$

$BM = PB \sin P$

$= 1000 \sin 15° = 258.2$

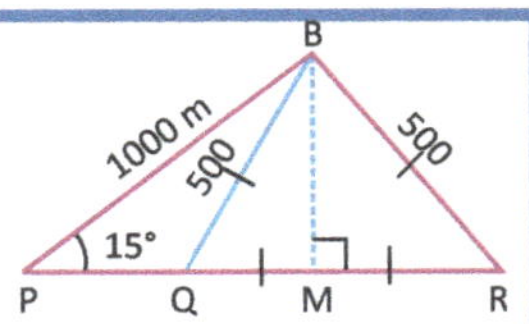

In $\Delta\ BMQ$, Pythagrean theorem

$QM^2 = BQ^2 - BM^2$

$QM = \sqrt{500^2 - 2588.82^2} = 427.8$

$QR = 2QM$

$QR = 2\ (427.8) = 855.6\ km = 8.56\ km$

The Cosine Law

The **cosine law** relates cosine of an angle to the three sides. It is a general form of the Pythagorean Theorem for any triangle where squares of sides is related with cosine of an angles.

Proof of Cosine Law

Consider a triangle ABC with altitude CD. The altitude divides the triangle in two right triangles as shown in the figure. Applying Pythagorean theorem in these triangles we can derive the cosine law.

Here $AC = b, BC = a, AB = c, CD = h, AD = x, AB = c - x$

Pythagorean Theorem in $\Delta\ ADC$

$b^2 = x^2 + h^2$ (1)

$\cos A = \dfrac{x}{b}$

$x = b\cos A$ (2)

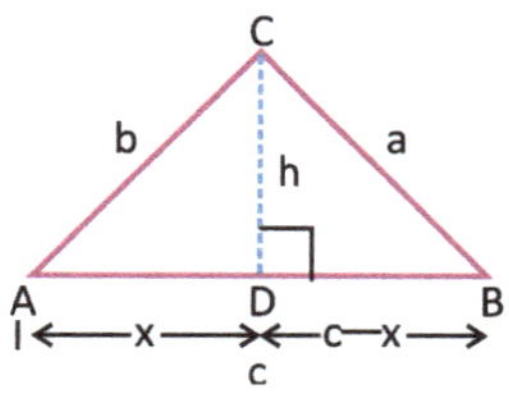

Pythagorean Theorem in ΔBDC

$a^2 = h^2 + (c - x)^2$

$a^2 = h^2 + c^2 - 2cx + x^2$

$a^2 = x^2 + h^2 + c^2 - 2cx$ from (1)

$a^2 = b^2 + c^2 - 2cx$

$a^2 = b^2 + c^2 - 2cb\cos A$ from (2)

Similar steps can be followed by considering altitude from other vertices to relate the squares of the sides with cosine law.

$a^2 = b^2 + c^2 - 2bc\ cos\ A$

$b^2 = c^2 + a^2 - 2ca\cos B$

$c^2 = a^2 + b^2 - 2ab\cos C$

Rearrange Cosine Law to find Angle

$a^2 = b^2 + c^2 - 2bc\cos A$

$2bc\cos A = b^2 + c^2 - a^2$

$\cos A = \dfrac{b^2 + c^2 - a^2}{2bc}$

$\cos B = \dfrac{a^2 + c^2 - b^2}{2ac}$

$\cos C = \dfrac{a^2 + b^2 - c^2}{2ab}$

Note:

In the SAS and SSS case cosine law is applied to solve the triangle

Example 12: Find a side length using cosine law in SAS Case

Solve ΔABC : $\hat{A} = 58°, b = 5.2,\ C = 4.1$

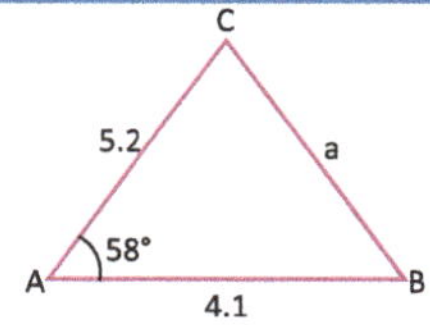

Solution 12: Find a side length using cosine law in SAS Case

Find side length using cosine law

$a^2 = b^2 + c^2 - 2bc\cos A$

$a^2 = 5.2^2 + 4.1^2$

$-2(5.2)(4.1)\cos 58°$

$a^2 = 21.25$

$a = \sqrt{21.25} = 4.6$

Now apply sine law to find angle

$\frac{\sin B}{b} = \frac{\sin A}{a}$

$\frac{\sin B}{5.2} = \frac{\sin 58°}{4.6}$

$\sin B = 0.9586$

$\hat{B}$

$= \sin^{-1}(0.9586)$

$\hat{B} = 73.46 = 73°$

Find the third angle

$\hat{C} = 180° - \hat{A} - \hat{B}$

$\hat{C} = 180° - 58° - 73°$

$\hat{C} = 49°$

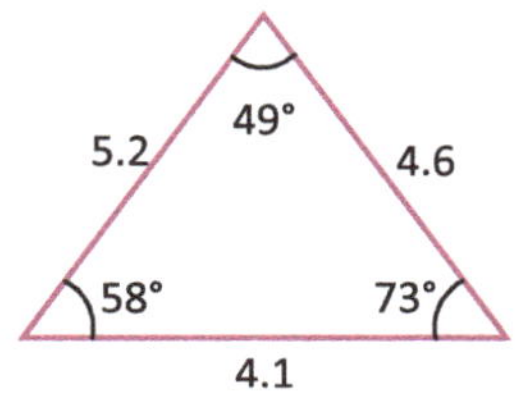

Explore Projection Formula SCAN ME

Application of Cosine Law

Example 13: Area of Triangle

A triangular yard, 7m by 8 m by 9 m is to be covered with sod. Find the cost if sod costs $1.75/m^2.

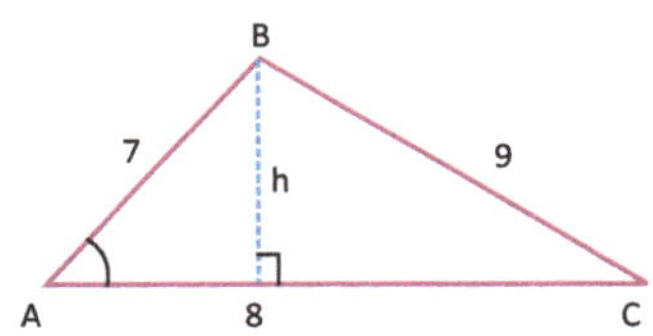

Solution 13: Area of Triangle

Three sides of a triangle are provided, SSS,

This configuration can be solved with cosine law.

Cosine Law to find $\hat{A}$

$\cos A = \frac{b^2 + c^2 - a^2}{2bc}$

$\cos A = \frac{7^2 + 8^2 + 9^2}{2(7)(8)} = \frac{2}{7}$

$\hat{A} = \cos^{-1}\left(\frac{2}{7}\right) = 73.4°$

Find height with trigonometric ratios

$h = 7\sin A = 7\, sin\, 73.4° = 6.7$

$Area = \frac{1}{2}(AC) \times h = \frac{1}{2}(8)(6.7) = 26.8\, m^2$

Cost of sod at rate $1.75/m^2

Cost $ $1.75\,(26.8) = \$47$

Alternate Method: Use Heron's formula to find area of ΔABC

$$A = \sqrt{s(s-a)(s-b)(s-c)}\,, s = \frac{1}{2}(a+b+c)$$

$$s = \frac{1}{2}\,(7+8+9) = 12$$

$$A = \sqrt{12(12-17)(12-8)(12-9)} = \sqrt{12(s)(4)(3)} = 12\sqrt{5}$$

$$Cost = 12\sqrt{5}(1.75) = \$\,21\sqrt{5} = 46.95 = \$47$$

Example 14: Minimum Perimeter

Dimensions of an obtuse triangle are $a = 5 - x, b = 4 + x \; and \; \angle C = 120°$.

- Find an expression for the length of the side c.
- Find the perimeter of the triangle for the minimum possible value of c correct to 3 significant figures.

Solution 14: Minimum Perimeter

Dimensions of an obtuse triangle are $a = 5 - x, b = 4 + x \; and \; \angle C = 120°$.

SAS configuration is given where the third side can be calculates by cosine law.

$$c^2 = a^2 + b^2 - 2ab\cos C = (5-x)^2 + (4+x)^2 - 2(5-x)(4+x)\cos 120°$$

$$c^2 = 25 - 10x + x^2 + 16 + 8x + x^2 + 20 + x - x^2$$

$\mathbf{c^2 = x^2 - x + 61}$ **expression for the third side c**

For minimum value of c, its square should also be minimum. Complete the square to find minimum.

$$c^2 = x^2 - x + 61$$

$$c^2 = x^2 - x + \left(\frac{1}{2}\right)^2 - \left(\frac{1}{2}\right)^2 + 61$$

$c^2 = \left(x - \frac{1}{2}\right)^2 - \frac{1}{4} + 61$, The minimum value if for $x = \frac{1}{2} = 0.5$

$$c = \sqrt{x^2 - x + 61} = \sqrt{0.5^2 - 0.5 + 61} = 7.79$$

$$a = 5 - x = 5 - 0.5 = 4.5; b = 4 + x = 4 + 0.5 = 4.5$$

Perimeter will be, $P = 7.79 + 4.5 + 4.5 = 16.79 = 16.8 \; units$

Example 15: Find Exact value of Sine Ratio

Three sides of triangle ABC are $AB = 6, BC = 7\ and\ AC = 11.$

Find exact value of $\sin A$.

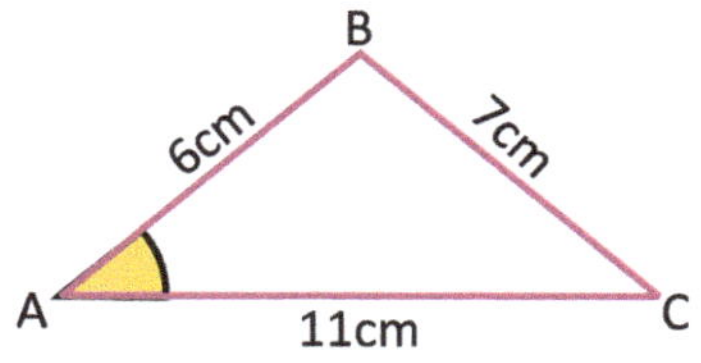

Solution 15: Find Exact value of Sine Ratio

Three sides of a triangle are provided, SSS. This configuration can be solved with cosine law.

Cosine Law to find $\cos A$

$$\cos A = \frac{b^2 + c^2 - a^2}{2bc}$$

$$\cos A = \frac{11^2 + 6^2 - 7^2}{2(11)(6)} = \frac{108}{2(11)(6)} = \frac{9}{11}$$

Consider a right triangle with

$$\cos A = \frac{9}{11}$$

Opposite side $= \sqrt{11^2 - 9^2} = 2\sqrt{10}$

$$\sin A = \frac{2\sqrt{10}}{11}$$

Example 16: Find Exact Value of Sine Ratio

In triangle ABC $AC = 6, BC = 5\ and\ \angle A = 45°.$

Find exact value of $\cos B$ where angle B is an obtuse angle.

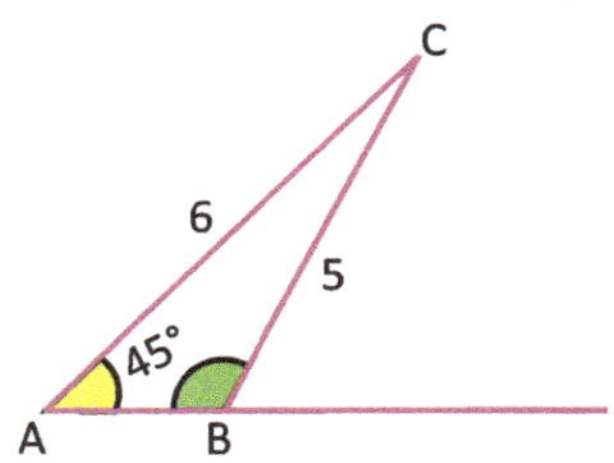

Solution 16: Find Exact vale of Sine Ratio

We are given the configuration SSA where sine law can be applied

$$\frac{\sin B}{6} = \frac{\sin 45°}{5}$$

$$\sin B = 6 \times \frac{\sin 45°}{5} = \frac{6}{5} \times \frac{\sqrt{2}}{2} = \frac{3\sqrt{2}}{4}$$

Sine is positive in quadrant I and quadrant II.

Since the angle, $\angle B$, is obtuse, cosine value will be negative.

Consider a right triangle for

$$\sin B = \frac{3\sqrt{2}}{4}$$

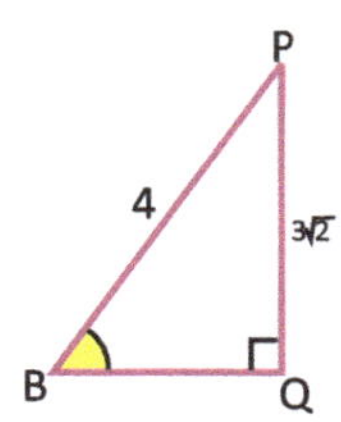

Adjacent side will be

$$\sqrt{4^2 - (3\sqrt{2})^2} = \sqrt{7}$$

$$\mathbf{\cos B = -\frac{\sqrt{7}}{4}}$$

Case I: Distance of an Inaccessible Object

We will derive a general formula for the solution of situations where observations are done to find height of inaccessible objects. Two points P & Q, d units apart in a straight line with the projection base of the object are selected as shown in the figure. Angle of elevation observed as α and β from these points can help to find the height, h, of the inaccessible object.

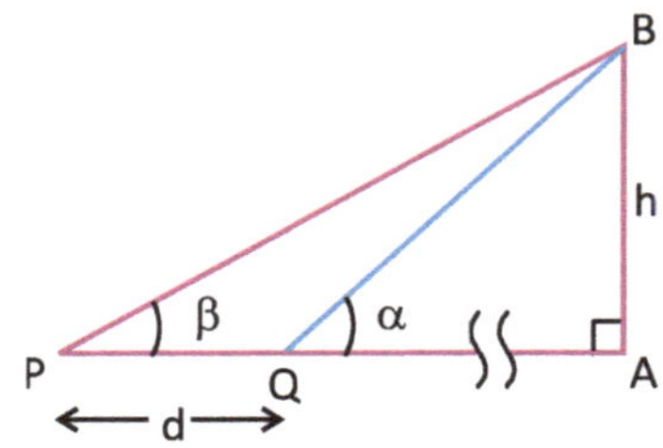

exterior angle is sum of remote interior angles $\angle PBQ = \alpha - \beta$ Sine ratio in ΔPBQ $\frac{QB}{\sin\beta} = \frac{d}{\sin(\alpha-\beta)}$ $QB = \frac{d\sin\beta}{\sin(\alpha-\beta)}$ (1)	Sine ratio in ΔQAB $\sin\alpha = \frac{h}{QB}$ $h = QB\sin\alpha$ **Substitute *QB* from (1)** $h = \frac{d\sin\beta\sin\alpha}{\sin(\alpha-\beta)}$

Example 17: Observation from Points in the same Line as projection

To find the altitude, h, of a promotional blimp angle of elevation was measured from two points, due west, 900 m apart. Find the height of blimp, to the nearest meter if the angle of elevation from the two points is 32° and 47° respectively.

Solution 15: Points in the same Line

Formula Application

$$h = \frac{d\sin\beta\sin\alpha}{\sin(\alpha-\beta)}$$

In our example,

$\alpha = 47°, \beta = 32°, d = 900$

$$h = \frac{900\sin 47°\sin 32°}{\sin(47-32°)} = 1347.6 = 1348\ m$$

Height of the blimp is 1348 m

Case II: Observation Points Not Aligned with Projection of Object

Three-dimensional situation solved with two triangles – one in the horizontal plane and the other in the vertical plane

Example 18: Bearing Angle

A promotional blimp was observed from two stations A and B located 0.4 km apart on Highway North. The bearing of the stationary blimp from A is 037.4° and its angle of elevation is 48.5°, while from B its bearing is 032.6°. Calculate the height of the balloon.

Solution 18: Bearing Angle

Promotional blimp P with projection Q on the ground, is shown along with the given information.

$d = AB = 400\ m,$

$\angle AQB = 37.4° - 32.6° = 4.8°$

Sine law in ΔABQ in horizontal plane

$$\frac{AQ}{\sin 32.6°} = \frac{400}{sin\ 4.8°}$$

$$AQ = \frac{400 \sin 32.6°}{\sin 4.8°}$$

In right ΔAPQ

$$\tan 48.5° = \frac{PQ}{AQ}$$

$$PQ = AQ \tan 48.5°$$

$$PQ = 400\frac{\sin 32.6°\ \tan 48.5}{\sin 4.8°}$$

$$PQ = 2911.m$$

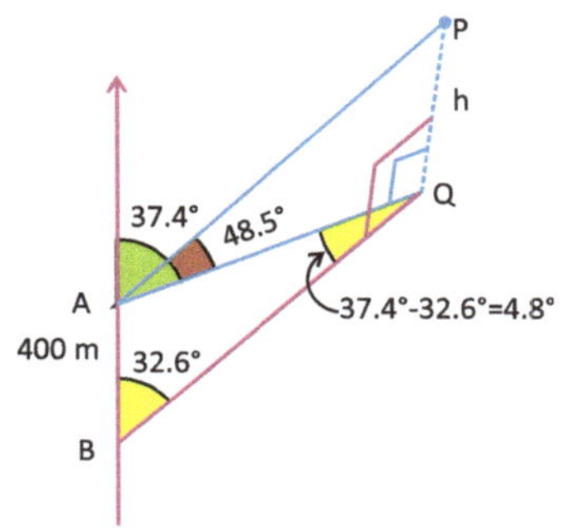

Example 19: 3D Application

Find the height, h of the balloon from the given illustration.

C is the projection of the balloon on the horizontal plane where A and B are observation points.

Solution 19: 3D Application

Consider horizontal and vertical plane separately to solve these triangles

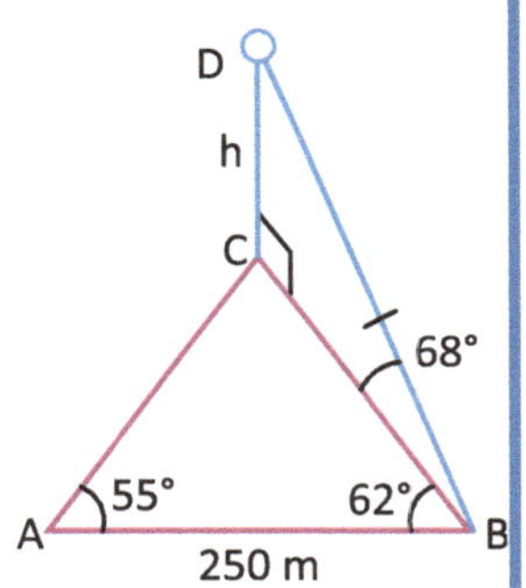

The Sine Law ΔABC in the horizontal plane

$$\angle ACB = 180° - (55° + 62°) = 63°$$

$$\frac{250}{\sin 63°} = \frac{CB}{\sin 55°}$$

$$CB = 250\frac{\sin 55°}{\sin 63°}$$

ΔBCD in vertical plane

$$\tan CBD = \frac{h}{CB} \Rightarrow h = CB \tan CBD$$

$$h = CB \tan 68° = 250\frac{\sin 55°}{\sin 63°} \times \tan 68° = 230$$

height of the balloon is 230 m.

Three-dimensional problems will normally involve three problems to be solve. One of these is in the horizontal plane and the other two in vertical plane, forming right triangle. Sine or combination of cosine and sine law will be applied to solve the horizontal plane. Separate figure may at times help.

Example 20: Find Angle of Elevation

Person X and Y are on the ground 25 m apart. The base of a tower stands such that the angle formed from the base of tower to person X to person Y is 50° and the angle formed from the base of tower to person Y to X is 60°. If the angle of elevation to the top of the tower from person X is 25°, what is the angle of elevation from person Y to the top of the tower? Figure illustrates the situation.

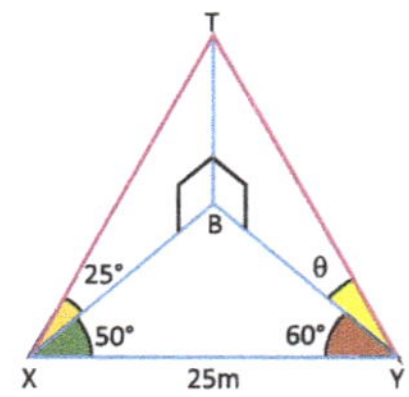

Solution 20: Find Angle of Elevation

Apply sine law to solve ΔXBY and tangent ratio to solve the triangles in vertical plane.

$\angle XBY = 180° - 50° - 60° = 70°$

$\frac{XB}{\sin 60°} = \frac{25}{\sin 70°}$. $\frac{BY}{\sin 50°} = \frac{25}{\sin 70°}$

$XB = 25\frac{\sin 60°}{\sin 70°}$ $BY = 25\frac{\sin 50°}{\sin 70°}$

$XB = 23.04$ $BY = 20.38$

$\Delta TBX : \tan 25° = \frac{TB}{XB} \Rightarrow TB = XB\tan 25°$

$TB = 2304\tan 25°\, TB = 10.74$

$\Delta TBY : \tan\theta = \frac{TB}{By} = \frac{10.74}{20.38}$

$\theta = \tan^{-1}\left(\frac{10.74}{20.38}\right) = 27.8°$

Answer: Angle of elevation is 27.8° from person Y to the top of the tower.

Example 21: Cosine Law Application

The observation deck of the Skylon Tower in Niagara Falls is 166 m above the Niagara River. A tourist in the observation desk notices two boats on the water. From the tourist's position, the bearing of boat A is 180° at an angle of depression on 40°. The bearing of boat B is 250° at an angle of depression on 34°. Calculate distance between the two boats to the nearest metre.

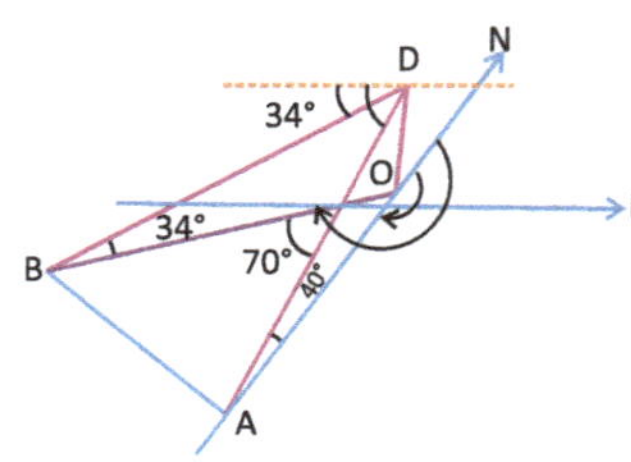

Solution 21: Cosine Law Application

Figure illustrates the situation.

$\angle AOB = 250 - 180 = 70°$

$OD = 166$ deck height

$$\tan 40° = \frac{OD}{AO} \Rightarrow AO = \frac{OD}{\tan 40°} = 1978.3$$

$$\tan 34° = \frac{OD}{OB} \Rightarrow OB = \frac{OD}{\tan 34°} = 246.1$$

$BA^2 = OB^2 + OA^2 - 2\,OB\,OA \cos 70°$

$= 197.8^2 + 246.1^2 - 2\,(197.8)(26.1)$

$= 66397.76$

$$BA = \sqrt{66398.36} = 257.68 = \mathbf{258\ m}.$$

Distance between the two boats: **258 m**

Example 22: Police Intercepts Smuggler

A smuggler leaves a private airfield at 06:00 and flies on a course of $N40°E$ at 200 km/h. The plane is detected by radar at a police airport, which is located 150 km northwest of airfield O. At 6.00 police airplane leaves its airport with the intention of intercepting the smuggler at 08:30 Determine the course and speed of the police airplane.

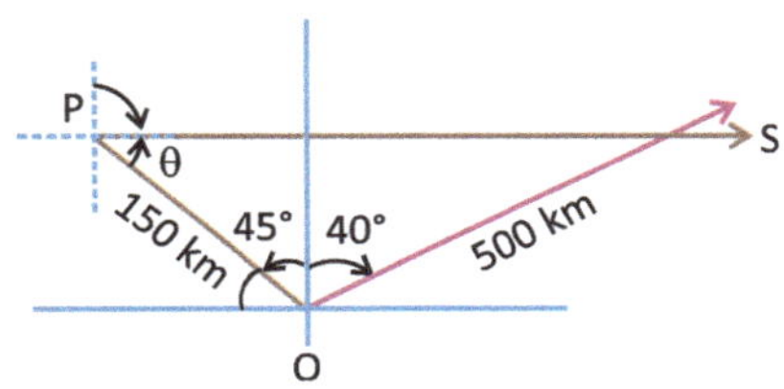

Solution 22: Police Intercepts Smuggler

From 6:00 to 8:30 time is 2.5 h

Distance travelled by smuggler at N40°E will be 2.5 (200) = 500 km

Figure illustrates situation

Apply cosine law since the configuration is SAS

$PS^2 = OP^2 + OS^2 - 2OP \times OS \times \cos P\hat{O}S$

$PS^2 = 150^2 + 500^2 - 2(150)(500)\cos 85°$

$PS^2 = 259429.63$

$PS = \sqrt{259426.63} = 509.34$

$$\therefore Speed = \frac{distance}{time} = \frac{509.34}{2} = 254.7\frac{km}{h}$$

Apply sine law in ΔPOS to find angle θ

$$\frac{\sin\theta}{500} = \frac{\sin 85°}{509.34}$$

$$\sin\theta = \frac{500}{509.54}\sin 85° = 0.9779$$

$\theta = \sin^{-1}(0.9779) = 77.9 = 78°$

Bearing Angle:

$$180° - 78° - 45° = \mathbf{057}°$$

Police aircraft will fly at speed of 254.7 km/h at a bearing of 057°.

Example 23: Triangles Cannot be Solved Independently

To find the height, h, of the tower OT, measurements are taken from the point A and B, 100 m apart making an angle of $60°$ with the base of the tower. Find the height, h, of the tower if the angle of elevation from A and B is $15°$ and $20°$ respectively. Round your answer to the nearest metre.

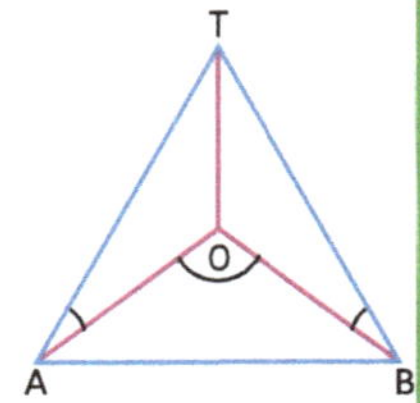

Solution 23: Triangles Cannot be Solved Independently

We have 4 triangles, none of them can be solved directly. We will write AO and BO in terms of h and then solve the triangle AOB.

$$\tan \angle TAO = \frac{h}{OA} \Rightarrow OA = \frac{h}{\tan \angle TAO}$$

$$OA = \frac{h}{\tan 15°} = 3.732h$$

$$OB = \frac{h}{\tan 20°} = 2.747h$$

$Given: \angle AOB = 60°, AB = 100\ m$

Apply Cosine Law in triangle AOB

$$AB^2 = OA^2 + OB^2 - 2(OA)(OB)\cos \angle AOB$$

$$100^2 = (3.732h)^2 + (2.747h)^2 - 2(3.732h)(2.747h)\cos 60°$$

$$10000 = 11.222h^2$$

$$h = \sqrt{\frac{10000}{11.222}} = 29.85 \approx 30\ m$$

height of tower: 30 m

Question

A triangle has three sides and three angles. To solve a triangle we need any three of these parameters. Three configurations with known sides and angles is given below:

SAS: Side – Angle – Side

SAA: Side – Angle – Angle

SSS: Side – Side – Side

In which of these configurations can sine law be applied to solve the triangle completely?

A. All of them
B. None of them
C. Only SAA
D. SAS and SSS
E. SAS and SAA

Communication

Explain why **cosine law** is used to solve a triangle if three sides are given or two sides with included angle is given.

Bearing Angle Application

Applications where both cosine law and sine law are applied

Example 24: Bearing Angle Application

A ship sails from A on a bearing of 065° for 10 km, then changes direction at B to a bearing of 110° for 12 km to reach at dock C.

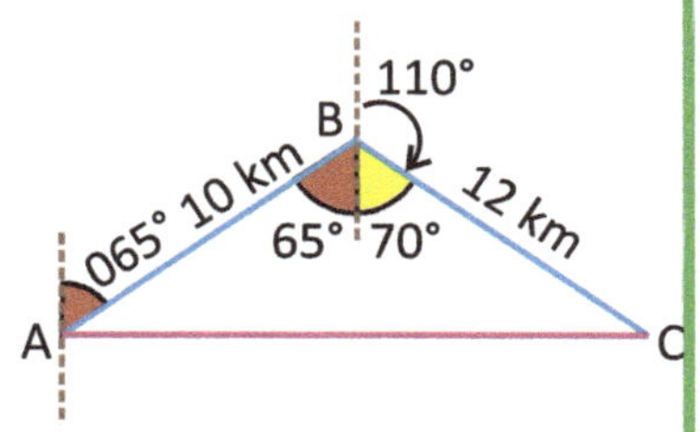

Find distance and bearing of C from A.

Solution 24: Bearing Angle Application

Figure illustrate the situation work out angles.

$\angle ABC = 65° + (180° - 110°)$

$\angle ABC = 65° + 70° = 185°$

SAS : Cosine Law Application

$AC^2 = AB^2 + BC^2 - 2\,AB \times BC \times \cos B$

$AC^2 = 10^2 + 12^2 - 2\,(10)(12)\cos 135°$
$= 412.7$

$AC = \sqrt{413.7} = 20.34$

Distance of dock C and A is 20.3 km.

Bearing of C: Sine Law Application

$$\frac{\sin A}{BC} = \frac{\sin B}{AC} \Rightarrow \sin A = \frac{BC}{AC}\sin B$$

$$\sin A = \frac{20.321}{10}\sin 135° = 0.4171$$

$\hat{A} = \sin^{-1}(0.4170) = 24.6°$

Bearing of C from A

$= 65° + 24.6° = 89.6°$

Bearing angle of dock C from dock A is 089.6°

Example 25: Distance and Bearing Angle

Anil leaves the marina on a 40 km sailboat race Lake Ontario intending to travel o a bearing of 355°. But an early morning fog settle, and by the time it clears, Anil has travelled 32 km on a bearing of 022°. Determine the shortest distance that Anil needs to travel to complete the race.

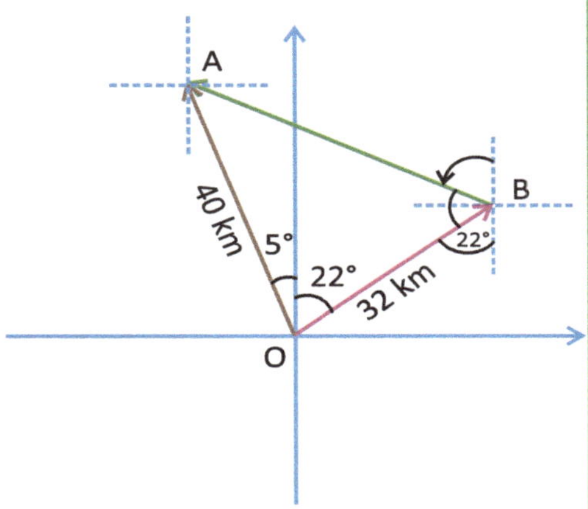

Solution 25: Distance and Bearing Angle

Diagram illustrates the situation where Anil reached at B instead of destination at A

Cosine law can be applied to solve the triangle AOB

$$AB^2 = 40^2 + 32^2 - 2(40)(32)\cos(5+22)°$$

$$AB^2 = 343.02$$

$$AB = \sqrt{343.02} = \mathbf{18.5\ km}$$

Apply sine law to find angle

$$\frac{\sin O\hat{B}A}{40} = \frac{\sin 27°}{18.5}$$

$$\sin O\hat{B}A = \frac{40}{18.5}\sin 27°, = 0.9816$$

$$O\hat{B}A = \sin^{-1}(0.9816) = 79°$$

Bearing angle is

$$180° + 22 + 79° = \mathbf{281°}$$

Anil needs to sail 18.5 km at bearing of 281°.

Example 26: Bearing Angle and Distance

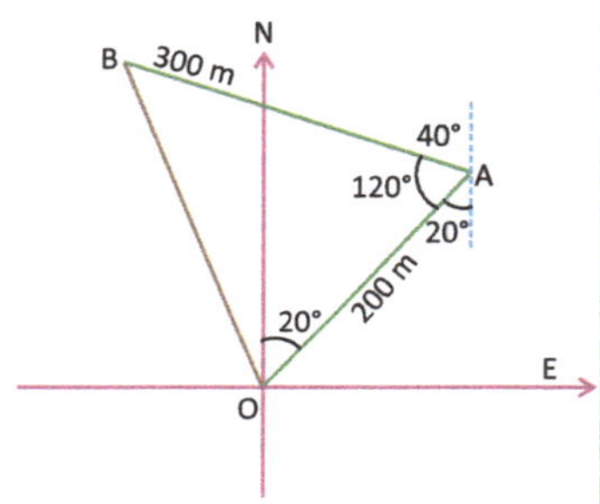

A surf-sailor at Tofino, west coast of Vancouver Island, sails 200 m on a bearing of 020°, changes direction to $N40°W$, sails 300 m, and stops.

How far from the original position should a friend sail to meet him? In what direction the friend sail?

Solution 26: Bearing Angle and Distance

Figure illustrates situation with angles calculated.

ΔOAB : SAS cosine law

$$OB^2 = OA^2 + AB^2 - 2\,OA.AB\cos A$$

$$OB^2 = 200^2 + 300^2 - 2\,(200)(300)\cos 120°$$

$$OB^2 = 190000$$

$$OB = \sqrt{190000} = 436\ m$$

Apply sine Law to solve for the vertex angle O in the triangle AOB.

$$\frac{\sin A\hat{O}B}{300} = \frac{\sin 120°}{436}$$

$$\sin A\hat{O}B = \frac{300}{436}\sin 120° = 0.5958$$

$$A\hat{O}B = \sin^{-1} 0.5958 = 36.57° = 37°$$

Bearing angle $(37° - 20° = 17°)$

N17°W or 343° bearing

Friend should sail a distance of 436 m at the bearing of 343°

Example 27: Sine Law Application

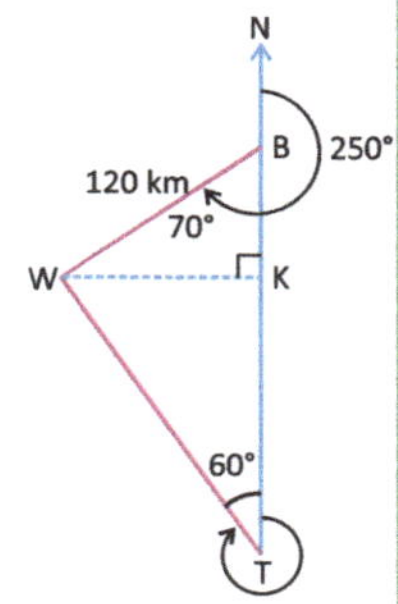

Highway 400 runs North from Toronto connecting many cities including Berry. Wasaga Beach is on a bearing of 250° from Berry and 300° from Toronto.

Find the shortest distance of Wasaga Beach to highway 400 if it is 120 km from Berry.

How far is Berry from Toronto?

Solution 27: Sine Law Application

Represent information on the figure as shown

Altitude, WK is the shortest distance.

$\Delta BWK : \sin 70° = \dfrac{WK}{WB}$

$WK = WB \sin 70°$

$WK = 120 \sin 70° = 112.8 = 113$

$\Delta TWB : \angle W = 180° - 70° - 60° = 50°$

$Sine\ Law : \dfrac{BT}{\sin 50°} = \dfrac{120}{\sin 60°}$

$BT = 120 \times \dfrac{\sin 50°}{\sin 60°} = 106.1 = 106$

Shortest distance of Wasaga beach from highway is 113 km and the distance between Berry and Toronto is 106 km.

SAT Test

To measure the height of a mountain, a surveyor measures angle of elevation of 23° and 32.5° from two points due south, 950 m apart. Find the height of the mountain.

a.	1100 m	b.	1200 m	c.	1300 m	d.	1400 m

Answer: B. 1200 m

Find the angle which the line passing through the two points (2, 5) and (5, 8) make with the x-axes.

a.	40°	b.	45°	c.	50°	d.	55°

Watch Video **3D Bearing Angle Sine Application** **SCAN ME**

Relative Velocity Application

Velocity is a vector quantity which has both magnitude and direction. It is represented by an arrow whose length is proportional to the magnitude and the head points at the direction. These arrows can be translated in a plane since both magnitude and direction remains same and the velocity is unaffected.

Example 28: Relative Velocity

An airplane is flying at speed of 500 km/h, on a bearing of 040°.

150 km/h wind is blowing from a bearing of 120°.

Determine the ground velocity of the airplane and the direction of the flight.

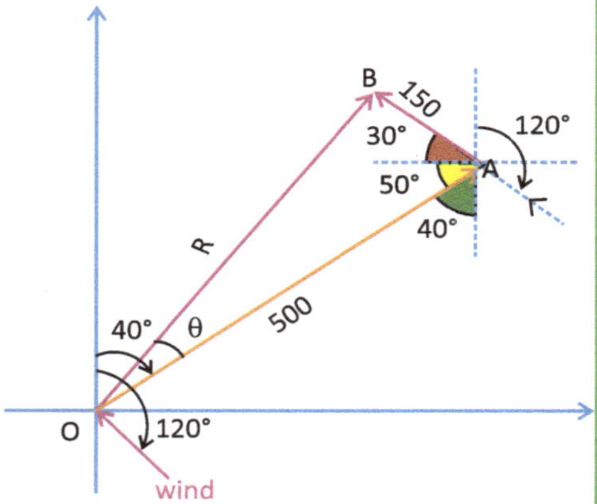

Solution 28: Relative Velocity

Figure provides the illustration of the situation.

In triangle OAB,

$OA = 500, AB = 150$

$\angle A = 30° + 50° = 80°$

Cosine Law: SAS

$OB^2 = 500^2 + 150^2 - 2(500)(150)\cos 80°$

$OB^2 = 246452.7$

$OB = \sqrt{246452.7} = 496.4$

Sine law for angle, θ

$\frac{\sin\theta}{180} = \frac{\sin 80°}{496.4}$

$\sin\theta = \frac{150}{496.4} \times \sin 80 = 0.2975$

$\theta = \sin^{-1}(0.2975) = 17.31°$

Bearing from North $40° - 17.31° = 22.7°$

Ground velocity of airplane in 496 km/h at bearing of 023°.

Solution 29: Distance Travelled by Planes

Two planes leave the same airport at the sometime in different directions. Plane A travels an unknown distance in the direction S20°E. Plane B travels an unknown distance in the direction N40E. The two planes are now 300 km apart and plane B is at a bearing of N30°E from plane A. How far did both planes travel to the nearest tenth of kilometre?

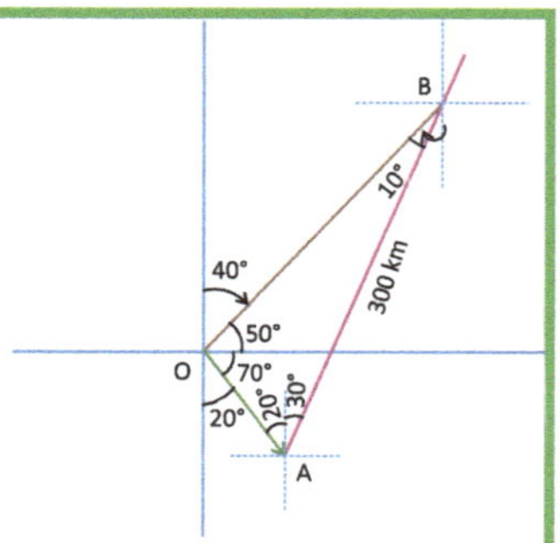

Solution 29: Distance Travelled by Planes

Figure illustrates the final position of the planes

Since combination of opposite side and angle combination is known sine law can be applied.

$\angle AOB = 180° = 40° - 20° = 120°$

$$\frac{\sin A\hat{O}B}{300} = \frac{\sin O\hat{B}A}{OA}$$

$$OA = \frac{300 \sin 10°}{\sin 120°} = 60.15$$

$$\frac{OB}{\sin 50°} = \frac{300}{\sin 120°}$$

$$OB = 300 \times \frac{\sin 50°}{\sin 120} = 265.36$$

Distance of plane A and B are 60.1 km and 265.4 km respectively.

Area of Triangle

Watch Video

SCAN ME

Example 30: Area of Triangle for Given SAS

Given ΔABC: $c = 12\ cm, b = 10\ cm,$ and $\angle A = 30°$.

Find the height of the triangle and then the area.

Solution 30

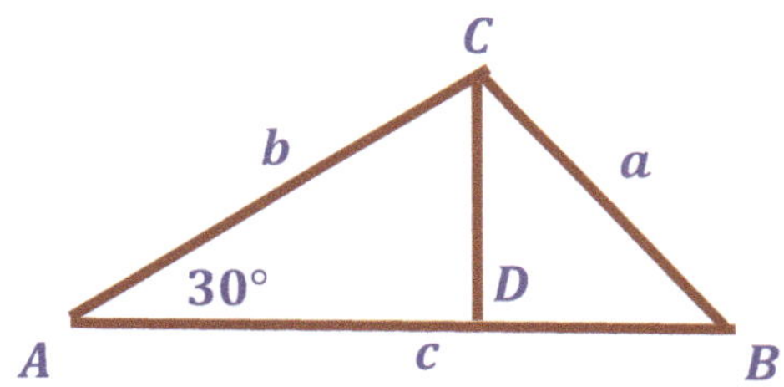

$$\sin A = \frac{CD}{AC} = \frac{CD}{b}$$

Height of the triangle $CD = b \sin A = 10 \times \frac{1}{2} = 5$

Area of $\Delta ABC = \frac{1}{2} base \times height$

Area of $\Delta ABC = \frac{1}{2} cb \sin A = \frac{1}{2} \times 12 \times 5 = 30cm^2$

Important Things to Note from above

- Height of the triangle is $\boldsymbol{b \sin A}$
- Area of the triangle is $\boldsymbol{\frac{1}{2} cb \sin A}$ when two sides and included angle is given.

Test 1: Sine Law and Cosine Law

Q1.	In $\Delta\, ABC, \angle B = 51°, \angle\, C = 63°, c =$ 15.2 Find a. A. 14.8 B. 15.2 C. 15.6 D. 15.9	Q2	In $\Delta LMN,\ \hat{L} = 43°,\ l = 16\ cm,\ n =$ $19\ cm.$ Find $\hat{N}$. A. 62° B. 58° C. 110° D. 126°
Q3	To find height of a tree, angles are measured from two points, 1000 feet apart, on the opposite side of the tree. Approximately, how tall is the tree if the angle of elevation from these points is 32° and 38° respectively? A. 31 feet B. 35 feet C. 400 feet D. 44 feet	Q4	Angles in a triangle are in the ratio of 1:3:8. The ratio of longest to the next shorter side can be A. $\sqrt{3} : 1$ B. $3 : \sqrt{2}$ C. $\sqrt{6} : 2$ D. $8 : 3$
Q5	The sides of a triangle are of integer length with consecutive numbers. If the perimeter of the triangle is 15 cm, find the smallest angle in the triangle. A. 32° B. 41° C. 58° D. 64°	Q6	For which of the following situations two triangles can be formed. A. $\hat{A} = 91°, \quad a = 10\ cm, \quad b = 7\ cm$ B. $\hat{A} = 45°, \quad a = 6.4\ cm,$ b=10 cm C. $\hat{A} = \ 30°, \quad a = 10\ cm, \quad b = 15\ cm$ D. $\hat{A}\ =\ 60\ , \quad a = 5\ cm \quad b = 6\ cm$
Q7	Mark is a landscaper who is creating a triangular planting garden the homeowner wants the garden to have two equal sides that contain an angle of 135°. Find perimeter of the garden if the longest side is 5 m. A. 14 m B. 13.2 m C. 12.1 cm D. 10.4 m 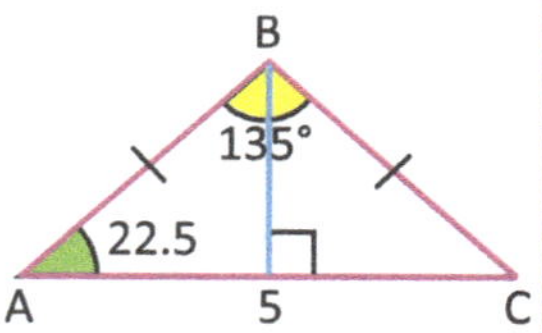 	Q8	Two ships A and B with underwater sonar have located the wreckage, W of an old ship on the sea floor. The ships are on the opposite sides of the wrecked ship W. When the distance AB between them is 8000 m, the angle of dispassion to the wreck is 18° and 56° respectively. Determine the depth of water above the wreck A. 2240 m B. 2160 m C. 2130 m D. 2119 m 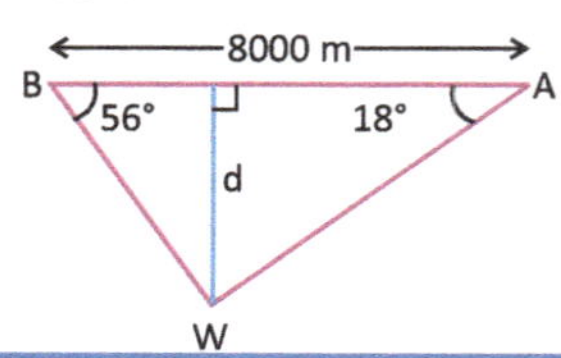

Q9. In a $\Delta ABC, a = 1\ cm,\ b = \sqrt{3}\ cm$ $and \angle A = 30°,\ find\ \angle B$

A. 30°
B. 60°
C. 60° or 120°
D. 30° or 150°

Q10 Sam walks 500 m south and then turns and travel 600 m in the direction N60°W. At what bearing will he need to walk to return straight to the starting point?

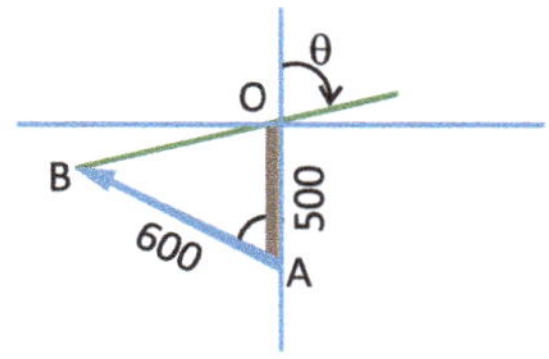

A. 021°
B. 030°
C. 069°
D. 078°

Q11 Person swims 300 m north from the dock. He then turns right 120° and swim 400 m. How far from the dock is he?

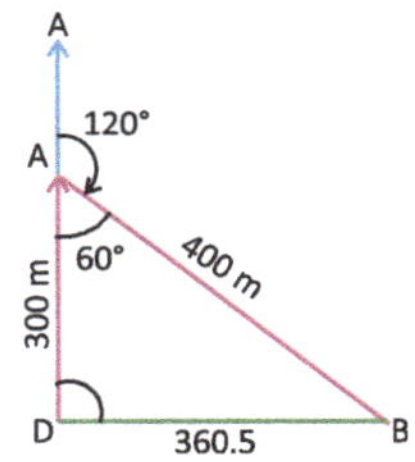

A. 360 m
B. 375 m
C. 470 m
D. 610 m

Q12 Person swims 300 m north from the dock. He then turns right 120° and swim 400 m.

Find the bearing of the person from the dock.

A. 036°
B. 043°
C. 074°
D. 084°

Q13 Triangle ABC with height of 12 m has an area of 126 m^2. Find its perimeter if $\angle ABC = 80°$ and the side length a= 17 m.

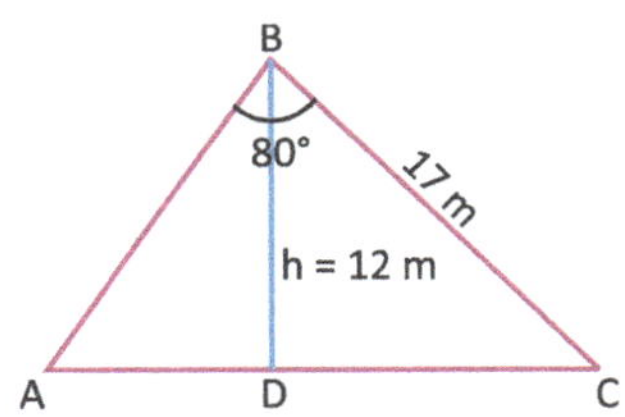

A. 51 m
B. 54 m
C. 56 m
D. 49 m

Q14 The bearing of a lighthouse, L, from a boat is 062°. The boat travels 500 m east and the bearing of the lighthouse how is 044°, as shown in the figure. Find the distance the boat needs to travel to be due south of the lighthouse.

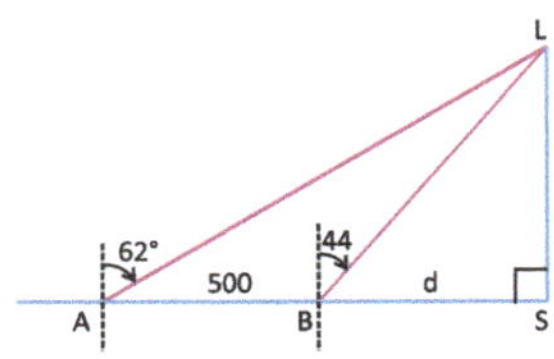

A. 546 m
B. 528 m
C. 257 m
D. 482 m

Q15 P lies on a bearing of 60° from Q. What is the bearing of Q from P?

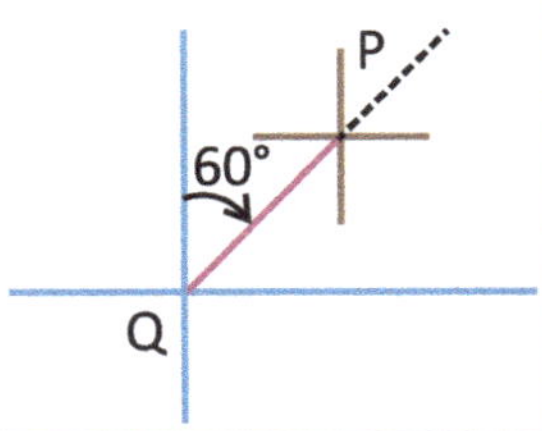

A. 60°
B. 240°
C. 210°
D. 150°

Q16 Find area of regular octagon inscribed in a circle of radius 10 cm

A. 320 cm^2
B. 283 cm^2
C. 242 cm^2
D. 141 cm^2

Q17	A triangle has side length of 20 cm and 11 cm. The angle opposite the later side is 29°. What can be the measure of the angle opposite the side of length 20 m. 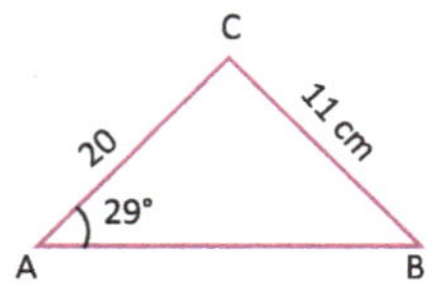 A. between 30°-40° B. between 50°-60° C. between 70°-90° D. greater than 90°	Q18	Find the length of the side AB in the given triangle where D is the midpoint of BC. A. 14.1 B. 14.8 C. 15.6 D. 16.2 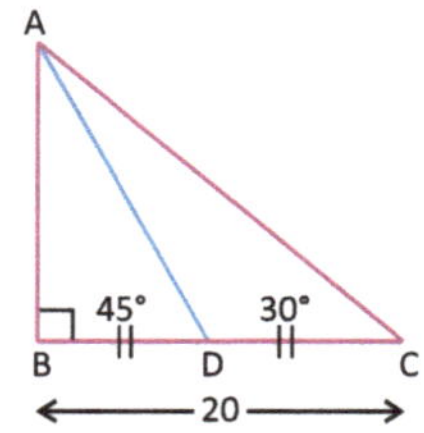
Q19	From point B, Ben uses a clinometer to determine an angle of elevation to the top of a cliff to be 38°. From point D, 68 m away from Ben, Anil estimates the angle between the base of the cliff, himself and Ben to be 42°, while Ben estimates the angle between the base of the cliff, himself and Anil to be 63°. How high is the cliff if Ben is 158 cm tall? A. 37 m B. 38 m C. 40 m D. 42 m 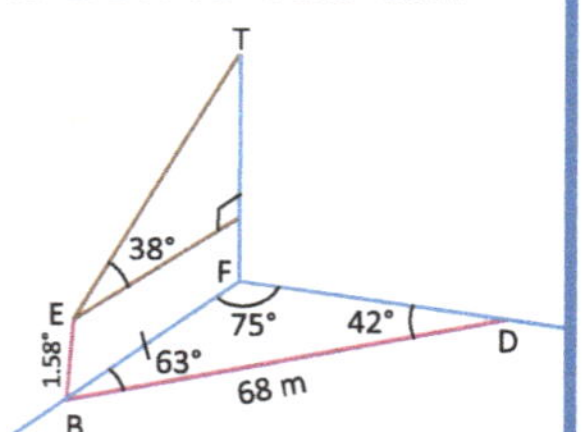	Q20	A line from the point P (30,0) intersects the circle $x^2 + y^2 = 169$ at points A and B. Determine the length of the chord AB. If the line makes an angle of 160° with the positive x-axis. 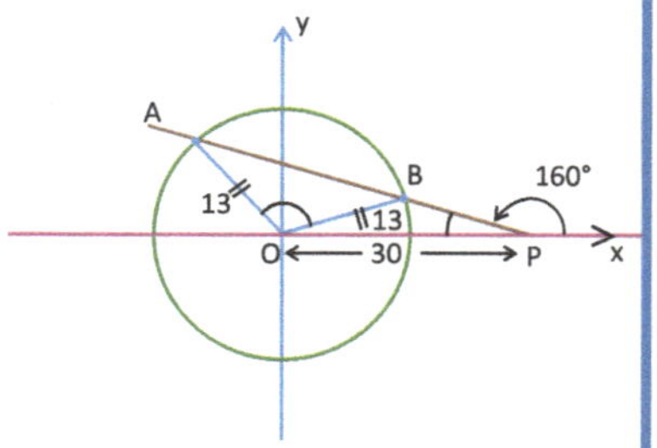 A. 14 B. 15 C. 16 D. 17
Q21	Find perimeter of regular octagon inscribed in a circle of radius 10 cm. A. 72 cm B. 61 cm C. 55 cm D. 48 cm 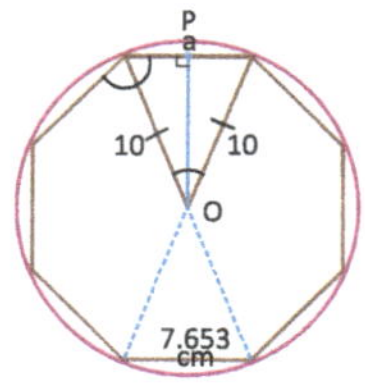		

Q22.	The confederation Bridge links Prince Edward Island with mainland New Brunswick. From one point on the Island, the angle of elevation of the highest point on the bridge is 25°. From a point 100 m. Closer to the bridge, the angle of elevation is 73°. Which of the following expression can be used to find the height of the bridge?

A. $100\frac{\sin 25°}{\sin 48°}\times \sin 73°$

B. $100\frac{\sin 25°}{\sin 73°}$

C. $100\frac{\sin 48°}{\sin 25°}\times \sin 73°$

D. $100\frac{\sin 76°}{\sin 25°}$

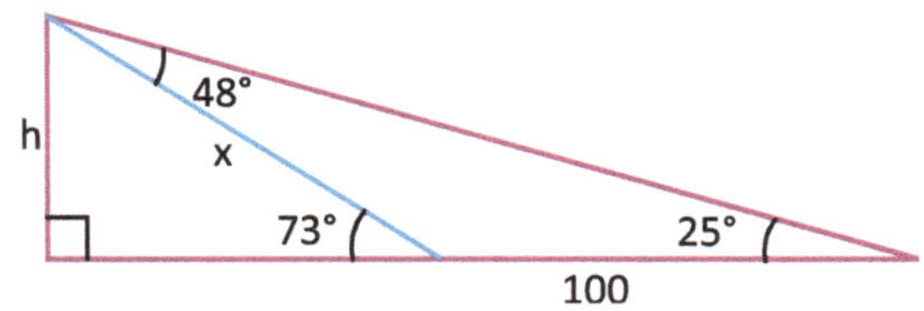

Explore

The figure shows semicircles with centre P and M. M is mid point of AP. Points A, M, P, and B are collinear. A small circle of radius r is drawn with centre O tangential to the semicircles as shown.

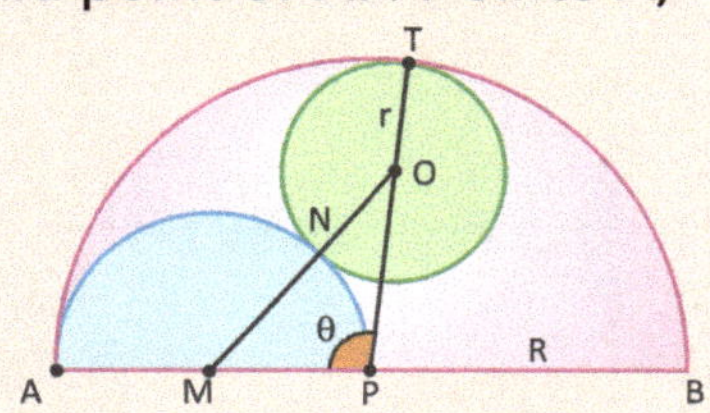

a) If the radius of bigger semicircle is R then prove that

$$\frac{R}{r} = \frac{3-\cos\theta}{1-\cos\theta}$$

b) Find the angle OMP for $R = 3r$.

c) What do you conclude from this example?

Answers Test 1: Sine Law and Cosine Law

Q1	C	Q2	D	Q3	B	Q4	C
Q5	B	Q6	C	Q7	D	Q8	C
Q9	C	Q10	C	Q11	A	Q12	C
Q13	C	Q14	B	Q15	B	Q16	B
Q17	D	Q18	a	Q19	b	Q20	C
Q21	B	Q22	A				

Q1	$a = 15.2 \times \dfrac{\sin 66°}{\sin 63°} = 15.6$	Q2	$\sin N = \dfrac{19}{16}\sin 43°, N = 54°$ $N_2 = 180 - 54 = 126$
Q3	$h\left(\dfrac{1}{\tan 32°} + \dfrac{1}{\tan 38°}\right) = 100$ $h = \dfrac{100}{2.88} = 34.7$	Q4	$1:3:8 \equiv 15° : 45° : 120°$ *are angles* $\dfrac{\sin 120°}{\sin 45°} = \dfrac{\sqrt{6}}{2}$
Q5	$x + (x+1) + (x+2) = 15$ $3x = 15 - 3 = 12$ $x = 4 \Rightarrow sides\ are: 4, 5, 6$ $\cos A = \dfrac{5^2 + 6^2 - 4^2}{2(5)(6)} = 0.75$ $A = \cos^{-1}(0.75) = 41.4°$	Q6.	$\hat{A} = 30°,\quad a = 10\ cm,\quad b = 15\ cm$ Calculate $b \sin A$ and check if $b \sin A < a < b$ $b \sin A = 15 \sin 30° = 7.5$ $7.5 < 10 < 15$ is true. So, 2 triangles are formed.
Q7.	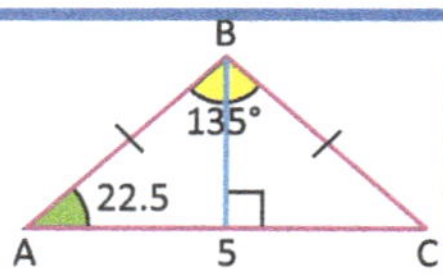 $\dfrac{AB}{\sin 22.5°} = \dfrac{5}{\sin 35°}$ $AB = \dfrac{5}{\sin 35°} \times \sin 22.5°$ $AB = 2.7$ $P = 5 + 2.7 + 2.7 = 10.4$	Q8.	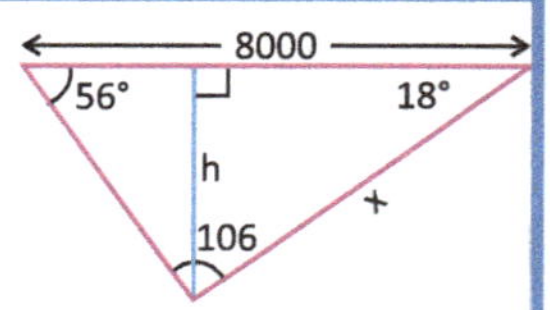 $\dfrac{x}{\sin 56} = \dfrac{8000}{\sin 106}$ $x = 8000 \times \dfrac{\sin 56°}{\sin 106°}$ $x = 6899.6$ $h = x \sin 18° = 6899.6 \sin 18° = 2132\ m$
Q9.	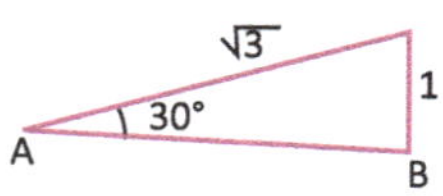 $\dfrac{\sin B}{\sqrt{3}} = \dfrac{\sin 30°}{1}$ $\sin B = \sqrt{3}\dfrac{1}{2} = \dfrac{\sqrt{3}}{2}$ $\angle B = \sin^{-1}(\frac{\sqrt{3}}{2}) = 60°\ or\ 120°$	Q10	$BO^2 = 500^2 + 600^2 - 2(500)(600)\cos 60°$ $BO^2 = 310000 \Rightarrow BO = 556.78$ $\dfrac{\sin BOA}{600} = \dfrac{\sin 6\,0°}{556.78}$ $\angle BOA = \sin^{-1}\left(600 \times \dfrac{\sin 60°}{556.78}\right) = 69°$

Q11	Diagram illustrates the situation. (notice the turn right by 120°) $\angle DAB = 180° - 120° = 60°$ $SAS : cosine\ Law$ $DB^2 = 300^2 + 400^2 - 2(300)(400)\cos 60°$ $DB^2 = 130000$ $DB = \sqrt{130000} = 360.5$	Q12	Sine law for bearing angle $\angle ADB$. $\frac{\sin D}{400} = \frac{\sin 60°}{360.5}$ $\sin D = \frac{400}{360.5}\sin 60° = 0.9609$ $\angle D = \sin^{-1}(0.9609) = 74°$

Solution 13

Triangle ABC with height of 12 m has an area of 126 m².

Find its perimeter if $\angle ABC = 80°$ and the side length $a =$ 17 m.

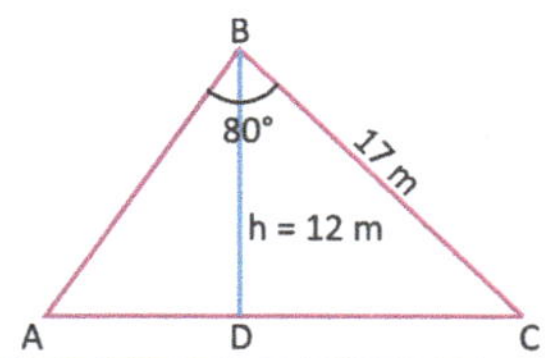

$Area = \frac{1}{2}(AC \times BD)$ $126 = \frac{1}{2}AC(12)$ $AC = \frac{126}{6} = 21$	Apply Sine Law $\frac{\sin A}{17} = \frac{\sin 80°}{21}$ $\sin A = \frac{17}{21}\sin 80°$ $\sin A = 0.7972$ $\hat{A} = 52.86$	$\hat{C} = 180° - 80° - 52.86 = 47.14$ $\frac{AB}{\sin 47.14°} = \frac{21}{\sin 80°}$ $AB = 21\frac{\sin 47.14°}{\sin 80°} = 15.63$ $P = 17 + 21 + 15.63 = 63.6$

Solution 14

The bearing of a lighthouse, L, from a boat is 062°. The boat travels 500 m east and the bearing of the lighthouse how is 044°, as shown in the figure. Find the distance the boat needs to travel to be due south of the lighthouse.

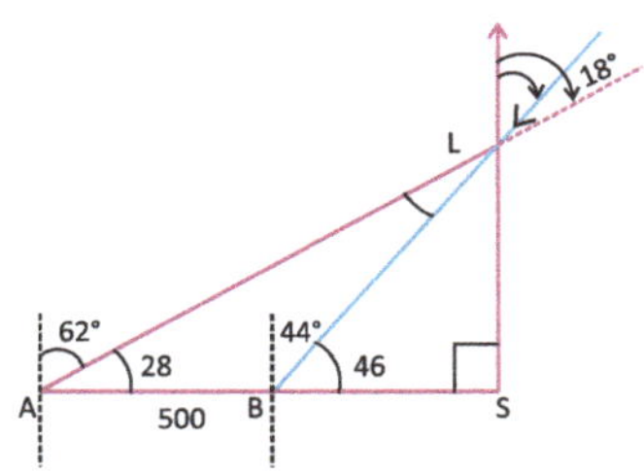

Understand steps to find angles using bearing angles Exterior angle: $46° = 28° + \angle ALB$ $\angle ALB = 46° - 28° = 18°$ Alternate Method: $\angle ALB = 180° - 28° - 46° = 18°$	$\frac{BL}{\sin 28°} = \frac{500}{\sin 8°}$ $BL = 500\frac{\sin 28°}{\sin 18°} = 760\ m$ $distance\ to\ move\ due\ south, LS = 760 \sin 46° = 546\ m$

Q15	P is on a bearing of 60° from Q. Q lines on the bearing of $180 + 60°$ $= 240$ from P 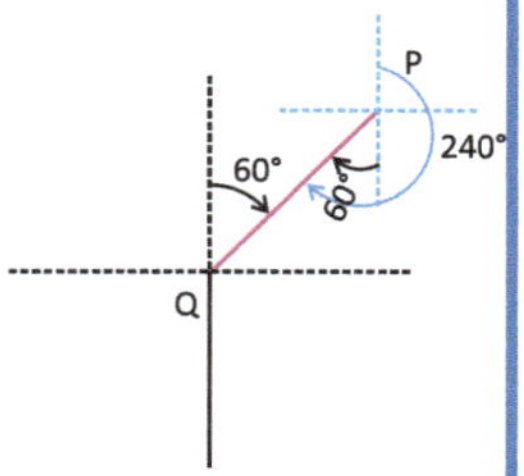 	Q16	$A = \frac{1}{2}$ base x height $h = OP = 10\cos 22.5°$ $a = 7.653$ $A = 8 \times \frac{1}{2}\,(7.653)(10\cos 22.5°)$ $A =$ 282.8 cm^2
Q17	$\frac{\sin B}{20} = \frac{\sin 29°}{11}$ $\sin B = \frac{20}{11}\sin 29° = 0.88147$ $\hat{B}_1 = 61.8° = 62°$ $B_2 = 180° - 62° = 118°$ Ambiguous case with two solutions 62 is not a given option Answer is d) 118°	Q18	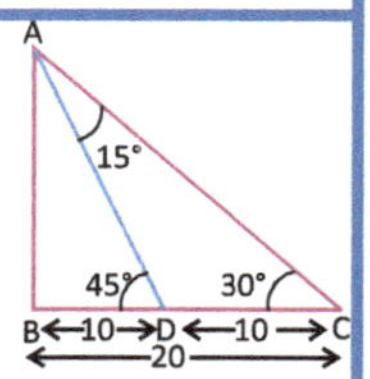 ΔADC First sine law $\angle DAC = 45° - 30° =$ $15°$ (external angle) $\frac{AD}{\sin 30} = \frac{10}{\sin 15}$ $AD = 10\frac{\sin 30}{\sin 15} = 19.3$ $AB^2 = 10^2 + 19.3^2$ $- 2(10)(19.3)\cos 45°$ $AB = \sqrt{199.54} = 14.12$
Q19	$\frac{BO}{\sin 42} = \frac{68}{\sin 75}$ $BO = 47.1\ m$ $h = 47.1\ m\tan 38° = 36.80$ $H = 36.80 + 1.58 = 38\ m$	Q20	Solution provided at the end
Q21	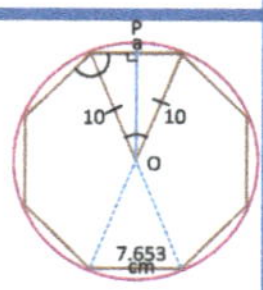 $\alpha = \frac{360°}{8} = 45°$ $SAS : cosine\ low$ $a^2 = 10^2 + 10^2 - 2(10)(10)\cos 45°$ $a = \sqrt{58.57} = 7.653$ $P = 8a = 8(7.653) = 61.22$ cm	Q22	$\frac{x}{\sin 25°} = \frac{100}{\sin 48°} \Rightarrow x = 100\frac{\sin 25°}{\sin 48°}$ $h = x\sin 73° = 100\frac{\sin 25°}{\sin 48°} \times \sin 73°$

Solution 20

A line from the point P (30,0) intersects the circle $x^2 + y^2 = 169$ at points A and B. Determine the length of the chord AB. If the line makes an angle of 160° with the positive x-axis.

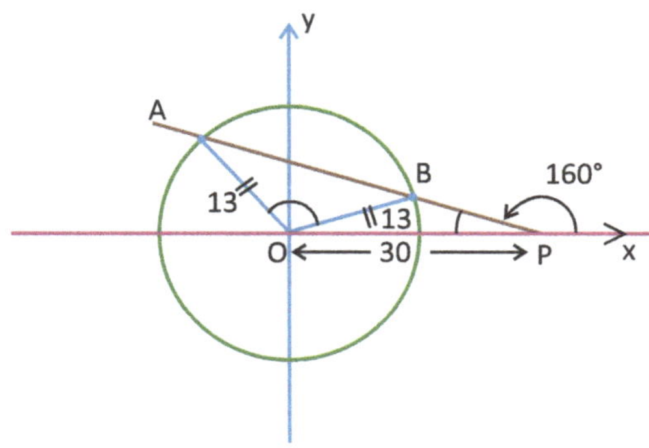

Solution 20:

$x^2 + y^2 = 169 = 13^2, r = \sqrt{13^2} = 13;$

$\angle OPB = 180° - 160° = 20°, OP = 30$

$$\frac{\sin A}{30} = \frac{\sin 20°}{13} \; in \; \Delta \, OAP$$

$$\sin A = \frac{30}{13} \times \sin 20° = 0.7892$$

$$\angle A = \sin^{1}(0.7892) = 52.1°$$

$$\angle AOB = 180° - 2\,(52.1) = 75.8°$$

$$\frac{AB}{\sin 75.8°} = \frac{13}{\sin 52.1}$$

$$AB = 13 \times \frac{\sin 75.8°}{\sin 52.1°} = 15.97$$

Canadian Math Olympiad, 1986

In the diagram line segment AB and CD are of length 1 while the angle ABC and CBD are 90° and 30° respectively. Find length AC.

Hint: Relate $\sin D$ in triangle BDC and ABD

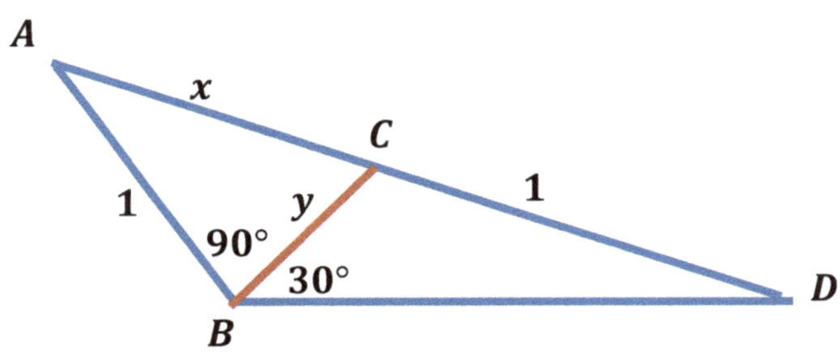

$AC = \sqrt[3]{2}$

Watch Video **Olympiad Challenge** SCAN ME

Circle – Triangle and Sphere – Cone Relation

Interesting questions to derive formula which can help understand relations between geometrical shapes better and improve performance. Also appreciate how we can apply our knowledge in real life applications.

STEM Application Challenge: Solution provided in Unit 7 page 541

Circle Inscribed in Equilateral Triangle

Formula Derivation

A circle, O, of radius r is inscribed in an equilateral triangle ΔABC.

Develop a relation between radius r of the inscribed circle and the side length a of the equilateral triangle.

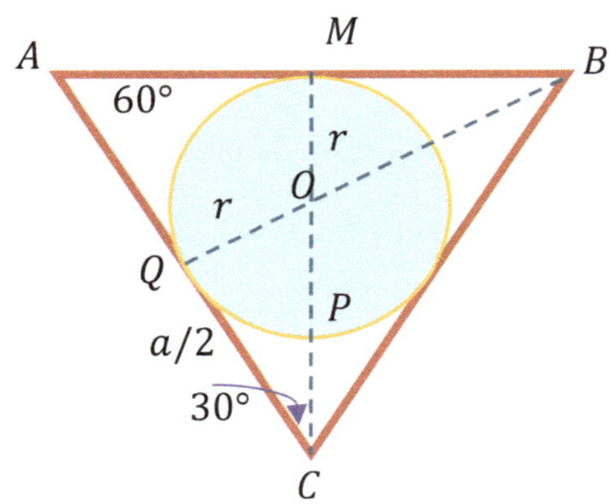

Solution

The paper cup for a snow cone has radius of R with cross-section of an equilateral triangle. Figure, not to the scale is shown. When a ball of radius r is dropped into the cup, the top of the ball is even with the top of the cup. Determine an expression for the distance from the bottom of the ball to the vertex of the cone in terms of R and r.

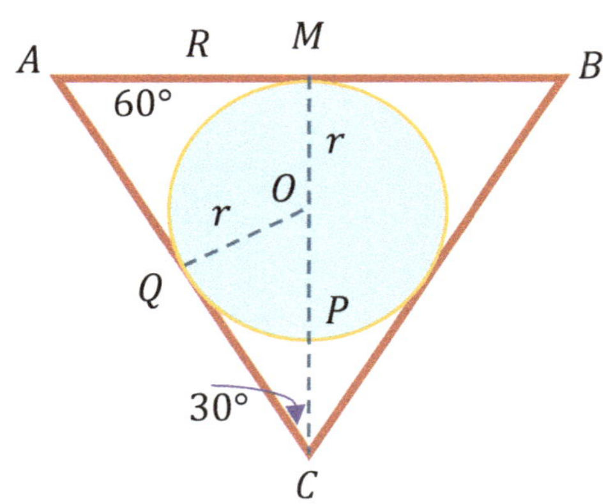

Solution

ΔAMC is a special triangle $30-60-90$

Chapter 13: Trigonometric Identities

What is an Equation?

An **equation** is a mathematical statement that asserts two expressions are equal. It's typically written in the form:

$f(x) = g(x)$

where $f(x)$ and $g(x)$ are algebraic expressions involving the **independent variable** x.

Equations can also involve **more than one variable**, such as x and y depending on the context.

Types of Equations

- If an equation is **true for every value** of the variable(s) within the domain of both expressions, it is called an **identity**.
  Example: $(x + y)^2 = x^2 + 2xy + y^2$
- If the equation is **only true for specific values**, it is called a **conditional equation**.
  Example: $2x + 3 = 7$ is true only when $x = 2$

Example 1: Equation and Identity

Explain why Set A of equations are **conditional equations** and equations of Set B are **identities**.

Set A: $2x - 1 = 5$ $\quad x^2 - 8x + 7 = 0$ $\quad x + y = 0$ $\quad x^2 = -1$

Set B: $3p + 5p = 8p$ $\quad x^2 - y^2 = (x + y)(x - y)$ $\quad \frac{x^2-1}{x+1} = x - 1, x \neq -1$

Solution 1: Equation and Identity

Set A:

Equation $2x - 1 = 5$ is true only for x = 3

Equation $x^2 - 8x + 7 = 0$ is true for x = 1 and x = 2

Equation $x + y = 0$ is true for infinite value but not all values of x and y

Equation $x^2 = -1$ is not true for any real number

All the above equations in the Set A are **conditional equation**; they are not true for every value of independent variable in the domain.

Set B:		
$3p + 5p = 8p$	$x^2 - y^2 = (x + y)(x - y)$	$\frac{x^2-1}{x+1} = x - 1, x \neq -1$
$3p + 5p = 8p$ Left Side (LS) terms are like terms which add to give Right Side (RS) for app values of p. This equation is true for all values and so it is an identity.	Expand and simplify RS $(x + y)(x - y)$ $= x^2 - xy + xy - y^2$ $= x^2 - y^2 = LS$ $RS = LS$ for all values of x and y	Factor LS $\frac{x^2 - 1}{x + 1}$ $= \frac{(x + 1)(x - 1)}{x + 1}$ $= x - 1, x \neq -1$ $=$ RS
Set B of equations are **identities** as they are true for every value of variable(s) for which it is defined.		

Understanding Trigonometric Identities

Trigonometric identities are equations that involve trigonometric functions and are **true for all values** of the variables where both sides are defined. These identities are essential tools in simplifying expressions, solving equations, and connecting algebra with geometry.

At the beginning of this chapter, you'll find a list of the **fundamental trigonometric identities**, which will be **proved in the next section**. These include:

- Three Reciprocal Identities
- Two Quotient (Ratio) Identities
- Three Pythagorean Identities
- Three Co-Function (Complementary Angle) Identities

SCAN ME

How Do We Prove an Identity?

To prove a trigonometric identity, we show that **one side of the equation is exactly equal to the other**. Typically, we:

1. Start with the more complex side
2. Use algebraic techniques (factoring, expanding, common denominators)
3. Apply known identities and trigonometric transformations
4. Simplify step by step until both sides match

Reciprocal identities: $\csc\theta = \frac{1}{\sin\theta}$ $\sec\theta = \frac{1}{\cos\theta}$ $\cot\theta = \frac{1}{\tan\theta}$

Quotient Identities: $\tan\theta = \frac{\sin\theta}{\cos\theta}$ $\cot\theta = \frac{\cos\theta}{\sin\theta}$

Pythagorean Identities: $\sin^2\theta + \cos^2\theta = 1$ $1 + \tan^2\theta = \sec^2\theta$ $1 + \cot^2\theta = \csc^2\theta$

Cofunction Identities: $\sin(90° - \theta) = \cos\theta$, $\cos(90° - \theta) = \sin\theta$, $\tan(90° - \theta) = \cot\theta$

Watch Video **Trigonometric Identities Master Concepts** **SCAN ME**

Proof of Fundamental Trigonometric Identities

The fundamental trigonometric identities can be proved by definition using a general point, $P(a,b)$, in the coordinate plane.

For any coordinate point, $P(a,b)$, $r = \sqrt{a^2 + b^2} > 0$.

The six trigonometric ratios are defined as:

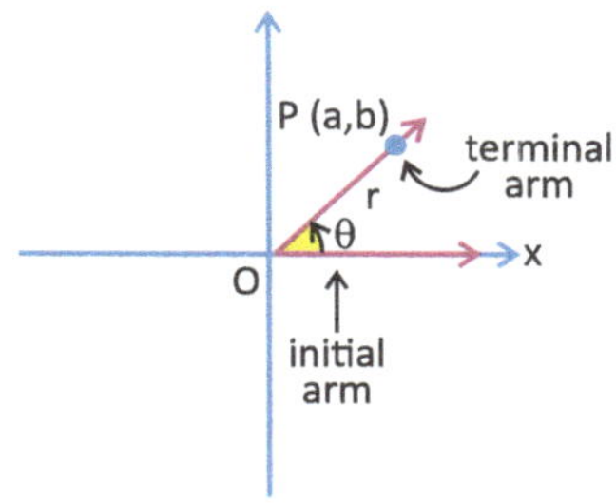

Primary Trigonometric Ratio:

$\sin\theta = \frac{b}{r}$ $\cos\theta = \frac{a}{r}$ $\tan\theta = \frac{b}{a}$

Secondary Trigonometric Ratio: $\csc\theta = \frac{r}{b}$ $\sec\theta = \frac{r}{a}$ $\cot\theta = \frac{a}{b}$

Reciprocal identities: $\csc\theta = \frac{1}{\sin\theta}$	$\sec\theta = \frac{1}{\cos\theta}$	$\cot\theta = \frac{1}{\tan\theta}$
$\csc\theta = \frac{r}{b} = \frac{1}{b/r} = \frac{1}{\sin\theta}$	$\sec\theta = \frac{r}{a} = \frac{1}{a/r} = \frac{1}{\cos\theta}$	$\cot\theta = \frac{a}{b} = \frac{1}{b/a} = \frac{1}{\tan\theta}$

Quotient Identities: $\tan\theta = \frac{\sin\theta}{\cos\theta}$	$\cot\theta = \frac{\cos\theta}{\sin\theta}$
$\frac{\sin\theta}{\cos\theta} = \frac{b/r}{a/r} = \frac{b}{a} = \tan\theta$	$\frac{\cos\theta}{\sin\theta} = \frac{a/r}{b/r} = \frac{a}{b} = \cot\theta$

Pythagorean Identities: $\sin^2\theta + \cos^2\theta = 1$	$1 + \tan^2\theta = \sec^2\theta$	$1 + \cot^2\theta = \csc^2\theta$
$b^2 + a^2 = r^2$ $\frac{b^2}{r^2} + \frac{a^2}{r^2} = \frac{r^2}{r^2}$ $\sin^2\theta + \cos^2\theta = 1$	$b^2 + a^2 = r^2$ $\frac{a^2}{a^2} + \frac{b^2}{a^2} = \frac{r^2}{a^2}$ $1 + \tan^2\theta = \sec^2\theta$	$b^2 + a^2 = r^2$ $\frac{b^2}{b^2} + \frac{a^2}{b^2} = \frac{r^2}{b^2}$ $1 + \cot^2\theta = \csc^2\theta$

Proof of Basic Trigonometric Identities Using General Point

Consider a general point, $P(a, b)$, in the coordinate plane.

For any coordinate point, $P(a, b)$, $r = \sqrt{a^2 + b^2} > 0$.

Primary Trigonometric Ratio: $\sin\theta = \frac{b}{r}$ $\quad \cos\theta = \frac{a}{r}$ $\quad \tan\theta = \frac{b}{a}$

Secondary Trigonometric Ratio: $\csc\theta = \frac{r}{b}$ $\quad \sec\theta = \frac{r}{a}$ $\quad \cot\theta = \frac{a}{b}$

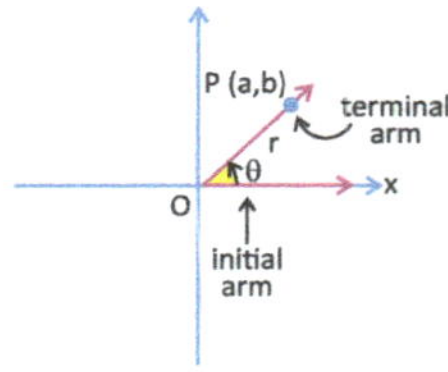

Example 1: Prove the following identities:

a.	$\cot\theta \ \sin\theta = \cos\theta$	b.	$\frac{1+\sin\theta}{\sin\theta} = 1 + \csc\theta$
c.	$\frac{1+\cot\theta}{\csc\theta} = \sin\theta + \cos\theta$	d.	$\sec^2\theta - \tan^2\theta = 1$

Solution 1: This is a very effective strategy to prove identities

a.	$\cot\theta \ \sin\theta = \cos\theta$ • Start from Left Hand Side (LHS) • Write ratios in terms of x, y, and r • Simplify and derive expression on RHS LHS $= \frac{x}{y} \times \frac{y}{r}$ $= \frac{x}{r} = \cos\theta = RS$	b.	$\frac{1+\sin\theta}{\sin\theta} = 1 + \csc\theta$ LHS $= \frac{1+\frac{y}{r}}{\frac{y}{r}} = \frac{r+y}{r} \times \frac{r}{y} = \frac{r+y}{y}$ $= \frac{r}{y} + 1 = \csc\theta + 1$ $= 1 + \csc\theta = RS$
c.	$\frac{1+\cot\theta}{\csc\theta} = \sin\theta + \cos\theta$ $LS: \frac{1+\cot\theta}{\csc\theta} = \frac{1}{\csc\theta} + \frac{\cot\theta}{\csc\theta}$ $= \frac{1}{r/y} + \frac{x/y}{r/y}$ $= \frac{y}{r} + \frac{x}{y} \times \frac{y}{r} = \frac{y}{r} + \frac{x}{r}$ $= \sin\theta + \cos\theta = RS$	d.	$\sec^2\theta - \tan^2\theta = 1$ $LS: \sec^2\theta - \tan^2\theta$ $= \left(\frac{r}{x}\right)^2 - \left(\frac{y}{x}\right)^2 = \frac{r^2 - y^2}{x^2}$ $= \frac{x^2}{x^2} = 1 = RS$

Guided Practice 1: Prove the following identities:

a.	$\frac{\sin\theta}{\tan\theta} = \cos\theta$	b.	$\frac{\tan\theta}{\cos\theta} = \frac{\sin\theta}{1-\sin^2\theta}$

What Is a Counterexample?

A counterexample is a **specific example** that shows a general statement is **not always true.**

Definition:
A counterexample is an example that disproves a statement or claim by providing a case where the statement fails.

Why Are Counterexamples Important?

In mathematics, we often test the validity of a statement or **conjecture**. If even one example exists where the statement does not hold, the statement is not universally true — and that single case is the counterexample.

Example:

Claim: "For all integers n, the number $n^2 + n + 41$ is prime."

Counterexample:
Try $n = 40$

$40^2 + 40 + 41 = 1600 + 40 + 41 = 1681 = 41 \times 41$

Since 1681 is not prime, this disproves the claim.
So, $n = 40$ is a counterexample.

A conjecture is an educated guess or a mathematical statement that appears to be true based on observations or patterns, but has not yet been proven

When to Use Counterexamples

- To disprove a general claim or **conjecture**
- To test the limits or scope of a mathematical rule
- To sharpen critical thinking by looking for exceptions
- In logic and proofs, to refute a statement without checking every case

Counter Example: Equation Not an Identity

Example 1: Counter Example

Provide counterexample to show that $\sin(A + B) = \sin A + \sin B$ is not an identity.

Solution 1: Counter Example

For identities, the equation is true for every real number in the domain of the functions. Counter example to show that the equation is not true for even one value is an effective way to answer at times.

To show that, $\sin(A+B) = \sin A + \sin B$ is not an Identity let's take $A = 30°$ and $B = 60°$

$$RS = \sin 30° + \sin 60° = \frac{1}{2} + \frac{\sqrt{3}}{2} = \frac{1+\sqrt{3}}{2}$$

$$LS = \sin(30° + 60°) = \sin 90° = 1$$

Since for $A = 30°$ and $B = 60°$ the equation $\sin(A+B) \neq \sin A + \sin B$ the given equation is not an identity.

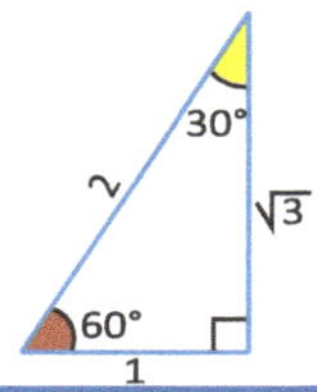

Guided Practice

Example 2: Identify Mistake

Provide counterexample to show that $sin\theta = \sqrt{1 - cos^2\theta}$ is not an identity

Hint: Try: $\theta = 60°, 120°, 240°, 300°$

Student's Activity

Prove **Complementary Angle or Cofunction Identity** with the help of right triangle ABC shown below.

$\angle A + \angle C = 90°$ or $\angle C = 90° - \angle A$

Cofunction Identity

$\sin(90° - \theta) = \cos\theta$

$\cos(90° - \theta) = \sin\theta$

$\tan(90° - \theta) = \cot\theta$

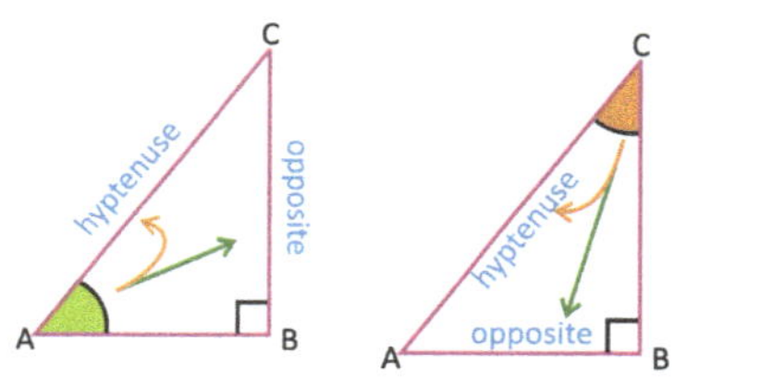

Hint:

$$\sin(A) = \frac{CB}{AC}$$

$$\cos C = \cos(90° - A) = \frac{CB}{AC}$$

Cofunction Identity Applications to Evaluate Expressions

Example: Evaluate each Expression

1.	$\frac{\sin 17°}{\cos 73°}$	2.	$\frac{\tan 23°}{\cot 67°}$
3.	$\tan 23° \tan 48° \tan 42° \tan 67°$	4.	$\tan 5° \tan 25° \tan 30° \tan 65° \tan 85°$

Express Trigonometric Ratio in Terms of Other Ratios

Pythagorean Identities: $\sin^2\theta + \cos^2\theta = 1 \quad 1 + \tan^2\theta = \sec^2\theta \quad 1 + \cot^2\theta = \csc^2\theta$

Rearranging $\sin^2\theta + \cos^2\theta = 1$

	Critical Understanding
$\sin^2\theta + \cos^2\theta = 1$ $\sin^2\theta = 1 - \cos^2\theta$ $\|\sin\theta\| = \sqrt{1-\cos^2\theta}$ **or** $\sin\theta = \pm\sqrt{1-\cos^2\theta}$ In quadrant I (all ratios are positive) So, for $0° < \theta < 90°$ You may write $\sin\theta = \sqrt{1-\cos^2\theta}$ Not otherwise.	$\sqrt{x^2} \neq x$ $\sqrt{x^2} = \|x\|$ You can prove the above statements with counter example. If $x = -7$, then $\sqrt{(-7)^2} = \sqrt{49} = 7 \neq -7$ So, $\sqrt{x^2} = x, x \geq 0$ Generally: $\sqrt{x^2} = \|x\|$

We can rearrange the Trigonometric Pythagorean identities as follows:

$\sin^2\theta + \cos^2\theta = 1$	$1 + \tan^2\theta = \sec^2\theta$	$1 + \cot^2\theta = \csc^2\theta$
$\cos\theta = \pm\sqrt{1-\sin^2\theta}$	$\sec\theta = \pm\sqrt{1+\tan^2\theta}$	$\csc\theta = \pm\sqrt{1+\cot^2\theta}$

$$\sin\theta = \pm\sqrt{1-\cos^2\theta} = \tan\theta\cos\theta = \frac{\tan\theta}{\sec\theta} = \pm\frac{\tan\theta}{\sqrt{1+\tan^2\theta}} = \frac{1}{\csc\theta} = \pm\frac{1}{\sqrt{1+\cot^2\theta}}$$

$$\cos\theta = \pm\sqrt{1-\sin^2\theta} = \frac{1}{\sec\theta} = \pm\frac{1}{\sqrt{1+\tan^2\theta}} = \pm\frac{\cot\theta}{\sqrt{1+\cot^2\theta}} = \pm\frac{\sqrt{\csc^2\theta - 1}}{\csc\theta}$$

Explain why $\cos\theta = \frac{1}{\sqrt{1+\tan^2\theta}}$ is not an identity

Hint: Use concept of counterexample.

Watch Video **Sine in Tangent Identity Strategy** SCAN ME

Example 1: Rearrange Pythagorean Identities

Write $\sin\theta, \cos\theta, and\ \sec\theta$ in terms of $\cot\theta$

$\csc^2\theta = 1+\cot^2\theta$ $\dfrac{1}{\sin^2\theta} = 1+\cot^2\theta$ $\sin^2\theta = \dfrac{1}{1+\cot^2\theta}$ $\sin\theta = \pm\dfrac{1}{\sqrt{1+\cot^2\theta}}$	$\cos\theta = \cot\theta\sin\theta$ $\cos\theta = \pm\cot\theta\dfrac{1}{\sqrt{1+\cot^2\theta}}$ $\cos\theta = \pm\dfrac{\cot\theta}{\sqrt{1+\cot^2\theta}}$	$\sec^2\theta = 1+\tan^2\theta$ $\sec^2\theta = 1+\dfrac{1}{\cot^2\theta}$ $\sec^2\theta = \dfrac{\cot^2\theta+1}{\cot^2\theta}$ $\sec\theta = \pm\dfrac{\sqrt{\cot^2\theta+1}}{\cot\theta}$
For $0° < \theta < 90°$ $\boldsymbol{\sin\theta = \dfrac{1}{\sqrt{1+\cot^2\theta}}}$	For $0° < \theta < 90°$ $\boldsymbol{\cos\theta = \dfrac{\cot\theta}{\sqrt{1+\cot^2\theta}}}$	For $0° < \theta < 90°$ $\boldsymbol{\sec\theta = \dfrac{\sqrt{\cot^2\theta+1}}{\cot\theta}}$

Guided Practice

Consider $0° < \theta < 90°$. Prove each of the following identities

$\sin\theta = \dfrac{\tan\theta}{\sqrt{1+\tan^2\theta}}$	$\cos\theta = \dfrac{1}{\sqrt{1+\tan^2\theta}}$	$\csc\theta = \dfrac{\sqrt{1+\tan^2\theta}}{\tan\theta}$
$\tan\theta = \dfrac{\sin\theta}{\sqrt{1-\sin^2\theta}}$	$\tan\theta = \dfrac{1}{\sqrt{\csc^2\theta-1}}$	$\sec\theta = \dfrac{\csc\theta}{\sqrt{\csc^2\theta-1}}$

Extra Practices

if $\cos\theta = 0.5$ for $0° < \theta < 90°$ Find:

a.	$\dfrac{\sqrt{1+\tan^2\theta}}{\tan\theta}$	b.	$\dfrac{\csc\theta}{\sqrt{\csc^2\theta-1}}$

Example 2: Pythagorean Trigonometric Identity with Secondary Ratios

If $\sec x + \tan x = p$, the find the value of:

a.	$\sec x$	b.	$\tan x$	c.	$\csc x$

Solution 2: Pythagorean Trigonometric Identity with Secondary Ratios

Given: $\sec x + \tan x = p$ (1)

Identity: $1 + \tan^2 x = \sec^2 x$

Rearrange Identity

$\sec^2 x - \tan^2 x = 1$

$(\sec x + \tan x)(\sec x - \tan x) = 1$

$p(\sec x - \tan x) = 1$

$\sec x - \tan x = \frac{1}{p}$ (2)

Add equation (1) and (2)

$$2\sec x = p + \frac{1}{p}$$

$$2\sec x = \frac{p^2+1}{p}$$

$$\mathbf{\sec x = \frac{p^2+1}{2p}}$$

Add equation (1) and (2)

$$2\tan x = p - \frac{1}{p}$$

$$2\tan x = \frac{p^2-1}{p}$$

$$\mathbf{\tan x = \frac{p^2-1}{2p}}$$

$$\csc x = \frac{1}{\sin x} = \frac{1/\cos x}{\sin x / \cos x} = \frac{\sec x}{\tan x}$$

$$\csc x = \frac{\frac{p^2+1}{2p}}{\frac{p^2-1}{2p}}$$

$$\mathbf{\csc x = \frac{p^2+1}{p^2-1}}$$

Challenge Problem

If $\csc x + \cot x = q$, the find the value of $\sec x$

$$\sec x = \frac{q^2+1}{q^2-1}$$

Practices

Q1	Prove $\sin^8 x + \cos^8 x = (\sin^2 x - \cos^2 x)(1 - 2\sin^2 x\cos^2 x)$
Q2	If $a\cos x + b\sin x = p$ and $-b\cos x + a\sin x = q$, prove that $a^2 + b^2 = p^2 + q^2$

Example 3: Extended Skills

If $\cos x + \sin x = \sqrt{2}\cos x$ show that $\cos x - \sin x = \sqrt{2}\sin x$

Solution 3:

$\cos x + \sin x = \sqrt{2}\cos x$

$\sin x = \sqrt{2}\cos x - \cos x$

$\sin x = \cos x\left(\sqrt{2} - 1\right)$ Factor and then multiply by $\left(\sqrt{2} + 1\right)$

$\left(\sqrt{2} + 1\right)\sin x = \cos x\left(\sqrt{2} - 1\right)\left(\sqrt{2} + 1\right)$ $\left(\sqrt{2} - 1\right)\left(\sqrt{2} + 1\right) = 2 - 1 = 1$

$\sqrt{2}\sin x + \sin x = \cos x$

$\sqrt{2}\sin x = \cos x - \sin x$

Example 4: Ratios in a Triangle

$A, B, and\ C$ are the interior angles of a triangle. Show that

$\tan\left(\frac{A+B}{2}\right) = \cot\frac{C}{2}$

Solution 4:

In any ΔABC, $A + B + C = 180°$ $A + B = 180° - C$ $\frac{A+B}{2} = \frac{180° - C}{2}$ $\frac{A+B}{2} = 90° - \frac{C}{2}$	$\tan\left(\frac{A+B}{2}\right)$ Cofunction Identity $= \tan\left(90° - \frac{C}{2}\right)$ $= \cot\frac{C}{2}$

✍ Practices

Q1	In a triangle ABC, $\sin(A + B - C) = \frac{1}{2}$ and $\cos(B + C - A) = \frac{1}{\sqrt{2}}$ find the angle A. angle $A = 67.5°$

Simplify Trigonometric Expressions: 4 Strategies

Simplify Trigonometric Expressions Using Algebraic Skills

Here are few Important Strategies that can be applied to simplify trigonometric expressions.

- Re-write expression in terms of sine and cosine
- Factor and then simplify
- Simplify taking common denominator
- Rationalize using conjugate of expression

Strategy 1: Re-write expression in terms of sine and cosine

Example 1: Simplify Expression	Example 2: Simplify Expression
$\tan\theta\cos\theta$	$\csc\theta\tan\theta\cos\theta$
Solution 1: $\tan\theta\cos\theta = \frac{\sin\theta}{\cos\theta}\times\cos\theta = \sin\theta$	**Solution 2:** $\frac{1}{\sin\theta}\times\frac{\sin\theta}{\cos\theta}\times\cos\theta = 1$

Example 3: Re-write in terms of sine and cosine to simplify			
a.	$\frac{\tan\theta}{\sec\theta}$	b.	$\frac{1+\tan\theta}{1+\cot\theta}$
Solution 3: Re-write in terms of sine and cosine to simplify			
a.	$\frac{\tan\theta}{\sec\theta} = \frac{\sin\theta/\cos\theta}{1/\cos\theta}$ $= \frac{\sin\theta}{\cos\theta}\times\frac{\cos\theta}{1}$ $= \sin\theta$	b.	$\frac{1+\tan\theta}{1+\cot\theta} = \frac{1+\frac{\sin\theta}{\cos\theta}}{1+\frac{\cos\theta}{\sin\theta}}$ $= \frac{\frac{\cos\theta+\sin\theta}{\cos\theta}}{\frac{\sin\theta+\cos\theta}{\sin\theta}} = \frac{\sin\theta}{\cos\theta} = \tan\theta$
✍ **Practice: Simply the following**			
a.	$\frac{\cot\theta-\cos\theta}{\cot\theta-\cos\theta}$	b.	$\frac{\csc\theta-1}{\csc\theta+1}$

Strategy 2: Factor and then Simplify

Example 4: Factor and then simplify			
a.	$\frac{\cos^2\theta + \cos\theta}{\sin^2\theta}$	b.	$\frac{1 + sin^2x}{1 - sin^4x}$
Solution 4: Factor and then simplify			
a.	$\frac{\cos^2\theta + \cos\theta}{\sin^2\theta}$ $= \frac{\cos^2\theta + \cos\theta}{\sin^2\theta} = \frac{\cos\theta(\cos\theta + 1)}{1 - \cos^2\theta}$ $= \frac{\cos\theta(\cos\theta + 1)}{(1 + \cos\theta)(1 - \cos\theta)}$ $= \frac{\cos\theta}{(1 - \cos\theta)}$	b.	$\frac{1 + \sin^2 x}{1 - \sin^4 x}$ $= \frac{1 + \sin^2 x}{1 - \sin^4 x}$ $= \frac{1 + \sin^2 x}{(1 + \sin^2 x)(1 - \sin^2 x)}$ $= \frac{1}{1 - \sin^2 x} = \frac{1}{\cos^2 x} = \sec^2 x$

Strategy 3: Common Denominator

Example 5: Simplify taking common denominator			
a.	$\frac{1}{1 - \sin x} + \frac{1}{1 + \sin x}$	b.	$\frac{1 - \sin\theta}{\cos\theta} + \frac{\cos\theta}{1 - \sin\theta}$
Example 5: Simplify taking common denominator			
a.	$\frac{1}{1 - \sin x} + \frac{1}{1 + \sin x} = \frac{(1 + \sin x) + (1 - \sin x)}{(1 - \sin x)(1 + \sin x)} = \frac{2}{1 - \sin^2 x} = \frac{2}{\cos^2 x}$ $= 2\sec^2 x$		
b.	$\frac{1 - \sin\theta}{\cos\theta} + \frac{\cos\theta}{1 - \sin\theta} = \frac{(1 - \sin\theta)^2 + \cos^2\theta}{\cos\theta(1 - \sin\theta)} = \frac{1 - 2\sin\theta + \sin^2\theta + \cos^2\theta}{\cos\theta\,(1 - \sin\theta)}$ $= \frac{1 - 2\sin\theta + 1}{\cos\theta\,(1 - \sin\theta)} = \frac{2 - 2\sin\theta}{\cos\theta\,(1 - \sin\theta)} = \frac{2(1 - \sin\theta)}{\cos\theta\,(1 - \sin\theta)} = \frac{2}{\cos\theta} = 2\sec\theta$		

Strategy 4: Rationalization

Example 6: Rationalize using conjugate of an expression

a.	$\sqrt{\dfrac{1+\cos x}{1-\cos x}}$	b.	$\dfrac{1}{\sec x - \tan x}$

Solution 6: Rationalize using conjugate of an expression

$$(\boldsymbol{a}+\boldsymbol{b})(\boldsymbol{a}-\boldsymbol{b}) = \boldsymbol{a}^2 - \boldsymbol{b}^2$$

a.	$= \sqrt{\dfrac{1+\cos x}{1-\cos x} \times \dfrac{1+\cos x}{1+\cos x}}$ $= \sqrt{\dfrac{(1+\cos x)^2}{1-\cos^2 x}}$ $= \sqrt{\dfrac{(1+\cos x)^2}{\sin^2 x}}$ $= \left\lvert\dfrac{1+\cos x}{\sin x}\right\rvert$	b.	$= \dfrac{1}{\sec x - \tan x} \times \dfrac{\sec x + \tan x}{\sec x + \tan x}$ $= \dfrac{\sec x + \tan x}{\sec^2 x - \tan^2 x}$ $= \sec x + \tan x$ $= \dfrac{1}{\cos x} + \dfrac{\sin x}{\cos x}$ $= \dfrac{1+\sin x}{\cos x}$

Practice to simplify following expression

a.	$\sqrt{\dfrac{1-\sin x}{1+\sin x}}$	b.	$= \dfrac{1}{\csc x - \cot x}$
c.	Simplify left side (LS) to show that it is equal to the right side (RS) $\sqrt{\dfrac{1-\sin x}{1+\sin x}} + \sqrt{\dfrac{1+\sin x}{1-\sin x}} = \dfrac{2}{\lvert\cos x\rvert}$		

Top Strategies to Prove Trigonometric Identities

1. Work on One Side

- Start with the more complicated side (LHS or RHS).
- Aim to transform it into the simpler or target form.
- Avoid working on both sides simultaneously.

2. Convert to Sine and Cosine

- Rewrite all trig functions (like tan, sec, cot) in terms of **sine and cosine**:
  - $\tan x = \frac{\sin x}{\cos x}$
  - $\sec x = \frac{1}{\cos x}$, $\csc x = \frac{1}{\sin x}$
- Helps simplify or find common denominators.

3. Use Fundamental Identities

- Always look out for:
  - $\sin^2 \theta + \cos^2 \theta = 1$
  - $1 + \tan^2 \theta = \sec^2 \theta$
  - $1 + \cot^2\theta = \csc^2\theta$
- Replace parts of the expression using these.

4. Get a Common Denominator

- If dealing with **sums or differences** of fractions, rewrite with a **common denominator**.

- This often leads to factoring or cancellation.

5. Factor Expressions

- Factor using algebraic techniques:
  - Take out common terms: $ax - ab = a(x - b)$
  - Use difference of squares: $a^2 - b^2 = (a + b)(a - b)$
  - Factor trinomials when possible.

6. Multiply by a Conjugate to Rationalize

- Especially helpful with **expressions involving 1 ± sin x or cos x**.

7. Think Creatively, but Logically

- Try substitution, rewriting, or even working backward mentally (but show forward work).

- Look for patterns or symmetry in the expression.

Exercise 1: Simplify Trigonometric Expressions with Solutions

Guided Practice: Solutions are provided on the next page

Q1	$3 - 3\sin^2 x$	Q2	$\cos^3 x + \cos x \sin^2 x$
Q3	$3\sin x + 7\cos x \tan x$	Q4	$\tan^2 x - 3 + \tan x + 2$
Q5	$(\cos\theta + \sin\theta)^2 + (\cos\theta - \sin\theta)^2$	Q6	$\dfrac{\sin\theta}{1+\cos\theta} + \dfrac{1+\cos\theta}{\sin\theta}$
Q7	$\dfrac{1-\sin^4 x}{1+\sin^2 x}$	Q8	$1 - \sin x \cos x \tan x$
Q9	$\dfrac{\sin\theta\,(1+\cos\theta)}{\cos^2\theta + \cos\theta}$	Q10	$\tan^2\theta\cos^2\theta + \cot^2\theta\sin^2\theta$

Solution Exercise 1: Simplify Trigonometric Expressions

Q1	$3 - 3\sin^2 x = 3\,(1 - \sin^2 x) = \;3\cos^2 x$
Q2	$\cos^2 x + \cos x \sin^2 x = \cos x(\cos^2 x + \sin^2 x) = \cos x$
Q3	$3\sin x + 7\cos x \tan x = 3\sin x + 7\sin x = 3\sin + 7\sin x = 10\sin x$
Q4	$\tan^2 x - \;3\tan + 2 = (\tan x - 1)\,(\tan x - 2)$
Q5	$(\cos\theta + \sin\theta)^2 + \;(\cos\theta - \sin\theta)^2$ $= \cos^2\theta + 2\sin\theta\cos\theta + \sin^2\theta + \cos^2\theta - 2\sin\theta\cos\theta + \sin^2\theta$ $= 2\,(\cos^2\theta + \sin^2\theta) = \;2$
Q6	$\dfrac{\sin\theta}{1+\cos\theta} + \dfrac{1+\cos\theta}{\sin\theta} = \dfrac{\sin^2\theta + \;(1 + cos\,\theta)^2}{(1+\cos\theta)\sin\theta} = \dfrac{\sin^2\theta + 1 + 2\cos\theta + \cos^2\theta}{\sin\theta\,(1+\cos\theta)}$ $= \dfrac{2 + 2\cos\theta}{\sin\theta\,(1+\cos\theta)} = \dfrac{2(1+\cos\theta)}{\sin\theta(1+\cos\theta)} = \dfrac{2}{\sin\theta} = 2\csc\theta$
Q7	$\dfrac{1-\sin^4 x}{1+\sin^2 x} = \dfrac{(1-\sin^2 x)(1+\sin^2 x)}{1+\sin^2 x} = 1 - \sin^2 x = \cos^2 x$

Q8	$1-\sin x\cos x\tan x = 1-\sin x\cos x\dfrac{\sin x}{\cos x} = 1-\sin^2 x = \cos^2 x$
Q9	$\dfrac{\sin\theta(1+\cos\theta)}{\cos^2\theta+\cos\theta} = \dfrac{\sin\theta(1+\cos\theta)}{\cos\theta(\cos\theta+1} = \dfrac{\sin\theta}{\cos\theta} = \tan\theta$
Q10	$\tan^2\theta\cos^2\theta+\cot^2\theta\sin^2\theta = \dfrac{\sin^2\theta}{\cos^2\theta}.\cos^2\theta+\cos^2\theta = 1$

Trigonometric Identity Proof and Test

Examples 1: Prove Identities (Technique: Basic Concepts)

a.	$\tan\theta\cos\theta = \sin\theta$	b.	$(1-\cos\theta)(1-\cos\theta) = \sin^2\theta$
c.	$\tan^2\theta-\sin^2\theta = \tan^2\theta\sin^2\theta$	d.	$\cot\theta+\tan\theta = \csc\theta\sec\theta$

Examples 2: Prove Identity (Technique: Convert to $\sin\theta$ and $\cos\theta$)

a.	$\dfrac{\tan^2\theta}{1+\tan^2\theta} = \sin^2\theta$	b.	$\tan\theta+\dfrac{1}{\tan\theta} = \dfrac{1}{\cos\theta\sin\theta}$
c.	$\dfrac{\cos\theta\sin\theta}{\tan\theta} = 1-\sin^2\theta$	d.	$\dfrac{\sin x+\tan x}{1+\cos x} = \tan x$

Example 3: Prove Identities and State Restrictions (Technique: Rationalization)

a.	$\dfrac{1-\cos\theta}{\sin\theta} = \dfrac{\sin\theta}{1+\cos\theta}$	b.	$\dfrac{\sec\theta-1}{\sec\theta+1} = \left(\dfrac{\sin\theta}{1+\cos\theta}\right)^2$

Examples 4: Prove Identity (Technique: Factoring)

a.	$\dfrac{\sin\theta-2\sin^3\theta}{2\cos^3\theta-\cos\theta} = \tan\theta$	b.	$\dfrac{\sin^4\theta-\cos^4\theta}{\cos^2\theta+\sin\theta\cos\theta} = \tan\theta-1$

Example 5: Prove Identities (Technique: Creative Thinking)

a.	$\frac{\tan\theta+\sec\theta-1}{\tan\theta-\sec\theta+1}=\frac{1+\sin\theta}{\cos\theta}$	b.	$\frac{\sin^2 x+2\cos x-1}{\sin^2 x+3\cos x-3}=\frac{\cos^2 x+\cos x}{-\sin^2 x}$

Solution Examples 1: Prove Identities (Technique: Basic Concepts)

1a.	$\tan\theta\cos\theta=\sin\theta$ $\boldsymbol{LS}$: $\boldsymbol{\tan\theta\cos\theta}$ $=\frac{\sin\theta}{\cos\theta}\times\cos\theta$ $=\boldsymbol{\sin\theta=RS}$	1b	$(1-\cos\theta)(1-\cos\theta)=\sin^2\theta$ $LS=(\boldsymbol{1-\cos\theta})(\boldsymbol{1-\cos\theta})$ $=1-\cos^2\theta$ $=\boldsymbol{\sin^2\theta=RS}$
1c	$\tan^2\theta-\sin^2\theta=\tan^2\theta\sin^2\theta$ $\boldsymbol{RS=\tan^2\theta\sin^2\theta}$ $=\tan^2\theta(1-\cos^2\theta)$ $=\tan^2\theta-\tan^2\theta\cos^2\theta$ $=\tan^2\theta-\frac{\sin^2\theta}{\cos^2\theta}\cos^2\theta$ $=\boldsymbol{\tan^2\theta-\sin^2\theta=LS}$	1d	$\cot\theta+\tan\theta=\csc\theta\sec\theta$ $\boldsymbol{RS=\csc\theta\sec\theta}$ $=\frac{1}{\sin\theta}\times\frac{1}{\cos\theta}$ $=\frac{\sin^2\theta+\cos^2\theta}{\sin\theta\cos\theta}$ $=\frac{\sin^2\theta}{\sin\theta\cos\theta}+\frac{\cos^2\theta}{\sin\theta\cos\theta}$ $=\frac{\sin\theta}{\cos\theta}+\frac{\cos\theta}{\sin\theta}$ $=\boldsymbol{\cot\theta+\tan\theta=LS}$

Examples 2: Prove Identity (Technique: Convert to $\sin\theta$ *and* $\cos\theta$)

2a.	$\frac{\tan^2\theta}{1+\tan^2\theta}=\sin^2\theta$ LS: $\frac{\tan^2\theta}{1+\tan^2\theta}$ $=\frac{\tan^2\theta}{\sec^2\theta}=\frac{\frac{\sin^2\theta}{\cos^2\theta}}{\frac{1}{\cos^2\theta}}$ $=\sin^2\theta=RS$	2b.	$\tan\theta+\frac{1}{\tan\theta}=\frac{1}{\cos\theta\sin\theta}$ $LS=\tan\theta+\frac{1}{\tan\theta}$ $=\frac{\sin\theta}{\cos\theta}+\frac{\cos\theta}{\sin\theta}$ $=\frac{\sin^2\theta+\cos^2\theta}{\sin\theta\cos\theta}=\frac{1}{\cos\theta\sin\theta}$ $=RS$

2c.	$\dfrac{\cos\theta\sin\theta}{\tan\theta} = 1 - \sin^2\theta$ $LS = \dfrac{\cos\theta\sin\theta}{\tan\theta}$ $\dfrac{\cos\theta\sin\theta}{\frac{\sin\theta}{\cos\theta}}$ $= \cos\theta\sin\theta \times \dfrac{\cos\theta}{\sin\theta}$ $= \cos^2\theta$ $= 1 - \sin^2\theta = RS$ $\tan\theta = \dfrac{\sin\theta}{\cos\theta}$	2d.	$\dfrac{\sin x + \tan x}{1 + \cos x} = \tan x$ $LS = \dfrac{\sin x + \tan x}{1 + \cos x}$ $= \dfrac{\sin x + \frac{\sin x}{\cos x}}{1 + \cos x}$ $= \dfrac{\frac{\sin x\cos x + \sin x}{\cos x}}{1 + \cos x}$ $= \dfrac{\sin x\,(\cos x + 1)}{\cos x\,(1 + \cos x)}$ $= \dfrac{\sin x}{\cos x} = \tan x = RS$

Example 3: Prove each identity and state restrictions

3a.	$\dfrac{1 - \cos\theta}{\sin\theta} = \dfrac{\sin\theta}{1 + \cos\theta}$	3b.	$\dfrac{\sec\theta - 1}{\sec\theta + 1} = \left(\dfrac{\sin\theta}{1 + \cos\theta}\right)^2$

Solution 3: Prove each identity and state restrictions

	$\dfrac{1 - \cos\theta}{\sin\theta} = \dfrac{\sin\theta}{1 + \cos\theta}$ Start from left hand side (LHS) $LS = \dfrac{1 - \cos\theta}{\sin\theta}$ **Rationalization** $= \dfrac{1 - \cos\theta}{\sin\theta} \times \dfrac{1 + \cos\theta}{1 + \cos\theta}$ $= \dfrac{1 - \cos^2\theta}{\sin\theta\ (1 + \cos\theta)}$ $= \dfrac{\sin^2\theta}{\sin\theta\,(1 + \cos\theta)}$ $= \dfrac{\sin\theta}{1 + \cos\theta} = RS$		$\dfrac{\sec\theta - 1}{\sec\theta + 1} = \left(\dfrac{\sin\theta}{1 + \cos\theta}\right)^2$ $LS = \dfrac{\sec\theta - 1}{\sec\theta + 1} = \dfrac{\frac{1}{\cos\theta} - 1}{\frac{1}{\cos\theta} + 1}$ $= \dfrac{1 - \cos\theta}{1 + \cos\theta}$ $= \dfrac{1 - \cos\theta}{1 + \cos\theta} \times \dfrac{1 + \cos\theta}{1 + \cos\theta}$ $= \dfrac{1 - \cos^2\theta}{(1 + \cos\theta)^2} = \dfrac{\sin^2\theta}{(1 + \cos\theta)^2}$ $= \left(\dfrac{\sin\theta}{1 + \cos\theta}\right)^2 = RS$

Example 3: Restrictions Logic

3a. $\dfrac{1-\cos\theta}{\sin\theta} = \dfrac{\sin\theta}{1+\cos\theta}$

Restrictions: denominator cannot be zero

$\cos\theta \neq -1 \Rightarrow \theta \neq 180° + 360n$

$\sin\theta \neq 0 \Rightarrow \theta \neq 360°n, n \in Z$

3b. $\dfrac{\sec\theta-1}{\sec\theta+1} = \left(\dfrac{\sin\theta}{1+\cos\theta}\right)^2$

Restrictions: denominator cannot be zero

$\cos\theta \neq -1 \Rightarrow \theta \neq 270° + 360n$

$\cos\theta \neq 0 \Rightarrow \theta \neq 90°(2n+1)$

Solution Examples 4: Prove Identity (Technique: Factoring)

4a. $\dfrac{\sin\theta - 2\sin^3\theta}{2\cos^3\theta - \cos\theta} = \tan\theta$

$$LHS = \frac{\sin\theta - 2\sin^3\theta}{2\cos^3\theta - \cos\theta}$$

$$= \frac{\sin\theta\,(1-2\sin^2\theta)}{\cos\theta\,(2\cos^2\theta - 1)}$$

$$= \frac{\sin\theta\,\big(1-2(1-\cos^2\theta)\big)}{\cos\theta\,(2\cos^2\theta - 1)}$$

$$= \frac{\sin\theta\,(1-2+2\cos^2\theta)}{\cos\theta\,(2\cos^2\theta - 1)}$$

$$= \frac{\sin\theta\,(2\cos^2\theta - 1)}{\cos\theta\,(2\cos^2\theta - 1)}$$

$$= \frac{\sin\theta}{\cos\theta} = \tan\theta = RHS$$

4b. $\dfrac{\sin^4\theta - \cos^4\theta}{\cos^2\theta + \sin\theta\cos\theta} = \tan\theta - 1$

$$LHS = \frac{\sin^4\theta - \cos^4\theta}{\cos^2\theta + \sin\theta\cos\theta}$$

$$= \frac{(\sin^2\theta + \cos^2\theta)(\sin^2\theta - \cos^2\theta)}{\cos\theta\,(\cos\theta + \sin\theta)}$$

$$= \frac{(\sin^2\theta - \cos^2\theta)}{\cos\theta\,(\cos\theta + \sin\theta)}$$

$$= \frac{(\sin\theta + \cos\theta)(\sin\theta - \cos\theta)}{\cos\theta\,(\cos\theta + \sin\theta)}$$

$$= \frac{(\sin\theta - \cos\theta)}{\cos\theta} = \frac{\sin\theta}{\cos\theta} - \frac{\cos\theta}{\cos\theta}$$

$$= \tan\theta - 1 = RHS$$

Important strategies applied above

Difference of Squares

$(a+b)(a-b) = a^2 - b^2$

$\sin^2\theta + \cos^2\theta = 1$

Challenge Problem: Evaluate Expression

If $10\sin^4 x + 15\cos^4 x = 6$ find the value of $8\sec^6 x + 27\csc^6 x$

Example 5: Prove Identities (Technique: Creative Thinking)

5a.	$\frac{\tan\theta+\sec\theta-1}{\tan\theta-\sec\theta+1}=\frac{1+\sin\theta}{\cos\theta}$ $LS=\frac{\tan\theta+\sec\theta-1}{\tan\theta-\sec\theta+1}$ $=\frac{\tan\theta+\sec\theta-(\sec^2\theta-\tan^2\theta)}{\tan\theta-\sec\theta+1}$ $=\frac{\tan\theta+\sec\theta-(\sec\theta+\tan\theta)(\sec\theta-\tan\theta)}{\tan\theta-\sec\theta+1}$ $=\frac{(\tan\theta+\sec\theta)(1-(\sec\theta-\tan\theta))}{\tan\theta-\sec\theta+1}$ $=\frac{(\tan\theta+\sec\theta)(1-\sec\theta+\tan\theta)}{\tan\theta-\sec\theta+1}$ $=\frac{(\tan\theta+\sec\theta)(\tan\theta-\sec\theta+1)}{\tan\theta-\sec\theta+1}$ $=\tan\theta+\sec\theta$ $=\frac{\sin\theta}{\cos\theta}+\frac{1}{\cos\theta}$ $=\frac{1+\sin\theta}{\cos\theta}=RS$	$\mathbf{1+\tan^2\theta=\sec^2\theta}$ $\mathbf{1=\sec^2\theta-\tan^2\theta}$ **Difference of Squares** $(a+b)(a-b)$ $=a^2-b^2$

Challenge Problem: Prove the Identity

$$\frac{\sin\theta+1-\cos\theta}{\sin\theta-1+\cos\theta}=\frac{1+\sin\theta}{\cos\theta}$$

Challenge Problem: Prove the Identity

$$\frac{2\sin^2x-2\sin^4x-1}{1-\sin^2x}=\sin^2x-\cos^2x-\tan^2x$$

Challenge Problem: Prove the Identity

If $\frac{\sin x}{\sin y}=p$ and $\frac{\cos x}{\cos y}=p$ find $\tan x$ and $\tan y$

Example 5: Prove Identities (Technique: Creative Thinking)

5b.

$$\frac{\sin^2 x + 2\cos x - 1}{\sin^2 x + 3\cos x - 3} = \frac{\cos^2 x + \cos x}{-\sin^2 x}$$

$\sin^2\theta + \cos^2\theta = 1$

$\sin^2\theta = 1 - \cos^2\theta$

$$LS = \frac{\sin^2 x + 2\cos x - 1}{\sin^2 x + 3\cos x - 3}$$

$$= \frac{1 - \cos^2 x + 2\cos x - 1}{1 - \cos^2 x + 3\cos x - 3}$$

$$= \frac{-\cos^2 x + 2\cos x}{-\cos^2 x + 3\cos x - 2}$$

$$= \frac{-\cos x(\cos x + 2)}{-(\cos x - 1)(\cos x + 2)}$$

$$= \frac{\cos x}{\cos x - 1}$$

Rationalize

$$= \frac{\cos x}{\cos x - 1} \times \frac{\cos x + 1}{\cos x + 1}$$

$$= \frac{\cos^2 x + \cos x}{\cos^2 x - 1}$$

$$= \frac{\cos^2 x + \cos x}{-(1 - \cos^2 x)}$$

$\sin^2\theta = 1 - \cos^2\theta$

$$= \frac{\cos^2 x + \cos x}{-\sin^2 x} = RS$$

Challenge Application

If $\sec x + \tan x = p$

then find $\frac{p^2-1}{p^2+1}$

Step-by-step solution provided later

If $\tan x = \frac{b}{a}$ then show that $\sqrt{\frac{a+b}{a-b}} + \sqrt{\frac{a-b}{a-b}} = \frac{2\cos x}{\sqrt{\cos^2 x - \sin^2 x}}$

Normal or Perpendicular Form of Line

Trigonometry and coordinate geometry meet beautifully in this concept.
Every straight line can be described not just by its slope or intercepts, but also by **how far it lies from the origin** and **the angle it makes with the x-axis**.

This relationship gives rise to the Normal (or Perpendicular) **Form** of a line — an equation that uses **trigonometric ratios** (cosine and sine) to express a line in terms of its **distance from the origin** and **orientation in the plane**.

$x\cos\alpha + y\sin\alpha = p$ is the equation of the straight line upon which the length of the perpendicular from the origin is p and this perpendicular makes and angle of α with $x - axis$. Here is a short proof of the equation of line in normal or perpendicular form.

Figure shows the line AB at a distance p from the origin at an angle α with x-axis. $p = OQ$.

From point $P(x, y)$ on the line construct

$PL = y, OL = x, LM \perp OQ$, and $PN \perp LM$.

In ΔOML, $\cos\alpha = \frac{OM}{ON} \Rightarrow OM = OL\cos\alpha$

$OM = x\cos\alpha$

In ΔPNL, $\sin\alpha = \frac{PN}{PL} \Rightarrow PN = PL\sin\alpha$

$MQ = PN = y\sin\alpha$

$$p = OQ = OM + MQ = x\cos\alpha + y\sin\alpha$$

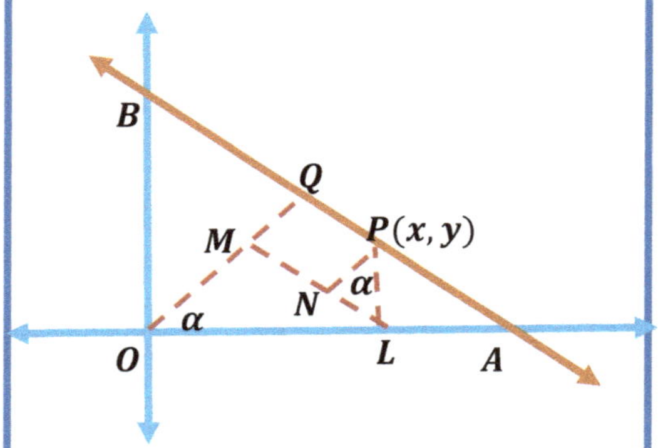

Examples

Find the equation of a line at a perpendicular distance of 5 units from the origin, making an angle of 60°with the x-axis.

$p = 5, \alpha = 60°$ Equation of line is $x\cos\alpha + y\sin\alpha = p$ $x\cos 60° + y\sin 60° = 5$	$x\frac{1}{2} + y\frac{\sqrt{3}}{2} = 5$ $x + \sqrt{3}y = 10$

Find the value of α and p if the equation $x\cos\alpha + y\sin\alpha = p$ is the normal form of line $\sqrt{3}x + y = -2$

$\sqrt{3}x + y = -2$ $-\frac{\sqrt{3}}{2}x - \frac{1}{2}y = 1$	$\cos\alpha = -\frac{\sqrt{3}}{2}; \sin\alpha = -\frac{1}{2}; p = 1$ $\alpha = 150°$

Test 1: Prove 30 Trigonometric Identities

Prove each of the following identities

Q1	$\sin x \tan x = \sec x - \cos x$	Q2	$\dfrac{\sin\theta - 1}{\cos\theta} = \tan\theta - \sec\theta$
Q3	$(1 - \cos x)\,(1 + \cos x) = \dfrac{1}{\csc^2 x}$	Q4	$1 - \sin x \cos x \tan x = \cos^2 x$
Q5	$\tan x + \cot x = \sec x \csc x$	Q6	$\cos^4 x - \sin^3 x = 1 - 2\sin^2 x$
Q7	$\sec^2 x - \sec^2 y = \tan^2 x - \tan^2 y$	Q8	$\sec^2 x - \tan^6 x$ $= 1 + 3\tan^2 x \sec^2 x$
Q9	$\dfrac{\sin x - \tan x}{1 + \cos x} = \tan x$	Q10	$\dfrac{\sin^2(-\theta) - \cos^2(-\theta)}{\cos(-\theta) - \sin(-\theta)} = \sin\theta - \cos\theta$
Q11	$1 - \dfrac{\cos^2\theta}{1 + \sin\theta} = \sin\theta$	Q12	$\dfrac{\cos^4\theta - \sin^3\theta}{\cos^2\theta} = 1 - \tan^2\theta$
Q13	$\sec^4 x + \tan^4 x$ $= 2\sec^2 x \tan^2 x + 1$	Q14	$\tan\theta = \dfrac{\sin\theta + \sin^2\theta}{\cos\theta\,(1 + \sin\theta)}$
Q15	$\sin^4\theta + \cos^4\theta = 1 - 2\sin^2\theta\cos^2\theta$	Q16	$\sec^4\theta - tan^4\,\theta = 1 + 2\tan^2\theta$
Q17	$\sin^6 x + \cos^6 x = 1 - 3\sin^2 x \cos^2 x$	Q18	$\dfrac{\sin^2\theta + 2\cos\theta - 1}{\sin^2\theta + 3 + \cos\theta - 3} = \dfrac{\cos\theta}{\cos\theta - 1}$
Q19	$(1 - \cos^2 x)(1 - \tan^2 x)$ $= \dfrac{\sin^2 x - 2\sin^4 x}{1 - \sin^2 x}$	Q20	$\dfrac{1 - \tan^2\theta}{1 + \tan^2\theta} = 2\cos^2\theta - 1$
Q21	$\tan\theta + \dfrac{1}{\tan\theta} = \dfrac{1}{\sin\theta\cos\theta}$	Q22	$\cot\theta - \tan\theta = \dfrac{2\cos^2\theta - 1}{\sin\theta\cos\theta}$
Q23	$\sec\theta - \tan\theta)^2 = \dfrac{1 - \sin\theta}{+\sin\theta}$	Q24	$\dfrac{\tan\theta - \cot\theta}{\sin\theta\cos\theta} = \tan^2\theta - \cot^2\theta$
Q25	$\dfrac{\tan\theta + \sin\theta}{\tan\theta - \sin\theta} = \dfrac{\sec\theta + 1}{\sec\theta - 1}$	Q26	$\tan^2 A - \tan^2 B = \dfrac{\sin^2 A - \sin^2 B}{\cos^2 A \cos^2 B}$

Q27	$\dfrac{\cot A + \tan B}{\cot B + \tan A} = \cot A \tan B$	Q28	$\sqrt{\dfrac{1+\sin x}{1-\sin x}} + \sqrt{\dfrac{1-\sin x}{1+\sin x}} = 2\sec x$
Q29	$\sqrt{\dfrac{1+\cos x}{1-\cos x}} = \left\lvert\dfrac{1+\cos x}{\sin x}\right\rvert$ $= \lvert\csc x + \cot x\rvert$	Q30	$\sqrt{\dfrac{1+\sin\theta}{1-\sin\theta}} = \lvert\sec\theta + \tan\theta\rvert$
Step-by-step solutions			

Solution Test 1: Prove 30 Trigonometric Identities

Q1	$RS = \sec x - \cos x = \dfrac{1}{\cos x} - \cos x = \dfrac{1-\cos^2 x}{\cos x} = \dfrac{\sin^2 x}{\cos x} == \sin x\ \tan x = LS$
Q2	$LS = \dfrac{\sin\theta - 1}{\cos\theta} = \dfrac{\sin\theta}{\cos\theta} - \dfrac{1}{\cos\theta} = \tan\theta - \sec\theta$
Q3	$1 - \cos^2 x = \sin^2 x = \dfrac{1}{\csc^2 x}$
Q4	$\text{LS}: 1 - \sin x\ \cos x\ \tan x\ = 1 - \sin x \cos x \frac{\sin x}{\cos x} = 1 - \sin^2 x = \cos^2 x$
Q5	$\tan x + \cot x = \dfrac{\sin x}{\cos x} + \dfrac{\cos x}{\sin x} = \dfrac{\sin^2 x + \cos^2 x}{\cos x \sin x} = \dfrac{1}{\sin x \cos x} == \sec x \csc x$
Q6	$\text{LS} = \cos^4 x - \sin^4 x = (\cos^2 x + \sin^2 x)(\cos^2 x - \sin^2 x) = \cos^2 x - \sin^2 x$ $= 1 - \sin^2 x - \sin^2 x = 1 - 2\sin^2 x = RS$
Q7	$\text{LS} = \sec^2 x - \sec^2 y = 1 + \tan^2 x - \ (1 + \tan^2 y) = \tan^2 x - \tan^2 y = RS$
Q8	$\text{LHS} = \sec^6 x - \tan^6 x = (\sec^2 x - \tan^2 x)\ (\sec^4 x + \sec^2 x \tan^2 x + \tan^4 x)$ $= \sec^4 x + \sec^2 \tan^2 x + \tan^4 x = (\sec^2 x - \tan^2 x)^2 + 3\tan^2 x \sec^2 x$ $= 1 + 3\tan^2 x \sec^2 x$

Watch Video **Trigonometric Challenge Identity** **SCAN ME**

Q9	$LS = \dfrac{\sin x + \tan x}{1 + \cos x} = \dfrac{\sin x + \sin x/\cos x}{1 + \cos x} = \dfrac{\sin x \cos x + \sin x}{\cos x\,(1 + \cos x)}$ $= \dfrac{\sin x}{\cos x} \tan x = RS$
Q10	$LS = \dfrac{\sin^2(-\theta) - \cos^2(-\theta)}{\cos(-\theta) - \sin(-\theta)} = \dfrac{(\sin(-\theta))^2 - (\cos(-\theta))^2}{\cos\theta + \sin\theta} = \dfrac{\sin^2\theta - \cos^2\theta}{\sin\theta + \cos\theta}$ $= \sin\theta - \cos\theta = RS$
Q11	$LS = 1 - \dfrac{\cos^2\theta}{1 + \sin\theta} = 1 - \dfrac{(1 - \sin^2\theta)}{1 + \sin\theta} = 1 - \dfrac{(1 + \sin\theta)(1 - \sin\theta)}{1 + \sin\theta}$ $= 1 - (1 - \sin\theta) = \sin\theta = RS$
Q12	$\dfrac{\cos^4\theta - \sin^4\theta}{\cos^2\theta} = \dfrac{(\cos^2\theta + \sin^2\theta)\ (\cos^2\theta - \sin^2\theta)}{\cos^2\theta} = \dfrac{\cos^2\theta - \sin^2\theta}{\cos^2\theta}$ $= \dfrac{\cos^2\theta}{\cos^2\theta} - \dfrac{\sin^2\theta}{\cos^2\theta} = 1 - \tan^2\theta$
Q13	$LS = \sec^4 x + \tan^4 x = \sec^4 x + \tan^4 x - 2\sec^2 x \tan^2 x + 2\sec^2 x \tan^2 x$ $= (\sec^2 x - \tan^2 x)^2 + \ 2\sec^2 x \tan^2 x = 1 + 2\sec^2 x \tan^2 x = RS$
Q14	$RS = \dfrac{\sin\theta + \sin^2\theta}{\cos\theta(1 + \sin\theta)} = \dfrac{\sin\theta}{\cos\theta} = \tan\theta = LS$
Q15	$LS:\ \sin^4\theta + \cos^4\theta = \sin^2\theta + \cos\theta + 2\sin^2\theta\cos^2\theta - 2\sin^2\theta\cos^2\theta$ $= (\sin^2\theta + \cos^2\theta)^2 - \ 2\sin^2\theta\cos^2\theta = 1 - 2\sin^2\theta\cos^2\theta = RS$
Q16	$LS:\sec^4\theta - \tan^4\theta = (\sec^2\theta + \tan^2\theta)\ (\sec^2\theta - \tan^2\theta) = \sec^2\theta + \tan^2\theta$ $= 1 + \tan^2\theta + \tan^2\theta = 1 + 2\tan^2\theta = RS$
Q17	$LHS:\sin^6 x + \cos^6 x = (\sin^2 x)^3 + (\cos^2 x)^3$ $= (\sin^2 x + \cos^2 x)\ (\sin^4 x - \sin^2 x\ \cos^2 x + \cos^4 x)$ $= \sin^4 x - \sin^2 x \cos^2 x + \cos^4 x$ $= \sin^4 x - \sin^2 x \cos^2 x + \cos^4 x + 2\sin^2 x \cos^2 x - 2\sin^2 x \cos^2 x$ $= (\sin^2 x + \cos^2 x)^2 - \ 3\sin^2 x \cos^2 x = 1 - 3\sin^2 x \cos^2 x = RS$

Q18	$LS\ : \dfrac{\sin^2\theta + 2\cos\theta - 1}{\sin^2\theta + 3\cos\theta - 3} = \dfrac{1-\cos^2\theta + 2\cos\theta - 1}{1-\cos^2\theta + 3\cos\theta - 3} = \dfrac{-\cos^2\theta + 2\cos\theta}{-\cos^2\theta + 3\cos\theta - 2}$ $= \dfrac{\cos^2\theta - 2\cos\theta}{\cos^2\theta - 3\cos\theta + 2} = \dfrac{\cos\theta(\cos\theta - 2)}{(\cos\theta - 2)\ (\cos\theta - 1)} = \dfrac{\cos\theta}{\cos\theta - 1} = RS$
Q19	$LS = (1-\cos^2 x)\ (1-\tan^2 x) = \sin^2 x\left(1 - \dfrac{\sin^2 x}{\cos^2 x}\right) = \sin^2 x\left(\dfrac{\cos^2 x - \sin^2 x}{\cos^2 x}\right)$ $= \dfrac{\sin^2 x\ (1 - 2\sin^2 x)}{1-\sin^2 x} = \dfrac{\sin^2 x - 2\sin^4 x}{1 - \sin^2 x} = RS$
Q20	$LS : \dfrac{1-\tan^2\theta}{1+\tan^2\theta} = \dfrac{1 - \dfrac{\sin^2\theta}{\cos^2\theta}}{1 + \dfrac{\sin^2\theta}{\cos^2\theta}} = \dfrac{\dfrac{\cos^2\theta - \sin^2\theta}{\cos^2\theta}}{\dfrac{\cos^2\theta + \sin^2\theta}{\cos^2\theta}}$ $= \dfrac{\cos^2\theta - \sin^2\theta}{\cos^2\theta} \times \dfrac{\cos^2\theta}{\cos^2\theta + \sin^2\theta} = \cos^2\theta - \sin^2\theta = 2\cos^2\theta - 1 = RS$
Q21	$LS :\ \tan\theta + \dfrac{1}{\tan\theta} = \dfrac{\sin\theta}{\cos\theta} + \dfrac{1}{\dfrac{\sin\theta}{\cos\theta}} = \dfrac{\sin\theta}{\cos\theta} + \dfrac{\cos\theta}{\sin\theta} = \dfrac{1}{\sin\theta\ \cos t\theta} = RS$
Q22	$LS: \cot\theta - \tan\theta = \dfrac{\cos\theta}{\sin\theta} - \dfrac{\sin\theta}{\cos\theta} = \dfrac{\cos^2\theta - \sin^2\theta}{\sin\theta\cos\theta} = \dfrac{\cos^2\theta - (1-\cos^2\theta)}{\sin\theta\cos\theta}$ $= \dfrac{2\cos^2\theta - 1}{\sin\theta\cos\theta} = RS$
Q23	$(\sec\theta - \tan\theta)^2 = \sec^2\theta - 2\sec\theta\tan\theta + \tan\theta$ $= \dfrac{1}{\cos^2\theta} - \dfrac{2.1}{\cos\theta} . \dfrac{\sin\theta}{\cos\theta} + \dfrac{\sin^2\theta}{\cos^2\theta}$ $= \dfrac{1 - 2\sin\theta + \sin^2\theta}{\cos^2\theta} = \dfrac{(1-\sin\theta)^2}{\cos^2\theta} = \dfrac{(1-\sin\theta)^2}{1-\sin^2\theta} = \dfrac{(1-\sin\theta)^2}{(1+\sin\theta)\ (1-\sin\theta)}$ $= \dfrac{1-\sin\theta}{1+\sin\theta} = RS$
Q24	$\dfrac{\tan\theta - \cot\theta}{\sin\theta\ \cot\theta} = \dfrac{\dfrac{\sin\theta}{\cos\theta} - \dfrac{\cos\theta}{\sin\theta}}{\sin\theta\ \cos\theta} = \dfrac{\sin^2\theta - \cos^2\theta}{\sin^2\theta\cos^2\theta} = \dfrac{\sin^2\theta}{\sin^2\theta\cos^2\theta} - \dfrac{\cos^2\theta}{\sin^2\theta\cos^2\theta}$ $= \dfrac{1}{\cos^2\theta} - \dfrac{1}{\sin^2\theta} = \sec^2\theta - \csc^2\theta = (1+\tan^2\theta) - (1+\cot^2\theta)$ $= \tan^2\theta - \cot^2\theta$

Q25	$LS = \dfrac{\tan\theta + \sin\theta}{\tan\theta - \sin\theta} = \dfrac{\dfrac{\sin\theta}{\cos\theta} + \sin\theta}{\dfrac{\sin\theta}{\cos\theta} - \sin\theta} = \dfrac{\sin\theta\left(\dfrac{1}{\cos\theta} + 1\right)}{\sin\theta\left(\dfrac{1}{\cos\theta} - 1\right)}$ $= \dfrac{\dfrac{1}{\cos\theta} + 1}{\dfrac{1}{\cos\theta} - 1} = \dfrac{\sec\theta + 1}{\sec\theta - 1} = RS$
Q26	$LS = \tan^2 A - \tan^2 B = \dfrac{\sin^2 A}{\cos^2 A} - \dfrac{\sin^2 B}{\cos^2 B} = \dfrac{\sin^2 A\,.\cos^2 B - \sin^2 B\cos^2 A}{\cos^2 A\cos^2 B}$ $= \dfrac{\sin^2 A - \sin^2 A\sin^2 B - \sin^2 B + \sin^2 A\sin^2 B}{\cos^2 A\cos^2 B} = \dfrac{\sin^2 A - \sin^2 B}{\cos^2 A\cos^2 B} = RS$
Q27	$\dfrac{\cot A + \tan B}{\cot B + \tan A} = \dfrac{\dfrac{\cos A}{\sin A} + \dfrac{\sin B}{\cos B}}{\dfrac{\cos B}{\sin B} + \dfrac{\sin A}{\cos A}} = \dfrac{\dfrac{\cos A\cos B + \sin A\sin B}{\sin A\cos B}}{\dfrac{\cos A\cos B + \sin A\sin B}{\sin B\cos A}} = \dfrac{\sin B\cos A}{\sin A\cos B}$ $= \cot A\tan B$
Q28	$\sqrt{\dfrac{1+\sin x}{1-\sin x}} + \sqrt{\dfrac{1-\sin x}{1+\sin x}} = \sqrt{\dfrac{1+\sin x}{1-\sin x} \times \dfrac{1+\sin x}{1+\sin x}} + \sqrt{\dfrac{1-\sin x}{1+\sin x} \times \dfrac{1-\sin x}{1-\sin x}}$ $= \sqrt{\dfrac{(1+\sin x)^2}{1-\sin^2 x}} + \sqrt{\dfrac{(1-\sin x)^2}{1-\sin^2 x}} = \sqrt{\dfrac{(1+\sin x)^2}{\cos^2 x}} + \sqrt{\dfrac{(1-\sin x)^2}{\cos^2 x}}$ $= \dfrac{1+\sin x}{\cos x} + \dfrac{1-\sin x}{\cos x} = \dfrac{1+\sin x + 1 - \sin x}{\cos x} = \dfrac{2}{\cos x} = 2\sec x$
Q29	$\sqrt{\dfrac{1+\cos x}{1-\cos x}} = \sqrt{\dfrac{1+\cos x}{1-\cos x}.\dfrac{1+\cos x}{1+\cos x}} = \sqrt{\dfrac{(1+\cos x)^2}{1-\cos^2 x}} = \sqrt{\dfrac{(1-\cos x)^2}{\sin^2 x}}$ $= \left\lvert\dfrac{1+\cos x}{\sin x}\right\rvert = \left\lvert\dfrac{1}{\sin x} + \dfrac{\cos x}{\sin x}\right\rvert = \lvert\csc x + \cot x\rvert = RHS$
Q30	$\sqrt{\dfrac{1+\sin\theta}{1-\sin\theta}} = \lvert\sec\theta + \tan\theta\rvert$ *follow above steps to prove the identity*

Test 2: Trigonometric Expressions with Some Solutions

Simplify Trigonometric Expression to match the correct expression

1 $\cos\theta\,(1+\sec\theta)\,(\cos\theta-1)$

a.	$\sin^2\theta$	b.	$-\sin^2\theta$
c.	$\cos^2\theta$	d.	$-\cos^2\theta$

2 $\sin^3\theta+\sin\theta\cos^2\theta-\cos\theta\tan\theta$

a.	$\sin\theta$	b.	$\cos\theta$
c.	$\tan\theta$	d.	0

3 $\sin(90°+x)-\cos(270°-x)+\sin(270°-x)$

a.	0	b.	$\sin x$
c.	$\cos x$	d.	$-\sin x$

4 $\sin^2\theta+\ 2\cos^2\theta-1$

a.	$\sin^2\theta$	b.	1
c.	$\cos^2\theta$	d.	-1

5 $\dfrac{\tan x\quad\cos x}{\tan x+\sin x}$

a.	$\dfrac{1}{\sec x\csc x}$	b.	$\dfrac{1}{\csc x+1}$
c.	$\dfrac{1}{\sec x+1}$	d.	$\dfrac{1}{\tan x+1}$

6 $\sin^2 x\left(1+\dfrac{1}{\tan^2 x}\right)$

a.	1	b.	$\sin x$
c.	$\cos x$	d.	0

7. $\dfrac{\tan x\quad\sin x}{\tan x+\sin x}$

a.	$\tan x+\sin x$	b.	$\tan x-\sin x$
c.	$\tan x+\sin x$	d.	$\cot x$

8. $\dfrac{\tan^2 x}{1+\tan^2 x}$

a.	$\tan^2 x$	b.	$\tan x$
c.	$\sin^2 x$	d.	$\cos x$

Challenge Application:

If $\boldsymbol{\csc x+\cot x=p}$ then show that

a.	$\csc x-\cot x=\dfrac{1}{p}$	b.	$\dfrac{p^2-1}{p^2+1}=\cos x$

Test 3: Prove Trigonometric

Prove the following identities

1	$\dfrac{1-\cos\theta}{1+\cos\theta}=(\csc\theta-\cot\theta)^2$	2	$\dfrac{1-\cos\theta}{\sin\theta}=\dfrac{\sin\theta}{1+\cos\theta}$
3	$\dfrac{\cos\theta}{1-\tan\theta}+\dfrac{\sin\theta}{1-\cot\theta}=\sin\theta+\cos\theta$	4	$\dfrac{\cos\theta}{1-\sin\theta}+\dfrac{\cos\theta}{1+\sin\theta}=2\sec\theta$
5	$\dfrac{\sin\theta}{\sec\theta+\tan\theta}+\dfrac{\cos\theta}{\csc\theta+\cot\theta-1}=1$	6	$\dfrac{\sin^2\theta+2\cos\theta-1}{\sin\theta+3\cos\theta-3}=\dfrac{1}{1-\sec\theta}$
7	$\dfrac{1+\cos\theta}{\sin\theta}+\dfrac{\sin\theta}{1+\cos\theta}=\dfrac{2}{\sin\theta}$	8	$\dfrac{1-\tan^2\theta}{\cot^2\theta-1}=\tan^2\theta$
9	$\dfrac{cos\theta}{1+\sin\theta}=\dfrac{1-\sin\theta}{\cos\theta}$	10	$\dfrac{\sec\theta-1}{\sec\theta+1}=\left(\dfrac{\sin\theta}{1+\cos\theta}\right)^2$
11	$\dfrac{1+\sec\theta}{\sec\theta}=\dfrac{\sin^2\theta}{1-\cos\theta}$	12	$\dfrac{1-\sin\theta}{1+\sin\theta}=(\sec\theta-\tan\theta)^2$

Challenge Application Solution:

If $\mathbf{\sec x+\tan x=p}$ then find $\frac{p^2-1}{p^2+1}$

$$\sec x+\tan x=p$$

$$\frac{1}{\cos x}+\frac{\sin x}{\cos x}=p$$

$$\frac{1+\sin x}{\cos x}=p$$

$$\frac{(1+\sin x)^2}{\cos^2 x}=p^2$$

$$\frac{p^2-1}{p^2+1}=\frac{\frac{(1+\sin x)^2}{\cos^2 x}-1}{\frac{(1+\sin x)^2}{\cos^2 x}+1}=\frac{(1+\sin x)^2-\cos^2 x}{(1+\sin x)^2+\cos^2 x}$$

$$=\frac{1+2\sin x+\sin^2 x-\cos^2 x}{1+2\sin x+\sin^2 x+\cos^2 x}$$

$$=\frac{2\sin x+\sin^2 x+\sin^2 x}{1+2\sin x+1}$$

$$=\frac{2\sin x+2\sin^2 x}{2+2\sin x}=\frac{2\sin x\,(1+\sin x)}{2(1+\sin x)}=\sin x$$

Watch Video **Trigonometric Identities** SCAN ME

Solution 7

7	$\dfrac{1+\cos\theta}{\sin\theta}+\dfrac{\sin\theta}{1+\cos\theta}=\dfrac{2}{\sin\theta}$ $LHS=\dfrac{1+\cos\theta}{\sin\theta}+\dfrac{\sin\theta}{1+\cos\theta}=\dfrac{(1+\cos\theta)^2+\sin^2\theta}{\sin\theta(1+\cos\theta)}$ $=\dfrac{1+2\cos\theta+\cos^2\theta+\sin^2\theta}{\sin\theta(1+\cos\theta)}$ $=\dfrac{1+2\cos\theta+1}{\sin\theta(1+\cos\theta)}=\dfrac{2+2\cos\theta}{\sin\theta(1+\cos\theta)}=\dfrac{2(1+\cos\theta)}{\sin\theta\ (1+\cos\theta}=\dfrac{2}{\sin\theta}=RHS$

Solution Test 2: Trigonometric Expressions

Express Trigonometric Expression in simpler form to match the simplified expression

1. B	$\cos\theta\ (1+\sec\theta)\ (\cos\theta-1)$ $=(\cos\theta+\cos\theta\ \sec\theta)\ (\cos\theta-1)$ $=(\cos\theta+1)\ (\cos\theta-1)$ $=\cos^2\theta-1=\ -(1-\cos^2\theta)$ $=\ -\sin^2\theta$	2 D	$\sin^3\theta+\sin\theta\cos^2\theta-\cos\theta\tan\theta$ $=\sin\theta\ (\sin^2\theta+\cos^2\theta)$ $-\cos\theta\dfrac{\sin\theta}{\cos\theta}$ $=\sin\theta-\sin\theta=0$
3 B	$\sin(90°+x)-\cos(270°-x)$ $+\sin(270°-x)$ $=\cos x+\sin x-\cos x$ $=\sin x$	4 A	$\sin^2 x+2\cos^2\theta-1$ $=\sin^2\theta+\cos^2\theta+\cos^2\theta\ -1$ $=1+\cos^2\theta-1=\cos^2\theta$
5 C	$\dfrac{\tan x\cos x}{\tan x+\sin x}$ $=\dfrac{\frac{\sin x}{\cos x}.\cos x}{\frac{\sin x}{\cos x}+\sin x}$ $=\dfrac{\sin x}{\sin x\left(\frac{1}{\cos x}+1\right)}$ $=\dfrac{\cos x}{1+\cos x}=\dfrac{1}{\sec x+1}$	6 A	$\sin^2 x\left(1+\dfrac{1}{\tan^2 x}\right)$ $=\sin^2 x+\sin^2 x\times\dfrac{1}{\frac{\sin^2 x}{\cos^2 x}}$ $=\sin^2 x+\sin^2 x\times\dfrac{\cos^2 x}{\sin^2 x}$ $=\sin^2 x+\cos^2 x=1$

Chapter 13: Test Trigonometry

Q1	Evaluate $\sin x \times \sec(90° - x)$
Q2	If $\csc x = 3p$ and $\cot x = \frac{3}{x}$, find the value of $9\left(p^2 - \frac{1}{p^2}\right)^2$
Q3.	If $\csc x + \cot x = p$ find the value of $\csc x - \cot x = p$
Q4.	If $\sec A = x + \frac{1}{x}$, prove that $\sec A + \tan A = 2x\ or \frac{1}{2x}$
Q5	If $\cos A = \frac{5}{12}$ where angle If A is an acute angle find the value of the expression $\frac{\tan A - \cot A}{\sin A\ \cot A}$
Q6	Point $\left(a, \frac{1}{2}\right)$ lies on the unit circle making an angle of θ with the x-axis. Find the value of $\tan\theta$
Q7	Prove the identity: $\tan^2\theta + \csc^2\theta - \cot^2\theta = \sec^2\theta$
Q8.	Prove the identity $$\frac{\sin^2\theta + 2\cos\theta - 1}{\sin\theta + 3\cos\theta - 3} = -\frac{\cos\theta + \cos^2\theta}{\sin^2\theta}$$
Q9.	A boat is located 20 km North and 14 km West of the lighthouse. Determine bearing of the boat from the lighthouse to the nearest degree.
Q10.	The angle of elevation to the top of a building from the foot of a tower is $30°$ and the angle of elevation of the top of the tower from the base of the building is $45°$. If the tower is $60\ m$ high find the height of the building.
Q11.	The angle of depression from the top of a tower to a point A on the even ground is $30°$. On moving a distance of $20\ m$ from the point A towards the foot of the tower to a point B, the angle of elevation to the top of the tower is $60°$. Find the height of the tower and its distance from the point A.
Q12	If $\frac{\sin x}{1-\cos x} + \frac{\sin x}{1+\cos x} = 4, 0° < x < 90°$ find x. $x = 30°$
Q13	Find the equation of a line at a perpendicular distance of 6 units from the origin, making an angle of $30°$with the x-axis. $\sqrt{3}x + y = 6$

Quotient Identity

$\tan\theta = \frac{\sin\theta}{\cos\theta} \qquad \cot\theta = \frac{\cos\theta}{\sin\theta}$

Reciprocal Identity

$\csc\theta = \frac{1}{\sin\theta}$; $\sec\theta = \frac{1}{\cos\theta}$; $\cot\theta = \frac{1}{\tan\theta}$

Pythagorean Identity

$\sin^2\theta + \cos^2\theta = 1$

$1 + \tan^2\theta = \sec^2\theta$

$1 + \cot^2\theta = \csc^2\theta$

Rearrange Pythagorean Identity

$\boldsymbol{\sin^2\theta = 1 - \cos^2\theta}$

$\sin^2\theta = (1 + \cos\theta)(1 - \cos\theta)$

$\cos^2\theta = 1 - \sin^2\theta$

$\cos^2\theta = (1 + \sin\theta)(1 - \sin\theta)$

$\boldsymbol{\sec^2\theta - \tan^2\theta = 1}$

$(\sec\theta + \tan\theta)(\sec\theta - \tan\theta) = 1$

$$(\sec\theta + \tan\theta) = \frac{1}{(\sec\theta - \tan\theta)}$$

$$(\sec\theta - \tan\theta) = \frac{1}{(\sec\theta + \tan\theta)}$$

$\boldsymbol{\csc^2\theta - \cot^2\theta = 1}$

$(\csc\theta + \cot\theta)(\csc\theta - \cot\theta) = 1$

$$(\csc\theta + \cot\theta) = \frac{1}{(\csc\theta - \cot\theta)}$$

$$(\csc\theta - \cot\theta) = \frac{1}{(\csc\theta + \cot\theta)}$$

Even Odd Identity

$\sin(-\theta) = -\sin\theta$

$\cos(-\theta) = -\cos\theta$

$\tan(-\theta) = -\tan\theta$

$\csc(-\theta) = -\csc\theta$

$\sec(-\theta) = -\sec\theta$

$\cot(-\theta) = -\cot\theta$

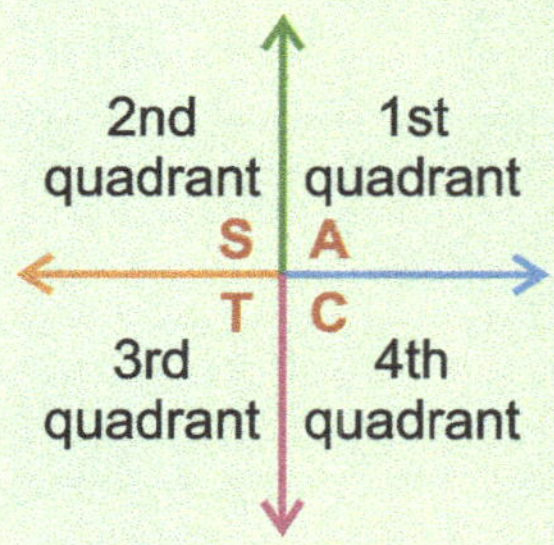

Co-Function Identity

$\sin(90° - \theta) = \cos\theta$

$\cos(90° - \theta) = \sin\theta$

$\tan(90° - \theta) = \cot\theta$

Related Acute Angle

$\sin(180° - \theta) = \sin\theta$

$\cos(180° - \theta) = -\cos\theta$

$\tan(180° - \theta) = -\tan\theta$

$\sin(180° + \theta) = -\sin\theta$

$\sin(180° + \theta) = -\cos\theta$

$\sin(180° + \theta) = \tan\theta$

$\sin(360° - \theta) = -\sin\theta$

$\sin(360° - \theta) = \cos\theta$

$\sin(360° - \theta) = -\tan\theta$

Coordinate Plane Point

$S^O_H \quad C^A_H \quad T^O_A$

$P(a,b), r = \sqrt{a^2 + b^2} > 0$

$\sin\theta = \frac{b}{r} \qquad \cos\theta = \frac{a}{r} \qquad \tan\theta = \frac{b}{a}$

$\csc\theta = \frac{r}{b} \qquad \sec\theta = \frac{r}{a} \qquad \cot\theta = \frac{a}{b}$

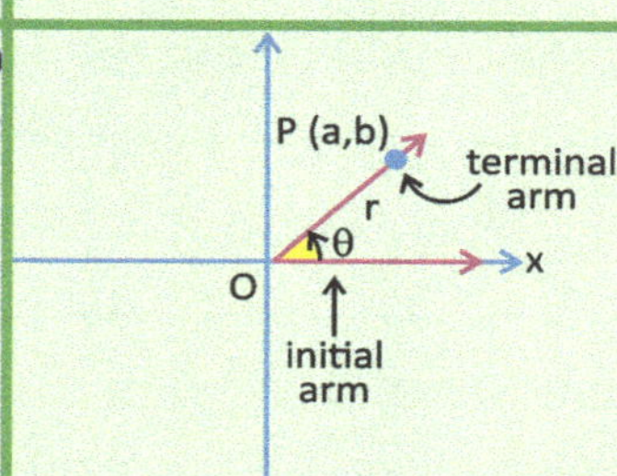

Assessment and Challenges

This section contains **25 structured tests** designed to assess understanding, accuracy, and problem-solving across algebra, geometry, and trigonometry. An **extended examination practice** set of 9**5 questions** is included to evaluate conceptual depth, procedural fluency, and application skills. The questions reflect the format, rigor, and progression commonly seen in **school examinations, AP/IB assessments, and competitive mathematics papers**. This section supports effective exam preparation through comprehensive coverage and graded challenge.

Student Spotlight: The Spognardi Family

We are deeply grateful for the guidance Anil provided to our children—Melisa, Joseph, and Nicholas—through their IB Math journey. His structured teaching, patience, and ability to simplify complex concepts not only helped them excel in high school but also gave them critical thinking skills that continue to support their success.

Today, Melisa is in medical school, while Joseph and Nicholas are thriving in university. Anil's mentorship extended far beyond exams—he prepared them for the challenges of higher education and inspired lasting confidence in their abilities.

— The Spognardi Family

Test 1: Basic Problem-Solving Skills

Answer the following questions and explain your strategy.

Q1.	Write the next three terms of the sequence: $1, 1, 2, 3, 5, 8,$
Q2.	A car travels at 60 mi/h. How many kilometres will it travel in 3 hours? 1 mi = 1.609 km 289.6 km
Q3.	The larger of two numbers is 4 more than the other. Twice the larger added to the smaller equals 86. Find the two numbers.
Q4.	Able is 7 years older than Ban. The sum of their ages is 41. How old will Able be after 7 years. 31 years
Q5.	Sum of three consecutive odd integers is 159. Find the integers.
Q6.	Unit conversion: $10\ m^3 = a\ cm^3$. Find the value of a. 10^6
Q7	Convert $4\frac{4}{5}$ out of 40 to percent. 12%
Q8	27 small cubs fit exactly inside a open top cubical box. How many of the cubes are touching the sides or the bottom of the box. 25
Q9.	A mixture contains almonds, raisins, and walnuts in the ratio of 6:5:1. If the weight of the mixture is 240 g, what is the weight of almond?
Q10.	Sam has $2.40 in nickels, dimes, and quarters. He has twice as many nickels as quarters, and twice as many quarters as dimes. How many of each does he have? Quarters = 6, Nickels = 12, Dimes = 3
Q11.	Sumit gave his Taxi driver $96.60, which included 15% tip. How much did the actual taxi ride cost, excluding the tip? $84
Q12	A salmon swims 100 m in 8 minutes downstream. Upstream, it would take the fish 20 minutes to swim the same distance. What is the speed of the salmon in still water? 8.75 m/min
Q13.	**Challenge Problem** If you research your ancestors back ten generations, how many people would you research? 2046

Test 2: Algebra and Geometry Essential

Stretch your skills with advanced arithmetic, algebra, and spatial reasoning – perfect for learners exploring AMC and real-world math.

<table>
<tr><td rowspan="4">Q1</td><td colspan="4">Write following numbers as specified</td></tr>
<tr><td>a.</td><td>2.5463 to the nearest hundredth</td><td>b.</td><td>7.8967 to one decimal place</td></tr>
<tr><td>c.</td><td>8.6420 to two significant figures</td><td>d.</td><td>4678 to two significant figures</td></tr>
<tr><td>e.</td><td>0.023567 to three significant figures</td><td>f.</td><td>1500 in standard form</td></tr>
<tr><td>Q2.</td><td colspan="4">$0.\overline{23} = \frac{p}{q}$ or 0.2323.... Find the value of p and q.</td></tr>
<tr><td>Q3.</td><td colspan="4">Evaluate $27^{1/3} + 16^{3/4} + 32^{4/5}$</td></tr>
<tr><td>Q4.</td><td colspan="4">Evaluate $\frac{3}{\sqrt{12}} - \frac{3}{\sqrt{3}}$.</td></tr>
<tr><td>Q5.</td><td colspan="4">Simplify to lowest terms $\frac{3^{-1}+3}{3^{-1}+1}$</td></tr>
<tr><td>Q6.</td><td colspan="4">Solve the equation $\frac{5}{6} = \frac{15}{2x^2}$</td></tr>
<tr><td>Q7a.</td><td colspan="4">Convert 16.234" to D° M' S" form</td></tr>
<tr><td>Q7b.</td><td colspan="4">Convert 26° 16' 30" to degrees.</td></tr>
<tr><td>Q8.</td><td colspan="4">700 marbles are among three friends Anil, Ben and Cindy. If the ratio of marbles is 4:3:7 respectively then find the number of marbles with Cindy.</td></tr>
</table>

Q9. Find the volume of a square based pyramid with base area $64\ m^2$ and four faces as equilateral triangles.

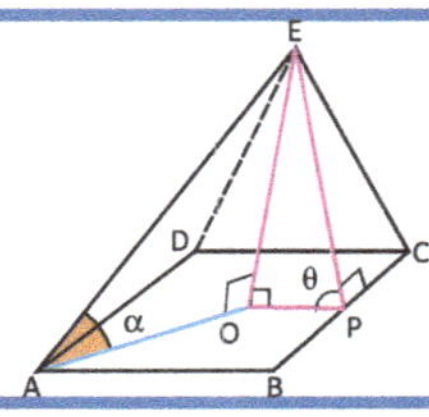

Q10. Find the range for parabola represented by the equation

$y = -5x^2 + 30x + 4.$

Hint: Range is all possible values of y. Find vertex (h, k) and notice that parabola opens down. Range is all real numbers where $y \leq k$.

Q11. Find tangent length, $\overline{PT}$ from $P(-6,8)$ to the circle $x^2 + y^2 = 4.$

Centre of the circle is at the origin.

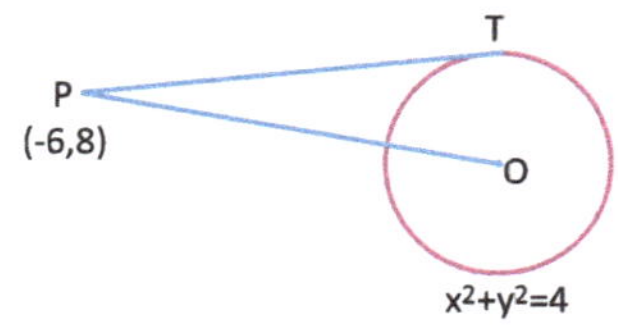

Q12. Solve following equations.

a.	$\frac{1}{3}x + 5 = 7$	b.	$9 - (x+1)^2 = 0$
c.	$\frac{12 - x}{x + 3} = 2$	d.	$2x^2 - 3x - 1 = 0$

Q13 Complete the equation using each digit from 1 to 5 once for each alphabet.

$$15 + (-A) - (-B) + C \times DE = 65$$

Q14 Domain of the function $f(x)$ is $-3 \leq x \leq 7$. Find the domain of $y = -2f(x - 1)$

Q15 If $f(x) = 2x - 1$ and $g(x) = x^2 - 4$ find the point(s) where $f(x) = g(x)$

Q16 Find all group of three consecutive integers for which product is equal to their sum.

Q17 Show that $2^{700} > 5^{300}$

Q18 $2^p - 2^q = 56$, find $p + q$.

Answers: Test 2: Algebra and Geometry Essentials

1a.	2.55	1b.	7.9	1c.	8.6	1d.	4.7×10^3
1e	2.36×10^{-2}	1f.	1.5×10^3	2.	$x = \frac{23}{99}$	3.	27
4.	$-\frac{\sqrt{3}}{2}$	5.	$\frac{5}{2}$	6	± 3	7a.	$16° + 14' + 2$
7b.	$26.275°$	8.	350 marbles	9	$85\sqrt{2}\,m^3$	10	$y \leq 49$
11	$4\sqrt{6}$	12a.	6	12b.	21	12c.	2
12d.	$x = \frac{3 \pm \sqrt{17}}{4}$	13	Try numbers	14	$-2 \leq x < 8$	15	$(-1,-3),(3,5)$
16	$\{-1,0,1\}; \{1,2,3\}; \{-1,-2,-3\}$			17	Hint: $2^7 = 128, 5^3 = 125$		

AP Challenge Test: Financial Application

Q1	Under certain law, the first 40,000 of earnings is subjected to a 20% tax, earnings greater than 40,000 and up to 100,000 are subjected to 30% tax, and earnings greater than 100,000 are taxed at 50%.
Q1a	Find the tax for the earning of 50,000 $11,000
Q1b	Find the tax for the earning of 100,000 $26,000
Q1c	Write piecewise function to model this situation in function notation.
Q1d	Sketch graph of this function.
Q1e	Calculate the tax due on the earning of 125,000. $38,500
Q1f	For what earning will the tax be 40,000. Estimate from your graph.
Q1g	For what earning will the tax be 40,000. Calculate algebraically. $128,000

Test 3: Equation of Lines and Applications

Q1. **Understand the Line – Explore Its Properties**

Use the graph of the given line to answer the following:

a. What is the **y-intercept**? (Where the line crosses the y-axis)

b. What is the **x-intercept**? (Where the line crosses the x-axis)

c. What is the **slope (m)** of the line? (Rate of change or steepness)

d. Write the **equation of the line** in **slope-intercept form**: $y = mx + b$

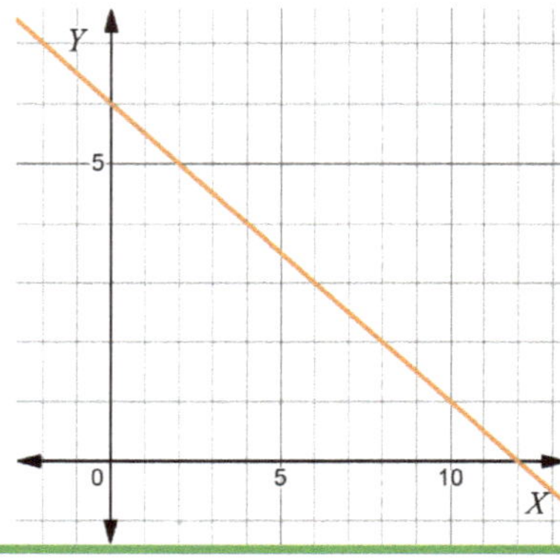

Q2 **Write the Equation of the Line**

Write the equation of each line in **slope-intercept form** $y = mx + b$:

a. The line has a **y-intercept of 3** and a **slope of 2**
(Use the form directly)

b. The line has a **y-intercept of -2** and an **x-intercept of 5**
(Hint: Use intercepts to find slope, then write the equation)

Q3 **Write the Equation of the Line**

Write the equation of a line that passes through the given pair of points. Express your answer in **slope-intercept form** $y = mx + b$.

a. Points: (−3, 4) and (2, −1)

b. Points: (0, 5) and (4, 1)

Q4 **Finding x-intercept and y-intercept**

Find x-intercept and y-intercept of the following equations

a. $2x - 4y - 3 = 0$

b. $x + 3y = 0$

Q5 Write equation of line that pass through the origin and point $(-3, -6)$

Q6. The line $5x - 2y + c = 0$ pass through the point $(2, -5)$. Find value of c.

Q7 Determine the value of $\boldsymbol{k}$ for the line $\boldsymbol{kx - 15y = 5}$ to be perpendicular to the line $2x - 3y + 4 = 0$

$k = -22.5$

Q8. **Rearranging to Slope-Intercept Form**

Rearrange each of the following linear equations into the form: $y = mx + b$. (Slope-Intercept Form)
Then, identify the **slope (m)** and **y-intercept (b)**.

a.	$2x - 3y - 6 = 0$	b.	$\frac{x}{3} - \frac{y}{4} = 1$

Q9

Finding Equations of Horizontal and Vertical Lines

Horizontal Line: A horizontal line has **zero slope (m = 0)** and runs **left to right**.
General Form: $y = k$
where k is a constant — the **y-coordinate** of all points on the line.
Vertical Line: A vertical line has an **undefined slope** and runs **up and down**.
General Form: $x = h$
where h is a constant — the **x-coordinate** of all points on the line.

a. Find equation of horizontal line through the point (2, −5)
b. Find the equation of vertical line through the point (7, 11)

Q10.

Write equation of the line shown in the graph

a.

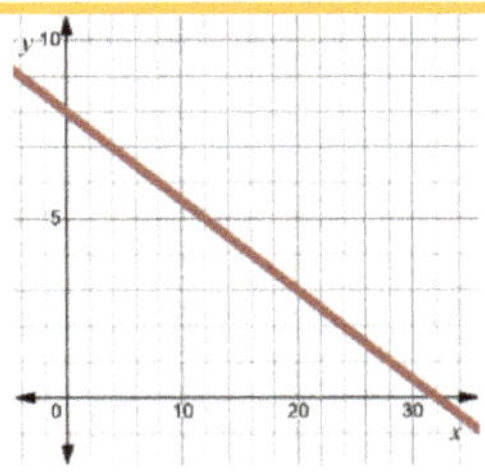

b.

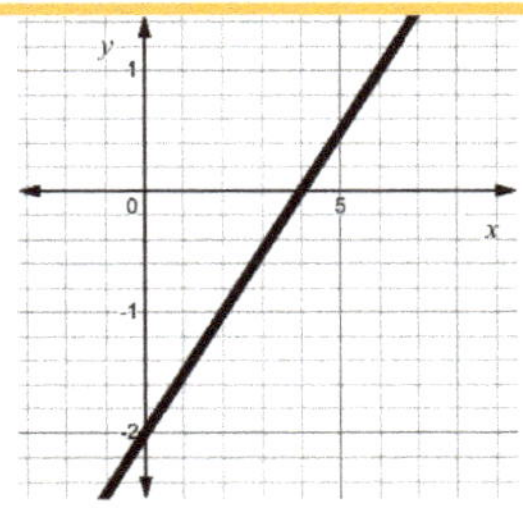

Q11

SAT Challenge

In the figure, one vertex of each square lies on the line as shown. If the length of the sides of the three squares is 4, 3, and p, what is the value of p?

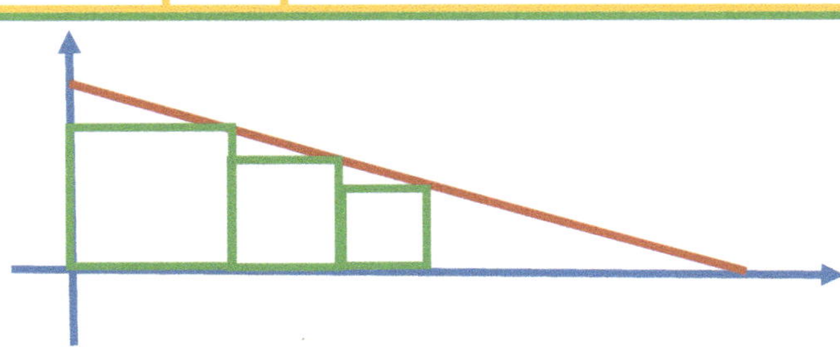

Q12.

Parallel and Perpendicular Lines

Use the concept of slope to answer the following questions. Remember:

- **Parallel lines** have the **same slope**.
- **Perpendicular lines** have slopes that are **negative reciprocals** of each other.
  a. Write the equation of a line **parallel** to $y = 2x + 5$ and passing through the point (3, −1).
  b. Write the equation of a line **perpendicular** to $y = -\frac{1}{2}x + 4$ and passing through the point (−4, 2).

Q13.

Write linear equation from the given data and then provide three other points on this line

x	-1	1	3			
y	-2	-1	0			

Q14

Determine the $x-$ and $y-$ intercepts of the line $y - 2 = \frac{5}{3}(x + 3)$. Use the intercepts to find the area of the triangle formed by the line and axes.

Q15

$\frac{x}{2} + \frac{y}{-3} = 1$. Sketch the line on the coordinate plane.
Write the equation in standard form and then find two points on this line.

Q16.	**Movie Rentals** A video rental store charges a one-time membership fee of \$10, plus \$4 for each movie rented. • Write a linear equation for the total cost C of renting xxx movies. • How many movies can be rented for \$42? • What does the slope and y-intercept represent in this context?
Q17	**School Bus Trip** A charter bus costs \$250 to hire, plus \$15 per student. • Write an equation for the total cost C in terms of the number of students $\boldsymbol{n}$. • If the total cost is \$625, how many students are on the trip? • Sketch a graph and label the intercept.
Q18	A 100-liter tank is leaking at a steady rate of 2 liters per minute. • Write a linear equation to show the amount of water left $\boldsymbol{W}$after $\boldsymbol{t}$ minutes. • How much water is left after 30 minutes? • After how many minutes will the tank be empty?
Q19	For what value of p will the following lines have infinitely many solutions: $(p-3)x+3y=p; px+py=12$ $p=6$
Q20	Determine the $x-$ and $y-$ intercepts of the line $y-2=\frac{5}{3}(x+3)$. Use the intercepts to find the area of the triangle with line and the axes.
Q21	$\frac{x}{2}+\frac{y}{-3}=1$. Sketch the line on the coordinate plane. Write the equation in standard form and then find two points on this line.
Q22	Find the equation of line perpendicular to $2x+5y=0$ that passes through the point (3, 6). $5x-2y-3=0$
Challenge Section	
Q23	Find the point closest to line $y=\frac{2}{3}x-4$ from the point $A(3,-4)$. $\left(\frac{27}{13},-\frac{34}{13}\right)$
Q24	Point $P(-1,4)$ is reflected on the $x-axis$ and then on the $y-axis$. Find the coordinates of the point after each reflection in sequence.
Q25	Find the mirror image of the point $P(-3,2)$ on the line $y=-3$
Q26	Find the mirror image of the point $P(8,6)$ on the line $y=x+2$
Q27	Find shortest distance of $P(3,6)$ from the line $y=x+1$ $\sqrt{2}$

Q28	The line $4x + 3y = 18$ pass through the point P(m, n) with intercepts A(a, 0) and B(0, b). Find coordinates of P if it divides segment AB in the ratio 1:2.
Q29	$7x + 8y + 9z = -4, 10x + 11y + 12z = -4, 13x + 14y - 16z = 89$ Above equations represent three planes. Find their point of intersection $(1, 2, -3)$
Q30	Find equation of the angle bisector(s) between the lines $3x + 4y - 5 = 0$ and $12x + 5y - 7 = 0$. Solution provided at the end
Q31	Find the locus of the points which moves so that its distance from the line $3x + 4y - 5 = 0$ is three times its distance from $12x + 5y + 2 = 0$.
Q32	Find equation of obtuse angle bisector between the lines $x + y - 6 = 0$ and $17x - y + 4 = 0$. $\mathbf{2x - 6y + 34 = 0; 12x + 4y - 26 = 0}$

Challenge Problem Solutions

Find equation of the angle bisector(s) between the lines $\mathbf{3x + 4y - 5 = 0}$ and $\mathbf{12x + 5y - 7 = 0}$.

Consider a point $P(x, y)$ on the angle bisector. It will be same distance from both the lines.

$$\left|\frac{3x + 4y - 5}{\sqrt{3^2 + 4^2}}\right| = \left|\frac{12x + 5y - 7}{\sqrt{12^2 + 5^2}}\right|$$

$$\frac{3x + 4y - 5}{5} = \pm\frac{12x + 5y - 7}{13}$$

$$13(3x + 4y - 5) = \pm 5(12x + 5y - 7)$$

Case 1: with $+ve$ sign

$$39x + 52y - 65 = +(60x + 25y - 35)$$

$$39x + 52y - 65 - (60x + 25y - 35) = 0$$

$$7x - 9y + 30 = 0$$

Case 1: with $-ve$ sign

$$39x + 52y - 65 = -(60x + 25y - 35)$$

$$39x + 52y - 65 + (60x + 25y - 35) = 0$$

$$99x + 77y - 100 = 0$$

Find the locus of the points which moves so that its distance from the line $\mathbf{3x + 4y - 5 = 0}$ is three times its distance from $\mathbf{12x + 5y + 2 = 0}$.

Solve:

$$\left|\frac{3x + 4y - 5}{\sqrt{3^2 + 4^2}}\right| = 3\left|\frac{12x + 5y + 2}{\sqrt{12^2 + 5^2}}\right|$$

Test 4: Word Problems: Linear Equations

<table>
<tr><td>Q1</td><td>Write an expression for the following phrases:<table><tr><td>a.</td><td>7 decreased from the number n</td></tr><tr><td>b.</td><td>Three times the sum of n and m.</td></tr></table></td></tr>
<tr><td>Q2</td><td>Cost of a pizza at Pizza House is $15.00 plus $1.25 for each topping.<table><tr><td>a.</td><td>Write an equation to represent the situation.</td></tr><tr><td>b.</td><td>Find the cost of pizza with 3 toppings.</td></tr><tr><td>c.</td><td>How many toppings can you have for $20?</td></tr></table></td></tr>
<tr><td>Q3</td><td>Find equation of a line passing through the points $(-2, -8)$ and $(1, -2)$.</td></tr>
<tr><td>Q4</td><td>A rectangle has length that is four times an unknown number increased by one, and width that is triple the same unknown number less two. If the perimeter is 19 cm, then determine the dimensions of the rectangle.</td></tr>
<tr><td>Q5</td><td>The sum of digits of two-digit number is 12. The number obtained by interchanging the two digits exceeds the given number by 18. Find the number</td></tr>
<tr><td>Q6</td><td>What is the solution of the equation $\frac{2}{3}x - 4 = 20$</td></tr>
<tr><td>Q7</td><td>Raisins and almonds are sold together in package of 360 g. The ratio of the mass of raisins to the mass of almonds is 3 to 5. Determine the mas of raisins in the package.</td></tr>
<tr><td>Q8.</td><td>A $860 TV is sold at a discounted price of $550.40. Find the percent discount on the TV.</td></tr>
<tr><td>Q9</td><td>Trees are planted in a row at a distance of 10 m. How many trees will be planted along a road of 1 km?</td></tr>
<tr><td>Q10</td><td>Mean or average of five numbers is 7. If the four numbers are 4, 6, 8, and 9, find</td></tr>
<tr><td>Q11</td><td>At a concert, adult ticket cost is $12 and children's ticket cost $8. If $960 was collected from sales of 90 tickets determine the number of children's tickets sold.</td></tr>
</table>

Q12	A cheetah notices an antelope 90 m away and begins to run after it. If cheetah runs at 20 m/s and the antelope runs at 10 m/s, how long will it take for the cheetah to catch the antelope?
Q13	A pigeon and a seagull begin to fly from the top of a building. Every second, the pigeon's altitude increases 3 m, while the seagull's altitude decreases 2 m. If the pigeon reaches an altitude of 66 m at the same time as the seagull reaches an altitude of 31 m, determine the height of the building.
Q14	Charles earns $50/h as a piano teacher and $25/h as a babysitter. If he worked for a total of 40 h last weekend and earned $1600, how much money did he earn babysitting.
Q15	The sum of squares of two numbers is 3538, and the difference of their squares is 2080. Determine the smallest number in this case.
Q16	Anil has 5% and 10% solution of hydrochloric acid. How much of 5% solution should be mixed to make 250 mL of 8% hydrochloric acid?
Q17	Peter rows 10 km downstream in 2 h and 8 km upstream in 4 h. Determine his average rowing speed.
Q18	Sam invested $2700, a part at 8% per annum and the rest at 10% per annum. In one year, the two parts earned equal amount of interest. How much did he invest at 10%.
Q19	A colony of 1920 is doubling every 10 days. How long will it take them to be 491520?
Q20	A used car dealer sells 5-year-old car for $4200. What was the original value of the car if the depreciation is 15% per year.

Answers: Test 4: Linear Equations and Applications

Q1	$n-7$; $3(n+m)$	Q2	5 toppings	Q3	$y = 2x - 4$	Q4	7, 5/2
Q5	57	Q6	36	Q7	135 g	Q8	36%
Q9	101	Q10	8	Q11	30	Q12	9 seconds
Q13	45 m	Q14	$400	Q15	-53	Q16	100 mL
Q17	3.5 km/h	Q18	$1200	Q19	80 days	Q20	$9466

Test 5: Quadratic Functions

Q1	Write characteristics of the parabola including intercepts, minimum value, increasing decreasing interval for $y = -x^2 - x + 6$
Q2	Write the number of real roots for the given relations 1, 0 a. $y = 2(x-3)^2$ b. $y = 3 - x + x^2$
Q3	Solve each equation using the indicated method a. $6x^2 + 5x = 1$, factorising b. $3x^2 - 6x + 1 = 0$, quadratic formula c. $-5x^2 + 30x + 10 = 0$, completing squares
Q4	For what value of k will the equation $3x^2 + 4x + k = 0$ have no zeros
Q5.	A parabolic bridge is 40 m wide. Its height is 8 m at a distance of 5 m from one end. Find the height of the bridge 12 m from the other end.
Q6	Determine two values of b so that expression $x^2 + bx + 15$ can be factored.
Q7	Height, h, in meters, of a ball thrown from the top of a building is modelled by the function, $h = -5t^2 + 25t + 30$ where the time, t, is in seconds. How long is the ball above 30 meters height?
Q8	Factor fully $y = x^4 - 4x^2 - 5$
Q9	Write equation of the following graph in standard form. a. 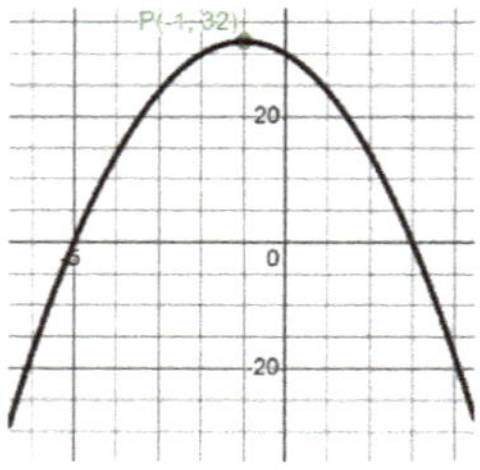 b. 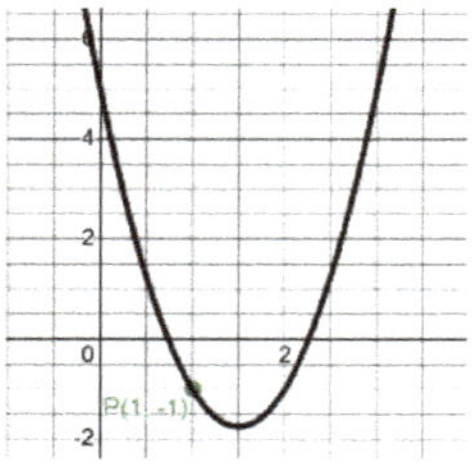

Q10	The line $y = 1 + mx$ is tangent to the graph of parabola $y = 2x - x^2$. Find possible values of m.
Q11	Two positive integers differ by 8 and the sum of their squares is 274. What are the numbers?
Q12	the product of two consecutive even integers increased by their sum is 98. Find the integers if they are less than zero.
Q13	an open box is made from a 10 cm by 20 com piece of cardboard by cutting squares with side $x\ cm$ from each corner and folding up the edges. The area of the resulting base is $96\ cm^2$. What is the length of the side of the cut out square.
Q14	Sam is twice as old as his sister. Four years hence, the product of their ages will be 160. Find their present age.
Q15	Factor: $x^2 - y^2 + 1 + 2x$
Q16	Write a quadratic equation with $y - intercept$ of 2 and roots at $-3 \pm \sqrt{5}$.
Q17	The area of an isosceles triangle is $60\ cm^2$ and the length of each one of its equal sides is $13\ cm$. Find its base.
Q18	The numerator of a fraction is three less than the denominator. If two is added to both numerator and denominator, then the sum of the new and the original fraction is $\frac{29}{20}$. Find the original fraction.
Q19	Find equation whose roots are squares of the roots of $4x^2 - 9x - 2 = 0$.
Q20	a two-digit number is such that the product of its digit is 14. When 45 is added to the number, the digits interchange their places. Find the number.
Q21	Write an equation to represent the given sequence. $2, 9, 22, 41, 66, \ldots$
Q22	$y = 18x^3 - 2x - 1$. Third finite-difference will be constant for the given cubic equation. Find the exact value of this constant finite difference.

Answers: Test 5: Quadratic Functions

Q2	1 and none	Q3a.	$-\frac{1}{2}, -\frac{1}{3}$	Q3b	$1 \pm \frac{\sqrt{6}}{3}$	Q3c	$0.76, 5.24$
Q4	$k > \frac{4}{3}$	Q5	15.36 m	Q6	8, 16	Q7	5 s
Q10	0 and 4	Q11	7, 15	Q12	-10, -12	Q13	2 cm
Q14	6, 12 years	Q16	$y = 0.5x^2 + 3x + 2$			Q17	10 cm
Q18	$\frac{7}{10}$	Q19	$y = 16x^2 - 97x + 4$			Q20	27
Q21	$3x^2 - 2x + 1$	Q22	3rd finite difference will be 3				

Test 6: Quadratic Applications

Q1	Describe each parabola by answering the following questions for each.

i.	Name Form of Equation	ii.	General Equation
iii.	Axis of Symmetry	iv.	Vertex
v.	Range	vi.	Direction of Opening
a.	$\boldsymbol{y = -5x(x-2) + 9}$	b.	$y = -2(x+1)^2 - 3$

Q2	Simplify fully

a.	$\frac{1}{2}\sqrt{8} - \frac{1}{4}\sqrt{96} + \frac{3}{5}\sqrt{50} - 9\sqrt{24}$
b.	$(2-\sqrt{5})(-2+3\sqrt{5})$

Q3	Solve. Express your final answer in proper form without using decimals. $x - 5 = 4x - (3x-2)(x+5)$
Q4	Express $y = \frac{3}{2}x^2 - 2x + 1$ in vertex form by completing the squares
Q5	Determine the number of zeros by using discriminant for $y = \frac{3}{2}x^2 - 2x + 1$
Q6	Find the value(s) of k so that the graph of $y = 3x^2 + (k-1)x + 12$ has one zero.

Q7	Show that the function $f(x) = (x - a)(x - b) - c^2$ will always have one or two zeros.
Q8	Explain the condition for the equation $a(x - p)^2 + q$ to always have 2 roots.
Q9	Determine the equation of parabola that passes through $\boldsymbol{P(3, -6)}$ if its zeros are $\boldsymbol{x = 2\ and\ 7}$ $y = -\frac{3}{2}x^2 - \frac{27}{2}x + 21$
Q10	Determine the equation of parabola that passes through $(-1, 4)$ if its zeros are $\boldsymbol{-3 \pm \sqrt{6}}$. $y = -2x^2 - 12x - 6$
Q11	Find the quadratic equation in standard form for the following roots. a. $\boldsymbol{-2\frac{1}{2}}$ b. $\boldsymbol{\frac{-3 \pm 2\sqrt{5}}{6}}$
Q12	A model rocket is shot into the air from the roof of a building. Its height, h, above the ground, measured in metres, can be modelled by the equation $h(t) = -5t^2 + 35t + 5$, where t is the time elapsed since liftoff in seconds. Note: $\boldsymbol{h(t)}$ represents height at any instant of time. This is function notation where the value of dependent variable is shown as related to the value of independent variable. $\boldsymbol{h(2)}$ means height at $\boldsymbol{t = 2}$. a. Determine the height of the building. Show your work. b. How long is the rocket in flight? c. Determine the greatest height reached by the rocket.
Q13	Find the equation of parabola with y-intercept of -8, image of $(4, -8)$, and has an x-intercept of -2, in standard form. $y = \frac{2}{3}x^2 - \frac{8}{3}x - 8$
Q14	A punter kicks a football. Its height, $h(t)$, in metres, t seconds after the kick is given by the equation $h(t) = -4.9t^2 + 18.24t + 0.8$. The height of an approaching blocker's hand is modelled by the equation $g(t) = -1.43t + 4.26$, using the same t. Can the blocker knock down the punt? If so, at what point will it happen? Yes, 0.18 s, 4 m
Q15	Without using discriminant, explain how can you tell how many roots the quadratic has.

Q16	A parabolic bridge is 40 m wide. The height of the bridge 5 m in from the outside edge is 10 m. a. Consider the ground to be the x-axis and assume the vertex lies on the y-axis. Determine the quadratic equation that models the parabolic bridge in standard form. Leave values found in fractions. A rough sketch may help. b. Determine the maximum height of the bridge. c. How far is one end of the bridge from the axis of symmetry? d. Suppose that one end of the bridge is at a point at the origin. Develop a new equation that represents the parabolic bridge. e. Describe the similarities and differences found in part a) and d).
Q17	An electronic store sells an average of 60 entertainment systems per month at an average of $800 more than the cost price. For every $20 increase in the selling price, the store sells one fewer system. What amount over the cost price will maximize revenue?
Q18	An Auto Collision Centre would like to construct a rectangular car-holding lot. The plan is to fence three sides of the lot, since the fourth side would already be bounded by the building. If there are 375 m of fencing available, what is the largest possible area of the car-holding lot?
Q19	Anita and Aditi decide to race each other, twice around a 400 m track. Amita ran 0.08 m/s slower than Aditi. If Aditi finished 20 seconds ahead of Amita, determine the speed of each runner.
Q20	Describe how will you determine the number of zeros from a quadratic function given in vertex form, $y = a(x-p)^2 + q$ Hint: Number of zeros depend on the product of a and q.
Q21	a hotel has 250 rooms. Management has determined that if they charge $150 per room on a Saturday night, they will book 210 rooms. For every $15 increase in room price, 10 less rooms are booked. What price should be charged to maximize their revenue? How many rooms will be booked?
Q22	Solve the cubic equation $$4x^3 - x - 12x^2 + 3 = 0$$

Q23 Consider a right triangle ABC. The length of the hypotenuse is 6 cm more than twice the smallest side. Find the length of the hypotenuse if the perimeter is 60 cm. . 26 cm

Q24 A projectile is launched of the top of a platform. The table gives the height of the projectile at different times during it's flight. Determine an equation in standard form that models these data point.

Tims (s)	0	1	2	3
Height (m)	11	36	51	56

Q25 Jim has difficult golf shot to make. His ball is 100 m from the hole. He wants the ball to land 5 m in front of the hole, so it can roll to the hole. A 20 m tree is between his ball and the hole, 40 m from the hole and 60 m from Jim's ball. With the base of the tree as origin, write an algebraic expression to model the height of the ball if it just clears the top of the tree.

Q26 A ball is dropped from a height of 20 m directly above the point where it will land. Its height after t seconds is given by $h = 20 - 5t^2$ (in metres). A lamppost of height 20 m stands 12 m away from the landing point. The light at the top of the lamppost casts a shadow of the ball on the horizontal ground. Find the position of the tip of the shadow from the base of the lamppost one second after the ball is released. 48 m

Q27 Expanding Circle

In a circle, the **rate of change of the surface area** with respect to the radius equals its **circumference**. This can be explored algebraically without using calculus by relating the **change in area** to the **circumference when the radius is increases by one.**

Prove that rate of change of area is equal to its circumference.

Q28 Challenge: **Factor $x^5 + x + 1$.**

$x^5 + x + 1$

$x^5 - x^2 + x^2 + x + 1$ Add and subtract x^2

$x^2(x^3 - 1) + (x^2 + x + 1)$

$x^2(x - 1)(x^2 + x + 1) + (x^2 + x + 1)$

$(x^2 + x + 1)(x^2(x - 1) + 1)$

$(x^2 + x + 1)(x^3 - x^2 + 1)$

Test 7: Pythagorean Power & Spatial Thinking

Use the Pythagorean Theorem to solve real-world problems, 3D challenges, and tricky geometry tasks. Let's apply and extend!

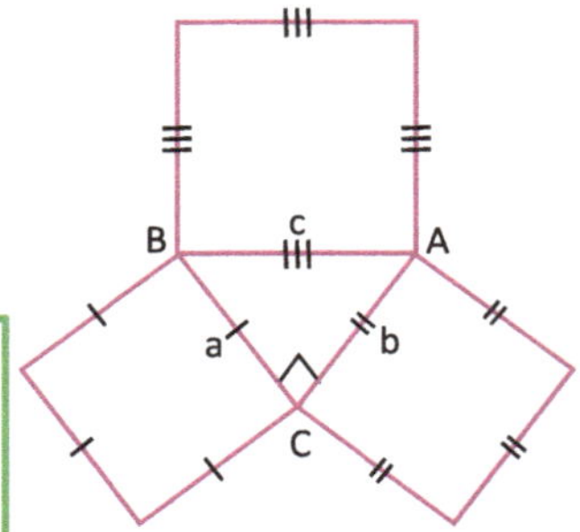

Concept Recap

Pythagorean Theorem

In a right $\Delta ABC,\ a^2 + b^2 = c^2$, where c is the hypotenuse.

$\boldsymbol{Sum\ of\ squares\ of\ two\ shorter\ sides\ =\ square\ of\ hypotenuse}$

Part A: Find the Missing Side

✍️ Solve for the missing value of $\boldsymbol{x}$. Round to the nearest tenth.

Q1	Determine the value of x in each diagram. Round your answer to nearest tenth, if required.				
a)	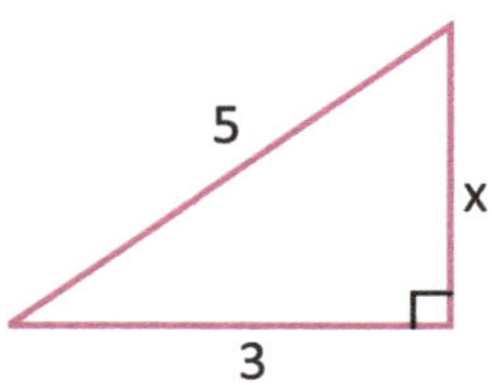	b)	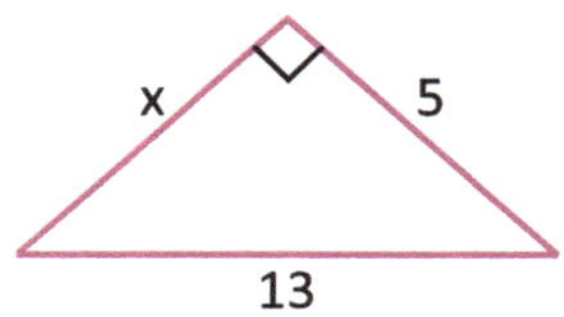	c)	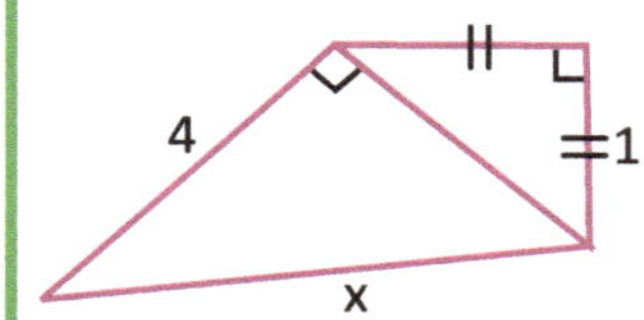
d)	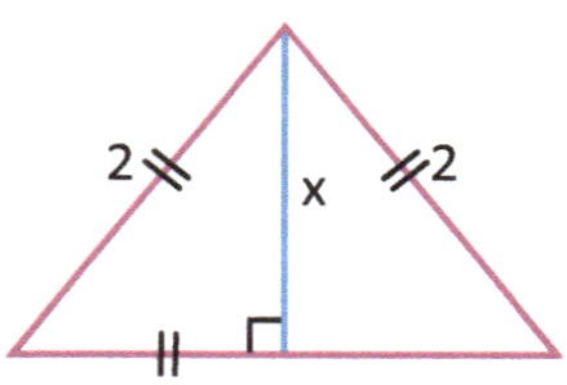	e)	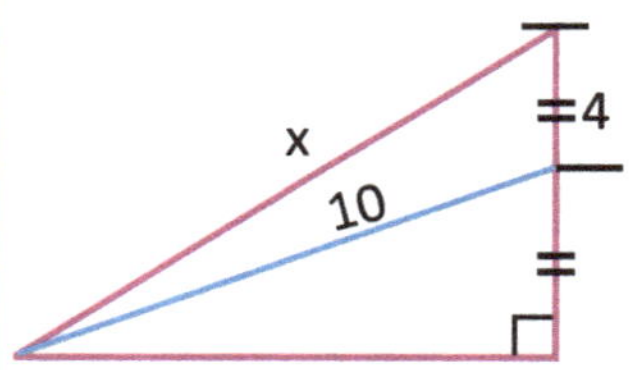	f)	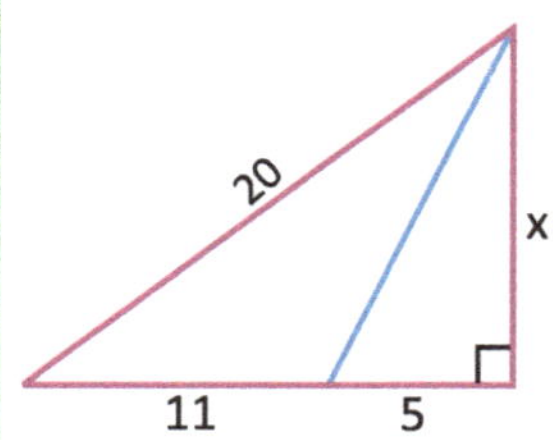

Part B: Word Problems and Applications

Q2	✈️ A kite flying at a height of 50 m is attached with a string of length 200 m. How far is the kite along the ground? Does your answer represent maximum distance? Explain.
Q3	📦 Find the longest diagonal length in a $9\ cm \times 8\ cm \times 6\ cm$ rectangular prism. Round your answer to three significant figures (3sf.) Can a $6''$ ruler be inserted in this box? (1 inch =2.54 cm)

Q4 An airplane flying at a content speed of 600 km/h at a height of 10 km crosses over the tracking radar at 11:00 am. Determine its distance from the radar at 11:05 am?

Write your answer to the nearest tenth of a mile.
Take 1 mile = 1.6 km

Part C: Challenge Zone

Q5 **Challenge**:

Suppose that m and n are positive integers with $\sqrt{7+\sqrt{48}} = m + \sqrt{n}$.

Find the value of $m^2 + n^2$.

Q6 Find the value of x in the given figure where $AB = CD$

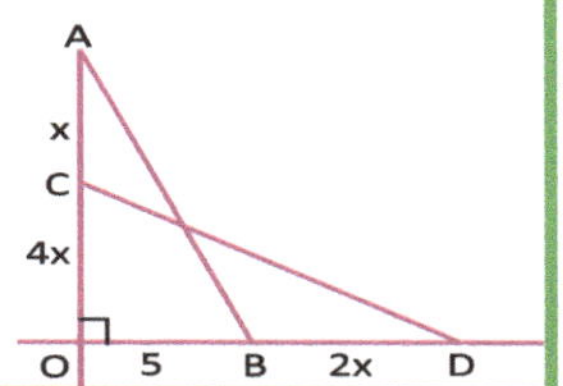

Q7 A cone has surface area of $96\pi\ cm$.

Find height of this solid cone if slant length is three times radius. (SA of cone $= \pi r^2 + \pi rl$)

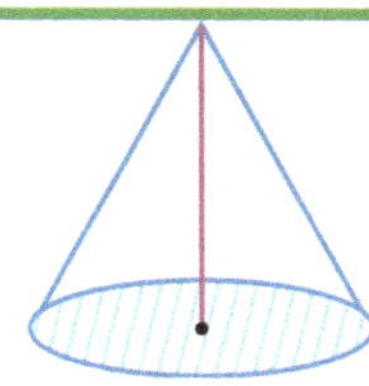

Q8 Find perimeter of the isosceles trapezoid whose parallel side are 20 and 8 cm and the height is 6 cm.

Write exact value as the answer.

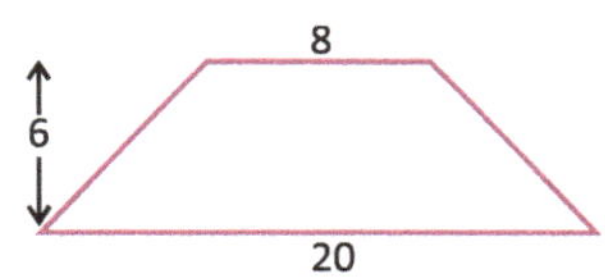

Q9 Find the value of base, b, in the given right triangle.

Explain how you can apply this result.

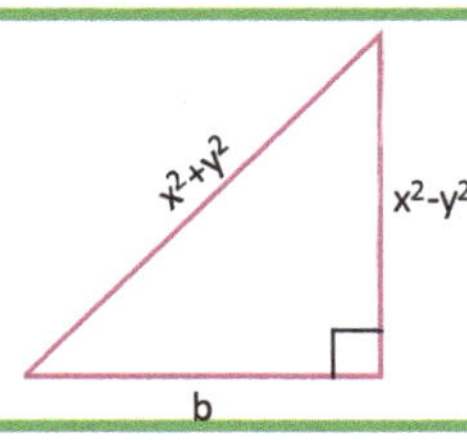

10 Find the area of an isosceles triangle with hypotenuse of 10 cm.

A. $5\sqrt{2}\ cm^2$
B. $10\sqrt{2}\ cm^2$
C. $25\ cm^2$
D. $50\ cm^2$

Test 8: Trigonometry with Special Triangles

Quiz 1: Working with Special Right Isosceles Triangle 45-45-90

Quick review of the concepts about the special right isosceles triangle. Consider an isosceles right triangle PQR with hypotenuse PR to answer the following questions. (Label the diagram)
Sketch and label the triangle to visualize the solution.

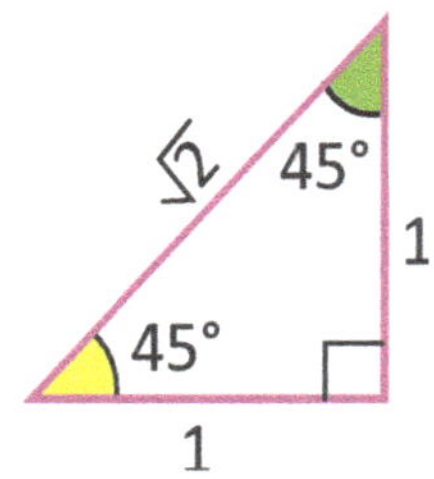

Q1. If PQ is 8 cm find the length of PR.
Q2. If PR is 10 cm find the length of PQ
Q3. Find the area of the triangle if PQ is 6 cm
Q4. Find the area of the triangle if PR = 7 cm
Q5. Find dimensions of ΔPQR if its area is 100 cm^2.
Q6. Write the length of PR, in simplified form, if shorter side length is $\sqrt{6}$
Q7. Write an expression for area of the ΔPQR if PR = c.
Q8. Write an expression for perimeter of the ΔPQR if PR = c

Quiz 2: Working with Special Triangle 30-60-90

Consider a right triangle LMN with ∠L = 30°, ∠M = 90°, and ∠N = 60° to answer the following questions. Sketch a triangle, write ratio of the sides and then answer orally.

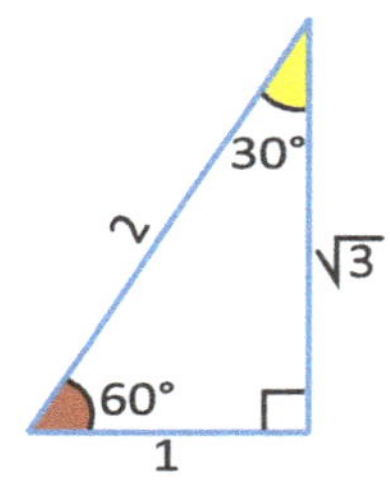

Q1. If length LM is 8 cm find the length of MN.
Q2. If length MN is 10 cm find the length of LN.
Q3. If length LN is 12 cm find the length of LM.
Q4. Find area of the ΔLMN if MN = 10 m.

STEM Skills Practice: Area with Special Triangles

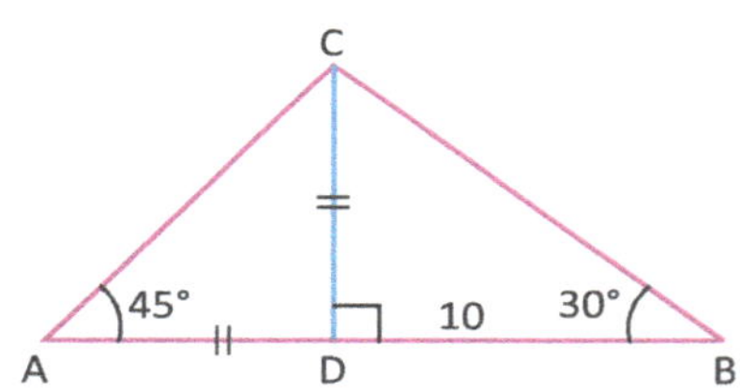

Find the area of the triangle ABC.
∠A = 45°, ∠B = 30°, DB = 10, CD = AD
Find CD, the altitude and then the area of ΔABC

Answers to Test 7: Pythagorean Power & Spatial Thinking

1a.	4	1b.	12	1c.	$3\sqrt{2}$	1d.	$\sqrt{3}$	1e.	$2\sqrt{37}$
1f.	9	2	194 m	3	13.4 cm	4.	31.9 mi	5.	13
6.	4	7.	$8\sqrt{3}$ cm	8.	$28 + 12\sqrt{2}$	9.	2ab	10.	c. 25

Test 9: Height and Distance Applications

Time Limit: Maximum 15 minutes **Total Questions: 12**

Q1	The shadow of a tower is $\sqrt{3}$ times its height. The angle of elevation of the top of the tower is							
	A	30°	B	45°	C	60°	D	75°
Q2	The angle of elevation of ladder leaning against a house is 60° and the foot of the ladder is 3 cm from the house. The length of the ladder is							
	A	3 m	B	$3\sqrt{3}$ m	C	$3\sqrt{2}$ m	D	6 m
Q3	A kite flying 400 m above the ground at an elevation of 60° is around d distance away. Find approximate value of d. (take $\sqrt{3} = 1.732$)							
	A	200 m	B	230 m	C	690 m	D	460 m
Q4	A 15 feet vertical post is broken at a certain height, h. It leans to meet the ground at an angle of 30°. Find the height at which the post is broken.							
	A	10 ft	B	5 ft	C	$5\sqrt{3}$ ft	D	$\frac{10}{\sqrt{3}}$ ft
Q5	The angle of elevation of a plane changes from a point on the ground from 60° to 30° in 15 seconds. If the plane is flying at the height of $1200\sqrt{3}$ m. Find the distance travelled by the plane during this time.							
	A	2000 m	B	2400 m	C	3000 m	D	$1200\sqrt{3}m$
Q6	The angle of elevation of an airplane from a point on the ground is 60°. After 30 seconds the elevation changes to 30°. If the airplane is flying at a height of $1800\sqrt{3}\ m$, find the speed of the plane in m/min.							
	A	3600	B	5400	C	7200	D	$3600\sqrt{3}$

Q7	Two planes are observed along the same vertical line. Their angle of elevation from a point on the ground, at that instance, in 60° and 30° respectively. If the height of the plane at lower altitude is 2.4 km find the distance between the planes at that instance.							
	A	1.2 km	B	2.4 km	C	3.6 km	D	4.8 km
Q8	The distance between the two street light poles across the road is $20\sqrt{3}m$. From a point somewhere between them their angle of elevation is $30°$ and $60°$ respectively. Find the height of the streetlight.							
	A	15 m	B	$10\sqrt{3}$ m	C	18 m	D	20 m
Q9	Distance between two parallel trees is $40\sqrt{3}\ m$. The angle of depression of the top of the second tree when seen from the top of the first is 30°. What is the height of the second tree if the height of the first is 100m?							
	A	40 m	B	60 m	C	50 m	D	$45\ m$
Q10	A helicopter, at an altitude of 1500 m is hovering over north-south highway. Two cars are spotted speeding over 150 km/h. A the angles of depression of these cars from helicopter is 30° and 60°. Find the cars if both are in same side travelling north							
	A	1000 m	B	$1000\ \sqrt{3}$	C	$2000\ \sqrt{3}$	D	$1500m$
Q11	A ladder is placed against a wall reaching its top. The foot of the ladder is at a horizontal distance of $4\ m$ from the wall. Ladder is making an angle of 70° with the ground. Suddenly, it starts slipping. Now it is making an angle of 20° with the ground. How much did the ladder slip?							
	A	4 m	B	6 m	C	7 m	D	1 m
Q12	The angle of depression of two ships from a light house are 30° and 45° respectively, towards east. If the ships are 200 m apart, find the height of the light house.							

	A	$\frac{200}{\sqrt{3}-1}$	B	$100\,(\sqrt{3}-1)$	C	$\frac{200}{\sqrt{3}+1}$	D	$200\frac{\sqrt{3}}{3}$

Answers Test 8: Quiz 1

Q1.	$8\sqrt{2}$	Q2.	$\frac{10\sqrt{2}}{2}$	Q3	18	Q4.	$\frac{49}{4}$
Q5	$10\sqrt{2}, 10\sqrt{2}, 20$	Q6	$2\sqrt{3}$	Q7	$\frac{c^2}{4}$	Q8	$c(1+\sqrt{2})$

Answers Test 8: Quiz 2

Q1.	$\frac{8\sqrt{3}}{3}$ cm	Q2.	20 cm	Q3	$6\sqrt{3}$cm	Q4.	$\frac{25\sqrt{3}}{2}$

Answers Test 9: Height and Distance Application

Q1.	A	Q2.	D	Q3	D	Q4.	B
Q5	B	Q6	C	Q7	D	Q8	A
Q9	B	Q10	B	Q11	C	Q12	C

An equilateral triangle and a hexagon have same perimeter. The area of the triangle is 6. Find the area of the hexagon.

A. 6
B. 9
C. 12
D. 18

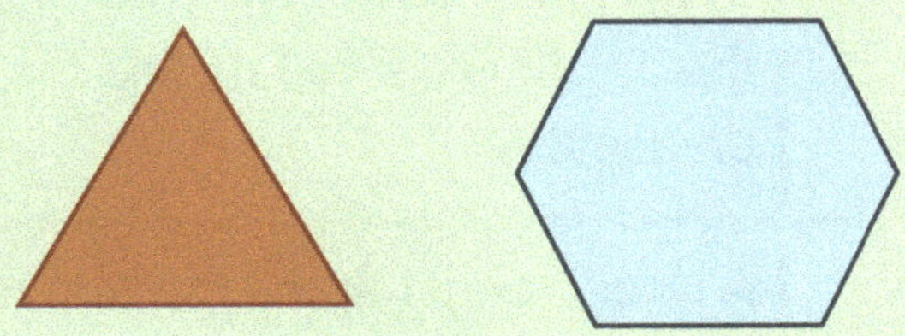

Hint: Equilateral triangle height, $h = \frac{\sqrt{3}}{2}a$

Answer: B

Test 10: Spatial Applications of Pythagorean Theorem

Note: The drawings are not to scale.

Q1	The vertices of the triangle ABC lie on the semicircle. AC forms the diagonal. Find the area of the semicircle not covered by the triangle ABC if AB is 12 cm and BC is 9 cm. 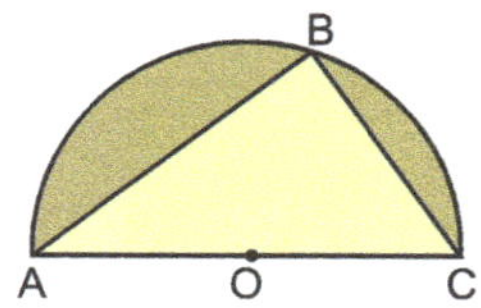
Q2	An equilateral triangle ABC is inscribed in the circle. Find area of the triangle ABC if the circle has radius of 10 cm. 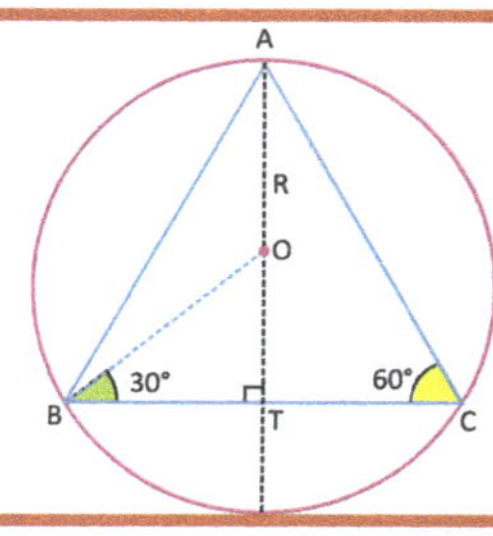
Q3	Circle $x^2 + y^2 = 25$ has radius of 5 units. Centre of the circle, O, is at the origin of a coordinate plane. PA is tangent to the circle at A from point P(20, 0). Find the length AP to the nearest centimetre. 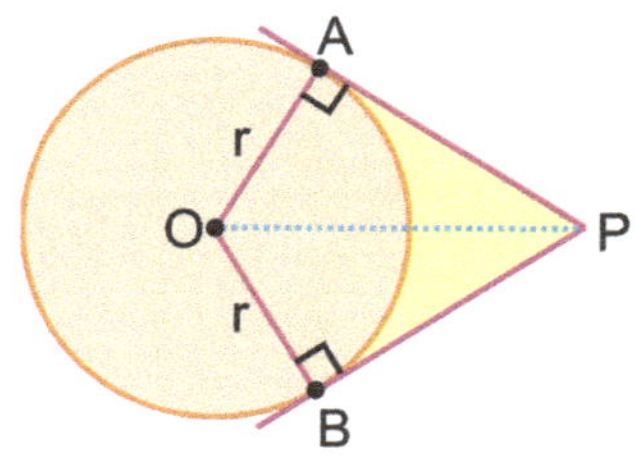

Q4	A wedge ABCDEF is shown in the figure. Base is 300 m by 400 m as shown. (drawing not to the scale). If the height FC is 100 m, find the length of the diagonal AF. 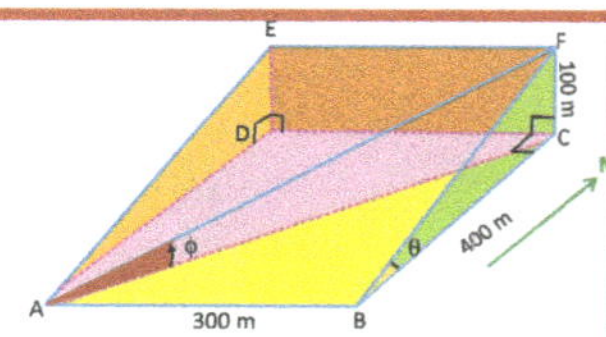
Q5	A forest ranger at the top of a tower sights two fires in opposite directions with angle of depression of $45°$ and $30°$. Find the hight of the tower if the estimated distance between the fires is 100 m. 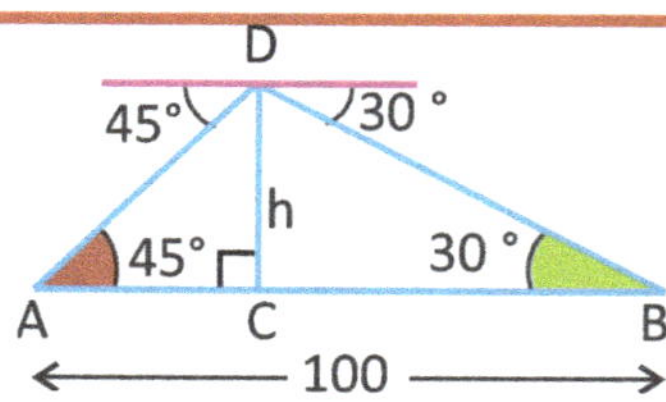 **Angle of depression is the angle that line of sight makes with horizontal line.**

Q6

A ladder is placed along the vertical wall making an angle of $60°$. Foot of the ladder is 10 ft from the wall. Then it slides from the point P to Q. At this instance the ladders rest at Q making an angle of $30°$ with the horizontal ground.

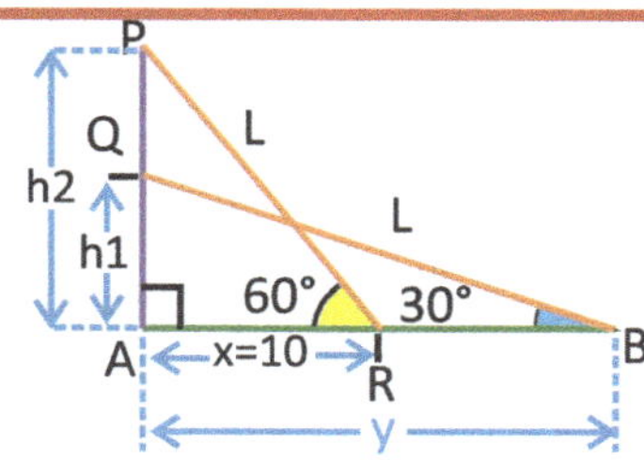

Find exact measure of

a.	height h_1	b.	height h_2
c.	ladder length L	d.	distance AB

Q7

Find the volume of rectangular based pyramid

ABCD is the rectangle with sides 8 cm and 6 cm. The slant length, AE, is 10 cm.

Find the height EO and then the volume of the pyramid shown in the figure.

Hint: First find AO, half the diagonal of base.

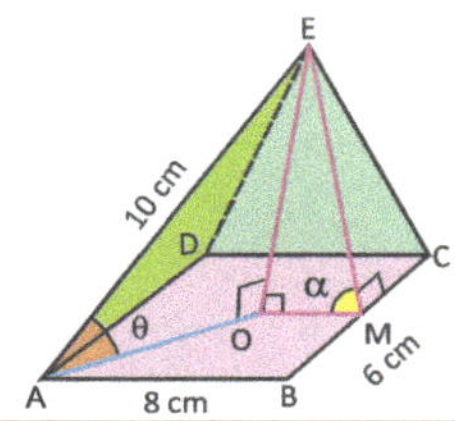

Q8

Find volume of a trough, 1 m long, with cross-section area of an isosceles trapezoid ABCD.

$\angle A = \angle B = 60°$

Hint: First find AE, using special triangle.

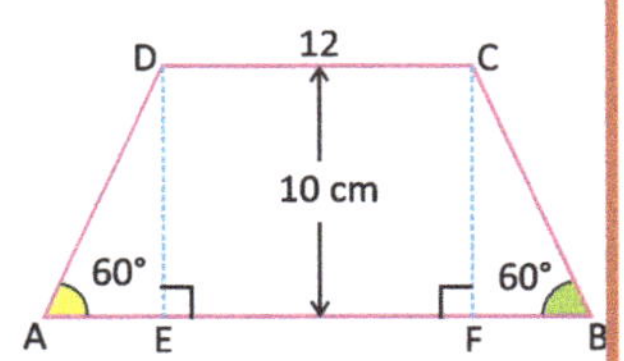

Watch Video **Review Complete Concept** **SCAN ME**

Answers Test 10: Spatial Application also check Video Solution

Q1.	$34.4\ cm^2$	Q2.	$75\sqrt{3}cm^2$	Q3	$5\sqrt{15}$	Q4.	$100\sqrt{26}m$
Q5	$50(\sqrt{3}-1)m$ $\approx 36.6\ m$	Q6	$10, 10\sqrt{3},$ $20, 10\sqrt{3}$	Q7	$138.6\ cm^3$	Q8	$0.0178\ m^3$

Test 11: Evaluate Trigonometric Expressions

Q1 Find exact value of following expressions and match with options.

a.	$2 \sin 30° \cos 30°$
b.	$\cos^2 30° - \sin^2 30°$
c.	$\cos 60° \cos 30° + \sin 60° \sin 30°$
d.	$\tan 60° \cos 60° \sin 60°$
e.	$\dfrac{\tan 60° - \tan 30°}{1 + \tan 60° \tan 30°}$
f.	$4\cos^2 30° - 2$

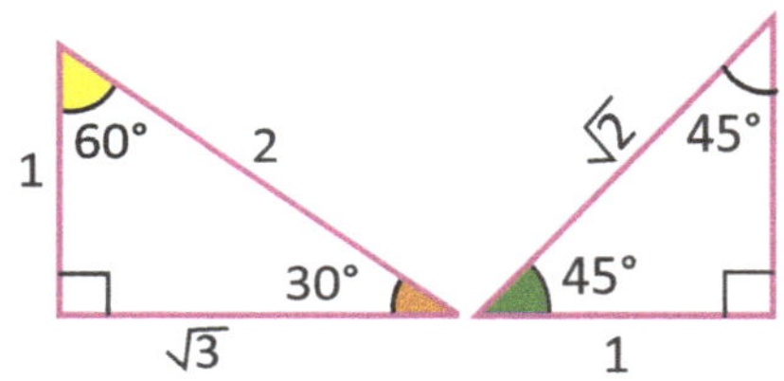

Match with Options

A	$\tan 30°$	B	$2\cos 60°$	C	$\sin^2 60°$
D	$\cos 60°$	E	$\sin 60°$	F	$\sin 90°$

Q2 Area of a triangle is half the produce of adjacent sides with sine of included angle. Find area of the triangle ABC if $AB = 10, AC = 12$ *and* $\angle BAC = 25°$.

$25.4\ unit^2$

Q3 A triangle has adjacent sides of length 8 cm and 12 cm. Find the included angle if the area is $44.8\ unit^2$

$69°$

Q4 a parallelogram has adjacent sides of length 6 and 9 cm respectively. If the included angle measures $62°$, find the area of the parallelogram.

$47.7\ unit^2$

Q5 Show that $\cos 60° \cos 30° - \sin 60° \sin 30° = \cos 90°$

Test 12: Trigonometry IB Practice

Q1 to Q3 refer to the three triangles given below

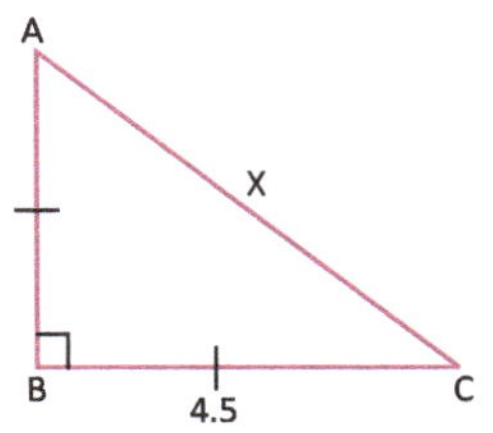

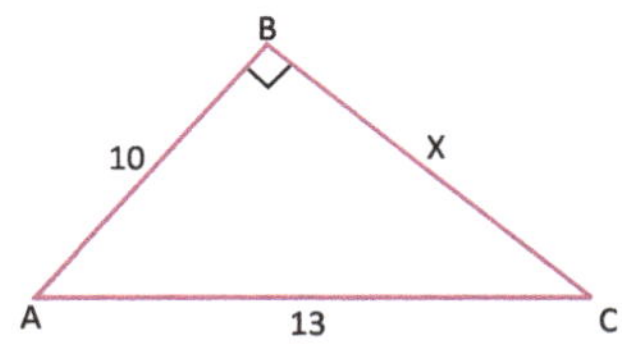

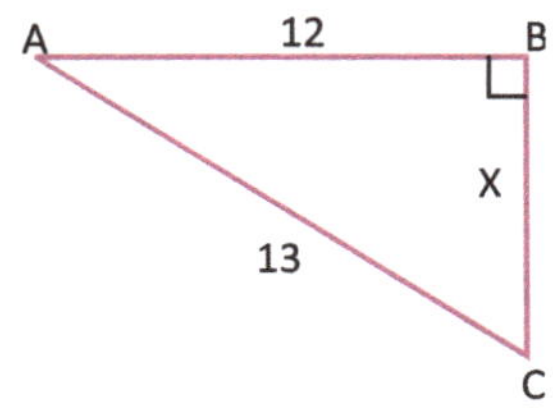

Q1	Use Pythagorean theorem to determine the value of x to the nearest tenth place.
Q2	Determine the primary trigonometric ratios for angle A in the above triangles to the nearest tenth of unit.
Q3	Determine angle measurement to the nearest degree for angle C in each given triangle.
Q4	If $\sin P = 0.8905$ find P to 3 significant figures.
Q5	If $\sec A = 2.5$ find A to 3 significant figures.
Q6	If $\sin R = \cos R$ find exact measure of angle R.
Q7	Arrange $\sin K$, $\cos K$, and $\tan K$ in increasing order if angle K is increasing from $50°$ to $70°$ in a right triangle.
Q8	From a point on the ground 30 m from the foot of the Peace Tower, the angle of elevation to the top of the tower is $72°$, Find the height of the tower.
Q9	A lighthouse is at the top of cliff 36 metres above the sea level. The angle of depression to the sight of a boat is $27°$. How far from the base of the cliff is the fishing boat, to the nearest metre?
Q10	In ΔABC, $BC = 6$, $AC = 8$, and $\angle B = 90°$. Find $\angle C$

Q11	In ΔPQR, $PR = 10$, $QR = 6$, and $\angle Q = 90°$. Find $\angle P$
Q12	In ΔMNP, $MP = 14$, $NP = 9$, $\angle N = 90°$. Find $\angle M$
Q13	In ΔDEF, $DE = 12$, $EF = 7$, $\angle E = 90°$. Find $\angle F$
	Find the measure of angle x in the given right triangles (Q14 – Q17).
Q14	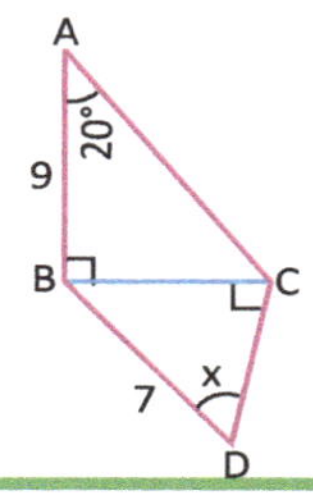
Q15	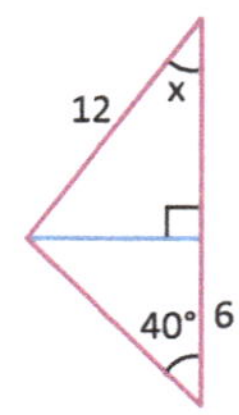
Q16	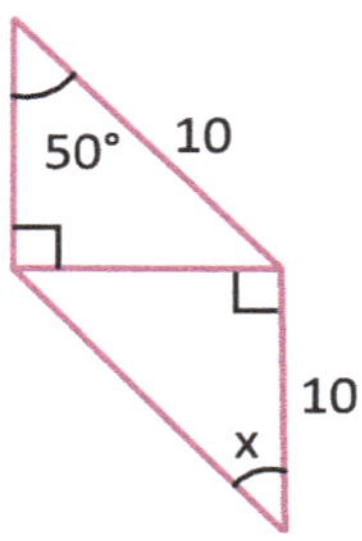
Q17	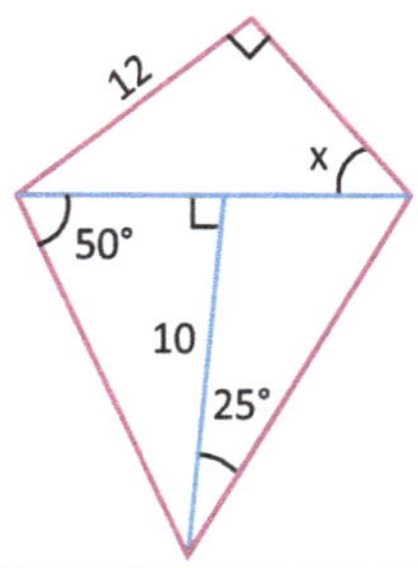
Q18	The Confederation Bridge spans the Northumberland Strait and joins New Brunswick and Prine Edward Island. From one point in the Strait, the angle of elevation of the highest point on bridge is $27°$. From a point 100 m closer to the bridge, the angle of elevation is $73°$. How high is the bridge, to the nearest metre?
Q19	Find the acute angle that the $2x - 3y = 6$ makes with the horizontal $x - axis$.
Q20	A chord of length 20 cm subtends an angle of 76° at the centre of the circle. Find the radius of the circle.

Application and Thinking

For the acute angles A and B, in a triangle ABC, $\sin B = \cos A$. If $A = 3p - 17$ and $B = 2p + 12$, what is the value of p?

Test 13: SAT Math 2 Trigonometry Applications

Q1	In a right triangle ABC where $\angle A$ and $\angle B$ are acute angle, $\sin A = 0.25$. Find $\cos B$. A. 0.25 B. 0.75 C. $\frac{\sqrt{15}}{4}$ D. 0.95	Q2	In a right triangle $ABC, \angle C = 90°$. If $\tan A = 4$ then $\tan B$ is A. 4 B. 2 C. 0.25 D. 0.5
Q3	In a right triangle $ABC, \angle C = 90°$ and $\tan A = 2.4$. Find value of $\sec B$. A. $1\frac{5}{12}$ B. $1\frac{1}{12}$ C. $2\frac{3}{5}$ D. $2\frac{1}{4}$	Q4	In a right triangle ABC, right angle at C. if $\tan = 1$, the value of $2 \sin A \cos B$ is A. 1 B. $\frac{1}{\sqrt{2}}$ C. $\frac{1}{2}$ D. 2
Q5	In a right triangle ABC, right angle at C, if $\tan A \frac{1}{\sqrt{3}}$ then evaluate $\sin A \cos B + \cos A \sin B$. A. 1 B. $1\frac{1}{2}$ C. $\sqrt{3}$ D. $\frac{\sqrt{3}}{2}$	Q6	In $\Delta\, ABC$, right angle at C, if $\sin A = \frac{a^2-b^2}{a^2+b^2}$ find $\tan A$ A. $\frac{a^2-b^2}{ab}$ B. $\frac{a^2-b^2}{2ab}$ C. $\frac{a^2+b^2}{ab}$ D. $\frac{a^2+b^2}{2ab}$
Q7	In a $\Delta\, ABC$, right angle at C, if $\cos A = \frac{a^2-b^2}{a^2+b^2}$ find $\tan B$ A. $\frac{a^2-b^2}{2ab}$ B. $\frac{2ab}{a^2-b^2}$ C. $\frac{a^2-b^2}{ab}$ D. $\frac{ab}{a^2+b^2}$	Q8	Which of the following is not defined for $\theta = 0^{\circ}$ A. $\sin\theta$ B. $\cos\theta$ C. $\sec\theta$ D. $\csc\theta$

Q9 Which of the following is not defined for $\theta = 90°$

A. $\frac{\cot\theta}{\sin\theta}$
B. $\frac{\sin\theta}{\sec\theta}$
C. $\sec\theta \sin\theta$
D. $\frac{\cos\theta}{\tan\theta}$

Q10 $\cot\theta = 1.376$ find θ to nearest degree.

A. 36°
B. 54°
C. 47°
D. 43°
E. 36°

Q11 $\csc\theta = 2$ find $\cos\theta$

A. $\frac{1}{2}$
B. $\frac{\sqrt{3}}{2}$
C. $\frac{1}{\sqrt{2}}$
D. 1

Q12 If $\sin\theta = \frac{a}{b}$ find $\sec\theta$

A. $\frac{b}{a}$
B. $\frac{b}{\sqrt{b^2-a^2}}$
C. $\frac{\sqrt{b^2-a^2}}{b}$
D. $\frac{b}{b^2-a^2}$

Q13 Find y in the figure

A. 9
B. 15
C. 12
D. 11

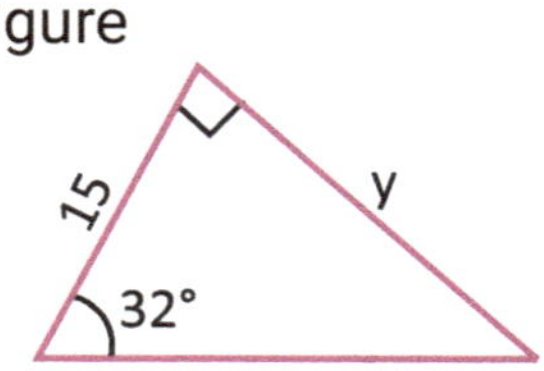

Q14 Find z in the figure

A. 16
B. 17
C. 18
D. 19

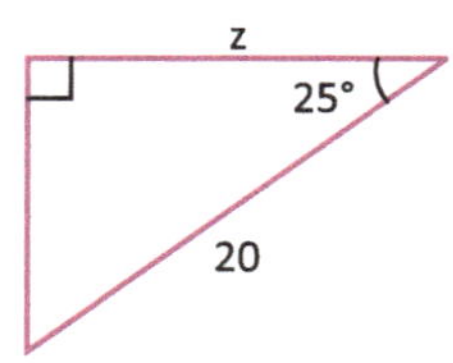

Q15 Find x to the nearest tenth.

A. 5.4
B. 4.6
C. 4.9
D. 6.1

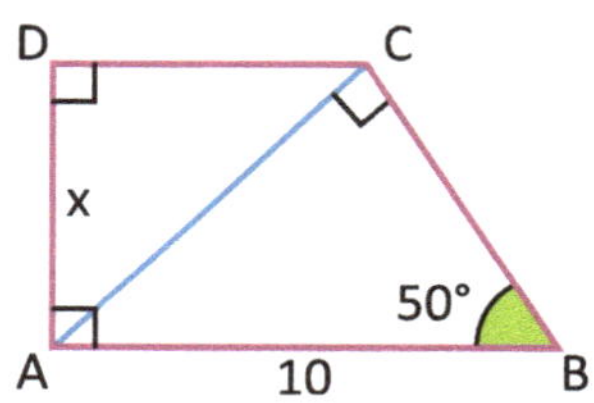

Q16 In triangle ΔABC, $AC = 125\ m$, $\angle A = 25°$, $\angle C = 30°$

Find value of h.

A. $32\ cm$
B. $56\ cm$
C. $48\ cm$
D. $39\ cm$

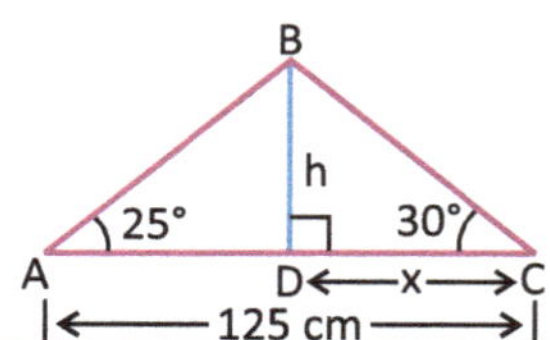

Q17 An airplane when flying at a height of 3125 m form the ground passes vertically below another plane at an instant when the angle of elevation of the two planes from the same point on the ground are 30° and 60° respectively. Find the distance between the two planes at the instant.

A. 9375 m
B. 5412 m
C. 5250 m
D. 7245 m

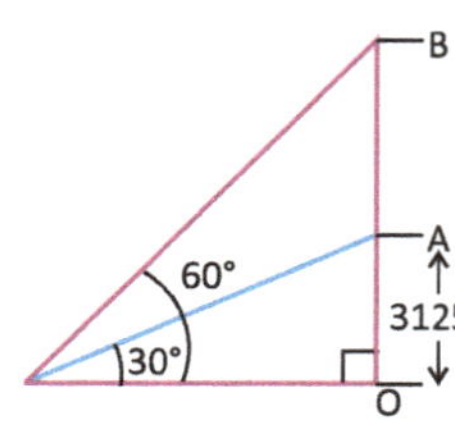

Explore Projection Formula

Projection formula relates sides of a triangle in terms of cosines.

In any triangle ABC show that:

$a = c\cos B + b\cos C\,;\;\; b = a\cos C + c\cos A\,; c = a\cos B + b\cos A$

Then prove that:

$$\frac{c - a\cos B}{b - a\cos C} = \frac{\sin B}{\sin C}$$

SCAN ME

Answers: Test 12: Right Triangles

Q10.	41.4°	Q11.	36.9°	Q12.	40.0°	Q13.	59.7°
Q14.	27.9°	Q15.	24.8°	Q16.	37.5°	Q17.	66.8°
Q18	60 m	Q19	33.7°	Q20	16.2 cm		

Test 13: Trigonometry Applications Answers:

Q1	A	Q2	C	Q3	B	Q4	A	Q5	A
Q6	B	Q7	A	Q8	D	Q9	C	Q10	A
Q11	B	Q12	B	Q13	A	Q14	C	Q15	C
Q16	C	Q17	A						

Challenge: Cylinder to Cone

A gold ingot in the shape of a cylinder is melted and the resulting molten metal molded into a few identical conical ingots. If the height of each cone is half the height of the original cylinder and the area of the circular base of each cone is one-fifth that of the circular base of the cylinder, then how many cone ingots can be made? 30

Challenge: Formula Derivation for Volume of Frustum

A frustum of a cone with base of radius R is formed when top of the cone is sliced with a plane parallel to the base. If the radius of the top is r and the height of the frustum is h then prove that the volume V of the frustum is given by the formula

$$V = \frac{1}{3}\pi h(R^2 + Rr + r^2)$$

Test 14: Height & Distance Applications

Q1. The top of a lighthouse is 90 m above the sea level. The angle of elevation from the deck of a sailboat to the top of the lighthouse is $26°$. Calculate the distance between the sailboat and the lighthouse.

Lighthouse Boat Distance SCAN ME

Q2. A ladder 12 feet long is leaning against a wall at a $74°$ angle. How far high up the wall does the ladder reach?

Inclination Angle Ladder Height SCAN ME

3. A ladder 12 feet long is leaning against a wall at a $74°$ angle. How far from the wall is the foot of the ladder?

Ladder from Wall SCAN ME

Q4. The angle of elevation to the top of the Niagara Falls from the boat which is 60 ft away is measured as $72°$. Find the height of the Niagara Fall to the nearest feet.

Niagara Fall Height SCAN ME

Q5 The Skylon Tower in Niagara Falls is about 526 feet high. From a certain distance, Anil measures the angle of elevation to the top of the tower to be $66°$. Then he walks away another 65 feet in the same direction and measures the angle of elevation again. Determine the measure of this angle.

Skylon Tower Angle SCAN ME

Q6. Brian is 100 m south of the CN Tower. He moves 50 m east to get a clear view of the tower and then measure the angle of elevation to the top as $75°$. Find the height of CN Tower to the nearest centimeter.

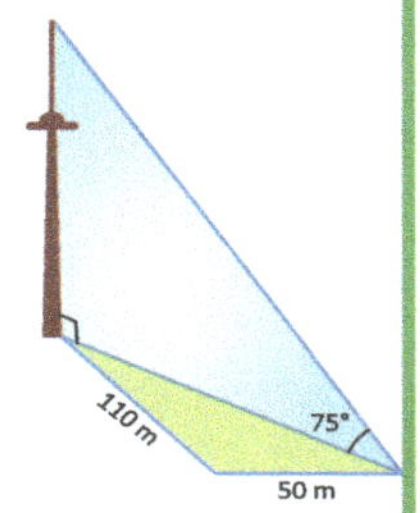

CN Tower Height SCAN ME

Q7 An aircraft travelling at 240 km/h is about to land at Billy Bishop Toronto City Airport in 3 seconds. Find the height of the plane above Lake Ontario if the angle of depression at this moment is $3°$.

Student Activity 1 Cofunction Identity Application

Consider a right triangle ABC.

BD of length z is the altitude of ΔABC

D divides hypotenuse AC in the ratio of $x : y$

Show that z is geometric mean of x *and* y.

Hint:

In the given triangle show that $z = \sqrt{xy}$

Remember $\tan A \times \tan C = 1$

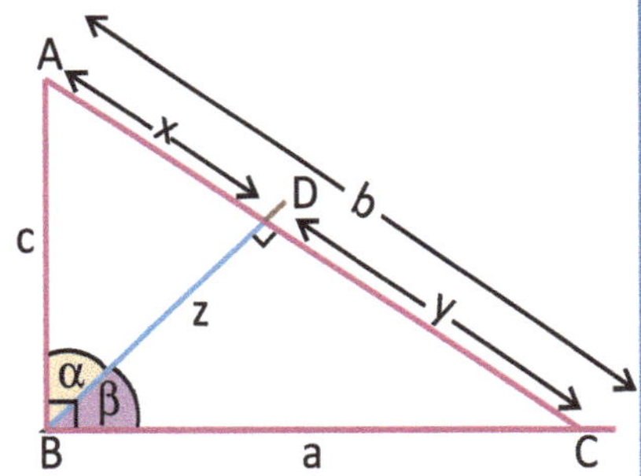

Challenge Problem

Triangle ABC with sides lengths of $a, b,$ and c is inscribed in a circle with radius R.

Prove the following relations:

$$2R = \frac{a}{\sin A} = \frac{b}{\sin B} = \frac{c}{\sin C}$$

$$R = \frac{abc}{4A}, A \text{ is area of triangle } ABC$$

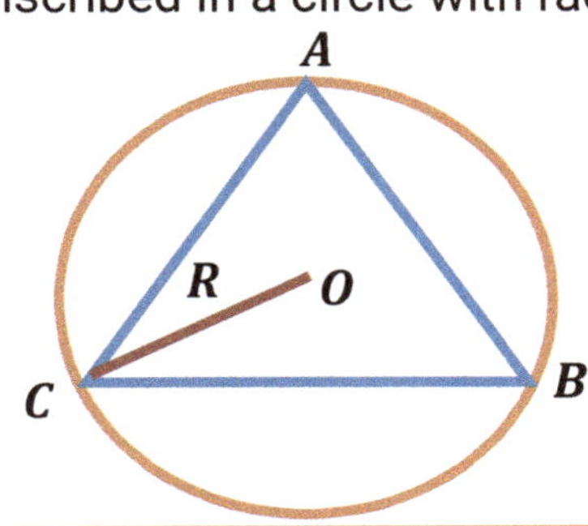

Test 15: Trigonometry Major

Enrich your Experience of Learning

Questions from Test Papers of my IB Math Students

Q1. If $tan\,\theta = \frac{5}{12}$

find $cos\,\theta$ and $sin\,\theta$

Q2. If $sin\,\theta = \frac{a}{b}$

find expressions for $tan\,\theta$.

Q3. Find the other two primary trigonometric ratios for each value of θ.

a.	$\sin\theta = \dfrac{8}{17}$	b.	$\cos\theta = \dfrac{19}{23}$	c.	$\tan\theta = \dfrac{20}{21}$
d.	$\sin\theta = \dfrac{P}{q}$	e.	$\cos\theta = \dfrac{9}{a+2}$	f.	$\tan\theta = \dfrac{x-y}{x+y}$

Consider the figure on the right to answer Q4 and Q5

Q4. Find the angles to 1 decimal place

(i) $\angle CAB$
(ii) $\angle DAC$
(iii) $\angle EAD$
(iv) $\angle FAE$

Note: This is how ancient Greeks constructed line segments of length $\sqrt{2}$, $\sqrt{3}$, $\sqrt{4}$.

Q5. How many triangles can be drawn without overlapping?

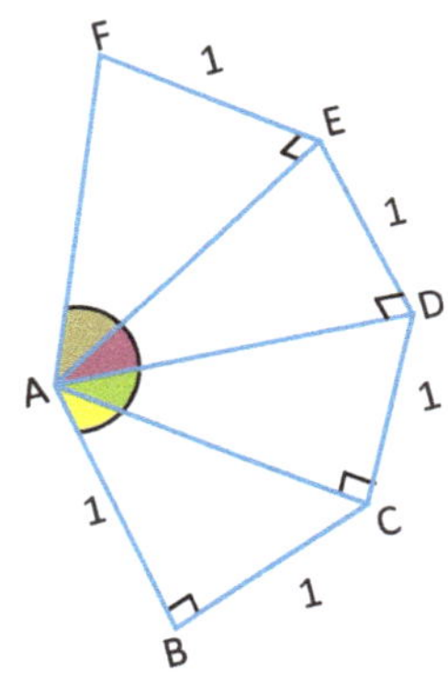

Q6. A box is resting inside a second box as shown in figure. How high is the lowest corner of the first box above the bottom of the second box?

A. 8.2
B. 8.9
C. 16.4 cm
D. 21.1 cm

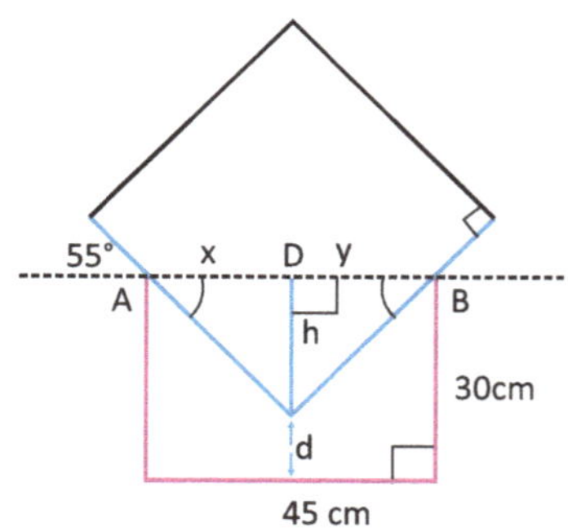

Information for Q7 and Q8

The length of rectangle ABCD is three times its width. Point M and N are the midpoints of the longer sides AB and DC respectively.

Q7 Find $\angle MAN$

A. 30°
B. 32°
C. 34°
D. 36°

Q8 If P is the midpoint of AD, find $\angle MPN$

A. 35°
B. 37°
C. 39°
D. 41°

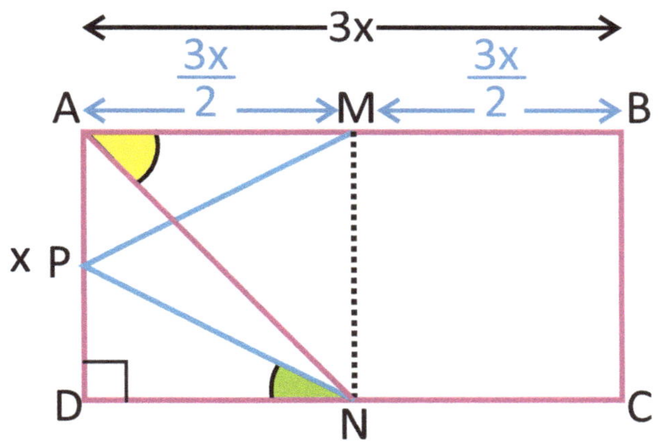

Q9. A funnel is placed in a cylindrical glass, as shown. If the glass is 14.5 cm tall and 7.6 cm in diameter, how high is the vertex of the funnel above the bottom of the glass?

A. 7.1 cm
B. 7.2 cm
C. 7.3 cm
D. 7.4 cm

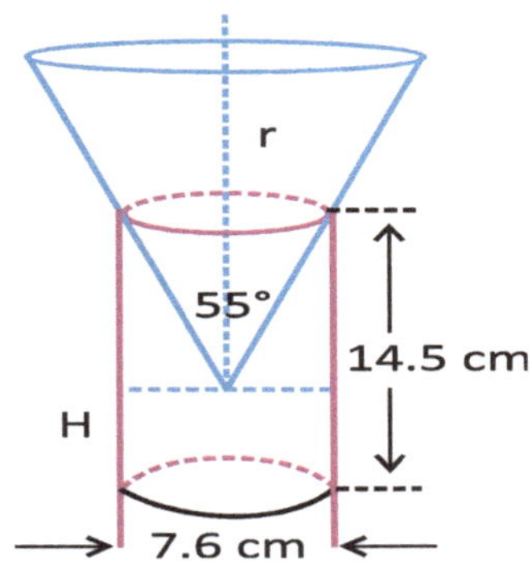

Q10. The angle of elevation of an aircraft from a point on the ground is 60°. After a flight of 30 seconds the angle of elevation become 30°. If it is flying at a constant height of $2000\sqrt{3}$m, find the speed of the aircraft.

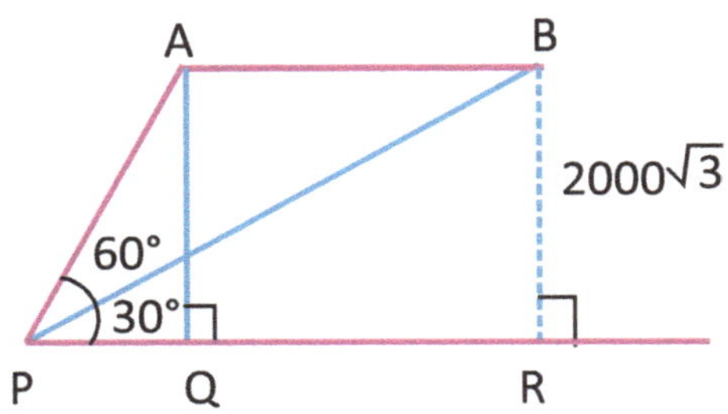

Q11.	At a point P, 10 m above the level of water in a lake, the angle of elevation of a cloud is 30°. The angle of depression of the reflection of the cloud in the lake, at P is 60°. Find the distance of the cloud from P.	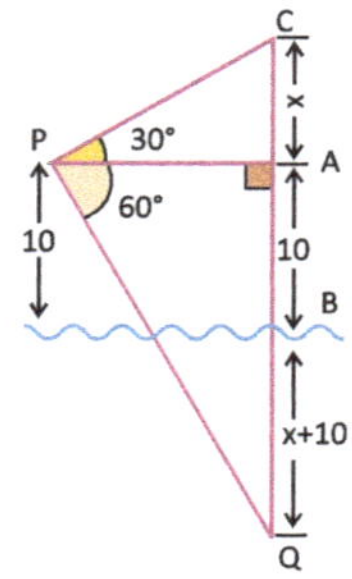
Q12.	Two ships are approaching a lighthouse from opposite directions. The angle of depression of the two ships from the top of the light house are 30° and 45°. If the distance between the two ships is 100 m, find the height of the lighthouse to the nearest meter.	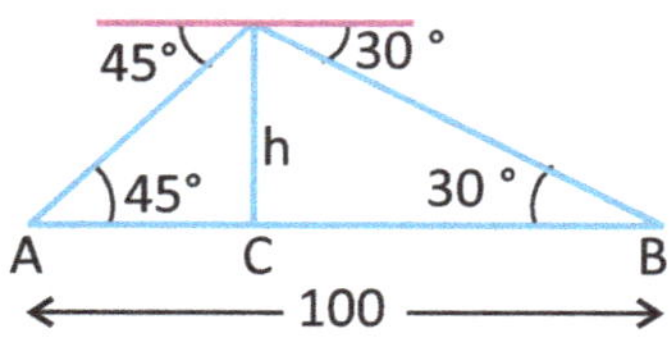
Q13	The shadow of a tower is found to be 24 m shorter when the sun's angle of elevation is 60° than when it is 30°. Find the height or the tower.	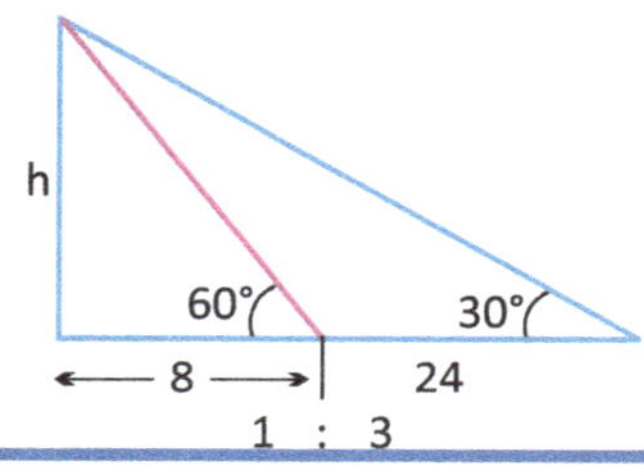
Q14.	A peregrine falcon is hovering 100 m above the grass land. Determine the are being scanned angle of depression is 5°.	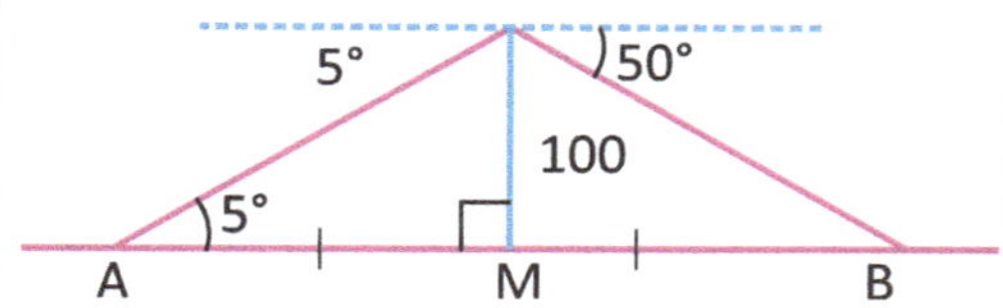
Q15	Find the equation of a line 10 units away from the origin whose perpendicular from origin makes an angle of $60°$.	

Answers Test 15

Q1	$\frac{12}{13}$	Q2	$\frac{a}{\sqrt{b^2 - a^2}}$	Q5	16	Q6	$8.9\ cm$
Q7	C	Q8	B	Q9	7.2 cm	Q10	$480 \frac{km}{h}$
Q11	$20\ m$	Q12	$37\ m$	Q13	$8\sqrt{3}$ m	Q14	$4.1\ km^2$

Test 16: Sine and Cosine Law Applications

Identify correct choice(s).

Q1	You can use sine law to solve triangle if you know the measures of A. a, b, and c B. a, b, and $\angle C$ C. c, b, and $\angle C$ D. a, c and $\angle B$ E. $\angle A, \angle B \; and \; c$	Q2	You can use cosine law for calculations if you know the measures of A. a, b, and c B. a, b, and $\angle C$ C. c, b, and $\angle C$ D. a, c, and $\angle B$ E. $\angle A, \angle B \; and \; c$
Q3	What is the minimum number of facts that you need to know to solve any triangle with sine/cosine law? A. 2 B. 3 C. 4 D. 5	Q4	How many triangles could be formed If $\angle A = 30°, a = 2.5 \; b = 5.0$ A. 0 B. 1 Right Δ C. 2 D. 1 or 2
Q5	How many triangles could be formed if $\angle A = 100°, a = a < b, b = 13.0$ A. 0 B. 1 Right Δ C. 2 D. 1 or 2	Q6	How many triangles could be formed: $\angle A = 50°$, a = 5.5, $\angle B = 45°$ A. 0 B. 1 Right Δ C. 2 D. 1 or 2
Q7	How many triangles could be formed if $\angle A = 70°, a = 1.51, b = 1.59$ A. 0 B. 1 Right Δ C. 2 D. 1 or 2	Q8	Determine measure of $\angle B$ to the nearest degree If $\angle A = 100°, a = 9, b = 6$ A. 40° B. 42° C. 41° D. 30°
Q9	Determine measure of side C to the nearest tenth of unit if $\angle A = 70°, \angle B = 50°, a = 5$ A. 4.6 B. 5.4 C. 4.1 D. 4.0	Q10	Determine measure of $\angle B$ to the nearest degree if $\angle A = 45°, b = 38.0, \; c = 24.5$ A. 100° B. 85° C. 40° D. 90°

Q11 Determine measure of smallest angle in ΔPQR if $p = 5, q = 6, r = 7$

A. 44°
B. 78°
C. 41°
D. 57°

Q12 A plane is

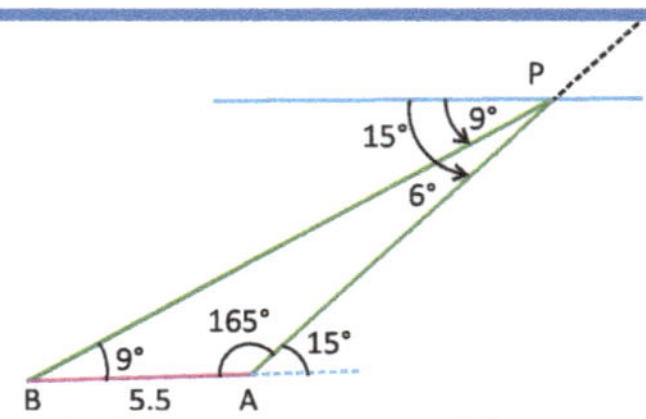

approaching a 5.5 km runway. The angles of depression at the end of the runway are 9° and 15°. How far is the plane from each end of the runway?

Q13 Two ships are approaching a harbour. One ship is 4.8 miles from the harbour on a bearing of 035°. The other ship of 8.2 miles away from the harbour on a bearing of 285°. How far apart are the two ships?

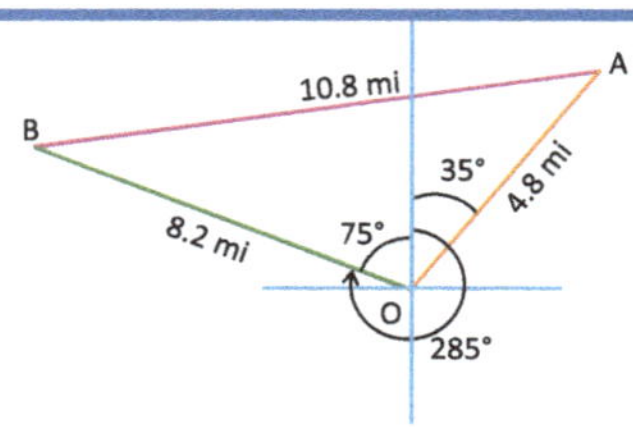

Q14 Part of a roof of a warehouse has a slope of 1/3, and the other part of the roof has a slope of 0.5. The width of the house is 30m. Find the height of the roof.

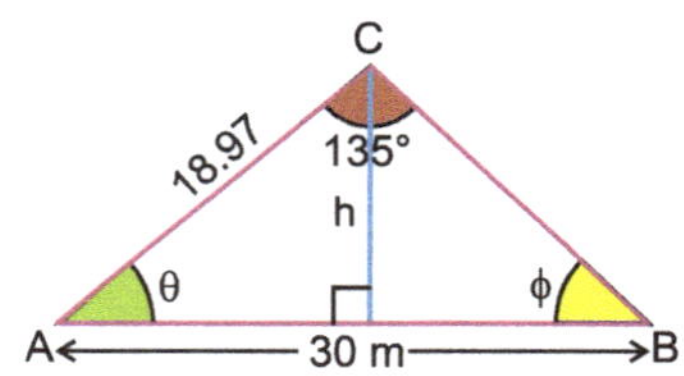

Q15 Two forest fire towers, A and B, are 20.3 km apart. The bearing from A to B is N70°E. The ranger in each tower observes a fire and radios the fire's bearing from the tower. The bearing from tower A and N25°E. From tower B, the bearing is N15°W. How far is the fire from tower A?

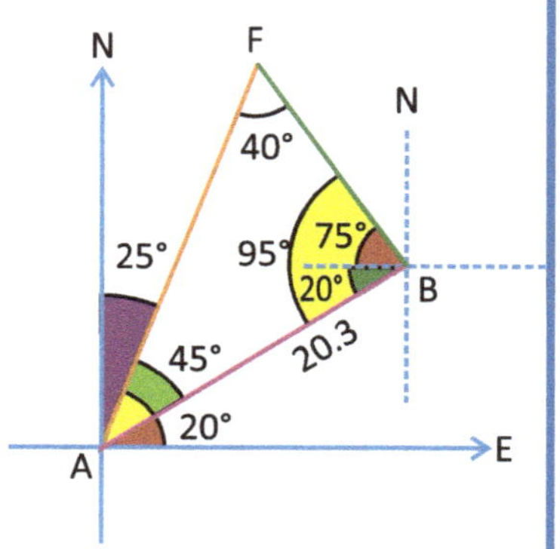

Q16 Two fire towers are located 100 km apart on high hill T and R. The bearing from T to R is north east. A fire F is observed from tower T at N10°E and from tower R at N76°W. The town of Pretty Valley, at point V, is on a bearing of N25°E from T and S70°W from tower R. The observers report the wind is blowing the fire directly towards Pretty Valley at a rate of 8 km/h. How many hours do the officials have to evacuate the town?

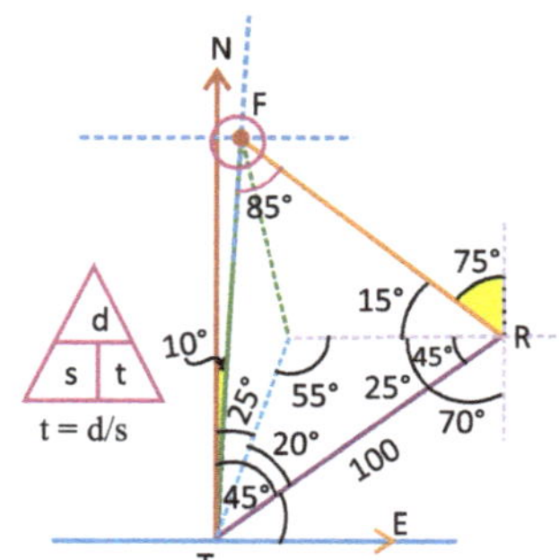

Q17	Three circles with centre A, B and C are mutually tangent and no circle lies inside another circle. 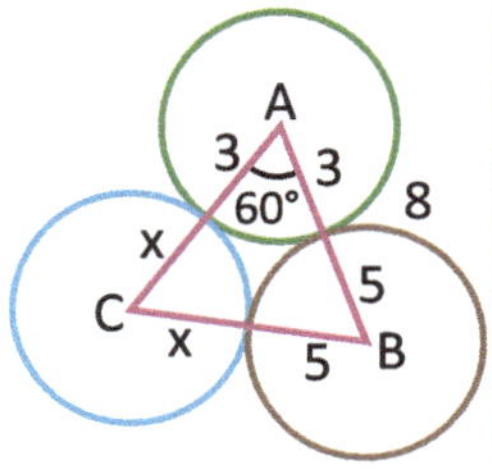 The circle with center A has radius 3 cm and the circle with center B has 5 cm radius. If the angle $\angle BAC$ is 60°, then determine the radius of third circle. Find also the area of triangle ΔABC	Q18	A falcon is spotted hovering by two friend Anil and Ben. The bird if due north at an angle of elevation of $40°$ for Anil. Ben is about a kilometer away, north-east of Anil. From his position the bird appears in the direction of $N65°W$. Estimate the height, h, of the falcon. 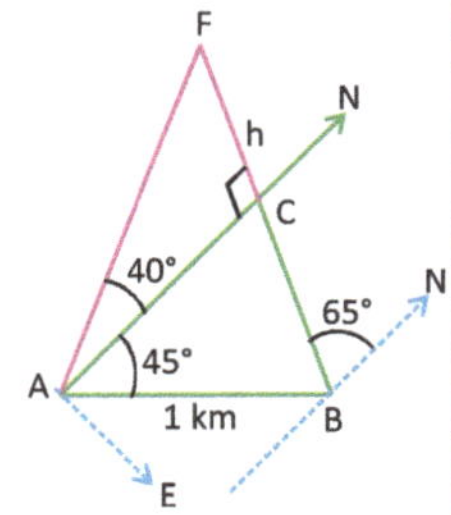

Answers Test 16

Q1	C, E	Q2	A, B, D	Q3	B	Q4	B	Q5	A
Q6	B	Q7	C	Q8	C	Q9	A	Q10	B
Q11	44°	Q12	13.6, 8.2	Q13	10 mi	Q14	135°	Q15	31.5 km
Q16	4h 7min	Q17	$2\ cm, 10\sqrt{3}cm^2$			Q18	$\frac{\sin 70° \tan 40°}{\sin 65°} = 870\ m$		

Communication Question – Ambiguous Case for SSA Condition

When two sides and a non-included angle (**SSA**) of a triangle are known, the given information does not always determine a unique triangle. This situation is called the **ambiguous case** of the sine law. **Explain why** this condition can lead to **no triangle, one triangle, or two possible triangles.**

Challenge Question

Prove that a n-sided regular polygon inscribed in a circle of radius r is $\mathbf{2rn \sin \frac{180}{n}}$.

Test 17: Trigonometry Final

Q1	A boat travels 11 km on a bearing of $080°$ from port A to the port B. Find the bearing angle at which it should return to port A from port B.
Q2	A tree is broken by the wind. The top hit the ground at an angle of $30°$ and at a distance of $30\ m$ from the trunk. Find the original height of the tree.
Q3	An aeroplane flying horizontally at a height of 1.5 km above the ground makes an angle of $60°$ at tracking Radar. After 15 seconds, its angle of elevation is $30°$. Calculate the speed of the aeroplane in km/h.
Q4	If $\cos x = \frac{2}{3}, 0° < x < 90°$ find the value of $\cot x$
Q5	If $\sin x = \frac{a^2-b^2}{a^2+b^2}, 0° < x < 90°$ find the value of $\tan x$
Q6	Evaluate: $\frac{\sec^2 53° - \cot^2 37°}{\csc^2 56° - \tan^2 34°} + 2\sin^2 36° \sec^2 54°$
Q7	Explain why the tangent ratio is positive in quadrant I and III.
Prove each of the following identities:	
Q8	$(\sec x - \tan x)^2 = \dfrac{1 - \sin x}{1 + \sin x}$
Q9	$\dfrac{1}{\sec x - \tan x} - \dfrac{1}{\cos x} = \dfrac{1}{\cos x} - \dfrac{1}{\sec x + \tan x}$
Q10.	$\dfrac{\tan x}{1 - \cot x} + \dfrac{\cot x}{1 - \tan x} = 1 + \csc x \sec x$
Q11	$\dfrac{\sec x + \tan x - 1}{\tan x - \sec x + 1} = \dfrac{\cos x}{1 - \sin x}$
Q12	$\sqrt{\dfrac{1 + \sin x}{1 - \sin x}} + \sqrt{\dfrac{1 - \sin x}{1 + \sin x}} = 2\sec x$
Q13	$\sin^2 x - \cos^2 x - \tan^2 x = \dfrac{2\sin^2 x - 2\sin^4 x - 1}{1 - \sin^2 x}$
Q14	**Prove that $\sec^2 x + \csc^2 x \geq 4$**

Test 18: SAT Math 2 Algebra

Q1 I have a box of chocolates with white, milk and dark chocolates. The ratio of white to not white is 3:4. The ratio of not dark to dark is 17:4. Find minimum number of milk chocolates in the box

a.	6	b.	7
c.	8	d.	9

Q2 $$\frac{18x^2-21x+5}{ax-2}=6x-3-\frac{1}{ax-2}$$

Is true for all values of $x \neq \frac{2}{a}$, where a is a constant. What is value of a?

a.	3	b.	6
c.	6	d.	9

Q3 Find the value x^3 if $4x^5 = 16^2$

a.	5.3	b.	8.0
c.	12.1	d.	27.9

Q4 $\frac{x+3y}{y} = 1\frac{2}{3}$, Find the value of $\frac{y}{x}$

a.	$\frac{3}{4}$	b.	$-\frac{3}{4}$
c.	$\frac{4}{3}$	d.	$-\frac{4}{3}$

Q5 For a function, $f(x) = x^2 + x + 1$, find the value of a if $f(a) = f(4)$

a.	-5	b.	-4
c.	-3	d.	6

Q6 If $x - y + xy = 0$, find possible integer value of $x^2 + y^2$

a.	2	b.	4
c.	8	d.	10

Q7 A line $y = mx + b$ has $x - intercept$ at $A(-4, 0)$ and $y - intercept$ at $B(0, b)$. What is the slope, m, if the area of the triangle AOB is $12\ unit^2$?

a.	2	b.	$\frac{3}{2}$
c.	$\frac{3}{4}$	d.	$-\frac{3}{4}$

Q8 The graph $f(x)$ is translated 4 units left and 2 units up. Find the equation of the resulting parabola if $f(x)$ is represented by the equation $x^2 - 2x + 2$.

a.	$x^2 - 6x + 4$	b.	$x^2 + 6x + 4$
c.	$x^2 + 6x + 12$	d.	$x^2 - 6x + 8$

Answers

1	c.	2	a.	3	c.	4	b.	5	a.	6	c.	7	b.	8	c.

Test 19: SAT Math 2 Algebra

Q1	When 4 times a number y is subtracted from 9, the result is 21. What number results when 4 times y is add to 30?	Q2	$9°C = 5(°F - 32)$ A temperature increase of $45°C$ will result in increase of $t°F$. Find t.
	a. 9 / b. 21 / c. 18 / d. 3		a. 45 / b. 81 / c. 25 / d. 77
Q3	Solve: $2x^3 - 6x^2 + x - 3 = 0$	Q4	Find the next term: 7, 16, 29, 46, 67
	a. -1 / b. -2 / c. 3 / d. 2		a. 90 / b. 92 / c. 93 / d. 94
Q5	If z varies jointly as x and square root of y, and $z = 6$ when $x = 4$ and $y = 9$, find z when $x = 9$ and $y = 4$.	Q6	Area of the regular hexagon inscribed in a circle is $6\sqrt{3}cm^2$. Find the radius of the circumscribing circle.
	a. 6 / b. 9 / c. 12 / d. 18		a. 1 / b. $\sqrt{3}$ / c. 2 / d. $\sqrt{3}/2$
Q7	The range of $f(x) = 2x(6 - x)$ is	Q8	For a function, $f(x) = x^2 + x - 2$, find the value of a if $f(a) = f(5)$
	a. $y \leq 18$ / b. $y \geq 18$ / c. $y \leq 36$ / d. $y \geq 36$		a. -7 / b. -6 / c. -5 / d. -4
Q9	$(9)^{2-x} = 27^{x-3}$ find the value of x	Q10	$2^{x+2} - 2^x = 96$ find the value of x
	a. 3 / b. 4 / c. 5 / d. 2		a. 3 / b. 4 / c. 5 / d. 2

Watch Video **SAT Math 2 Algebra** **SCAN ME**

Q11	x^2+y^2 $= 12, xy$=−13 find $(x-y)^2$			
	a.	38	b.	36
	c.	49	d.	25

Q12	$2m^2 - 6m + 8 = 0$. What is sum of all m that satisfy this equation?			
	a.	16	b.	8
	c.	32	d.	4

Q13	Peter has mean, M, of his quiz score. If he scores 11 on the next quiz, his mean will increase by 1. If he scores 11 on each of the next three quizzes, his mean will increase by 2. Find value of M			
	a.	5	b.	6
	c.	7	d.	8

Q14	A cylindrical storage tank is one-fourth full. When 3L is removed, the tank is one-fifth full. How many liters does the tank hold?			
	a.	$1000L$	b.	$80L$
	c.	$60L$	d.	$50L$

Q15	If $f(x) = \sqrt{3x} + 3$, what is $f(f^{-1}3)$?			
	a.	3	b.	6
	c.	1	d.	0

Q16	$18p^7 + 33p^6 + 14p^5 = 0$ is true for			
	a.	$p = -3/2$	b.	$p = -5/4$
	c.	$p = -7/6$	d.	$p = 2/3$

Q17	$2x - 3y = 18; x + 2y = -5$ What is product of x and y?			
	a.	-12	b.	-4
	c.	62	d.	-62

Q18	$h(x) = 2x - 1, g(x) = 2h(x) + 3$ Find solution for $h(x) = g(x)$.			
	a.	-2	b.	-1
	c.	0	d.	1

Q19	$h(x) = \frac{\sqrt{\pi - x}}{x^2 - 9}$ find the largest integer for which $h(x)$ is defined.			
	a.	1	b.	2
	c.	3	d.	4

Q20	If $\frac{x^{a^2}}{x^{b^2}} = x^{16}, x > 1$, and $a + b = 2$ What is the value of $a - b$?			
	a.	2	b.	4
	c.	8	d.	16

Q21	If measure of an exterior angle of a triangle is greater than 50°, what is the greatest number of sides it can have?			
	a.	6	b.	7
	c.	8	d.	9

Q22	Sum of number n and its reciprocal is $\frac{15}{6}$. Find the possible value of five times this number.			
	a.	5	b.	$\frac{5}{2}$
	c.	$\frac{5}{3}$	d.	30

Q23	$x^2 + y^2 = 25$ represents circle with center at the origin and radius of 5 units. Point $(a, 0)$ lies on the tangent to the circle at $(-3, 4)$. Find a.	Q24	Sum of number n and its reciprocal is $\frac{29}{10}$. Find the possible value of the number

a.	$-\frac{5}{3}$	b.	$-\frac{25}{3}$
c.	$-\frac{25}{4}$	d.	$\frac{25}{4}$

a.	5	b.	$\frac{5}{2}$
c.	$\frac{5}{3}$	d.	$\frac{5}{4}$

Q25	Find perimeter of a sector cut out with radius 5 and central angle of $60°$.	Q26	The length of an arc of a circle is 10 cm. The corresponding area is 50 cm^2. Find the radius in cm of the circle.

a.	12.6	b.	14.8
c.	13.4	d.	15.2

a.	10	b.	8π
c.	5π	d.	8

27 A circle is inscribed in the right triangle as shown in the figure. The circle touches the hypotenuse and divides it in the lengths a and b. Find the area of the right triangle.

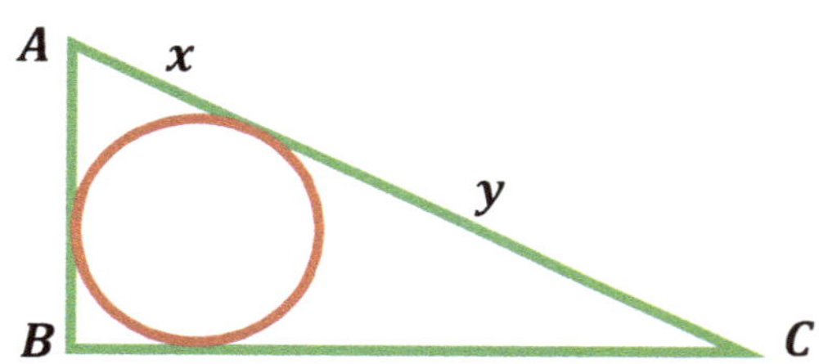

a.	$0.5xy$	b.	$0.25xy$
c.	xy	d.	$\sqrt{2}xy$

SCAN ME

Answers: Test 19 SAT Math 2

1	c.	2	b.	3	c.	4	b.	5	b.	6	a.	7	a.	8	b.
9	c.	10	c.	11	a.	12	b.	13	d.	14	c.	15	a.	16	c.
17	a.	18	b.	19	b.	20		21	b.	22	b.	23	b.	24	b.
25	d.	26	a.	27	c.										

Test 20: Lines and Quadratic Functions TIPS

Q1	Solve the linear system: $6(x+1)-2(y-2)=4$ $\frac{2}{3}x-\frac{1}{4}y=-1$	Q2	the following three lines intersect at one point. Find the value of k. $3x+5y=10, x+2y=4$ $5x+ky=10$
Q3	If $a+b=12, b+c=14$ and $c+a=16$, find $a+b+c$	Q4	Solve the system $2xy+3=4y$ and $3xy+2=5y$.
Q5	A parabola $y=ax^2+k$ pass through the points $(-1,3)$ and $(3,-13)$. Find the value of a and k.	Q6	If $b-2c=3, -3b+7c=-10$, and $a+2b-5c=13$, what is the value of a?
Q7	Sam threw a ball. After 3 seconds, the ball was at the highest point, 18 m above the ground. Two seconds later, the ball had fallen 8 m. Algebraically determine the equation for the height, h, that models the function with respect to time, t.	Q8.	An open box can be created by cutting congruent squares from each corner of a cardboard that is of 20 cm by 30 cm and folded up the sides. Determine the dimensions of the squares that must be cut to create a box with a volume of 1008 cm^3.
Q9	A survey done by a website reported that for a \$15 monthly fee, they could obtain 2000 new customers. It also reported that for every \$0.50 decrease in the fee, they would obtain an additional 100 new customers. What fee will result in maximum revenue?	10	determine range of the quadratic function. 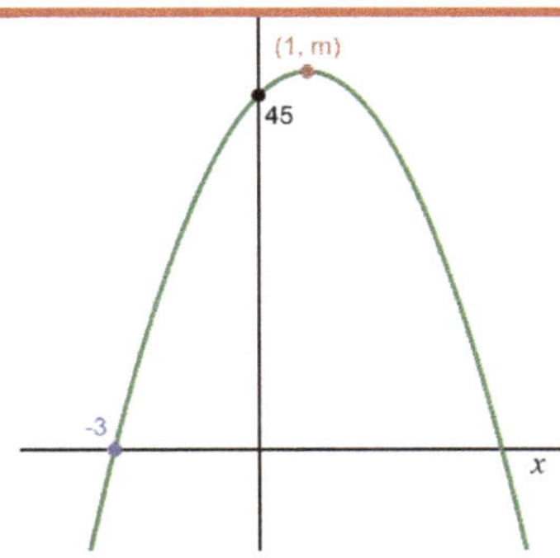

Answers: Test 20

Q1	$x=3, y=12$	Q2	$k=5$	Q3	21	Q4	$(3.5, 2.5)$
Q5	$a=-2, k=5$	Q6	$a=6$	Q7	$h(t)=-2(t-3)^2+18$		
Q8	3 cm by 3 cm	Q9	\$12.50	Q10	$y \leq 48$		

Test 21: Challenge Section: Integrated Approach

SAT Math 2 Challenge: Relates Areas

Q1 Find the area of shaded region

Area of square $PQRS$ is $625\ cm^2$

Area of rhombus $QRTU$ is $500\ cm^2$

Find area of shaded region in centimetre square.

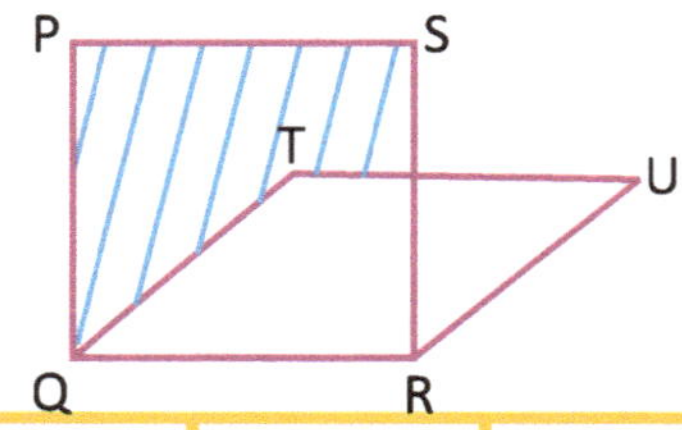

a.	200	b.	225	c.	275	d.	300

IB Math Challenge: Tangent to Circle

Q2 Find equation of line tangent to the unit circle at $P\left(\frac{1}{2}, \frac{\sqrt{3}}{2}\right)$.

AP Math Challenge: Functions

Q3 Sketch the graph of function $\boldsymbol{f(x) = \sqrt{1 - x}}$. Write its domain and range.

AP Math Challenge: Rate of Change

Q4 **Rate of change at a point $x = a$** is given by the expression:

$$\frac{f(a+h) - f(a)}{h}$$

Above expression provides slope of the tangent line if value of h approaches zero.

Find rate of change at $x = a$ on the function $f(x) = x^2$.

Evaluate the rate of change, using your simplified expression for $x = 3, h = 0.001$.

AP Math Challenge: Parallel and Perpendicular Lines

Q5 ABCD is a rectangle. Coordinate of the points are $A(1,14), B(-3,8), C(a,0)$ and $D(m,n)$. Find the coordinated of the points C *and* D.

Q6. **Olympiad Challenge:** If $\boldsymbol{x + \frac{1}{x} = 4}$ then find the value of

a.	$x^2 + \frac{1}{x^2}$	b.	$x^3 + \frac{1}{x^3}$
c.	$x^4 + \frac{1}{x^4}$	d.	$x^5 + \frac{1}{x^5}$

Solution 1: Area of Shaded Region	Let V be on QP so that $TV\ \ QP$
Option C: $\mathbf{275\ cm^2}$	VQ = height = $20\ cm$
	$PV = PQ - VQ = 25 - 20 = 5\ cm$
Area of square $PQRS = 625\ cm^2$	$\Delta\ VTQ$ is right angle
Side lengths $QR = \sqrt{625} = 25\ cm$	$VT^2 = QT^2 - QV^2$
Rhombus side lengths = $25\ cm$	$VT^2 = 25^2 - 20^2$
Area of rhombus $QRTU = 500$	$VT^2 = 625 - 400$
Base × height = 500	$VT^2 = 225$
Height $= \frac{500}{base} = \frac{500}{25} = 20$	$VT = \sqrt{225} = 15$

Area of shaded region = Area of rectangle $PSWV$ + Area of ΔVTQ

$$= 25 \times 5 + \frac{1}{2}\ 20 \times 15 = 125 + 50 = \mathbf{275\ cm^2}$$

Short Solution IB Math Challenge: Tangent to Circle

Find equation of line tangent to the unit circle at $P\left(\frac{1}{2}, \frac{\sqrt{3}}{2}\right)$.

Tangent will be perpendicular to the radius from origin, O, to the point $P\left(\frac{1}{2}, \frac{\sqrt{3}}{2}\right)$.

Slope of OP is $\frac{\frac{\sqrt{3}}{2}}{\frac{1}{2}} = \sqrt{3}$. Slope of tangent will be $m_{\perp} = -\frac{1}{\sqrt{3}}$

Equation of tangent: $y - \frac{\sqrt{3}}{2} = -\frac{1}{\sqrt{3}}\left(x - \frac{1}{2}\right)$

Student Activity: Write the tangent equation in standard form.

Short Solution 3: AP Math Challenge Domain and Range

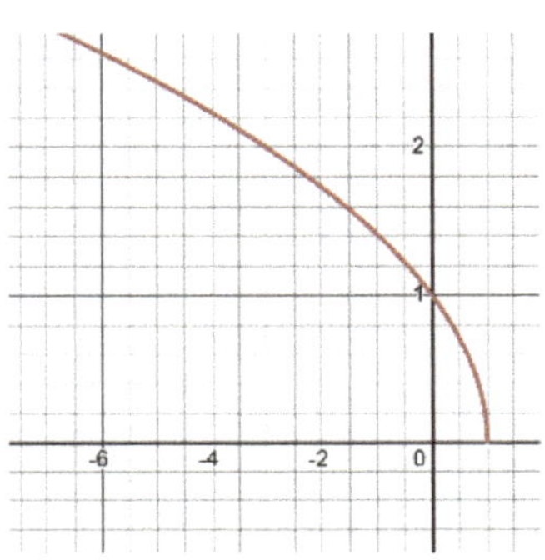

$\boldsymbol{f(x) = \sqrt{1 - x}}$.

$1 - x \geq 0 \Rightarrow 1 \geq x \Rightarrow x \leq 1$

Domain: $x \leq 1$

To sketch the graph, find $f(x)$ for $x = \{1, 0, -1, -3\}$

Range: $y \geq 0$ as is very clear from the graph.

Short Solution

Rate of change at a point $x = a$ is:

$$\frac{f(a+h)-f(a)}{h}$$

$f(x) = x^2 \Rightarrow f(a) = a^2$ and $f(a+h) = (a+h)^2 = a^2 + 2ah + h^2$

Rate of Change

$$\frac{f(a+h)-f(a)}{h} = \frac{(a^2+2ah+h^2)-a^2}{h} = \frac{2ah+h^2}{h} = \frac{h(2a+h)}{h} = 2a+h$$

Evaluate the rate of change for $x = 3, h = 0.001.$

Rate of Change $= 2a + h = 2(3) + 0.001 = 6.001 \approx 6$

Short Solution: Coordinates of Rectangle

ABCD is a rectangle. $A(1,14), B(-3,8), C(a,0)$ and $D(m,n)$.

Find the coordinated of the points C *and* D.

BA is perpendicular to BC

$$m_{BA} = -\frac{1}{m_{BC}} \Rightarrow \frac{14-8}{1-(-3)} = -\frac{a-(-3)}{0-8}$$

$$\frac{6}{4} = -\frac{a+3}{-8} \Rightarrow \frac{3}{2} = \frac{a+3}{8} \Rightarrow 12 = a+3$$

$a = 9 \Rightarrow$ Coordinates $C(9,0)$

BA is parallel to CD

$A(1,14), B(-3,8), C(9,0)$ and $D(m,n)$.

Student Activity

Show that coordinates of $D(13,6)$.

Video Solution 6

If $x + \frac{1}{x} = k$

Then show that:

SCAN ME

$x^2 + \frac{1}{x^2} = k^2 - 2$	$x^3 + \frac{1}{x^3} = k^3 - 3k$	$x^4 + \frac{1}{x^4} = (k^2-2)^2 - 2$

Test 22: Challenge Review

Q1 Given a rectangle ABCD with side lengths of $AD = x$ and $AB = y$
Point P is mid-point of length DC and point Q is the midpoint of BC.
Show that the area of the triangle APQ is $\frac{3xy}{8}$.

SCAN ME

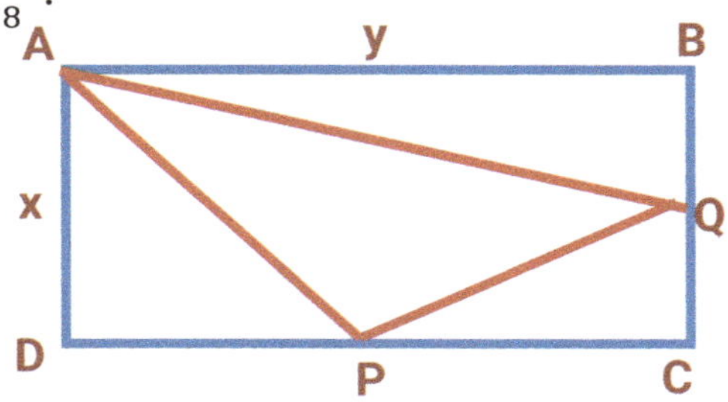

Q2 An open-top square-based prism has surface area, A, of $54m^2$. Find the area of its base, in meter square for maximum volume.

a.	27	b.	$\frac{27}{\sqrt{2}}$	c.	18	d.	$18\sqrt{2}$

Q3. Find the maximum volume, V, in metre cube, of an open-top square-based prism with surface area, A, of $54m^2$.

a.	27	b.	$27\sqrt{2}$	c.	$27\sqrt{3}$	d.	$18\sqrt{3}$

TIP: Open Top Prism Dimensions: $2x \times 2x \times x$; Area of base = $\frac{1}{3}A$.

Q4 $a_1x^2 + b_1x + c_1$ and $a_2x^2 + b_2x + c_2$

Show that the condition for the above equations to have

Q4a **both common roots is**

$$\frac{a_1}{a_2} = \frac{b_1}{b_2} = \frac{c_1}{c_2}$$

Q4b **only one common root is:**

$$(a_1b_2 - a_2b_1)(b_1c_2 - b_2c_1) = (a_2c_1 - a_1c_2)^2$$

SCAN ME

Watch Video **Quadratic Equations One Common Root**

Q5	Suppose that m and n are positive integers with $\sqrt{7+\sqrt{48}} = m + \sqrt{n}.$ Show that $m^2 + n^2 = 13$	Q6	Given $a, b, c, and\ d$ are positive integers. If $a^5 = b^4, c^3 = d^2$, and $c - a = 19$. Determine value of $d - b$. Answer: 757
Q7	Write the following in increasing order $2^{555}, 3^{444}, 4^{333}, 5^{222}$	Q8	Find the Larger Number $\sqrt{3\sqrt{2\sqrt{5}}}\ or\ \sqrt{2\sqrt{5\sqrt{3}}}$
Q9	Given $m\ and\ n$ are positive integers. $\frac{9^n \cdot 3^2 \cdot 3^n - 27^n}{3^{3m} \cdot 2^3} = \frac{1}{27}$ Determine value of $m - n$. Answer: 1	Q10	Show that $\frac{1}{2\sqrt{n+1}} < \sqrt{n+1} - \sqrt{n},$ $n \epsilon Z, n > 0\ (n\ is\ positive\ integers)$ Solution on Next Page

Test 23: Challenge: Simultaneous Equations

Q11	$\begin{cases} x^2 + y^2 = 5 \\ \frac{1}{x^2} + \frac{1}{y^2} = \frac{5}{4} \end{cases}$ $(\pm1, \pm2), (\pm2, \pm1)$	Q12	$\begin{cases} x^2 + y^2 = 10 \\ \frac{1}{x} + \frac{1}{y} = \frac{4}{3} \end{cases}$ $(1,3), (3,1), (0.605, -3.105), (-3.105, 0.605)$
Q13	$\begin{cases} x^2 - y^2 = 24 \\ \frac{1}{x+y} + \frac{3}{x-y} = \frac{11}{12} \end{cases}$ $x = \pm4, y = \pm1$	Q14	$\begin{cases} x^2 - 4xy + y^2 = 1 \\ \frac{y}{x} + \frac{x}{y} = \frac{17}{4} \end{cases}$ $x = \pm4, y = \pm1$

Try solving before watching the video! Even partial progress helps you grow.

Watch Video **Challenge Simultaneous Equations** **SCAN ME**

STEM Project Solution: Rationalization Inequality

Show that

$$\frac{1}{2\sqrt{n+1}} < \sqrt{n+1} - \sqrt{n}, n\epsilon Z, n > 0\ (n\ is\ positive\ integers)$$

Steps #1: Rationalization

$$\sqrt{n+1} - \sqrt{n}$$

$$= \sqrt{n+1} - \sqrt{n} \times \frac{\sqrt{n+1} + \sqrt{n}}{\sqrt{n+1} + \sqrt{n}}$$

$$= \frac{(n+1) - n}{\sqrt{n+1} + \sqrt{n}} = \frac{1}{\sqrt{n+1} + \sqrt{n}}$$

Proved

$$\sqrt{n+1} - \sqrt{n} = \frac{1}{\sqrt{n+1} + \sqrt{n}}$$

Step #2

$$\sqrt{n+1} + \sqrt{n+1} > \sqrt{n+1} + \sqrt{n}$$

$$2\sqrt{n+1} > \sqrt{n+1} + \sqrt{n}$$

$$\frac{1}{2\sqrt{n+1}} < \frac{1}{\sqrt{n+1} + \sqrt{n}}$$

Therefore

$$\frac{1}{2\sqrt{n+1}} < \sqrt{n+1} - \sqrt{n}$$

Extended: Substitute $\boldsymbol{n} = \mathbf{1}$ and then show that

$$\frac{1}{\sqrt{2}} < 2\sqrt{2} - 2$$

Challenge: Equation of Tangent Line to a Circle

Find the equation of the tangent to the circle $x^2 + y^2 - 8x - 10y - 128 = 0$ at the point $(-8, 10)$.

$12x - 5y + 146 = 0$

Test 23: Algebra: Advanced Skills Video Support

Sum of Reciprocals

If $x + \frac{1}{x} = k$

Then show that:

$x^2 + \frac{1}{x^2} = k^2 - 2$	$x^3 + \frac{1}{x^3} = k^3 - 3k$	$x^4 + \frac{1}{x^4} = (k^2 - 2)^2 - 2$

Special Case

If $x + \frac{1}{x} = 1$ then $x^3 = -1$	If $x + \frac{1}{x} = -1$ then $x^3 = 1$

Practice Question

If $x + \frac{1}{x} = 4$ then find the value of

a.	$x^2 + \frac{1}{x^2}$	b.	$x^3 + \frac{1}{x^3}$
c.	$x^4 + \frac{1}{x^4}$	d.	$x^5 + \frac{1}{x^5}$

Watch Video **Algebra Reciprocals Part 2** SCAN ME

Purpose of the Challenge Section

These questions are designed to bridge **conceptual understanding and algebraic mastery**.

They go *beyond routine practice* and help students **connect patterns, structure, and logic** — essential skills for **STEM and competition math**.

Test 24: Advanced Algebra Skills Multiple Choice Test

Q1	If $x + y = 2z$ find the value of $\frac{x}{x-z} + \frac{z}{y-z}$							
	a.	-1	b.	1	c.	2	d.	2
Q2	If $x^2 + 1 = 2x$ find the value of $x^{40} + x^{39} + x^{38} + \cdots. + x^2 + x + 1$							
	a.	20	b.	21	c.	40	d.	41
Q3	If $x + y + z = 6$ find the value of $(x-1)^3 + (y-2)^3 + (z-3)^3$							
	a.	0	b.	1	c.	2	d.	3
Q4	If $\frac{x}{3} + \frac{3}{x} = 2$ then find the value of x^3							
	a.	3	b.	6	c.	9	d.	27
Q5	If $a = q + r, b = p + r, c = p + q$ and $p^3 + q^3 + r^3 - 3pqr = 4$ then find the value of $a^3 + b^3 + c^3 - 3abc$.							
	a.	2	b.	4	c.	8	d.	12
Q6	If $\sqrt{a^2 + b^2 + ab} + \sqrt{a^2 + b^2 - ab} = 1$ then find the value of $(1 - a^2)(1 - b^2)$.							
	a.	$\frac{1}{4}$	b.	$\frac{3}{4}$	c.	$\frac{5}{4}$	d.	4
Q7	If $\frac{a^2-bc}{a^2+bc} + \frac{b^2-ca}{b+ca} + \frac{c^2-ab}{c^2+ab} = 1$ then find the value of $\frac{a^2}{a^2+bc} + \frac{b^2}{b^2+ca} + \frac{c^2}{c^2+ab}$							
	a.	0	b.	1	c.	2	d.	3
Q8	If $\frac{1}{1+a} + \frac{1}{1+b} + \frac{1}{1+c} = 2$ then find the value of $\frac{a}{1+a} + \frac{b}{1+b} + \frac{c}{1+c}$							
	a.	0	b.	1	c.	2	d.	3

Q9	If $a+b+c=0$ then find the value of $\frac{a^2+b^2+c^2}{a^2-bc}$							
	a.	0	b.	1	c.	2	d.	3
Q10	If $a^3+b^3=9, a+b=3$ then find the value of a^4+b^4							
	a.	17	b.	27	c.	32	d.	81

Algebra 10 Best Substitution Strategies **SCAN ME**

Answers

Q1	b.	Q2	d.	Q3	a.	Q4	d.	Q5	c.
Q6.	b.	Q7	c.	Q8	b.	Q9	c.	Q10	a.

Challenge Section with Video Solution:

If $x=7+4\sqrt{3}$. Prove that $\sqrt{x}=2+\sqrt{3}$ Hence show that $\sqrt{x}+\frac{1}{\sqrt{x}}=4$
If $x+\frac{1}{x}=5$. Prove that $x^2=5x-1$. Hence show that $\frac{4x}{3x^2-5x+3}=\frac{2}{5}$
If $x+x^2+x^3=1$ show that $x^3+\frac{1}{x}=2$
If $x-y=\frac{x+y}{7}=\frac{xy}{4}$ show that $\frac{x}{y}=\frac{4}{3}$
Find the value of $\frac{1}{1+11^{2025}}+\frac{1}{1+11^{-2025}}$

Golden Ratio **SCAN ME**

STEM Project: Mersenne Numbers

A **Mersenne Number** is a special number of the form:

$$M_n = 2^n - 1, n \text{ is positive integer}$$

They appear in cryptography, computer science, and the search for **large prime numbers**. Research how mathematicians use **supercomputers** to test whether $M_n$is prime — such primes are called **Mersenne Primes**.
Explore questions such as:

- Why are Mersenne primes important for digital security?
- What is the largest known Mersenne prime today?
- How does the **Lucas–Lehmer test** help identify them?

Challenge: Write a short report or create a presentation explaining how Mersenne numbers connect math to **modern technology and computing power**.

Example 1: $M_5 = 2^5 - 1 = 31$

- 31 is a Mersenne Prime, because 31 itself is prime.
- **Use in computing:** Binary systems rely on powers of 2. Since $2^5 = 32, 31 = 32 - 1$ is often used in programming to create masks or limits — for instance, 5-bit registers can store numbers up to 31.
- **STEM Insight**: This connection between powers of two and Mersenne numbers helps in data compression, cryptography, and processor design.

Example 2: $M_7 = 2^7 - 1 = 127$

- 127 is also a Mersenne Prime.
- **Application:** In early computer systems, the 7-bit ASCII code used 127 as the maximum representable value — all letters, digits, and symbols fit within this range.
- **Mathematical Significance:** $M_7 = 127$is small enough to test easily by hand, yet it represents the same pattern mathematicians use when searching for extremely large Mersenne primes with supercomputers.

Linear Law: Extended Problems Supported by Video Playlist

Convert non-linear equations to the transformed linear form

$\boldsymbol{Y = mX + c}$, where m and c are constants. Graph of transformed equation Y vs X will be a straight line.

SCAN ME

In each of the following questions find the value of Y and X

Q1	Write $Ax + By = xy$ in two different Linear Forms ($Y = mX + c$). **Solution 1:** $y = B\frac{y}{x} + A$ $Y = y, X = \frac{y}{x}$ **Solution 2:** $x = Ay + B$ $Y = x, X = \frac{x}{y}$
Q2	Write $y = \frac{c}{kx+d}$ in the Linear Forms $Y = mX + c$ $Rearrange{:}\frac{1}{y} = \frac{k}{c}x + \frac{d}{c};\ Y = \frac{1}{y}, x = x$
Q3	Write $y = ax + \frac{k}{x}$ in the Linear Forms $Y = mX + c$ $Rearrange{:}\ yx = ax^2 + \frac{b}{x};\ Y = yx, x = x^2$
Q4	the variable x and y are related by the equation $ay = x + bx^2$. When a graph of $\frac{y}{x}$ is plotted against x, the resukting line has slope of 5 and an intercept on the $\frac{y}{x} - axis$ of 0.5. Calculate the value of a and b. Sketch the graph $Rearrange{:}\frac{y}{x} = \frac{b}{a}x + \frac{1}{a};\ a = 2, b = 10$
Q5	Write $y = p(x)^q$ in the Linear Forms $Y = mX + c$ $Rearrange{:}\ \log y = q\log x + \log p;\ Y = \log y, x = \log x$
Q6	Write $y = Ae^{-kx}$ in the Linear Forms $Y = mX + c$ $Rearrange{:}\ \ln y = -kx + \ln A;\ Y = \ln y, x = x$

Lattice Points or The Points with Integer Values

Lattice points in the Cartesian coordinate system are the points where both coordinates are integers. For example, (1, 3), (−2, 5), and (0, 0) are all lattice points, while (0.5, 5) or (−2, ½) are not. Lattice points often appear in geometry and number theory problems. They form the grid intersections on a coordinate plane, which makes them essential in studying area, slope, and shapes made of integer coordinates.

In this section, we explore how to count lattice points on line segments and within polygons. One important tool you will learn is **Pick's Theorem**, which connects the area of a polygon with integer-coordinate vertices to the number of lattice points inside it and on its boundary.

Example: Finding Lattice Points in a Right Triangle

Consider a triangle with vertices at (0, 0), (6, 0), and (0, 4). We will determine:
(a) The number of lattice points lying on each side of the triangle.
(b) The number of lattice points strictly inside the triangle.

We will use the **gcd method** to count boundary points and **Pick's Theorem** to find the number of interior points.

Step 1: Identify vertices and sides

Vertices: A(0,0), B(6,0), C(0,4)

Step 2: Counting lattice points on each side

For any line segment connecting two lattice points (x_1, y_1) and (x_2, y_2), the number of lattice points (including both endpoints) is given by:
Points = $\gcd(|x_2 - x_1|, |y_2 - y_1|) + 1$

→ Side AB: (0,0) → (6,0): gcd(6,0) + 1 = 6 + 1 = 7 points

→ Side AC: (0,0) → (0,4): gcd(0,4) + 1 = 4 + 1 = 5 points

→ Side BC: (6,0) → (0,4): gcd(6,4) + 1 = 2 + 1 = 3 points

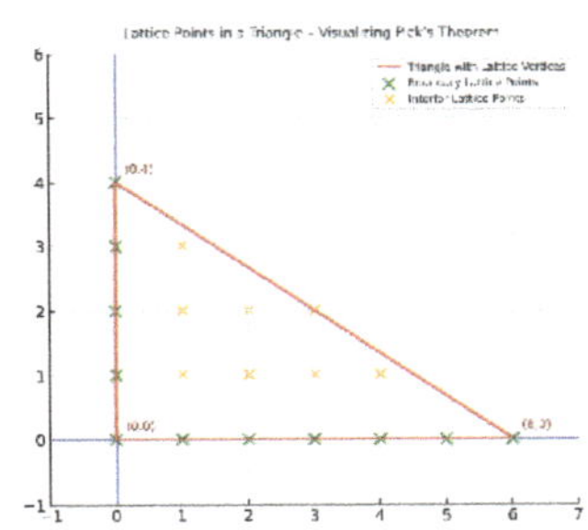

Step 3: Total boundary lattice points

If we simply add the counts from each side, we count the three vertices twice. Hence, subtract 3 to remove duplicates.
B = (7 + 5 + 3) − 3 = 12 boundary points.

Step 4: Area of the triangle Base = 6, Height = 4 Area = ½ × 6 × 4 = 12 square units. **Step 5: Apply Pick's Theorem** **Pick's Theorem: A = I + B/2 − 1** Here, A = 12 and B = 12. ⇒ 12 = I + (12/2) − 1 ⇒ 12 = I + 6 − 1 ⇒ I = 7	**Step 6: Final Answer** Boundary points (B) = 12 Interior points (I) = 7 **Interpretation** There are 12 lattice points on the edges (including vertices) and 7 lattice points inside the triangle. This makes a total of 19 lattice points that belong to or lie within the triangle.

This example demonstrates how **geometry, arithmetic, and number theory** come together in understanding **lattice points**. The concept can be extended to rectangles, polygons, and even 3D structures where integer-coordinate points play a key role in modeling grids, crystal structures, and computer graphics.

Guided Practice

Q1. Identify the lattice points among: (a) (3, 4), (b) (−2.5, 3), (c) (0, −1), (d) (2√2, 1).

- Short Description: Check whether both coordinates are integers.
- Solution: Lattice points are (3, 4) and (0, −1). (−2.5, 3) and (2√2, 1) are not lattice points because at least one coordinate is not an integer.

Q2. How many lattice points lie on the x-axis between x = −3 and x = 4 (inclusive)?

- Short Description: Points on the x-axis have y = 0; count integer x from −3 to 4.
- Solution: There are 4 − (−3) + 1 = 8 lattice points: (−3,0), ..., (4,0).

Q3. How many lattice points lie on the y-axis between y = −5 and y = 5?

- Short Description: Points on the y-axis have x = 0; count integer y from −5 to 5.
- Solution: There are 5 − (−5) + 1 = 11 lattice points: (0, −5), ..., (0, 5).

Q4. Which quadrants contain lattice points with both coordinates positive?

- Short Description: Positive x and positive y correspond to one quadrant.
- Solution: Quadrant I only (x > 0, y > 0).

Q5. Is (√3, √2) a lattice point? Explain.

- Short Description: A lattice point requires both coordinates to be integers.
- Solution: No, because √3 and √2 are irrational; hence not integers.

Q6. Sketch the line x + y = 4 and find all lattice points on it.

- Short Description: Solve $x + y = 4$ in integers.
- Solution: All integer solutions (x, y) with $x = 0..4$ and $y = 4 - x$: (0,4), (1,3), (2,2), (3,1), (4,0).

Q7. Find all lattice points on the axes for the circle $x^2 + y^2 = 25$.

- Short Description: Points on axes have one coordinate 0. Check (±5,0), (0,±5).
- Solution: The four axis lattice points are (5,0), (−5,0), (0,5), (0,−5).

Q8. How many lattice points lie inside or on the square with vertices (0,0), (0,3), (3,3), (3,0)?

- Short Description: Count integer pairs (x,y) with $0 \le x \le 3$, $0 \le y \le 3$.
- Solution: There are $(3+1)\times(3+1) = 16$ lattice points on or inside the square.

Q9. For rectangle with vertices (1,1), (1,4), (5,1), (5,4), how many lattice points lie inside (not on sides)?

- Short Description: Count interior integer pairs: $x = 2..4$ and $y = 2..3$.
- Solution: Interior count = $3 \times 2 = 6$ points: (2,2), (3,2), (4,2), (2,3), (3,3), (4,3).

Q10. Triangle with vertices (0,0), (6,0), (0,4): (a) lattice points on each side (b) total interior lattice points.

- Short Description: Use gcd for boundary counts and Pick's Theorem for the inside $A = I + B/2 - 1$.
- Solution: (a) Sides:
  • (0,0)→(6,0): gcd(6,0)+1 = 6+1 = 7 points.
  • (0,0)→(0,4): gcd(0,4)+1 = 4+1 = 5 points.
  • (6,0)→(0,4): $\Delta x=-6$, $\Delta y=4 \Rightarrow$ gcd(6,4)+1 = 2+1 = 3 points.
  Boundary $B = 7 + 5 + 3 - 3 = 12$ (subtract 3 to avoid counting vertices twice).
  Area $A = 1/2\cdot6\cdot4 = 12$. Pick: $12 = I + 12/2 - 1 \Rightarrow 12 = I + 6 - 1 \Rightarrow I = 7$.
  (b) Interior lattice points = 7.

Q11. Number of lattice points on the line segment from (2,3) to (10,9).

- Short Description: Use count = gcd(|Δx|, |Δy|) + 1 with $\Delta x=8$, $\Delta y=6$.
- Solution: gcd(8,6) + 1 = 2 + 1 = 3 points: (2,3), (6,6), (10,9).

Q12. Determine all lattice points on the segment joining (−4,2) and (2,−1).

- Short Description: Compute step via gcd; parametrize points.
- Solution: $\Delta x = 6$, $\Delta y = -3 \Rightarrow$ gcd(6,3)=3 $\Rightarrow$ 4 lattice points.
  Step = (6/3, −3/3) = (2, −1). Points: (−4,2), (−2,1), (0,0), (2,−1).

Communication and Thinking: Lattice Points

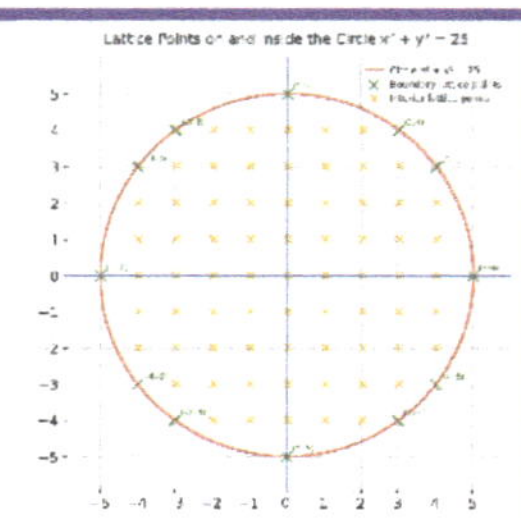

Show that there are 12 lattice points on the circle and 73 inside the circle with equation $(x-6)^2+y^2=25$

Hint: Pick's Theorem: A = I + B/2 − 1

SCAN ME

STEM Challenge Problems: Lines

Image After Reflection on a Slant Line

Determine the coordinates of the image of the point $P(3,6)$ which is reflected in the mirror placed along the straight line $y=x+1$.

💡 Hint Equation of line perpendicular to $y=x+1$ through $P(3,6)$: $y=-x+9$

Point of intersection of the two lines is $M(4,5)$

This is midway between point and its image (x',y')

$4=\dfrac{3+x'}{2}\Rightarrow x'=8-3=5$	$5=\dfrac{6+y'}{2}\Rightarrow y'=10-6=4$

Coordinates of the mirror image of point $P(3,6)$ reflected on $y=x+1$ is $P'(5,4)$

STEM Project: Foot of Perpendicular (Projection)

Prove that the projection $P(x_p,y_p)$ from $A(x_0,y_0)$ to the line $ax+by+c=0$ is

$$x_p=\frac{b^2x_0-aby_0-2ac}{a^2+b^2},y_p=\frac{a^2y_0-abx_0-2bc}{a^2+b^2}$$

STEM Project: Mirror Image

Prove that the image $P'(x_0',y_0')$ from $P(x_0,y_0)$ to the line $ax+by+c=0$ is

$$x_0'=\frac{(b^2-a^2)x_0-2aby_0-2ac}{a^2+b^2},y_0'=\frac{(a^2-b^2)y_0-2abx_0-2bc}{a^2+b^2}$$

STEM Project Thinking and Application

The Golden Rectangle and the Golden Ratio (φ)

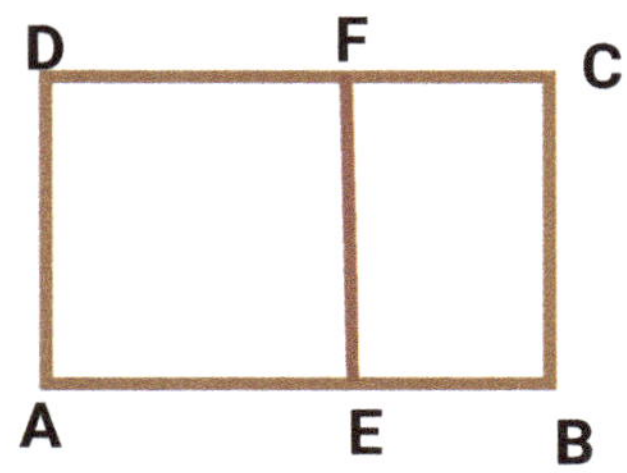

A **golden rectangle** is a special type of rectangle where the **ratio of the length to the width is**

$$\varphi = \frac{1+\sqrt{5}}{2}$$

This number, φ (pronounced "phi"), is known as the **golden ratio**, and it appears in nature, art, architecture, and mathematics.

Key Property:

When a square is removed from a golden rectangle, the remaining rectangle is **similar** to the original — meaning the ratios of their sides remain the same.

Given:

- Rectangle **ABCD** is a golden rectangle.
- Point **E** lies on side **AB** such that **AE** is the side of the square constructed inside the golden rectangle.

To Prove:

$$\frac{AE+EB}{AE} = \frac{AE}{EB} = \varphi = \frac{1+\sqrt{5}}{2}$$

Hint:

Let $AB = x \ and$

Watch Video **Golden Ratio** SCAN ME

The Fibonacci Connection to the Golden Ratio

Fibonacci Sequence: 1, 1, 2, 3, 5, 8, 13, 21, 34, ... where each term is the sum of the two before it.

Ratios that "chase" φ

If you take the ratio of consecutive Fibonacci numbers F(n+1)/F(n), these values get closer and closer to the golden ratio φ = (1 + √5) / 2 ≈ 1.61803.

🌱 UOI – Patterns in Nature: From Numbers to Growth

Mathematical patterns shape growth in nature—from spirals in shells to seeds in sunflowers. **Think about**: Where do you see ratios or symmetries? **Try this:** Capture a pattern and identify its ratio. **Reflect:** How does math reveal beauty in nature?

STEM Challenge: Similar Right Triangles & Geometric Mean

Explore how altitudes, ratios, and proofs uncover deep relationships in geometry—and lead to powerful formulas for problem-solving.

Interesting questions to derive formula which can help understand relations between geometrical shapes better and improve performance. Also appreciate how we can apply our knowledge in real life applications.

Part A: Identity in Right Triangles

In a right triangle ΔABC, sides $a, b, and\ c$ form the triangle with hypotenuse $AB = c$. Let point D lie on AB, and altitude $CD = p$ is drawn from the right-angle C to hypotenuse AB.

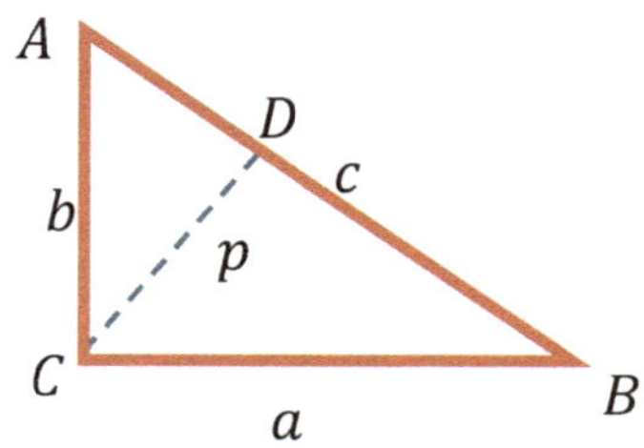

Challenge:

Prove the identity: $\frac{1}{p^2} = \frac{1}{a^2} + \frac{1}{b^2}$

Strategy: Use area of triangle in two ways

Part B: Geometric Mean Theorem (Altitude Form)

In a right triangle ΔABC, altitude CD divides the hypotenuse into two segments of length m and n as shown in the figure.

Prove that the length of altitude p is geometric mean of m $and\ n$.

In other words, prove that: $p^2 = mn$ or $\mathrm{p} = \sqrt{mn}$

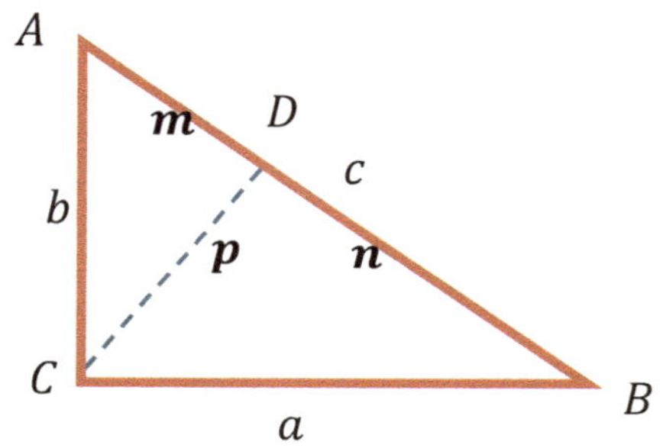

Example: Geometric Mean

In a right triangle ΔABC, altitude CD divides the hypotenuse into two segments of length 8 and 18 as shown in the figure.

$\angle ACB = \angle ADC = 90°$

Find the value of variable x.

Answer: $x = 9$

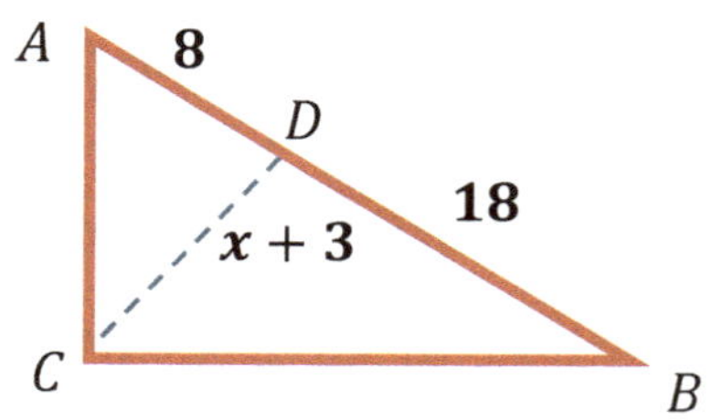

Test 25: AMC-Style Reflection Questions: Geometric Mean and Altitude

Objective: Deepen your understanding of right triangles, altitudes, and their relationships using geometric mean, area arguments, and algebraic reasoning. Each question builds critical problem-solving skills for math competitions such as AMC.

1. In triangle ABC, the altitude divides hypotenuse into two segments of length 4 and x. If the altitude is 6, find all possible values of x.
2. If △ABC has sides 5, 12, and 13, find the length of the altitude to the hypotenuse.
3. Prove that the **reciprocal of the square of the altitude** equals the **sum of reciprocals of squares of legs** — again — using **area methods**.
4. Challenge: In triangle ABC, the altitude from C to hypotenuse is 10. If the legs are in the ratio 3:4, find the lengths of all sides.

Watch Video **Geometric Mean** SCAN ME

STEM Project: Relating Equation of Line and Trigonometry

$$\frac{x - x_1}{\cos\theta} = \frac{y - y_1}{\sin\theta} = r$$

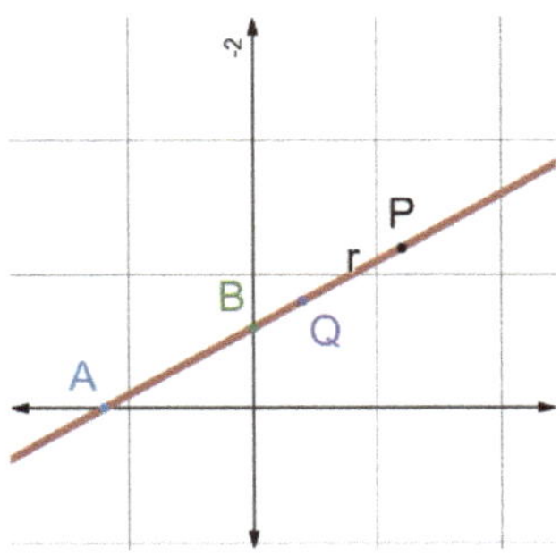

This is equation of line passing through the point $\boldsymbol{Q}(x_1, y_1)$ and making an angle θ with the positive direction of the $x - axis$, where r is the distance of the point $P(x, y)$ from the point (x_1, y_1).

Write a proof for the above equation of the line.

This equation is the distance form of a line.

Distance form of a line can be rearranged to get the parametric form as:

$x = x_1 + r\cos\theta$ and $y = y_1 + r\sin\theta$

Derive the parametric form of the equation of the line.

Practice Problem:

A straight line is drawn through the point $\boldsymbol{Q}(2, 5)$ ans is inclined at an angle of $30°$ with the x-axis. Find coordinates of two points on it at a distance of 4 from **Q**.

Solution: Circle Inscribed in Equilateral Triangle Solution

Formula Derivation

A circle, O, of radius r is inscribed in an equilateral triangle ΔABC.

Develop a relation between radius r of the inscribed circle and the side length a of the equilateral triangle.

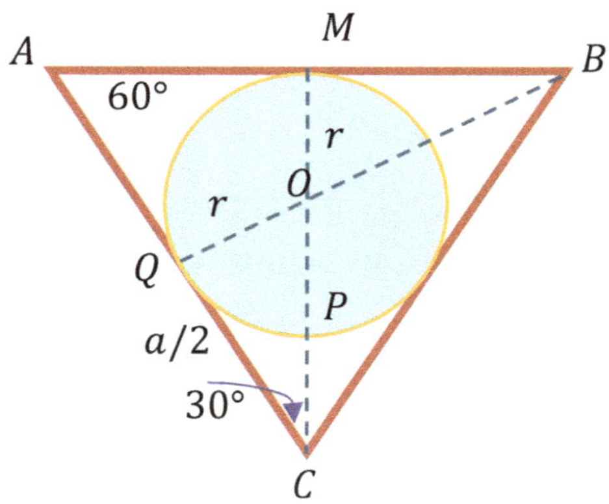

Solution

Consider the figure to represent the situation.

Line BQ is drawn through the centre connecting the vertex B and the mid-point Q of the side AC

$AC = a$ so $QC = \frac{a}{2}, OQ = r$

ΔOQC is a special $30° - 60° - 90°$ triangle

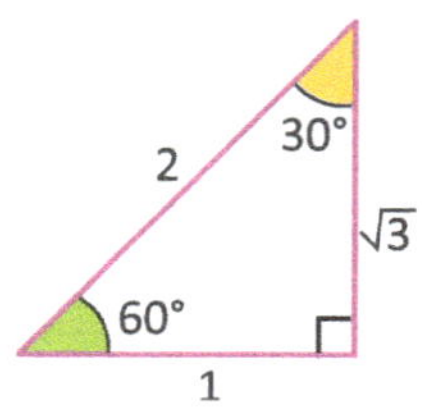

Ratio of the legs is $\sqrt{3}$

$$\frac{QC}{OQ} = \sqrt{3}$$

$$\frac{a/2}{r} = \sqrt{3}$$

$$r = \frac{a}{2\sqrt{3}}$$

$$r = \frac{a}{2\sqrt{3}}$$

Cone and Inscribe Sphere Application

The paper cup for a snow cone has radius 10 cm and a vertical angle of $60°$. When a bass of radius r is dropped into the cup, the top of the ball is even with the top of the cup. Find the radius r *in cm*.

a.	5	b.	3.2	c.	2.9	d.	2.5

Solution: Real Life Model: Ice-cream Cone Solution

The paper cup for a snow cone has radius of R with cross-section of an equilateral triangle. Figure, not to the scale is shown. When a ball of radius r is dropped into the cup, the top of the ball is even with the top of the cup. Determine an expression for the distance from the bottom of the ball to the vertex of the cone in terms of R and r.

Solution

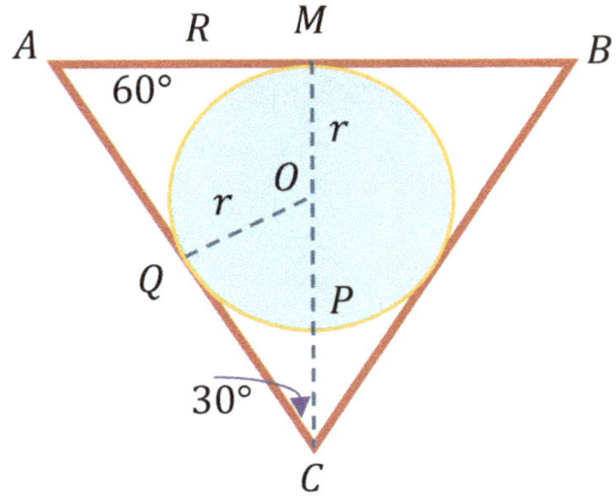

Consider the figure to represent the situation.

$AM = R$ and $OM = OQ = OP = r$

Line MC is drawn through the centre connecting the vertex C and the mid-point M.

$MC = r + r + PC$

$MC = 2r + PC$

We need to find the distance PC in terms of R and r.

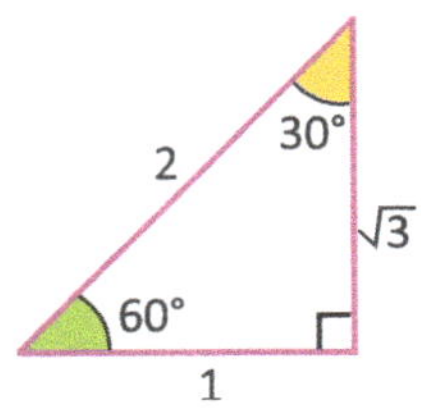

ΔAMC is a special triangle $30 - 60 - 90$

Ratio of the legs is $\sqrt{3}$

$$\frac{MC}{AM} = \sqrt{3}$$

$$\frac{2r + PC}{R} = \sqrt{3}$$

$$2r + PC = R\sqrt{3}$$

$$PC = R\sqrt{3} - 2r$$

$$PC = R\sqrt{3} - 2r$$

Triangle and Inscribe Circle Application

A circle of radius 5 cm is inscribed in an equilateral triangle. Find the perimeter, in cm, of the triangle.

a.	11.5	b.	10.4	c.	10.1	d.	9.6

Exam Thinking & Advanced Applications

Rediscover Algebra and Trigonometry

> This exam practice set includes carefully selected challenge problems that test conceptual understanding, problem-solving strategies, and application across topics. These questions reflect the style and rigor of school exams, AP/IB assessments, and advanced mathematics challenges.

Knowledge: 60	Communication:30	Applications: 96	Thinking: 76

Q1 1/c	Can square-root of a number be less than the number itself. Provide reason.
Q2 4/k	If $x = -\frac{3}{2}, y = -\frac{1}{4}, z = \sqrt{2}$, evaluate $x + y \times z^2 \div (x - y)$ $\frac{x^2z^2 + 4xy}{8y^2 - x}$
Q3 1/c	Find the one-digit place value in 3^{2025}. Explain your strategy.
Q4 2/k	Solve Equation: $2\|2x - 4\| - 1 = 7$
Q5 4/k	Evaluate Expression: a. $\frac{2^{-3} - 3^{-2}}{2^{-2} + 3^{-1}}$ b. $1 + \cfrac{1}{1 + \cfrac{1}{1 + \cfrac{1}{1 + 1}}}$
Q6 4/k	Simplify Expression: $\frac{(2 + h)^3 - 8}{h}$ Rationalize Numerator: $\frac{\sqrt{5x + h} - \sqrt{5x}}{h}$
Q7 2/k	Sum of number, n and its reciprocal is $\frac{15}{6}$. Find n.

Q8 2/a	The perimeter of a right triangle is 72 cm. The lengths of the sides are in the ratio of 3:4:5. Find the area of triangle.
Q9 2/a	Time period, in seconds, for a complete swing of a pendulum is given by the formula $T = \pi\sqrt{\frac{L}{9.8}}$ where L is length in metres. a) Rearrange the equation to make L as the subject. b) Find length L for the time-period of $10\ seconds$ correct to 3 s.f.
Q10 3/k	Round answer to one decimal place. a. 42 is what percent of 120? b. 14 is 35% of which number? c. What is $12\frac{3}{4}\%$ of 84?
Q11 2/k	Amy gave $73.60 for the taxi ride, which included 15% tip. What was the actual cost of the ride?
Q12 2/a	P is 3 times Q If P is increased by 10% and Q is decreased by 30% then find percent change in product of P and Q.
Q13 2/a	Show that the square of an odd number is always one more than the multiple of four.
Q14 3/t	n is an integer such that $2n - 1 \leq 7$ and $\frac{6n}{5+n^2} > 1$. Find all possible values of n
Q15 2/t	Arrange the following in increasing order: $2^{5555}, 3^{4444}, 4^{3333}, 5^{2222},$
Q16 2/c	Find the number of seconds in 6 weeks. Show that these are $10! = 1 \times 2 \times 3 \times 4 \times 5 \times 6 \times 7 \times 8 \times 9 \times 10$
Q17 4/a	For the following velocity – time graph a) Describe the motion of the object b) Sketch a possible acceleration – time graph c) Sketch a possible distance – time graph with correct shape of non-linear portion. 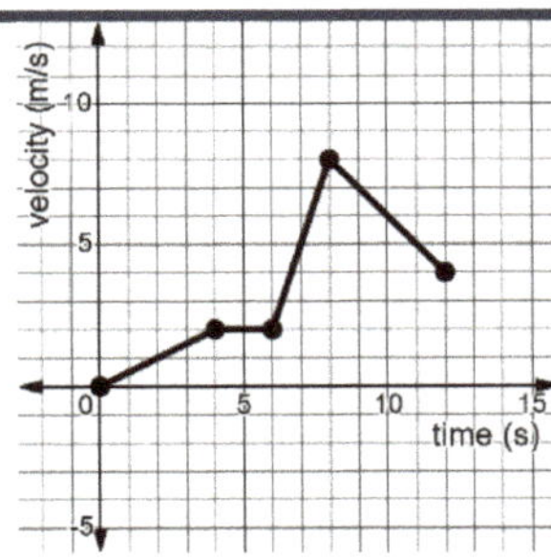

Q18 2/t	Any six digit number of the form "$abcabc$" is always divisible by what number? Hint: Write number in expanded form and use factoring.
Q19 2/t	If $\frac{2}{3}P:\frac{3}{4}Q:\frac{3}{2}R$ find the ratio of $P:Q:R$.
Q20 2/k	The three vertices of a rectangle $ABCD$ are $A(-1,-1), B(2,1), and\ C(6,-5)$ find the coordinates of the fourth vertex D.
Q21 2/a	Anil made four trips to Niagara Falls along the same route. His average speed on the first trip was 60 km/h and in each subsequent trip his speed was half of the previous trip. Find the average speed of the four trips. **16 km/h**
Q22 2/a	Find equation of tangent line at the point $(-3,-4)$ on the circle $x^2+y^2=25$.
Q23 2/k	Determine whether the points $(-2,2)$, $(2,5)$, and $(6,8)$ are collinear. Justify your answer.
Q24 3/t	The points $A(-2,3)$ and $B(m,n)$ form a segment AB. $3x-2y+5=0$ if the right bisector of segment AB. Find the coordinates of B. $B(4,2)$
Q25 2/t	$A(2,3)$ and $B(11,18)$ form a line segment. Find coordinates of two points that divide the segment AB in three equal parts.
Q26 2/k	Find equation of line perpendicular to $2x+3y=15$ with same $x-intercept$.
Q27 3/a	Find area of the triangle formed by the line $2x+3y=18$ and the coordinate axes.
Q28 2/c	Consider quadratic function $f(x)=kx-8=2x^2$. Find the range of k for $f(x)$ to have two solutions.
Q29 3/t	Find line whose distance from the line $x+y-5=0$ is twice as compared to the distance from $x-7y+3=0$.

Q30 3/c

Prove that $\sqrt{2}$ is not a rational number.

Then prove that $\sqrt{2}+\sqrt{3}$ is not a rational number.

Q31 4/k

Simplify the following radicals

a.	$3\sqrt{12}-2\sqrt{36}+\sqrt{24}$	b.	$2^3 \times 6^{-2} \div 3^{-2}$
c.	$(3\sqrt{2}-\sqrt{3})(2\sqrt{6}+\sqrt{2})$		$(3\sqrt{2}-5)^2$

Q32 8/k

Write domain and range of the following functions

a.	$y=\dfrac{x+1}{2x-1}-1$	b.	$f(x)=-2\sqrt{-x+1}+3$
c.	$g(x)=-2x+4, x=\{-1,2,5\}$		$h(x)=\frac{1}{2}(x+1)^2$

Q33 4/a

Let $f(x)=x^2-2x$ and $g(x)=2x-3$, determine

a.	$f(-2)-g(-2)$	b.	$f(x+1)+f(-x)$
c.	$g(f(-2))$		$f(g(x))$

Q34 6/k

Factor completely

	$x^3+3x^2-4x-12$	b.	x^2-8x-y^2+16
c.	x^4-3x^2-4	d.	$8x^3+27y^3$
e.	$2(x+a)^2+3(x+a)+1$	f.	$3x^{3/2}+6x^{1/2}-9x^{-1/2}$

Q35 8/a

Solve equations. Write restrictions. Check for extraneous roots

a.	$\sqrt{3x+4}+9=2x$	b.	$3\sqrt[3]{x^2}=\sqrt{6}$
c.	$\dfrac{x-2}{x+3}=\dfrac{x-4}{x+5}$	d.	$\dfrac{1-3x}{x^2-1}+\dfrac{x}{x-1}=0$

Q36 12/a

Sove equations

a.	$-\frac{1}{2}(x+3)^2+8=0$	b.	$14x^2-19x-3=0$
c.	$x^4-5x^2+4=0$	d.	$2^{2x}-12(2^x)+32=0$
e.	$\left(x-\frac{1}{x}\right)^2-7\left(x-\frac{1}{x}\right)^2+12=0$		$2\pm\sqrt{5}, \frac{3\pm\sqrt{13}}{2}$

Q37 2/a

Two numbers are in the ratio of 2:5.
If 16 is added to both the numbers, the ratio becomes 1:2. Find the numbers.

Q38 2/a	Find all group of three consecutive integers whose product is equal to their sum.
Q39 2/a	Write the quadratic function $f(x) = ax^2 + bx + c$ that will represent the following pattern. Explain your strategy. 2, 7, 16, 29, 46, 67,
Q40 2/a	The cost $C(n)$, of operating a cement-mixing truck is modelled by the function $C(n) = 2.2n^2 - 66n + 700$, where n is the number of minutes the truck is running. What is the minimum cost of operating the truck?
Q41 2/t	Find the prime number of pairs of positive integers (x, y) where x and y are prime numbers and $x^2 - 2y^2 = 1$
Q42 3/t	The obtuse triangle ABD, $\angle ABD = 120°$. Point C is on the side AD such that $CD = 1$ and $\angle CBD = 30°$. If $CD = AB = 1$, find the length AC. Draw a diagram to show your solution. $\boldsymbol{AC = \sqrt[3]{2}}$
Q43 3/t	Sketch graph of absolute value function: $$\lvert y\rvert = 2\lvert x+1\rvert - 3$$
Q44 3/t	Solve quadratic reciprocal challenge $x^2 + y^2 = 5 \text{ and } \frac{1}{x^2} + \frac{1}{y^2} = \frac{5}{4}$
Q45 3/t	y is inversely proportional to a^2. When $a = 10, y = 4$ a is directly proportional to x^2. When $x = 2, a = 24$ Write y as a function of x in the simplest form $\boldsymbol{y = \frac{100}{9x^4}}$
Q46 3/t	Find the sum of x and y if $$\frac{16^{x+2y}}{8^{x-y}} = 32 \text{ and } \frac{32^{x+3y}}{16^{x+2y}} = \frac{1}{8}$$ $\boldsymbol{x = -17, y = 2, x + y = -15}$
Q47 2/a	Find altitude of the right triangle whose legs are of length 6 cm and 8 cm. **4.8**
Q48 3/t	Show that there are 12 lattice points on the circle and 73 inside the circle with equation $x^2 + (y-6)^2 = 25$

Q49 2/c	Make t as subject in the equation $s = ut + \frac{1}{2}at^2$
Q50 3/c	Write family of quadratic function which has the roots $\frac{3\pm2\sqrt{5}}{2}$
Q51 3/t	Find the equation whose roots are the squares of the reciprocal of the roots of $x^2 + 6x - 2 = 0$
Q52 3/a	A high school is planning to build a new fiels surrounded by a running track. The track coach wants three laps around the track to be 2000 m. The footbal coach wants the rectangle infield area to be as large as possible. Can both coaches be satisfied?
Q53 2/t	The temperature of a lake is $24°$C at the surface and decreases by $1.2°$C for every 6 m of depth. Write a function that represents the water temperature at a depth of d metres.
Q54 2/k	The point $P(4, 16)$ lies on the graph of $y = x^2$. Find the image of this point on the transformed graph of $y = -0.5(x + 2)^2 + 1$
Q55 2/a	Domain of the function $f(x)$ is $-12 \le x < 18$. Find domain of transformed function $y = 3 - 2f(2x - 1)$.
Q56 3/t	If $x + \frac{1}{x} = 4$ find the value of $x^3 + \frac{1}{x^3}$.
Q57 3/t	If $\sqrt{6 - 4\sqrt{2}} = p - q\sqrt{2}$, find value of If $p + q$
Q58 3/a	Find radius of the circle inscribed in an equilateral triangle with perimeter of 24 cm.
Q59 4/c	Three circles of radii 2, 3, and 4 cm are mutually tangent. Find the area enclosed between the circles.
Q60 3/a	Determine area of equilateral triangle that encloses circle of radius 10 cm.

Draw labelled diagram for Q38 – Q43	
Q61 3/a	A ship sails north from port A in a strong wind and arrives at port B, that is 80 km east and 50 km north of its original location. Find the bearing of A from B.
Q62 3/a	A ship sails north from port A on a bearing of $055°$ for 10 km, then changes direction at B to a bearing of $120°$ for 12 km to reach port C. Find the distance and the bearing of C from A. $\mathbf{18.13\ km, 091.9°}$
Q63 3/a	A tower lies due north of my house. I walk a kilometer to the point B from where the tower is due west with an angle of elevation of $17°$. Determine the height of the tower if the point B is $N48°E$ of my house. 211 m
Q64 4/t	From a point A, the summit of two mountains B and C are on a bearing of $024°$ and $058°$, and the angle of elevation are $8.5°$ and $9.3°$ respectively. If B is $850\ m$ above A, and C is due east of B find: a) The horizontal distance from B to A. b) The height of C above A.
Q65 3/a	Solve the possible triangles: $\angle A = 41°, a = 6.5, b = 9.2$.
Q66 3/a	The posts of a hockey goal are 2 m apart. A player attempts to score by shooting the puck along the ice from a point 6.2 m from one post and 7.8 m from the other. Within what angle θ must the shot be made?
Q67 3/c	Solve: $3x + 5 - 2\|x\| = 3$
Q68 2/c	Prove that $\sin x = \sqrt{1 - \cos^2 x}$ is not an identity.
Q69 2/k	If $\cos x = \frac{a^2-b^2}{a^2+b^2}, 0° < x < 90°$ find the value of $\sin x \tan x$
Q70 2/k	If $\csc^2 \theta = \frac{169}{144}$, find the value of $\frac{\sin\theta - \cos\theta}{\sin\theta + \cos\theta}, 0° < \theta < 90°$.
Q71 2/a	If $\tan^2 \theta + \cot^2 \theta = 2$, find the value of $\sec\theta \csc\theta$.
Q72 2/a	Evaluate $\frac{3\sin 62°}{\cos 28°} - \frac{\tan 42°}{\cot 48°}$ Hint: Co-function Identity

Q73 4/t	Find area of triangle(s) DEF where $d = 17. e = 18, \angle D = 63°$ Note: There are 2 possible triangles.
Q 9/t	**Challenge Questions** 74 — If $a\cos\theta - b\sin\theta = c$ prove that $a\sin\theta + b\cos\theta = \pm\sqrt{a^2+b^2-c^2}$ 75 — If $\cos\theta + \sin\theta = \sqrt{2}\cos\theta$ prove that $\cos\theta - \sin\theta = \sqrt{2}\sin\theta$ 76 — If $\tan\theta + \sin\theta = m$ and $\tan\theta - \sin\theta = n$ prove that $m^2 - n^2 = 4\sqrt{mn}$
Q77 2/t	If $\frac{a}{\sin\theta} = \frac{b}{\cos\theta}$ find the value of $\sin\theta\cos\theta$ Hint: $\frac{a}{\sin\theta} = \frac{b}{\cos\theta} = k;\ \sin^2\theta + \cos^2\theta = 1$
Q78 2/t	If $\sec x + \tan x = p$ show that $\sin x = \frac{p^2-1}{p^2+1}$
Q79 2/t	Simplify Expressions a. $4 + (\tan x - \cot x)^2$ b. $\frac{1 - 2\sin x\cos x}{\sin x - \cos x}$
Q80 2/t	$(\sin x + \sec x)^2 + (\cos x + \csc x)^2 = (1 + \csc x\sec x)^2$ Hint: Expand and simplify
Q81 2/a	If $\tan x + \sec x = x$, find $\tan x$ a. $\frac{x^2-1}{2x}$ b. $\frac{x^2+1}{2x}$ c. $\frac{x-1}{x+1}$ d. $\frac{1-x^2}{2x}$
Q82 4/c	Show that there are 12 lattice points on the circle and 73 inside the circle with equation $x^2 + (y-6)^2 = 25$

Q 9/k 9/a	Prove the following identities

83	$\dfrac{1+\sin\theta}{1-\sin\theta} = (\sec\theta + \tan\theta)^2$	84	$\dfrac{1+\sin\theta+\cos\theta}{1-\sin\theta+\cos\theta} = \dfrac{1+\sin\theta}{\cos\theta}$
85	$\sqrt{\dfrac{\sec x - 1}{\sec x + 1}} = \csc x - \cot x$	86	$\dfrac{\sin^2\theta + 2\cos\theta - 1}{\sin^2\theta + 3\cos\theta - 3} = \dfrac{1}{1-\sec\theta}$
87	$\sec^2 x + \csc^2 x = \sec^2 x \cdot \csc^2 x$		
88	$\cos^4 x - \cos^2 x = \sin^4 x - \sin^2 x$		
89	$\sin^6 x + \cos^6 x + 3\sin^2 x \cos^2 x = 1$		
90	$\dfrac{\tan\theta}{1-\cot\theta} + \dfrac{\cot\theta}{1-\tan\theta} = 1 + \tan\theta + \cot\theta$		
91	$\frac{\tan\theta}{1-\cot\theta} + \frac{\cot\theta}{1-\tan\theta} = 1 + \sec\theta\csc\theta$		
92	$\sin^2 x - \cos^2 x - \tan^2 x = \dfrac{2\sin^2 x - 2\sin^4 x - 1}{1-\sin^2 x}$		

Q93 3/t	*if* $\tan x = \dfrac{b}{a}$ *then show that* $\sqrt{\dfrac{a+b}{a-b}} + \sqrt{\dfrac{a-b}{a+b}} = \dfrac{2\cos x}{\sqrt{\cos^2 x - \sin^2 x}}$
Q94 2/a	Solve $(2\sin^2 x - 1)(\cos x - 1) = 0, 0^\circ \le x \le 90^\circ$
Q95 3/c 3/t	**Bonus Question**

If $\dfrac{\cos^4 A}{\cos^2 B} + \dfrac{\sin^4 A}{\sin^2 B} = 1$ *prove that*

a.	$(\cos^2 A - \cos^2 B)^2 = 0$
b.	$\sin^2 A = \sin^2 B$
c.	$\sin^4 A + \sin^4 B = 2\sin^2 A \sin^2 A$

Hint:
Write all terms in cosine.
Expand and simplify.
Solve step by step to get all three.

Beyond Procedures: A Self-Assessment for STEM Readiness

These problems are designed to assess how well you reason, communicate ideas, and apply concepts in unfamiliar situations. They reflect the thinking style required in advanced mathematics and STEM disciplines.

Q1	**Communication and Reasoning** (a) Define *rational numbers* clearly. (b) Using the method of contradiction, prove that $\sqrt{6}$is not a rational number.
Q2	**Prove by Mathematical Induction:** Prove that $(9^n - 2^n)$ is divisible by 7 for all positive integers n. Suggested structure: Step 1: Verify the statement for $n = 1$. Step 2: Assume the statement is true for $n = k$. Step 3: Show that it is true for $n = k + 1$.
Q3	**Recursive Sequences** Let $f_0 = \frac{x}{x+1}$ and $f_{n+1} = f_0(f_n)$ for $n = 0, 1, 2, 3, \ldots$ show that $f_n = \frac{x}{nx+1}$ Hint: Compute $f_1(x)$explicitly, simplify, and observe the pattern in $f_2(x), f_3(x), \ldots$.
Q4	**Algebraic Skills** Given: $x(x-6) = 0$ Find: $x^9 - 1153x^5 - 198x^2 + 2x$
Q5	**Apply Strategy for Difference of Squares** $(x^2 - 5x - 5)(x^2 - 5x + 3) = 9$ Hint: Try strategy other than expand and simplify
Q6	**Complex Roots** Find square root of complex number $8 + 6i$, where $i = \sqrt{-1}$. Hint: Let $\sqrt{8 + 6i} = x + yi$. Now square both sides, equate real and imaginary parts and solve for $x \text{ and } y$. Write both the roots.

Q7

Extended Skills: Exponential Growth and Approximation

The number $\boldsymbol{e}$ can be approximated using expressions of the form $\left(1+\frac{1}{n}\right)^{n}$, where $\boldsymbol{n}$ is a large positive integer.

- Evaluate $\left(1+\frac{1}{n}\right)^{n}$ for several increasing values of $\boldsymbol{n}$, such as $\boldsymbol{n = 10, 100, 1,000, 10,000}$.
- Use your results to estimate the value of $\boldsymbol{e}$, correct to four decimal places.

Q8

The velocity–time graph of a particle moving along a straight line is shown.

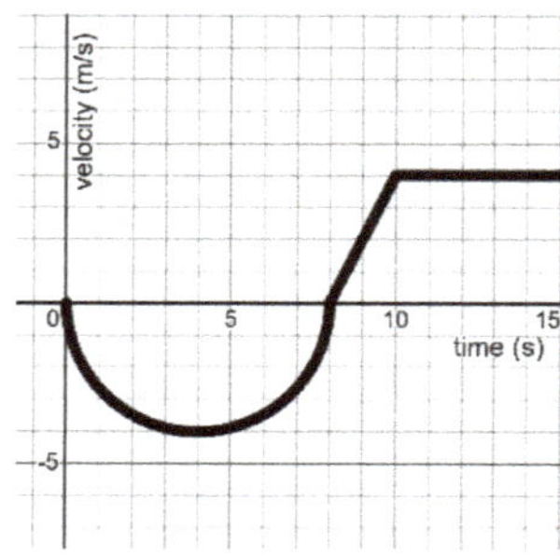

a) Analyze the motion of the particle over the given time interval.
b) Sketch a corresponding distance–time graph.
c) Identify the interval(s) during which the particle is **speeding up**.

Hint: A particle is speeding up when the product of its velocity and acceleration is positive.

Q9

Transformation of Transformed Function

The function $h(x) = 3 - \sqrt{x+5}$ is transformed to $g(x) = -5 + \sqrt{2x+5}$

a) Explain the transformation from $h(x)$ to $g(x)$.
b) Find image of $(11, -1)$ on $g(x)$

Q10

Three-Dimensional Modeling

A cylinder has radius R and height H. A cone is inscribed in the cylinder with its vertex at the center of the base. A sphere is placed inside the cylinder so that it touches the cone and the base of the cylinder, as shown in the figure.

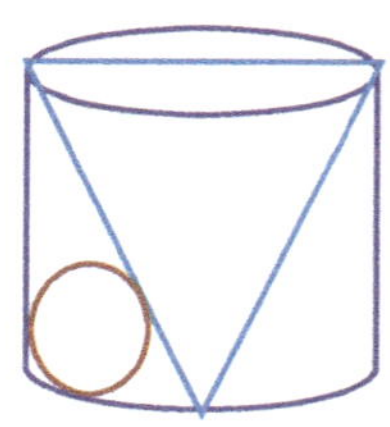

Show that the radius r of the sphere is given by:

$$r = \frac{R + H - \sqrt{R^2 + H^2}}{2}$$

Q11

Trigonomtery Identity Challenge

If $\cot x\,(1 + \sin x) = 4m$ and $\cot x\,(1 - \sin x) = 4n$, prove that

$$(m^2 - n^2)^2 = mn$$

Student Activity Challenge

Students Activity 1

Karan is flying a kite with an angle of elevation of θ. A gust of wind makes the kite rise d m higher and changes the angle of elevation by α. Show that the fixed length, L, of the string is given by the expression $L = \frac{d}{sin(\theta+\alpha) - sin\,\theta}$

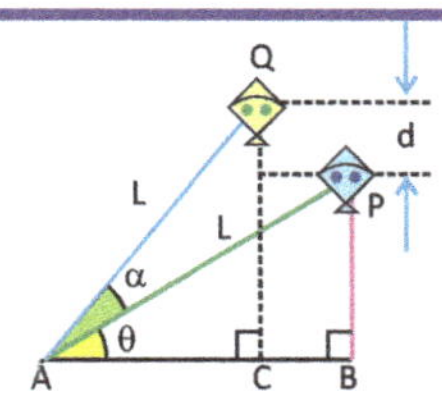

Student's Activity 2

Find the acute angle of intersection between the straight lines $x + y = 3$ and $2x - 3y + 2 = 0$.

Hint:

$slope, m = tan\,\theta$, where θ is the angle that the line makes with the x – axis.

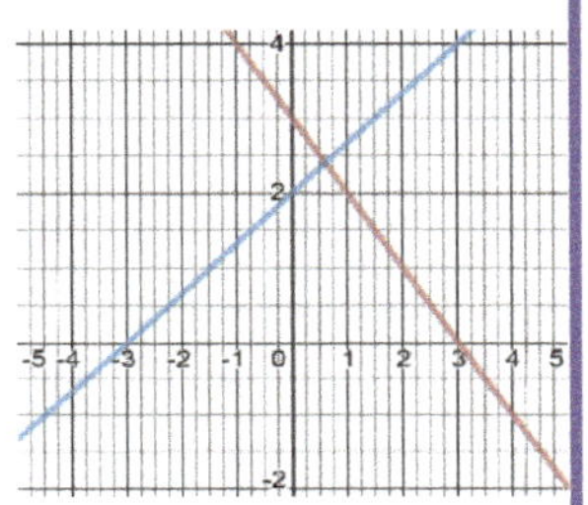

Hint: External angle is sum of remote angles

Student's Activity 3

Given Δ ABC with AD ⊥ BC.

Prove: $b^2 = a^2 + c^2 - 2ac\cos B$

Hint: $\cos B = \frac{x}{c}$

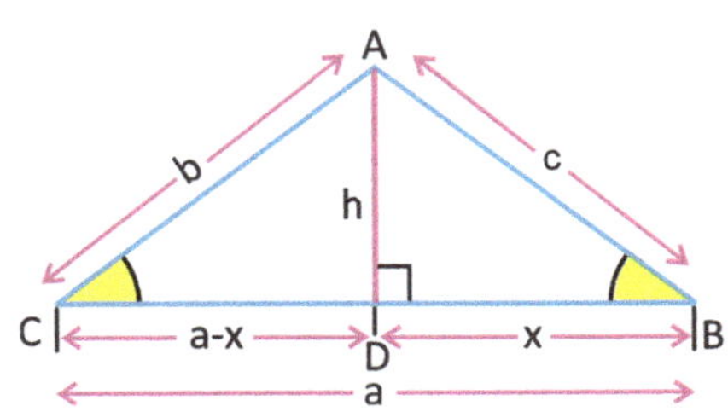

Note: This is proof of Cosine Law

$$b^2 = a^2 + c^2 - 2ac\cos B$$

Student's Activity 4

Describe the situation shown in the figure.

CD represents height, h, of an object above the ground.

Show that the height h

$$h = \frac{d \sin \alpha \sin \beta}{\sqrt{\sin^2 \beta - \sin^2 \alpha}}$$

Hint:

Write AD and BD in terms of h

$\angle BAD = 90°$, apply Pythagorean theorem in ΔABD.

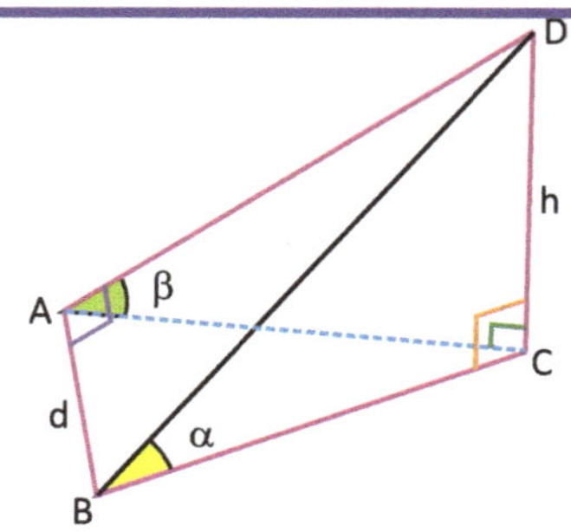

Student's Activity 5

The figure shows semicircles with centre P and M. M is mid point of AP. Points A, M, P, and B are collinear. A small circle of radius r is drawn with centre O tangential to the semicircles as shown.

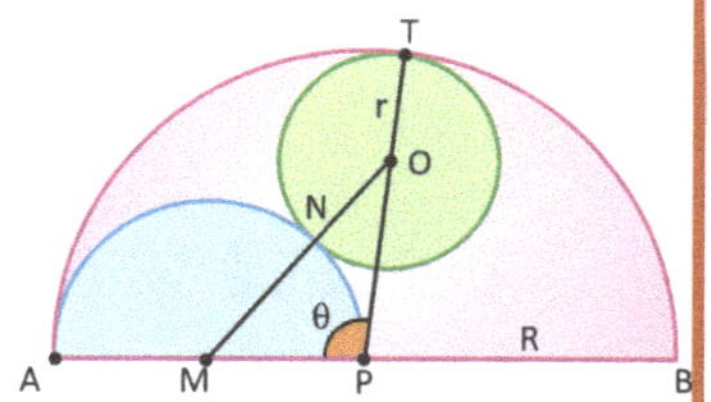

a) If the radius of bigger semicircle is R then prove that $\frac{R}{r} = \frac{3-\cos\theta}{1-\cos\theta}$.
b) Find the angle OMP for $R = 3r$.
c) What do you conclude from this example?

Student's Activity 5

A triangle ABC with area A is inscribe in a circle of radius R. Prove that

$$R = \frac{abc}{4A}$$

Watch Video **Triangle Inscribed in a Circle** SCAN ME

Student Activity: Approximate Height of a Shed

Objective:

Projects are designed to foster creative thinking and help students connect Science, Technology, Engineering, and Mathematics (STEM) concepts to real-life situations.

Project Description:

While trekking, Anil Kumar comes across a small shelter or shed. Based on quick observations and some rough calculations, he estimates the height of the shed to be approximately 12 feet.

Your task is to:

Explain a method Anil could have used to estimate this height using mathematical reasoning, basic tools, and scientific thinking.

Project Instructions:

- **Design a Procedure:**
  Describe at least one method that Anil could use to estimate the height. You can use basic tools like:
  - A stick or walking pole
  - Shadows
  - Smartphone features
  - Right triangle concepts
  - Estimation based on personal height
- **Show the Math Behind It:**
  Include a diagram or sketch and apply one or more of the following:
  - Similar triangles
  - Trigonometry (e.g., using $\tan\theta$ = height/distance)
  - Proportional reasoning

💡 Reflection Questions:

- What assumptions did you make?
- What might cause errors in your estimation?
- Can technology (e.g., phone clinometer app) give better results?
- Would your method work at different times of the day?

✅ Deliverables:

- Description of method
- Mathematical explanation
- Final estimated height
- Sketch or diagram
- Sources of error and how to improve

Explore: Circle tangent to two semicircles

The figure shows semicircles with centre P and M. M is mid point of AP. Points A, M, P, and B are collinear. A small circle of radius r is drawn with centre O tangential to the semicircles as shown.

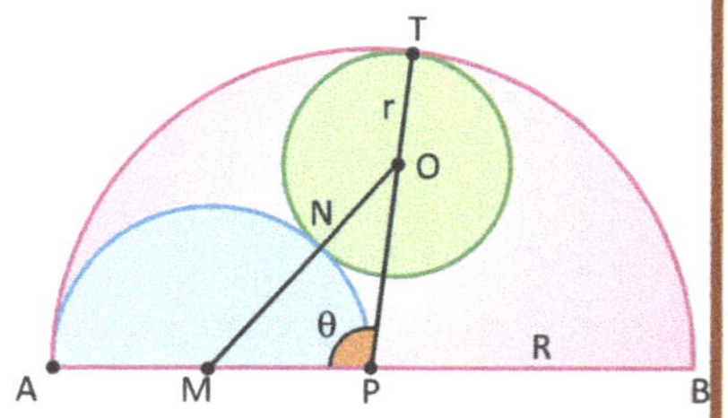

a) If the radius of bigger semicircle is R then prove that $\frac{R}{r} = \frac{3-\cos\theta}{1-\cos\theta}$.

b) Find the angle OMP for $R = 3r$.

c) What do you conclude from this example?

Solution: Circle tangent to two semicircles

Consider the triangle MPO.

$$MP = \frac{R}{2}, MO = \frac{R}{2} + r, PO = R - r$$

Cosine Law for the triangle MPO

$$\left(\frac{R}{2} + r\right)^2 = \left(\frac{R}{2}\right)^2 + (R - r)^2 - 2\left(\frac{R}{2}\right)(R - r)\cos\theta$$

$$\frac{R^2}{4} + Rr + r^2 = \frac{R^2}{4} + R^2 - 2Rr + r^2 - R(R - r)\cos\theta$$

$$3Rr - R^2 + R(R - r)\cos\theta = 0$$

$$3r - R + R\cos\theta - r\cos\theta = 0$$

$$3r - r\cos\theta = R - R\cos\theta$$

$$r(3 - \cos\theta) = R(1 - \cos\theta)$$

$$\frac{R}{r} = \frac{3 - \cos\theta}{1 - \cos\theta}$$

If $R = 3r$ then substituting in $\frac{R}{r} = \frac{3-\cos\theta}{1-\cos\theta}$ we get

$$\frac{3r}{r} = \frac{3 - \cos\theta}{1 - \cos\theta}$$

$$3 = \frac{3 - \cos\theta}{1 - \cos\theta}$$

$$\cos\theta = 0 \Rightarrow \theta = 90°$$

We conclude that the angle $\angle MPO = 90°$

Video Links Reference Resource

Anil Kumar YouTube Channel Links:
https://www.youtube.com/@MathematicsTutor

Links to some videos have been provided to explore and learn more.

BEDMAS	Achilles' Paradox	Repeated Decimals
Pythagorean Theorem	Multilayer Fractions	Prime Factorization applications
Arithmetic Lesson	Solve Word Problems	Percent Change Tricks
Rational Indices	Rationalize Denominator	Timeless Age Puzzle

Distance Speed Time Problems	Functions and Function Notations	Domain and Range
Mirror Image	Linear Systems with 3 Equations	Absolute Value Concept
Discriminant Analysis	Quadratic Exam Review	Quadratic Linear System Extended
Multiple Challenge Equations	Golden Ratio	Layered Difference Method

Advanced Algebra Reciprocal	Trinomial Squares:	Algebra Substitution Skills:
Algebra Reciprocals Part 2	Factor Polynomial	Ball trajectory
$\sqrt{3}$ on Number Line	Regular Polygon	Rectangle to Octagon
Trigonometry Solve Multiple Triangles	Inclination Angle Ladder Height	Exact Value of Trig Expression

Niagara Fall Height	Skylon Tower Angle	CN Tower Height
Angle of Elevation	Special Triangle Applications	SAT Circle & Triangle
SAT Math 2 Algebra	3D Applications Sine Law	3D Bearing Angle
Lattice Points Circle	Projection Formula	Trigono-metric Identities

Epilogue: Rediscovering Mathematics

You have now journeyed through Foundations, Algebra, Geometry, and Trigonometry — the pillars of mathematics that support all advanced study in STEM. Each section was designed not only to prepare you for school exams, but also to build the creative and analytical habits that mathematicians, scientists, and engineers rely on every day.

Mathematics is not just about solving for x or proving a theorem. It is about asking questions, seeing connections, and applying patterns to real-world challenges. Whether you were exploring the Sine Law in navigation, Quadratics in optimization, or Mersenne Numbers in number theory, you stepped into a tradition of discovery that stretches back centuries and continues to shape the future.

As you close this book, remember:

- Every solved problem is a building block toward confidence and mastery.
- Every challenge question is an invitation to think differently.
- Every STEM project is a chance to see math come alive in our world.

Mathematics, like nature, never truly ends. What begins here with algebra and trigonometry extends to calculus, vectors, probability, and beyond. Carry forward your curiosity, your persistence, and your creativity — and you will rediscover mathematics again and again, at every stage of your journey.

A Final Note from the Author

This book was written not just as a reference, but as a companion in learning. The problems, tests, and challenges are here to guide you, but the real rediscovery happens when you make mathematics your own — by asking "what if?", by exploring new ideas, and by daring to connect math with your own passions.

Mathematics has been my lifelong journey. I invite you to make it yours.

— Anil Kumar Khandelwal
Founder, Global Math Institute

Closing Poem – The Path of Numbers

By Anil Kumar Khandelwal

Numbers are rivers, flowing wide,
Equations the bridges, we learn to ride.
From roots that square to angles that soar,
Each step reveals there's always more.

Algebra whispers, transform and see,
Trigonometry guides through land and sea.
Beyond each problem, a hidden door—
Opens to patterns, and truths to explore.

So carry the torch, let curiosity lead,
Math is a language the world will need.
Not an ending, but a place to start—
A journey of numbers, with wonder at heart.

www.ingramcontent.com/pod-product-compliance
Lightning Source LLC
LaVergne TN
LVHW070906210926
848482LV00017B/1001